EXPERIMENTAL NEUTRON
RESONANCE SPECTROSCOPY

Contributors to This Volume

Robert C. Block

Lowell M. Bollinger

F. W. K. Firk

W. M. Good

E. Melkonian

M. S. Moore

Ernest R. Rae

EXPERIMENTAL NEUTRON RESONANCE SPECTROSCOPY

Edited by **J. A. HARVEY**

Oak Ridge National Laboratory
Oak Ridge, Tennessee

1970

ACADEMIC PRESS   New York and London

ACADEMIC PRESS, INC.
111 Fifth Avenue, New York, New York 10003

United Kingdom Edition published by
ACADEMIC PRESS, INC. (LONDON) LTD.
Berkeley Square House, London W1X 6BA

LIBRARY OF CONGRESS CATALOG CARD NUMBER: 75-84151

PRINTED IN THE UNITED STATES OF AMERICA

TO
DAD

CONTENTS

IV. Gamma Rays from Neutron Capture in Resonances

Lowell M. Bollinger

V. Measurements on Fissile Nuclides

M. S. Moore

LIST OF CONTRIBUTORS

Numbers in parentheses indicate the pages on which the authors' contributions begin.

ROBERT C. BLOCK, Department of Nuclear Science, Rensselaer Polytechnic Institute, Troy, New York (155)

LOWELL M. BOLLINGER, Argonne National Laboratory, Argonne, Illinois (235)

F. W. K. FIRK, Yale University, New Haven, Connecticut (101)

W. M. GOOD, Oak Ridge National Laboratory, Oak Ridge, Tennessee (1)

E. MELKONIAN, Columbia University, New York, New York (101)

M. S. MOORE,* Idaho Nuclear Corporation, Idaho Falls, Idaho (347)

ERNEST R. RAE, Nuclear Physics Division, Atomic Energy Research Establishment, Harwell, Great Britain (1, 155)

* Present address: Los Alamos Scientific Laboratory, Los Alamos, New Mexico

PREFACE

The original intention of this book was to cover the experimental techniques of neutron cross-section measurements by the time-of-flight method up to neutron energies of $\sim$10 keV. This energy limit was selected since a comprehensive survey for higher energy neutrons was available. [J. L. Fowler and J. B. Marion, "Fast Neutron Physics," Wiley (Interscience), 1960 and 1963.] However, great improvements of pulsed neutron sources and experimental equipment have been made in recent years and the time-of-flight method now excels up to the MeV neutron energy region. Several conferences have been held during the past decade on various subjects covered in this book, such as intense neutron sources, neutron time-of-flight methods, nuclear structure studies with neutrons, neutron cross sections, and technology, etc.; and there are many publications of detailed results. However, no unified presentation of the subject is available. As such, all chapters of this book have comprehensive bibliographies for the reader who desires supplemental information.

Chapter 1 deals mainly with the characteristics of time-of-flight spectrometers using pulsed electron and positive-ion accelerators, presently the most versatile of all neutron spectrometers. Other neutron sources continue to be used since they sometimes have unique characteristics which are valuable for particular experiments. For example, the "fast-chopper" spectrometer (a mechanical rotor at a high-flux steady-state reactor) was the major source of neutron resonance parameters one to two decades ago. This spectrometer is useful for transmission measurements upon samples available only in small quantities and is also used for spectral measurements of gamma rays following the capture of low energy neutrons. The use of a nuclear explosion as a pulsed neutron source is invaluable for measuring fission and capture cross sections of highly radioactive nuclides and of nuclides which are available only in minute quantities. For certain experiments on a single low-energy neutron resonance a crystal spectrometer at a high-flux reactor may have advantages. However, only the pulsed accelerator spectrometer is reviewed in detail in this chapter since it can be used for measurements over the entire neutron resonance energy region.

The other chapters cover the experimental techniques, such as detectors, data acquisition equipment, methods of analysis, etc., which are used for neutron cross section measurements, and the interpretation and the significance of the results. Chapter 2 on neutron total cross sections contains a brief treatment of neutron resonance theory which is needed in the analysis of total cross section data. Parameters of resonances obtained from transmission measurements many years ago were promptly used by theoreticians for two important discoveries—the Porter–Thomas distribution of neutron widths and the Wigner distribution of resonance spacings. Total cross section measurements are still the greatest source of resonance data today. Scattering measurements, which are used to determine angular momenta and parities of resonances, and capture measurements, which can be used as a sensitive technique for detecting very weak resonances, are reviewed in Chapter 3. Chapter 4 on gamma-ray spectra from the capture of neutrons in resonances deals with this powerful technique for obtaining information on both the capturing and final states and also for learning about the neutron capturing reaction. The detailed and varied experiments which have been performed on the complicated fission process are discussed in the final chapter. The recent discovery of intermediate structure in subthreshold fission can be interpreted in terms of a second minimum in the nuclear potential well. This final chapter contains a comprehensive summary of the parameters of the resonances of the fissile nuclides.

The editor would like to thank the authors and publisher for both their patience and cooperative efforts in trying to bring out a volume as up to date as the rapidly changing state of the art would allow.

EXPERIMENTAL NEUTRON RESONANCE SPECTROSCOPY

I

PULSED ACCELERATOR
TIME-OF-FLIGHT SPECTROMETERS

ERNEST R. RAE
ATOMIC ENERGY RESEARCH ESTABLISHMENT
HARWELL, GREAT BRITAIN

W. M. GOOD
OAK RIDGE NATIONAL LABORATORY
OAK RIDGE, TENNESSEE

I. INTRODUCTION

A. Scope of Chapter

It is well known that the resonance level spacing is a nuclear quantity that decreases in a systematic fashion with increasing mass number except in the region of magic mass numbers where the spacings become large in comparison to those of neighboring mass regions. This resonance spacing may be in the order of 1 MeV for light nuclei, of several kiloelectron volts for medium mass number nuclei, and less than 1 eV for heavy nuclei. For the purpose of the present chapter we define the resonance energy region to cover the energy range where only capture, fission, and elastic scattering are energetically possible, i.e., inelastic scattering is excluded. Hence, we consider resonance neutrons as those having energies from $\sim$1 eV to $\sim$100 keV. The only technique that has proved successful for studying this entire energy region is that of neutron time-of-flight using a continuous energy spectrum neutron source. From the optical analogy it is appropriate to speak of such a source as a "white spectrum" source.

It is obvious from the relatively shorter times involved, that the higher a given neutron's energy, the more precisely its flight time for a given distance must be determined, or in time-of-flight parlance the shorter must be the neutron burst. For this reason mechanical means of producing short bursts of neutrons, namely choppers, for higher energies must eventually be replaced by electromagnetic devices. Such devices are, of course, pulsed accelerators, any one of several types, and the neutrons are produced in a charged-particle nuclear reaction, in which the duration of the neutron burst is almost the same as the duration of the beam of charged particles on the neutron-producing target. In accordance with the opening paragraph, this chapter is concerned with continuous-spectrum neutron sources. It is convenient in the following discussion to classify continuous-spectrum neutron sources into three distinguishable types: (a) bounded, continuous-spectrum sources which are characterized by well-defined upper and lower energy limits. For these sources the neutron producing target is thin and the neutron energy spectrum is a consequence of target thickness; and the instrument most commonly associated with such a source is the pulsed Van de Graaff. Next (b) boil-off or Maxwellian-type continuous sources. For these the neutron spectrum is intrinsically boil-off or Maxwellian in character with a neutron yield which rises to a maximum at some considerable fraction of 1 MeV. Such spectra result from the bombardment of a high-Z target by any type of particle if the bombarding energy is above, say, 20 MeV. Examples of instruments of this type are the electron linac and the positive ion cyclotron. Finally, (c) moderated continuous spectra in which the original intrinsic spectrum is changed to a $1/E$ spectrum by moderator action or the original boil-off spectrum shifted to lower effective temperature by placing a moderator after the target.

Almost any instrument can be utilized in the form of (c), but to be effective, the primary beam energy and/or current must be high. Any instrument producing an intrinsic boil-off spectrum can be used without moderation as in (b) to study interactions above a few hundred kiloelectron volts. In such cases, it is advantageous to use deuterons because with deuterons the direct reaction component serves to enrich the high-energy part of the neutron spectrum which may frequently be of interest. An instrument constructed along these lines for neutron research at higher energies, e.g., above 100 keV, has recently been given special emphasis at Karlsruhe. It should be noted that for producing the highest neutron energies, however, charge exchange reactions with protons are most useful since a cyclotron can accelerate protons to twice the energy of deuterons and the neutron can take the whole of the proton energy. From our definition of resonance neutrons, instruments of type (b), although working on the time-of-flight principle, are outside the scope of this chapter.

The instruments to be considered in this chapter fall either into type (a) or type (c). Type (a) is limited to the pulsed Van de Graaff; type (c) has as examples the linacs, the 400-MeV synchrocyclotron (Nevis), and others. As an instrument for the laboratory devoted exclusively to neutron physics, the electron linac has emerged as the most versatile of all instruments. It has emerged in place of the Nevis-type synchrocyclotron because the higher relative cost of the latter makes it necessary for its activities to be shared between high-energy physics and neutron physics with some loss of flexibility. It has emerged in preference to the Van de Graaff because the latter is restricted to neutron energies above 2 keV, and the electron linac is superior to the Van de Graaff for most cross-section measurements. The present chapter, therefore, will be devoted in greatest detail to the linac. Other instruments, especially the 400-MeV synchrocyclotron and pulsed Van de Graaff, are included essentially as special purpose devices, useful if available, but not suitable exclusively as instruments for investigation in resonance neutron physics.

B. Resolution and Intensity

There are two properties, related to some extent, that must be considered in a discussion of time-of-flight instruments, as indeed they must be considered in a critical discussion of any measuring instrument. These two properties are resolution and intensity.

Clearly, it is only possible to speak of the flight time of clusters of neutrons, i.e., bursts of finite duration. The measurement of the flight time of a neutron associated with a given burst over any fixed distance, is uncertain by an amount Δt_{inst} which is composed, in addition to the burst duration, of the jitter in the detector response and two other time uncertainties to the extent that either or both the length of the source and the length of the detector are appreciable compared to the length of the flight path. If these four components are identified by the subscripts 1, 2, 3 and 4 respectively, then the total instrumental time uncertainty has the root-mean-square value

$$\Delta t_{inst} \approx \Big(\sum_{i=1}^{4} \Delta t_i^2 \Big)^{1/2} \tag{1}$$

The burst duration and the jitter in the detector are independent of the flight time. The burst duration must be defined in terms of particles of the same energy separated by the burst duration Δt_{burst}. However, different energy particles which start together will also arrive at different times, and these will be separated by an interval which depends upon their mean flight time t. In particular, two particles which originate simultaneously from the same point, but which are separated in energy by ΔE, will arrive after a mean flight time t separated by a time interval

$$\Delta t_{(\Delta E)} = \tfrac{1}{2}(\Delta E/E)t \tag{2a}$$

In the case of pulsed beams, the time intervals $\Delta t_{(\Delta E)}$ and Δt_{burst} are, of course, not distinguishable; thus associated with a burst duration Δt_{burst} is an effective energy difference

$$\Delta E_{eff} = 2(\Delta t_{burst}/t)E$$

This effective energy difference is the minimum real-energy difference that is distinguishable by bursts of duration Δt_{burst}, and is conveniently designated as the energy resolution ΔE_{res}. It is evident that several instrumental components must contribute to a general energy resolution in correspondence with their respective contributions to the general time uncertainty.

A more convenient quantity for instrumental comparison purposes than energy resolution is the logarithmic energy resolution,

$$\Delta E/E = 2\,\Delta t/t = 2(\Delta t/L)/(t/L) \tag{2b}$$

because it is equal to a purely instrumental quantity $\Delta t/L$ divided by the reciprocal velocity at the energy E: L is the flight distance. Since the present chapter is concerned with pulsed sources for time-of-flight, the time uncertainties and corresponding energy resolutions will be restricted to that of the burst duration Δt_{burst} alone. In the case of the pulsed Van de Graaff in which the target thickness is negligible, it follows that the quantity $\Delta t_{burst}/L$ is the measure of the resolution of the instrument. In pulsed accelerators utilizing moderated targets, the spread in time due to the moderation process of the emerging neutrons must be considered. Rainwater *et al.* (1960) have shown that this time spread is equivalent to a length uncertainty in the flight path, and so introduces a time uncertainty which is proportional to $1/\sqrt{E}$. This time uncertainty can be neglected in the high-energy limit where $\Delta t/L$ once again becomes an energy independent quantity. We shall therefore refer to $\Delta t/L$ (where Δt is the root-mean-square burst length from the neutron source of the accelerator) expressed in microseconds or nanoseconds per meter as the "resolution" of the spectrometer, bearing in mind that this quantity is directly related to the observed resolution through Eq. (2b) only in the high-energy limit.

The neutron intensity would ideally be expressed as neutrons per unit energy interval, per unit time, and per unit solid angle at the detector. Such a quantity can be estimated from the properties of the neutron producing reaction in the case of the pulsed Van de Graaff. However, this figure is seldom quoted in the literature for spectrometers employing a white continuous spectrum. The more normal method of expressing the neutron intensity is to give the instantaneous rate of emission of neutrons of all energies into 4π during the machine pulse. For such instruments it is possible, in general, to give only rough estimates of the energy dependent flux at the detector from the total instantaneous flux since this clearly depends on the neutron spectrum which, in turn, depends on the reaction used and on the thickness, composition, and disposition of the moderator, if any.

II. PULSED ACCELERATOR SPECTROMETERS WITH MODERATED CONTINUOUS NEUTRON SPECTRA

A. General Considerations

As we stated in the introduction to this chapter, any pulsed accelerator capable of initiating a neutron-producing reaction can be used for neutron time-of-flight measurements provided the peak neutron intensity is sufficiently high. The earliest machines used in this way were mainly cyclotrons accelerating deuterons which were used to bombard beryllium targets. The electron linac has become the most popular machine for this purpose because of its relatively low cost and great flexibility. The electron beam is stopped in a heavy target to produce *bremsstrahlung*, and the *bremsstrahlung* then eject neutrons from the target by photoneutron and photofission processes. The 400-MeV synchrocyclotron at Columbia University has been used intensively as a pulsed neutron source, utilizing the spallation reaction induced by the energetic protons in a lead target. The 150-MeV synchrocyclotron at Harwell is also equipped for this work.

A spectrometer, however, consists not only of the pulsed accelerator but also of a specially designed neutron-producing target and moderating system, a flight path or paths, detector systems, and time-of-flight electronics. All these features are important, and in some measure determine the types of experiment which can be pursued effectively with the equipment. Target design is especially important for the electron accelerators, since the nuclear photoeffect is a relatively inefficient mechanism for neutron production in that several giga electron volts (GeV) of electron energy must be dissipated in the target for each neutron produced. A boosted fission-type source, such as that used at Harwell, is more efficient, since the thermal energy to be dissipated per neutron emitted is ∼500 MeV. The spallation reaction induced by energetic protons in heavy elements is, however, the most efficient mechanism of all for neutron production, the heat generated being in this case only about 50 MeV per neutron for 1 GeV protons. Thus in the limit, where target cooling would determine the maximum pulsed-source intensity, high-energy protons would provide the ultimate pulsed neutron source (Bartholomew and Tunnicliffe, 1966). In practice, however, this limit is still well beyond the capability of current technology.

The neutron-producing targets to which we have been referring are invariably associated with a moderator which enhances the low-energy part of the neutron spectrum. The moderator is usually a slab of hydrogenous material placed close to the pulsed source. Fast neutrons from the source are slowed down by collisions with the hydrogen atoms in the moderator and escape with lower energy. The effect of the moderator on the neutron spectrum of a typical pulsed neutron source is illustrated in Fig. 1 which shows the moderated and unmoderated spectra from the Harwell Neutron Booster source (Rae, 1967). The unmoderated spectrum is in this case a degraded fission neutron spectrum which falls off rapidly with decreasing neutron energy below 100 keV. The presence of the moderator (in this case a thin-walled aluminium box containing water placed close to the source between the source and detector) attenuates the fast-neutron component and produces a low-energy tail to the spectrum of the well-known "slowing down" type. In the case of a very thick moderator this tail has a neutron-energy dependence close to $1/E$, but, in practice, the

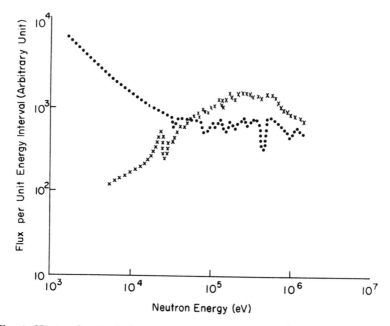

Fig. 1. Neutron booster leakage spectra. ✕, without water and boron; ●, with water and boron; normalized using electron input to booster (Rae, 1967, p. 845).

thickness and material of the moderator are chosen to maximize the moderated neutron flux for a given spectrometer resolution. Considerations of this type (Ribon and Michaudon, 1961) lead to the use of materials like polyethylene with a thickness of 2 to 3 cm. In the example shown in Fig. 1 (2 cm of water) the spectrum has the form $1/E^{0.78}$ rather than $1/E$ in the low-energy region due to the leakage from the moderator surface. It will be observed that the presence of the moderator leads to an enhanced flux below about 60 keV, the enhancement being very great in the resonance region.

The slowing down of the fast neutrons by collisions with moderator atoms takes a finite time, and has a spread, or uncertainty Δt_m, associated with it. This timing uncertainty plays an important role in the choice of a pulsed neutron source and in the optimization of a time-of-flight experiment. The interaction of this timing uncertainty with other parameters in the comparison of different spectrometer systems will be considered in detail later in the chapter, but it is useful at this stage to note that Δt_m varies with neutron energy as $1/\sqrt{E}$ and has a full width at half-maximum value ~ 50 nsec at a neutron energy of 1 keV or 500 nsec at 10 eV. Thus, although a short burst is desirable, in principle, for good energy resolution in a time-of-flight spectrometer, it is pointless to aim at a burst width for a moderated source which is much shorter than Δt_m for the energy range of interest, especially if in so doing the neutron intensity of the source is reduced.

The pulsed neutron target with its moderating slab or slabs must be enclosed in a thick biological shield, usually of the order of 2 meters of concrete, the shield being pierced by holes or channels through which emerge the neutron beams which fly down the flight path or paths. These are normally tubes which are either evacuated or filled with a gas like helium of very small cross section for resonance neutrons. At the end of the flight paths the detector systems are sited which may be either neutron detectors for use in transmission or scattering experiments, or devices to detect reaction products such as γ-rays or fission fragments. These detectors, in the simplest case, produce a digital pulse which indicates the time at which an event occurs. The timing interval between the target pulse and the event determines the energy of the neutron causing the event, and the object of the time-of-flight instrumentation is the recording of these time intervals and the construction of a frequency distribution for the events, as a function of the time interval. In a more complicated experiment, the event may require a more sophisticated descriptor composed of a number representing the time of flight, together with a second number representing some other feature of the event. This might be the amplitude of the pulse in a γ-ray detector, the angle of scattering in a scattering event, the presence or absence of a sample in the beam, in the case of a transmission experiment, or possibly a combination of more than one of these features. In this case the whole descriptor may represent a binary word of 30 to 40 bits.

For a simple experiment, recording only the time-of-flight of the event, the basic electronics would consist of a fast scaler counting a train of "clock pulses." The scaler is started at the time of the neutron burst, or at some known time thereafter, and stopped when the event occurs. The number recorded in the scaler is then called the channel number of the event, and relates directly to the time-of-flight. The contents of the scaler, or encoder, are then transferred to a store of some sort, the scaler content is adjusted to allow for the transfer time and restarted ready for the next event. This process can be repeated several times during the pulsing cycle.

Many time-of-flight systems transfer the encoder contents initially to a fast buffer store or derandomizer (dead time $\lesssim 1$ μsec) from which they are transferred more slowly to the main store. The fast buffer normally has a content of a few words which is sufficient to minimize the loss of data when two events occur in quick succession.

The clock pulses occur at intervals chosen for the time channels required, anywhere between a few nanoseconds and a few microseconds, and may often be programmed so as to give a number of very narrow channels followed by a number of wider channels, and so on, in order to achieve something approaching a constant value of ΔE over the whole energy range studied.

The main memory may be a core store with a relatively fast cycle time (a few microseconds) or it may be magnetic tape with a relatively slow writing speed (few hundred microseconds). The advantages of the former are fast writing speed, higher count rates with the same buffer, and immediate display of data. The advantages of the latter are low cost, permanence of data, and very much larger number of channels available. Core store memories in excess of 4000 channels are unusual, whereas quite inexpensive magnetic tape systems can give many millions of channels if necessary. This is essential if more complex experiments involving long descriptor words are being performed. Data recorded on magnetic tape are later analyzed, in blocks, in a digital computer.

Modern systems are normally computer based, with both memory blocks for prompt analysis of simple experiments and magnetic tape or disk units for more complex experiments, available to the experimenter.

B. Early Work

Before considering in more detail some of the major pulsed accelerator spectrometer installations, it is interesting to look back for a moment at the development of pulsed accelerator sources, over the past few decades. An excellent review of the development of Pulsed-Accelerator Slow-Neutron Velocity Selectors with emphasis on developments in the United States, was given by Havens (1955) in a paper at the first Geneva Conference on the Peaceful Uses of Atomic Energy. Havens' review, together with the paper by Wiblin (1955) on the development of Linear Electron Accelerator Spectrometers in the United Kingdom, should be consulted for a detailed account of much of the work on pulsed neutron sources up to 1955. Here, we shall present only an outline of the early developments of pulsed accelerators for resonance neutron time-of-flight work.

1. PULSED CIRCULAR ACCELERATORS

The first pulsed accelerator time-of-flight spectrometer was reported by Alvarez (1938) who modulated the power amplifier drive to the Berkeley 36-in. cyclotron. Neutrons were produced from deuterons on a beryllium target and a paraffin moderator during a period of about 4 msec. The resolution available was thus very poor, but for the first time it was possible to study the effects of thermal energy neutrons free from fast-neutron background, and the $1/v$ dependence of the boron cross section on neutron velocity was confirmed qualitatively with this apparatus for energies below 0.025 eV.

The next step forward was the use of a modulated ion source which was first applied to a Cockcroft–Walton accelerator by Fertel et al. (1940) and was used with the Cornell 16-in. cyclotron by Baker and Bacher (1941). This enabled the neutron burst length to be reduced to 50–100 μsec with corresponding improvement in resolution, and extension of the useful neutron energy to a few electron volts.

The technique of pulsing the ion source was later applied in a greatly improved manner to the 36-in. cyclotron at Columbia University (Rainwater and Havens, 1946; Rainwater et al., 1947; Havens and Rainwater, 1951) where the pulse length was reduced to about 2 μsec, and measurements were made with a flight path of 6 meters. The nominal resolution, defined as the duration of the neutron burst divided by the flight path length, was thus about 330 nsec/meter. The neutron production rate during the pulse was 6×10^{13} neutrons/sec.

The Brookhaven 60-in. cyclotron spectrometer, described by Havens (1955), used an even faster system of ion-source modulation and achieved a burst of less than 1 μsec. With a flight path of 4.3 meters, the nominal resolution was about 200 nsec/meter. This pulsed cyclotron gave a peak neutron intensity during the pulse $\sim 3 \times 10^{14}$ neutrons/sec, and the spectrometer could undoubtedly have been used at much higher resolution by employing a longer flight path, but no such improvement has been reported, due, no doubt, to the development of more powerful pulsed sources. All of these pulsed cyclotrons accelerated deuterons and used beryllium targets.

Another circular accelerator which is useful as a pulsed neutron source is the betatron which is intrinsically a pulsed accelerator. A circulating electron current, established by injection from an electron gun, is accelerated by the increasing magnetic flux through the orbit. During acceleration the orbit radius remains constant, and to obtain short pulses of X-rays when the desired electron energy has been reached, the orbit must be rapidly shifted so that it impinges on a target placed close to the equilibrium orbit. This method of producing short pulses has the advantage that the total X-ray yield is independent of the duration of the X-ray pulse, and depends only on the magnitude of the circulating electron current.

In the case of the 100-MeV General Electric Betatron (Yeater et al., 1957) the orbit shifting was accomplished by rapidly contracting the orbit so that the electrons struck a square laminated natural uranium target of 1.9-cm side and 0.8 cm thick with its outer edge placed 5 cm inside the equilibrium orbit, and produced an X-ray burst which was approximately gaussian, with a width at half-amplitude of 0.11 μsec. The uranium target performed the dual function of producing an intense X-ray pulse from the electron beam, and of partially absorbing the X-rays to give intense pulses of neutrons due to various photonuclear reactions with the uranium nuclei. Uranium was used because, having a high Z, it has a high photonuclear cross section which peaks at the relatively low energy of 15 MeV so that the target could be of small physical size, while giving a high neutron yield. An additional advantage was that in a heavy nucleus like uranium, the photofission process is strongly excited, which augments the neutron yield. The uranium target was partially surrounded by a hydrogenous moderator to increase the yield of resonance neutrons for time-of-flight work. According to Yeater et al. (1957) the best mean neutron flux obtained with the betatron was 8×10^{10} neutrons/sec. Since the repetition rate was 60 pps (pulses per second), and the width at half-height of the pulse 0.11 μsec, this corresponds to a peak neutron emission rate in the pulse of the order of 10^{16} neutrons/sec. The flight paths used with the betatron spectrometer had lengths of 7 and 20 meters, respectively. The nominal resolution with the 20-meter flight path was better than 6 nsec/meter.

2. Electron Linear Accelerators

In principle, the electron linac offers several advantages over circular machines as a pulsed neutron source. First, for a given peak neutron yield it is less costly than the equivalent circular machine, since no large magnet is involved in its design. Second, the pulse length and repetition frequency can easily be changed, and finally the linac, being linear, automatically provides an extracted beam, so that many flight paths may be used simultaneously around a target placed some distance from the accelerator itself, whereas most circular machines use internal targets and a single flight path. It is true that the use of multiple flight paths introduces certain difficulties not encountered in the single flight path arrangements. For example, it is not possible to have all of the flight paths normal to a plane neutron source, so that some flight paths may have excessive length uncertainty Δl due to this feature. A careful allocation of the experiments to the flight paths, however, can minimize this drawback. Again, because of the geometry, it is difficult to combine the use of a large number of flight paths with optimum moderation efficiency so that a loss of up to a factor of 2 may have to be accepted in neutron intensity in return for the multiplicity of experiments. These difficulties, however, do not detract seriously from the advantages of the electron linac as a flexible pulsed neutron source, and the number of large electron linac installations now in use would seem to indicate that this is the logical machine to use where an installation is being set up primarily for time-of-flight resonance neutron spectroscopy.

The first suggestion that the electron linear accelerator might prove to be very suitable as a pulsed neutron source, was made by Cockcroft (1949). The 3-MeV accelerator first used at Harwell (see, for example, Merrison and Wiblin, 1952) to test this suggestion, was similar to that described by Fry *et al.* (1948) and was driven by a single magnetron operating at 3000 MHz and giving a 2-μsec pulse of rf power at a peak rate of 2 MW. The electron beam (peak current 125 mA) was accelerated to 3.5 MeV. Neutrons were produced by stopping the electrons in a water-cooled platinum X-ray target 0.02 in. thick beyond which was placed a thin aluminum box containing compressed beryllium metal of dimensions 38 $\times$ 15 $\times$ 15 cm which was large enough to absorb two-thirds of the energetic photons. Beryllium was used because of its low photoneutron threshold, and the low energy of the accelerator. The peak neutron production rate was 2.5 $\times$ 10^{12} neutrons/sec. This pulsed neutron source was used with a 10-meter flight path and had a nominal resolution of 200 nsec/meter. Its resolution was thus comparable with the Brookhaven 60-in. pulsed cyclotron (Section II,B,1) though the peak neutron yield was lower by two orders of magnitude.

Another early electron linear accelerator which was used for neutron time-of-flight work was the one constructed at Yale University (Schultz and Wadey, 1951). The accelerator gave a peak beam current of 30 mA at 5 MeV electron energy, with a variable pulse length of 1 to 2 μsec, and a peak neutron production rate of 3 $\times$ 10^{12} neutrons/sec, which was very similar to the original Harwell machine. The Yale University machine used a gold X-ray target and beryllium neutron convertor. A later modification increased the electron energy to 10 MeV and the peak current to 100 mA with a peak neutron output of 10^{14} neutrons/sec. The flight-path length could be varied from 5 to 15 meters and the nominal resolution was about 140 nsec/meter.

In 1952, the 3-MeV accelerator at Harwell was replaced by a more powerful machine producing electrons of up to 15 MeV. This machine was described in detail by Bareford and Kelliher (1953), and its use as a neutron spectrometer by Wiblin (1955). The accelerator driven by a single 2 MW magnetron provided initially 25 mA peak electron current in a 2-μsec pulse at a repetition rate of 400 pps and with an electron energy of 14 MeV. Under these running conditions, the duration of the neutron pulse was found to be about 1 μsec, and the peak neutron output was about 3 $\times$ 10^{14}

neutrons/sec using a uranium target 2 in. in diameter and 2 in. long, surrounded by a Plexiglass moderator one inch thick. An important feature of this spectrometer was the first use of multiple flight paths, separate flight paths being set up for fission, scattering, capture, and total cross-section measurements of lengths varying from 3 meters (for fission cross-section measurements) to 60 meters for transmission work, five separate 100-channel analyzers being used to take data. Nominal resolution varied from 380 nsec/meter for the early fission cross-section work at 3 meters, to 17 nsec/meter for transmission work at 60 meters.

This 15-MeV accelerator was later improved by fitting a separate fast modulator on the electron gun (Firk *et al.*, 1958) which applied a large triangular pulse of amplitude 50 kV to the gun cathode. By biasing the pulse transformer which applied this triangular pulse to the gun cathode, it was possible to produce neutron pulses as short as 50 nsec. Since the gun-modulator pulse could be adjusted in phase relative to the rf pulse to give optimum neutron output, and since the lowering of the duty cycle of the machine by using short pulses reduced the loading on the rf system and increased the electron energy, it was found that the pulse length could be reduced from 1 μsec to 100 nsec for a loss of only a factor of 2 in total neutron output. Under these conditions the neutron production rate in the pulse had risen to about 10^{15} neutrons/sec. In this condition the machine gave a nominal resolution $\sim$2 nsec/meter with a 60-meter flight path (Firk *et al.*, 1960).

C. Current Technology and Applications

1. Cyclotrons

Although the field of resonance neutron spectroscopy is currently dominated by the electron linac installations, the highest instantaneous pulsed neutron intensities are, in fact, provided by cyclotrons; and it seems likely, as mentioned earlier, that the ultimate pulsed neutron source must finally be provided by a high-energy proton accelerator because of the thermal efficiency of the spallation reaction at high energies. The latter consideration is only of future interest since current technology does not allow us to approach the heat transfer limit, but the high pulse intensity of the cyclotrons is important at the high-energy end of our field of interest, say, at 100 keV neutron energy. The isochronous cyclotron at Karlsruhe (Cierjacks *et al.*, 1966, 1968), for instance, accelerates deuterons to 50 MeV and the pulsed current reaching the target is 3 A. This yields a neutron production rate of 6 × 10^{17} neutrons/sec in a pulse length of about 1 nsec. In addition, the Karlsruhe cyclotron has a very high repetition rate (20,000 pps in 1966 with an anticipated increase to 160,000 pps). It therefore follows from the short pulse, high repetition rate, and long flight paths (up to 180 meters) used with this facility, that its main usefulness is with unmoderated targets in the energy region well above 100 keV where the moderation time uncertainty is still 5 nsec.

Another cyclotron installation which provides even higher pulsed neutron intensities is the synchrocyclotron at Harwell. This accelerator was modified during 1968 (Taylor, 1969) to give pulses of 150-MeV protons with a duration of 4 nsec and a current again of 3 A. From a heavy target this yields a neutron production rate of 3×10^{19} neutrons/sec (cf. Havens, 1966), and repetition rates of up to 1600 pps for deuterons and 800 for protons are available. This accelerator is equipped with three underground flight paths of lengths up to 100 meters and with many detector stations, and will certainly become an important spectrometer installation for both high- and low-energy work, since the pulse length in this case is tailored to the moderation time spread at 100 keV. The use of this spectrometer in the low-energy field is, however, at an early stage.

The only cyclotron installation fully active in the field of resonance neutron spectroscopy, and which makes use of the spallation reaction with high-energy protons, is the 385-MeV synchrocyclotron at the Nevis site of Columbia University. This machine has been used for resonance work for over a decade (Rainwater et al., 1960), and currently provides pulses of 385-MeV protons of duration 25 nsec with a pulse current of 330 mA. This yields a neutron production rate of 1.6×10^{19} neutrons/sec in the pulse (Havens, 1966) which is half as great as that obtained with the Harwell machine, but which is sustained for a much longer time, thus increasing its usefulness at lower neutron energies. Its repetition rate is very low (60 pps), but plans exist for increasing both the repetition rate and the pulse current (Rainwater, 1966). The reason that the peak neutron production rate is less than a factor of 2 below that of the Harwell accelerator while the pulse current lies almost a factor 10 below that machine lies in the higher proton energy which increases the neutron yield considerably as shown in Fig. 2 (Havens, 1966). The main characteristics of these and earlier spectrometers based on circular accelerators are shown in Table I.

When a synchrocyclotron is used as a pulsed neutron source, the principle of the pulsing is that the circulating bunched proton beam is suddenly deflected to strike a heavy target. Because of the bunching of the protons, the beam hits the target in a time considerably shorter than the circulation time of the beam, but which is nevertheless dependent on that time, and hence on the size of the accelerator. The Columbia synchrocyclotron is

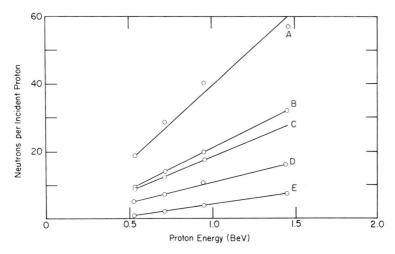

FIG. 2. Neutron production by high energy protons. A: U (4 in. × 24 in.); B: Pb (8 in. × 24 in.); C: Pb (4 in. × 24 in.); D: Sn (4 in. × 24 in.); E: Be (4 in. × 36 in.) (Havens, 1966, p. 568).

used primarily for meson physics, but is fitted with an electrostatic deflection system, a heavy moderated target, and a flight path so that it can be used for neutron time-of-flight work. The neutrons are produced when the circulating proton beam is deflected downward to strike the water-cooled

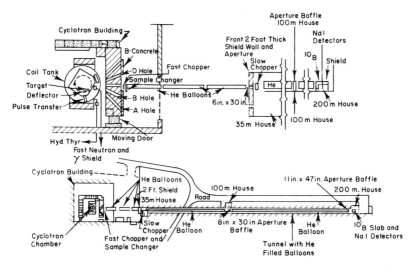

FIG. 3. Plan view of the Nevis time-of-flight system including the synchrocyclotron and the 200-meter flight path (Rainwater, 1966, p. 393).

TABLE
SPECTROMETERS BASED

Spectrometer installation	Particle accelerated	Target	Energy (MeV)	Current in pulse (mA)	Neutron production rate in pulse (neutrons/ sec)	Pulse duration (nsec)
Columbia 36-in. cyclotron	Deuterons	Be	8	10	6×10^{13}	2000
Brookhaven 60-in. cyclotron	Deuterons	Be	21	1.2	3×10^{14}	1000
General Electric betatron	Electrons	U	100	30	2×10^{16}	110
Columbia 170-in. synchrocyclotron	Protons	Pb	385	330	1.6×10^{19}	20
Karlsruhe 90-in. isochronous cyclotron	Deuterons	U	50	3000	6×10^{17}	1
Harwell 110-in. synchrocyclotron	Protons	W	150	3000	3×10^{19}	4

lead target. Figure 3 shows the general layout of the synchrocyclotron spectrometer, and Fig. 4 shows in more detail the deflection electrodes and the lead target with its polyethylene moderator. This whole assembly can be moved remotely from outside the vacuum system for optimum positioning. At the appropriate point in the machine cycle, when the orbit radius has reached 69 in., the beam is suddenly deflected downward by the graphite electrostatic deflectors. These are $\frac{3}{4}$ in. thick and 3 in. wide, and subtend 0.71 rad of arc at 69 in. radius, and are supported by swivel-mounted copper pipes on lucite insulators so that they can be adjusted for proper height from outside the vacuum chamber, and moved out of the way when the machine is used for high-energy physics. The polyethylene moderator ($\sim$4 cm thick) is situated below the lead target, and the flight path is so collimated that only the moderator is seen by the detectors. The shield wall of the synchrocyclotron is of steel, 10 ft thick, and the collimated path for the neutron beam through it is filled with helium and has dimensions $2\frac{3}{4}$ in. $\times$ 9 in. Immediately outside the shield wall, the transmission sample changer and a fast chopper are situated; a second slow chopper is sited at 35 meters from the source. These choppers can be phased

I

ON CIRCULAR ACCELERATORS

Repetition rate (pps)	Flight path length (meters)	No. of flight paths	Best nominal resolution (nsec/ meter)	No. of timing channels	Reference
60	6	1	330	64	Rainwater et al. (1947) Rainwater and Havens (1946) Havens and Rainwater (1951)
60	4.3	1	200	64	Havens (1955)
60	7 or 20	1	6	200	Yeater et al. (1957)
60	37 or 200	1	0.1	16,000	Rainwater et al. (1960) Havens (1966) Liou et al. (1968)
20,000 160,000	56 or 200	1	0.005	—	Cierjacks et al. (1966)
up to 800	8 to 100	3	0.04	16,000	Taylor (1969)

with the accelerator pulses, and are used to remove from the neutron pulse fast neutrons and γ-rays which would adversely affect the background conditions for low-energy studies ($\lesssim 300$ eV) with the 200-meter flight path, and also to minimize overlap effects between cycles due to very slow neutrons from the moderator (Rainwater, 1966). This latter problem is more usually solved by the insertion of a filter of ^{10}B which has a $1/\sqrt{E}$ cross-section dependence at low energies, with a very large cross section (4000 b) at thermal energy.

2. APPLICATION OF CYCLOTRONS

Most recent cross-section work at Columbia has been concerned with high-resolution measurements of total cross sections with the 200-meter flight path. The nominal resolution of the system is 0.13 nsec/meter which is extremely good. This high resolution is only meaningful, as was pointed out earlier, at energies above the point where the moderator time spread Δt_m is of the same order as the accelerator pulse length, 25 nsec. This means above ~ 5 keV.

The detector used in these transmission measurements is notable for its

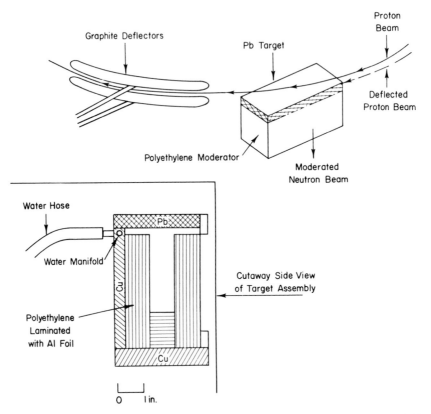

Fig. 4. Deflection system and target for Nevis synchrocyclotron (Rainwater, 1966, p. 393).

size and efficiency. It is of the ^{10}B-NaI type first introduced by Rae and Bowey (1953) in which the NaI crystals are used to detect the 480-keV γ-ray from the reaction ^{10}B(n,αγ)^{7}Li. The Columbia detector is interesting because of its large size and the fact that it allows two alternative geometries best suited to higher and lower energy neutrons. The detector is shown in Fig. 5, and it will be noted that the ^{10}B slab has the dimensions 30 cm × 120 cm × 4 cm and contains 15 kg of ^{10}B. The thickness of the ^{10}B is such that up to ∼10 keV, virtually all neutrons entering the slab are captured. The four large NaI crystals used to detect the γ-rays from the (n,αγ) reaction can be placed either above the boron slab, and outside the neutron beam as in most other detectors of this type or else directly behind the slab. The latter geometry achieves very high efficiency (close to 50%) because of the large solid angle subtended by the crystals at the slab, but can only

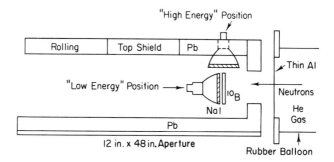

FIG. 5. The 200-meter detector: ^{10}B $\sim$ 12 in. $\times$ 48 in. $\times$ 1$\frac{1}{2}$ in. (Rainwater, 1966, p. 395).

be used for low-energy neutrons, and with good suppression of the γ-rays and fast neutrons from the cyclotron.

Some figures on the timing and energy resolution of the Columbia system, also taken (like the figures) from Rainwater (1966) are given in Table II below. It will be noted that because of moderator effects the resolution of 0.2 nsec/meter represents the effective resolution at about 1 keV (assuming

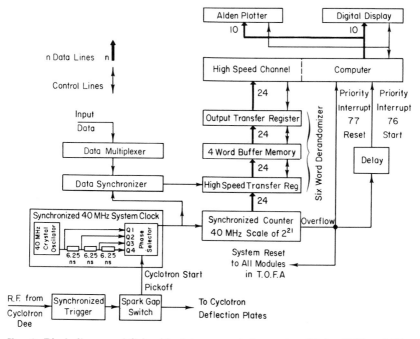

FIG. 6. Block diagram of Columbia data accumulation system (Hahn, 1968, p. 105).

TABLE II

TIMINGS AND ENERGIES OF INTEREST FOR THE SYSTEM

E	Microseconds flight time for:				Energy resolution for:		Doppler width Δ (eV) for $kT = 1/40$ eV for:	
	3.7 cm	7 m	35 m	200 m	1 nsec/m	0.2 nsec/m	A = 50	A = 200
4 eV	1.34	254	1265	7230	0.0002 eV	0.00004 eV	0.089	0.0447
10 eV	0.85	160	800	4560	0.0009 eV	0.0002 eV	0.1415	0.0708
40 eV	0.42	80	400	2280	0.007 eV	0.0014 eV	0.283	0.1415
100 eV	0.27	50.8	253	1446	0.028 eV	0.006 eV	0.447	0.224
400 eV	0.134	25.4	126	723	0.222 eV	0.044 eV	0.894	0.447
1000 eV	0.085	16.0	80.0	456	0.878 eV	0.176 eV	1.415	0.708
4 keV	0.042	8.0	40.0	228	7.02 eV	1.40 eV	2.83	1.415
10 keV	0.027	5.08	25.3	144.6	27.7 eV	5.5 eV	4.47	2.24
40 keV	0.013	2.54	12.6	72.3	222 eV	44.3 eV	8.94	4.47
100 keV	0.008	1.60	8.00	45.6	878 eV	176 eV	14.15	7.08
400 keV	0.004	0.80	4.00	22.8	7.02 keV	1.40 keV	28.3	14.15
1 MeV	0.003	0.508	2.53	14.5	27.7 keV	5.54 keV	44.7	22.4
4 MeV	0.001	0.254	1.26	7.23	222 keV	44.3 keV		
10 MeV	0.001	0.160	0.80	4.56	878 keV	176 keV		

negligible timing channel width and detector jitter) while 1 nsec/meter represents the effective resolution at about 50 eV. We should also note that the Doppler broadening of resonances at room temperature exceeds the instrumental resolution up to ~5 keV.

The time-of-flight electronics used by the Columbia group since 1968 are centered on an ASI 6050 computer with magnetic-tape and disk backing stores and with card-reader typewriter plotter and display unit. The computer has a 16-k fast core store ($k \equiv 1024$) with 1.9-μsec cycle time and 24-bit word length. When used for time-of-flight data acquisition, the clock speed is 40 MHz giving 25-nsec channels. A block diagram of the timing system, taken from Hahn et al. (1968) is shown in Fig. 6. The data go into a multiplexer which can handle up to eight independent detectors and gives a 3-bit detector identification code. The timing information is joined to the detector code to give a 24-bit descriptor which is transferred to a 6-word fast derandomizing buffer which has a dead time of 80 nsec— thence to the computer which can accept data at a rate corresponding to its cycle-time of 1.9 μsec.

As stated above, the major contribution of the Columbia group has been in the field of very high resolution resonance studies (Garg et al., 1965; Wynchank et al., 1968) mainly in the study of neutron total cross sections.

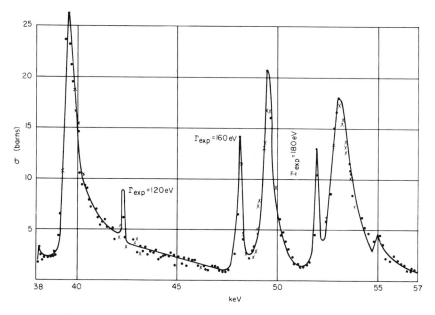

FIG. 7. High resolution total cross section of vanadium (Havens, 1966, p. 586).

A section of a high-resolution measurement on vanadium at an energy ~50 keV is shown in Fig. 7 (Havens, 1966) which shows two narrow resonances for which the observed width is compatible with an effective resolution of ~0.5 nsec/meter. Since this run was taken on the 200-meter flight path with timing channels of 100 nsec width, the observed resonance widths are determined essentially by the timing channels.

Another resonance study technique, unique to Columbia, which has been useful in resonance spin assignments, is the application of the self-indication method to high-resolution studies. Figure 8 shows a section of a self-indication measurement on silver (Rainwater, 1966) in which two curves are displayed. The higher curve (D or detector foil only) represents the counts observed in a plastic scintillator γ-ray detector placed close to a silver foil in the neutron beam, as a function of time-of-flight. The lower curve represents the yield when a second foil of the same material is placed in transmission. The use of data of this type in resonance analysis is illustrated in Fig. 9a (Liou *et al.*, 1968) which shows the analysis of the 34.6 eV resonance in ^{171}Yb. The various curves in the Γ vs $g\Gamma_n$ plane represent functional relationships derived from the various types of resonance data, as explained in the legend. In Fig. 9b the curves are concurrent, indicating

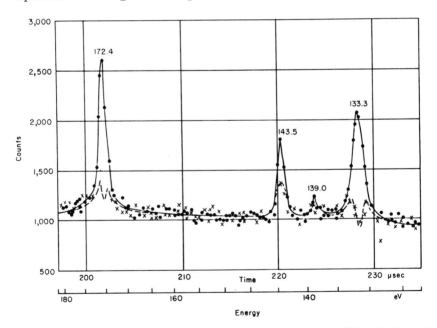

Fig. 8. Self-indication data on silver. $\bullet$, D only; $\times$, D and T; $1/N = 103.3$; $\tau = \frac{1}{4}$ (Rainwater, 1966, p. 395).

the set of parameters Γ_n, Γ, and J for which all the relationships are satisfied simultaneously. In addition to the high-resolution resonance work, the Columbia group are also active in the field of fission physics, studies having been made of fission-fragment energy and mass distribution as a function of neutron energy and in the study of ternary fission (Melkonian and Mehta, 1965; Mehta and Melkonian, 1966; Mehta *et al.*, 1967).

3. ELECTRON LINEAR ACCELERATORS

a. Survey. Since the days of the early projects described in Section II,B the traveling-wave electron linac has been intensively developed. The rf power source now universally used is the klystron, a pulsed power-amplifier for centimeter wavelengths. A simplified explanation of its action is that a high-current electron beam is bunched in velocity as it traverses a resonant cavity excited by a low-power external source. The electron beam then traverses a drift tube in which the velocity modulation is converted to a current modulation; the beam then passes through a second resonant cavity which it excites to produce a large gain in rf power. Klystron amplifiers as used in electron linac technology can now produce peak power ratings of tens of megawatts, and can operate at mean power ratings of several tens of kilowatts. Being amplifiers, many of these tubes can be driven from the same stable rf supply, and can be accurately adjusted in phase so that very long accelerators with many sections driven by many klystrons are possible. The outstanding accelerator in this connection is in the SLAC project at Stanford. This largest of all *S*-band accelerators is 2 miles long and is driven by 240 klystrons each producing a peak power of over 20 MW. The resultant electron beam has an energy of 20 GeV, and is used for high-energy physics studies. An interesting popular account of the history of this fantastic project is given by Dupen (1966) and Neal (1968).

The accelerators used as pulsed neutron sources are, of course, of more modest energy, which ranges from 40 to 150 MeV. As shown in Fig. 10 from Leiss (1966), very little is gained in neutron production per megawatt by increasing the electron energy above 100 MeV, and as high-energy machines are long, and involve expensive buildings and shielding, the trend in this field is toward higher peak beam currents, with flexibility in pulse length and repetition frequency. The combination of high beam current and short pulses can be achieved most efficiently with an electron linac in terms of hardware cost by using the accelerating waveguide as a store for the rf energy. A fraction of this energy is then transferred to a high-current electron beam which is passed down the waveguide for a time which is short compared with the filling time of the waveguide structure. Under these conditions the peak power in the beam, and hence the peak neutron produc-

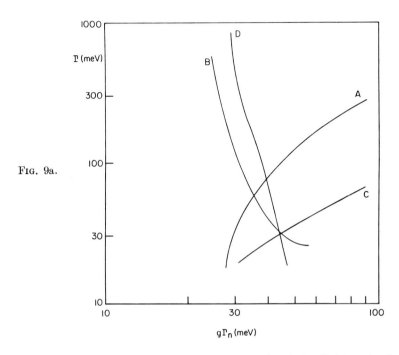

FIG. 9a.

FIG. 9. Analysis for the 34.6 eV resonance in ¹⁷¹Yb. The implied functional relationships depend on the choice of the assumed angular momentum J of the compound
nucleus resonance. (a) shows the analysis curves assuming $J = I + \frac{1}{2} = 1$, while (b)
shows the corresponding plots assuming $J = I - \frac{1}{2} = 0$. Inspection shows excellent
agreement of the four curves for $J = 0$ and the inconsistency of curve A with the common intersection point of curves B, C, D for $J = 1$. Curve A results from the self-
indication "$D + T$" to "D only" resonance area ratio. Curve B is based on the "D only"

tion rate, can exceed the peak rf power by factors of up to ∼20 at least;
whereas under steady state conditions, in long pulses, the peak beam power
does not exceed ∼70% of the peak rf power even with optimum design.
Such large factors of enhancement in beam power for pulses of length applicable in the resonance neutron spectroscopy field, say 10–20 nsec duration,
have only been achieved recently. In order to obtain these factors it is
necessary to have a large amount of energy stored in the guides, i.e., a long
filling time, and this requires either rather long or rather large diameter
waveguide sections or both. The diameter of the waveguide can be
increased by going from S-band (10 cm) to L-band (23 cm) rf supplies,
and this is clearly advantageous for short-pulse high-current work. The
new Oak Ridge electron linac project (ORELA) makes use of a 140-MeV

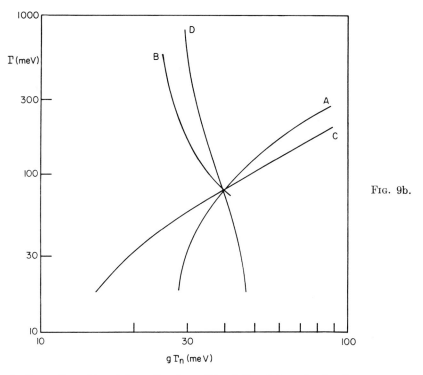

FIG. 9b.

peak. Curve C combines information from "*D* only" and from the flat detector transmission measurements. Curve D is based on flat detector transmission measurements
only. It is the difference in slopes of the curves that determines the degree of localization
of the intersection point and thus $g\Gamma_n$ and Γ separately. An additional self-consistency
test (not shown) is that $\Gamma_\gamma = \Gamma - \Gamma_n$ should be near the average value $\langle\Gamma_\gamma\rangle$ for levels in
the given isotope. This test further rules out the solution given by the intersection point
of curves B, C, and D in (a) (Liou *et al.*, 1968, pp. 11–12).

electron linac which gives 15 A in short pulses up to ∼24 nsec for a total
klystron peak power of ∼100 MW (Harvey, 1969). This represents a
stored energy enhancement factor of 20.

   Even higher stored energy enhancement is possible with the use of low-
frequency standing-wave cavity accelerators, such as the PHERMEX
accelerator at Los Alamos (Venable, 1964; Boyd *et al.*, 1965). This 50 MHz
cavity accelerator is driven by nine 1 MW amplifiers, yet so great is the
stored energy in the huge high Q cavities (inside diameter 4.6 meters)
which have a filling time of ∼1 msec, that 28-MeV electron pulses are
produced having a duration of 0.1 to 0.2 μsec and current ∼70 A, i.e., a
peak beam power ∼2000 MW as compared with the rf drive of 9 MW, an
enhancement factor ∼200. Each pulse is composed of a train of 6 nsec

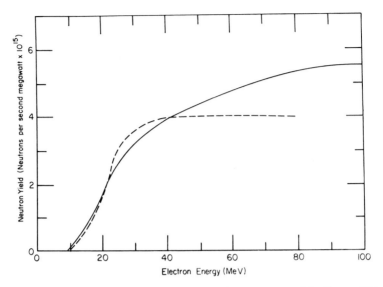

FIG. 10. Neutron yield per megawatt from uranium target. —, MacGregor (GA and ARCO); ——Baldwin, Gaerttner, and Yeater (GE) (Leiss, 1966, p. 612).

micropulses separated by 20 nsec, so that the enhancement factor in a single micropulse is even greater, ∼600. As originally used for flash radiography, PHERMEX was operated in a single-pulse mode, but various proposals for modification to produce a repetitively pulsed neutron source were made by Motz (1965). He suggested that the machine as it stood could, by modification of the injector system, be made to operate repetitively at up to 15 pps, but higher repetition frequencies would involve extensive injector development work, and an improved cooling system for the cavities.

A modified PHERMEX type accelerator, in providing ∼6000 MW of power in a 6 nsec pulse at 28 MeV would provide an intrinsic pulsed neutron flux of ∼10^{19} neutrons/sec as compared with the figure of 4×10^{18} for ORELA when used with a tantalum target. It is unlikely, however, that the cavity accelerator could compete in repetition rate, and the minimum pulse length of ORELA at 2.3 nsec would certainly give it the advantage for fast neutron work. Where PHERMEX would really score would be at lower neutron energies below 1 keV, say, where the moderation-time jitter is ∼100 nsec, and where PHERMEX can still maintain a target pulsed power of 2000 MW and a neutron production rate of 3×10^{18} neutrons/sec for a pulse of 100 nsec duration. This has not yet been achieved, but another method of enhancing the neutron yield from longer pulses for use at lower energies is the use of a neutron multiplier or booster. Such a device (essentially a small subcritical fast reactor) which currently gives a production rate of $\sim 10^{18}$ neutrons/sec in a 100 or 200 nsec pulse has been in use with the Harwell electron linac (Neutron Project) for a decade or more (Poole and Wiblin, 1958). It gives a factor of 12 enhancement of neutron yield over a plain natural uranium target of the type used in many linac installations and a factor of enhancement of at least 20 over a water-cooled tantalum target such as is in use at ORELA. The Harwell booster, which will be described later, has a relaxation time of 80 nsec and so is particularly useful for neutron energies below 1 keV.

Another device which is used as a booster and which has been extensively employed for neutron time-of-flight spectroscopy is the IBR pulsed reactor at Dubna (Blokhin *et al.*, 1962; Malyshev, 1966). This plutonium-fuelled, small, fast reactor is pulsed mechanically in reactivity and can be used in one of two modes. In its primary mode it is briefly pulsed to a prompt critical condition which produces a 40-μsec pulse with a neutron intensity $\sim 3 \times 10^{18}$ neutrons/sec. This mode is used for research work with thermal neutrons, mainly in the solid state field, and the mean power rating when used in this way is limited to 6 kW by cooling problems. It can also, however, be pulsed mechanically to a subcritical condition with a gain of 200, and in this mode it is driven by a 30-MeV microtron to give a pulse length of 4 μsec. With this duration of pulse the system is still optimized for very low energies, $\lesssim 1$ eV, but it has been used successfully with very long flight paths (up to 1 km) for resonance region work, including the first experiments with neutrons polarized by a dynamically polarized proton filter (Shapiro, 1966). This machine is currently being converted to operate at higher powers and to be driven by a 45-MeV electron linac (Shapiro, 1968). Proposals for even more powerful long pulse boosters have been made by Poole (1966) and Beyster and Russell (1966).

The present generation of high-current, klystron-driven, traveling-wave electron linacs dates from around 1957 onward, and the main characteristics of a selection of machines which are used as pulsed neutron sources is given in Table III along with those of the earliest linac spectrometers. It is noticeable that the electron energy has tended to rise over the past decade, while pulse currents have risen from 100 mA to ~ 15 A. Most of the well-established installations have been designed for a wide variety of applications, including electron and photon irradiation, electron scattering, pulse radiolysis, photonuclear studies, reactor physics, and so on, in addition to neutron spectroscopy, so that in many cases the number of flight paths and instrumentation available for the study of neutron interactions is limited.

Since our interest is in the use of the electron linac for resonance neutron

Spectrometer installation	Frequency (MHz)	Number of klystrons	Total rf power (MW)	Beam current (mA)	Beam energy (MeV)	Pulse duration (nsec)	Max prf (pps)	Target
Harwell	3000	1 (mag)	2	125	3.5	2000	200	Pt/Be
Yale (cavity accelerator)	600	8 (triodes)	0.5	100	10	1000	90	Au/Be
Harwell	3000	1 (mag)	2	25	14	1000	400	U
				100	15	100	400	
Saclay	3000	3	15	0	55			
				650	40	100	1000	U
				2000	50	10	1000	U
Kurchatov Institute (Moscow)	3000	1	20	0	32	50–600	120	U
				500	25			
Harwell	2998	7	56	0	55			
				550	35	1700	500	U
				650	40	100	500	^{235}U booster
				2000	50	10	750	U
Livermore	3000	3	15	0	50			
				250	30	20–2000	360	U
San Diego	1300	3	30	0	46			
				750	23	4500		
				2000		10	720	—
Yale	1300	5	50	0	60			
				700	40	4500		
				1600		100	500	—
RPI (Troy)	1300	9	90	0	100			
				800	60	4500		
				2000		10	720	Ta
Tokai-Mura (Japan)	2857			(200)	2–20	4000		Pb
						40	300	
Toronto	2856	2	40	0	43			
				500	27	2000		W
				500	35	<10	960	
NBS (Washington)	1300	12	120	0	250			
				680	130	7000		
				(2000)		1.5	720	(Cd)
Geel	2856	2	66	0	66			
				350	43	2000	250	
				1200	59	100	880	U
				3600	64	10	1000	
San Diego	1300	4	70	0	80			
				700	50	7–4500	720	—
				3500	40	50	720	
Oak Ridge	1300	4	100	500	140	1000	1000	Ta
				15,000	140	2.3–24	1000	
Livermore	3000	6	120	0	200			
				700	80	3000	600	U
				10,000	140	5	660	
							(1800)	

ON ELECTRON LINEAR ACCELERATORS

Neutron production rate in pulse (neutrons/sec)	Flight path length (meters)	No. of flight paths	Best nominal resolution (nsec/meter)	No. of timing channels	Reference
2.5×10^{12}	10	1	200	32	Fry et al. (1948)
					Merrison and Wiblin (1952)
10^{14}	5–15	1	70	500	Schultz and Wadey (1951)
					Schultz et al. (1956)
2×10^{14}	3–60	4	2	500	Bareford and Kelliher (1953)
10^{15}					Wiblin (1955)
					Firk et al. (1958)
10^{17}	5–200	6	0.05	400,000	Leboutet et al. (1957)
4.5×10^{17}				(4×10^{9})	Roland et al. (1967)
3×10^{16}	2–110	7	0.5	10,000	Vladimirsky and Sokolovsky (1958)
					Voronkov et al. (1962)
					Malyshev (1966)
10^{18}	5–300	12	0.5	650,000	Poole and Wiblin (1958)
4.5×10^{17}	100	1	0.1	(7×10^{10})	Poole and Robinson (1964)
					Austin and Fultz (1959)
2×10^{16}	10–60	4	0.3	—	Fultz (1969)
—	19–50	2	0.2	—	Sund et al. (1964)
				2000	
2×10^{17}	—	—	—		Harvey (1963)
10^{17}					Gaerttner et al. (1962)
5×10^{17}	8–250	4	0.04	8000	Tatarczuk and Block (1967)
2×10^{15}	10–50	2			
			0.8	500	Takekoshi (1965)
—	up to 60	3			McNeill (1967)
			<0.16	—	
					Harvey (1963)
					Schwartz et al. (1968)
$\sim 10^{18}$	up to 40	—	0.04	16,000	
6×10^{16}	30–400	9	0.025	$(16,000)$	Allard and Salome (1966)
3×10^{17}					
1.2×10^{18}					
$(\sim 10^{17})$	19–220	3	0.03	—	Adcock et al. (1967)
$(\sim 4 \times 10^{17})$					
$(\sim 10^{17})$	10–80	11	0.03	$\sim 10^{6}$	Harvey (1969)
4×10^{18}					
3×10^{17}	16–250	10	0.02	—	Fultz (1969)
8×10^{18}					

spectroscopy and since space does not permit us to examine each of the installations in Table III in detail, we shall restrict ourselves to looking more closely at a few examples of installations used primarily for resonance neutron spectroscopy, and then go on to considering briefly the types of experiments carried out with linac pulsed sources, in general. The choice of examples is difficult since many of the laboratories listed have made extensive contributions to the literature of our subject, but we have selected as our first example the Saclay Laboratory which is a well-developed facility making extremely good use of a rather small accelerator, a straightforward natural uranium target, and a sophisticated data-accumulation system. The second example is the Neutron Project at Harwell which utilizes the booster target mentioned earlier, and which has a unique programmed beam-handling system. The third and final example is the ORELA Project at Oak Ridge which will provide the most intense short neutron pulses currently available from an electron linac for use in the upper part of our energy range.

b. Saclay. The current accelerator at Saclay has an unloaded energy of 55 MeV and a pulse width variable from 10 nsec to 2 μsec. It is operated typically with a burst width of 100 nsec and under these conditions gives 650 mA of 40-MeV electrons at repetition rates up to 1000 pps. The maximum current available in a 10-nsec pulse is 2 A. We note that when running with 100-nsec pulses the peak beam power of the accelerator (26 MW) approaches twice the peak rf power and is maintained for 20% of the waveguide filling time ($\sim$500 nsec). If we allow for the loss of about half of the rf power in the copper of the waveguides, this means that $\sim$70% of the stored energy is extracted. The neutron target of the Saclay accelerator is a cylinder of natural uranium 3 cm in diameter and 10 cm long which is water cooled and has a maximum power rating of 4 kW. The cylinder has a re-entrant conical hole into which the electron beam is directed, so that the heat produced by the beam is spread over a considerable area on the surface of the cone. Figure 11 (Michaudon, 1964) shows a section through the target which illustrates clearly the disposition of the two moderator slabs and the large quantity of ^{10}B and B_4C placed between them. The object of the boron is to "decouple" the two moderator slabs and prevent the migration of neutrons between the two slabs which would lengthen the neutron pulse. The relaxation time of the uranium target is of the order of the transit time of a primary neutron, i.e., $\sim$3 nsec, so that the timing uncertainty is controlled by the accelerator pulse length and the moderator timing uncertainty. A remotely controlled target changer has been installed in association with a separate storage chamber. Up to ten different targets can be stored and exchanged.

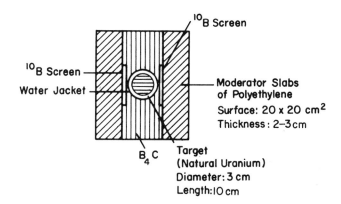

FIG. 11. Schematic section of Saclay neutron target. The electron beam is perpendicular to the section (Rae, 1967, p. 844).

Figure 12 shows the present (1969) layout of the accelerator, (Michaudon, 1969) target cell and flight paths. Six flight paths are provided, four of them having several different detector stations arranged along their length. These flight paths range from 15 to 200 meters in length and additional shorter flight paths are available within the building itself.

The electronics and data handling system used with the Saclay linac is sophisticated and extensive. Each of the six main flight paths is equipped with a 64-k channel time-of-flight encoder ($k \equiv 1024$), and the channel widths can be varied across the spectra so that the number of channels actually used to cover a given energy range can be optimized. The minimum channel width is normally 50 nsec, but one encoder is capable of a minimum width of 10 nsec. The coded information is first fed to derandomizing buffers and can then be utilized in one of the following ways:

(i)   Each event can be recorded on a 1-in. sixteen-track magnetic tape. The recording speed of magnetic tape, as mentioned earlier, is rather low, but this type of recording is of special interest for multiparameter experiments (for example, capture gamma-ray studies) and descriptor words of up to 32 bits are possible.

(ii)   The events can be integrated in a 4-k word ferrite-core store of a "bloc memoire" Intertechnique BM 96.

(iii)   The events can be integrated in the 24-k word (16-k words for accumulation) memory of a CAE computer.

At definite preset time intervals the contents of the computer memory are dumped on to an IBM $\frac{1}{2}$-in. seven-track magnetic tape, as are the contents of the memories of the BM 96. The CAE 510 computer can handle up to

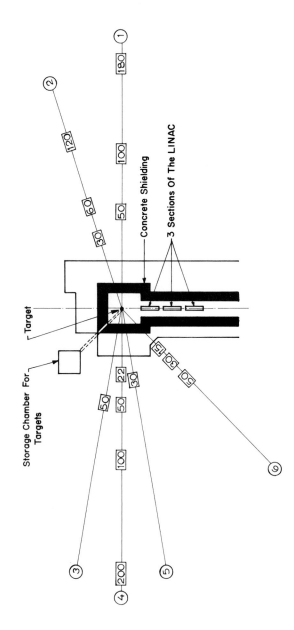

FIG. 12. General layout of the Saclay LINAC used as a time-of-flight neutron spectrometer. The detector stations are represented by rectangles. The length of the flight path is written in meters inside each of them. Short flight paths (from 5 to 18 m) are available in the building itself. (Michaudon, 1969.)

seven different experiments, and coding facilities exist so that at these pre-set time intervals the conditions of the experiments can be changed (for example, samples can be changed).

c. *Harwell.* One of the first, and still probably the largest, of the time-of-flight installations based on an electron linac is the Harwell Neutron Project. This project was built around an electron linac consisting orig-inally of six sections of S-band waveguide, each driven by a 6-MW klys-tron, the whole machine giving electron pulses of up to 750 mA at 25 MeV, with a range of pulse lengths and repetition frequencies. The neu-tron-producing target was unique in that it utilized neutron multiplica-tion in a small subcritical assembly of ^{235}U. This target gives an enhance-ment of a factor of 12 in neutron flux over a plain uranium target such as was employed on the Saclay accelerator, at the cost of introducing a relaxa-tion time in the multiplying assembly of 80 nsec. A second unique feature of the installation was the use of pulsed deflecting magnets which permit successive electron pulses to be directed onto various different target posi-tions thereby permitting several (up to three) different electron targets to be used simultaneously, each having its own selected repetition frequency and pulse length. This feature, coupled with the provision of several dif-ferent target cells and a large array of flight paths, led to great versatility in experimental arrangements.

The electron linac has been improved since its installation in 1958. It now consists of seven 1-meter waveguide sections, each driven by a klystron of peak power 7 or 8 MW and the unloaded energy of the accelerator is now 55 MeV. The accelerator is conservatively designed, and its perform-ance on 100-nsec and 10-nsec pulses is similar to that of Saclay, although the repetition rate is limited to 500 pps. It has, however, a much higher mean power and can deliver 500 mA in pulses up to 1.7 μsec in length giving a maximum mean power of 20 kW. These long pulses are useful in solid state studies and in activation analysis work. The accelerator operates on a three-shift 7-day/week basis, and produces a useful beam for 90% of its scheduled operating time.

The layout of the accelerator and target cells is shown schematically in Fig. 13. When the beam is undeflected it travels along about 40 ft of drift tube to the main pulsed neutron source, the Neutron Booster, in Cell I. Cell II contains the main beam deflecting magnets (the initial deflectors can be pulsed as mentioned earlier) and also two target positions, one equipped with a pneumatic "rabbit" for irradiation work. Cells III and IV are situated some distance from the main building and contain several target positions used for a variety of experimental work. Any one of the targets in these cells can be run together with the booster and with the

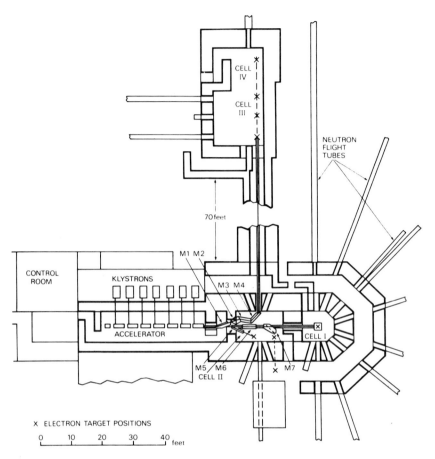

FIG. 13. Layout of target cells of Harwell neutron project (Harwell Electron Linac Brochure—1967).

pulsed target in Cell II under the programmed pulse deflection system, the pulse length and repetition rate of each target being independently selected within the restrictions of the accelerator's performance. Cells III and IV are particularly useful in that their deflection system can be locked off and work carried out safely on the target assemblies while the accelerator continues to provide beams in Cells I and II.

Some details of the construction of the booster target are shown in Fig. 14. The electron beam is stopped in a small cell through which passes a continuous flow of mercury. This permits the dissipation of up to 5 kW of electron power in a volume not much larger than 1 cc. The mercury *bremsstrahlung* target is backed by a cylinder of ²³⁵U, and both together are surrounded by a set of concentric hollow cylinders of ²³⁵U to form the booster assembly. All the ²³⁵U components are canned in stainless steel and cooled by mercury. The fission power rating of the system is 2.5 kW, and with a pulse current of 500 mA of 40 MeV electrons, the neutron production rate in the pulse is ∼10¹⁸ neutrons/sec.

The estimated factor of 12 enhancement of neutron output over a natural uranium target is made up of a factor of 10 multiplication of neutrons together with a factor 1.6 improvement given by a ²³⁵U convertor over one of natural uranium because of the increased photofission. The resulting factor is 12 rather than 16 because of the use of a mercury *bremsstrahlung* target which is less efficient than one of natural uranium. The multiplication of 10 was chosen as a compromise between neutron enhancement on the one hand, and long relaxation time and high delayed neutron fraction on the other. The use of plutonium-239 rather than uranium-235 would have permitted higher multiplication for the same relaxation time, or alternatively a narrower pulse for the same multiplication and with a smaller delayed neutron fraction, but its use was ruled out at the time by lack of availability, and safety considerations. In fact, the booster has now been in use for over 10 years and has been almost trouble free—only one assembly element having required replacement in this time, due to swelling caused by a small leak developing in the canning.

The booster is surrounded on three sides by a 1-cm thick layer of ¹⁰B

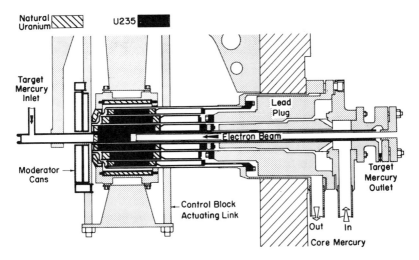

Fɪɢ. 14. Harwell neutron target (Booster) (Rae, 1967, p. 844).

element, followed by 2.5-cm thick slabs of polyethylene which serve as moderators. The function of the ^{10}B is similar to that in the Saclay target, and is to decouple the moderator slabs from the target proper and prevent the return of slow neutrons which would prolong the neutron pulse. The whole assembly is provided with a lead biological shield 12 cm thick which must be raised to surround the booster before access to the cell is possible. On a smaller scale Fig. 15 shows the layout of the project and the range of flight paths available for various target positions. There are no less than twelve flight paths arranged around the booster, and these vary in length from 5 to 300 meters. A further array of flight paths of length up to 200 meters is available for use with air- or water-cooled targets in Cells II, III, and IV.

The time-of-flight electronics and data handling system of the Harwell Project are based primarily on a 1-in. magnetic-tape recording system, and ten recorders are available with 64-k channel encoders, derandomizing buffers and pulse width programming units if required. Used in this way the minimum channel width is 62.5 nsec (16 MHz clock) but channels of width down to 1 nsec are available with the use of time expander circuits. The recorders may be used for single or multiparametric work including coding for sample changing, and so forth, with a descriptor length of up to 16 bits. Tape speeds of up to 16 in./sec can be used to accept counting rates per experiment of up to 3200 words/sec, although counting rates are normally much lower. The data tapes are analyzed by playing them back into a PDP-4 computer of 20-k word fast-core capacity (18-bit word length) of which 16-k words are available for data storage. In an average experiment a 14-bit descriptor is used and the tape is analyzed in a single 16-min pass through the reader.

The PDP-4 computer is equipped with the usual paper tape and typewriter equipment, display with light pen, slave display with automatic camera facilities, DEC-tape for systems program storage, and IBM $\frac{1}{2}$-in. magnetic-tape units for fast input-output use. The PDP-4 is used for basic data reduction and inspection, and can carry out simple operations such as averaging and integrating data, listing selected data, summation of runs, and so on. The basic data from the PDP-4 are transferred daily via $\frac{1}{2}$-in. IBM magnetic tape to a disk archive on the main Harwell IBM 360/65 computer, where final computation of cross-section data, fitting of theoretical curves, and extraction of nuclear parameters takes place.

A second smaller PDP-4 computer with 8-k words of fast store and also equipped with display and with two $\frac{1}{2}$-in. IBM compatible tape units is used for on-line recording of high count-rate experiments and multiparameter experiments where a longer word length is required. This system allows

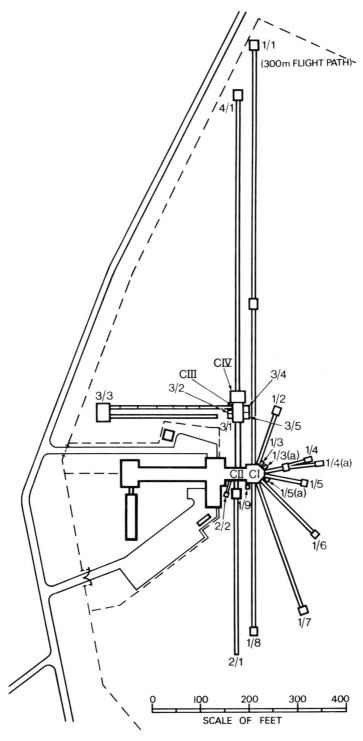

Fig. 15. Layout of Harwell experimental area (Harwell Electron Linac Brochure—1967).

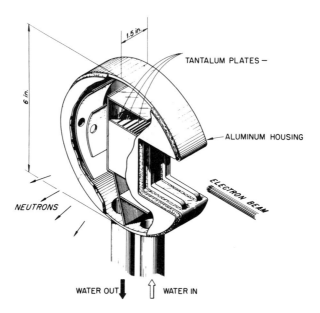

TANTALUM PLATES —

ALUMINUM HOUSING

ELECTRON BEAM

NEUTRONS

WATER OUT   WATER IN

FIG. 16. ORELA neutron target showing tantalum plates 1.114-in.-wide, 2-in.-high, with thicknesses from 0.06 to 0.31 in. (Harvey, 1969).

descriptors of up to 36 bits to be used, the descriptor words being written on to $\frac{1}{2}$-in. IBM tape. Simultaneous display of certain features of the data is provided for monitoring purposes.

d. Oak Ridge. Perhaps the most advanced project of all those devoted to resonance neutron work is that now starting operation (early 1969) at Oak Ridge National Laboratory. The Oak Ridge Electron Linac (ORELA) is an *L*-band machine of nominal energy 140 MeV driven by 4 klystrons each supplying 25 MW of peak rf power at a frequency of 1300 MHz. Pulse lengths are variable down to 2.3 nsec, and currents of over 15 A of electrons can be accelerated for pulses of duration up to 24 nsec with repetition rates of up to 1000 pps. For longer pulses the current will drop off as the pulse length begins to approach the filling time of the waveguides ($\sim$2 μsec) so that for 1-μsec pulses (essentially steady state operation) the anticipated current is about 500 mA and the energy 100 MeV.

For neutron time-of-flight work the electron beam is stopped in a water-cooled tantalum target capable of accepting a beam power of 50 kW. This neutron producing target is of an unusual design as shown in Fig. 16. The water-cooled tantalum target is surrounded by a collar of water, which serves as part of the cooling circuit, and also as moderator. The efficiency of this geometry for the moderator is high since nearly half of the neutrons

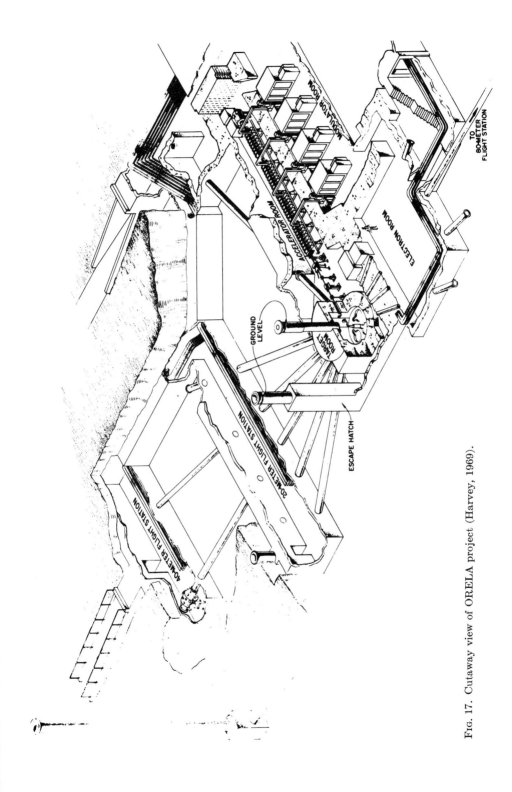

Fig. 17. Cutaway view of ORELA project (Harvey, 1969).

emitted by the target enter the moderator. It is intended that for fast neutron experiments the flight paths will be collimated to "look" only at the tantalum; while for resonance neutron experiments, the collimation system will contain a shadow cone so that only the moderator is seen. The tantalum target is less efficient for neutron production than a natural uranium target by at least a factor of 2, and the peak neutron-production rate in short bursts is estimated at 4×10^{18} neutrons/sec.

The neutron target is placed at the center of a concrete walled cell 12 ft in diameter and 10 ft high. This cell can be evacuated to a pressure of 1/100 atmosphere so as to reduce neutron scattering, and to avoid the production of excess radioactive and noxious gases where the electron beam passes through the air, as it does with this target design. The air pumped out of the target cell is exhausted through a 50-ft high stack and monitored for radioactivity.

Eleven flight tubes radiate from this target cell and several detector stations are available on each tube. These detector stations are all underground and their distances from the neutron target range from 10 to 80 meters. A general cutaway view of the project is shown in Fig. 17. The offices and laboratories are the only parts of the project to be located above ground. Ten laboratories containing the electronic equipment for the experiments surround a central data room which contains the computers for data acquisition and the control console for the accelerator (Harvey, 1969).

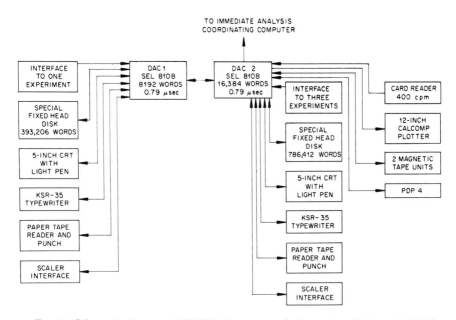

FIG. 18. Schematic diagram of ORELA data accumulation system (Betz *et al.*, 1969).

The data-acquisition system of the ORELA Project, initially capable of dealing with four simultaneous experiments, is different from those used in earlier linac installations. It is intended that the system should provide hundreds of thousands of channels, too many to be provided by direct-core memory, and yet operate at count rates (up to 10^4 events/sec) which are too high to permit the efficient use of magnetic tape. The system chosen is based on updating totalized data on a magnetic-disk store with a fast computer. A schematic diagram of the system is given in Fig. 18 from Betz *et al.* (1969), and shows that two fast SEL 810B computers are used, each with a fixed head disk. Between them they can service up to four experiments, simultaneously. The computers can perform preliminary operations on the data in addition to carrying out the updating function. For example, they can eliminate time-of-flight data between resonances, and can adjust time-of-flight gate widths as a function of neutron energy in order to minimize the number of channels required in an experiment and avoid the recording of purely background information. They can also display selected areas of data and carry out the usual functions of listing and plotting.

4. Applications of Electron Linear Accelerators

In contrast to the relatively narrow field explored with the cyclotrons, electron linac installations have been used to study all aspects of resonance-energy neutron interactions. Total, scattering, capture, and fission cross sections have been studied by many groups, and more detailed studies of the fission and capture processes have been the subject of many investigations. References to much of the recent work can be found in Nuclear Structure Studies with Neutrons (North Holland, 1966), Neutron Cross-Section Technology (USAEC CONF 660303, 1966) and Neutron Cross Sections and Technology (NBS Spec. Publ. 299, 1968).

a. Cryogenic Techniques. Many examples will be found in the above general references of high-resolution resonance and average total cross-section measurements, including the use of separated isotopes which are now available in large enough quantities to permit the use of neutron beams up to, say, 2 cm in diameter. An interesting development is the use, pioneered at Saclay, of samples cooled to liquid-nitrogen temperatures in order to reduce the Doppler broadening of the resonance structure (Le Pipec *et al.*, 1961). For heavy nuclei with high-resolution spectrometers, this effect is more severe than the experimental resolution broadening below a neutron energy of ~ 1 keV, and minimizing the Doppler effect permits the observation of many additional small resonances in this energy range. Figure 19 shows an example of the effect of cooling the target. A close doublet in ^{239}Pu is seen to be clearly resolved at liquid-nitrogen temperature and not at all resolved at room temperature (Michaudon, 1968a). Another

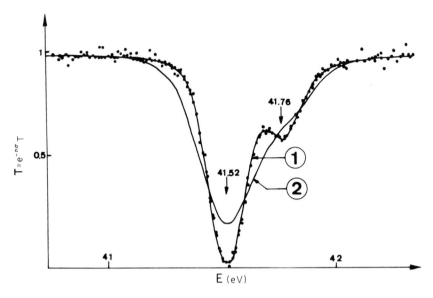

Fig. 19. Transmission curve of ^{239}Pu—cooled sample. (1) $\Delta = 0.012E^{1/2}$ ($T \simeq 80°K$), (2) $\Delta = 0.0208E^{1/2}$ ($T \simeq 300°K$) (Michaudon, 1968a).

interesting technique is the use of polarized neutron beams and polarized targets in transmission experiments to determine resonance spins. This has been successfully done for very low-energy neutrons using a cobalt crystal as polarizer and monochromator, in conjunction with a polarized target, at the Naval Research Laboratory, Washington, D.C. (Stolovy, 1960) and by Sailor's group at Brookhaven (Marshak *et al.*, 1962). More recently the difference in the hydrogen cross sections for singlet and triplet scattering interactions has been used as a basis for producing polarized neutrons over an energy range up to tens of kiloelectron volts. Shapiro (1966) reported the use of a 44% polarized neutron beam obtained from the Dubna pulsed reactor by transmission of the beam through the dynamically polarized protons in a crystal of LMN (La$_2$Mg$_3$(NO$_3$)$_{12}$·24H$_2$O) at a temperature of about 1°K. This polarized neutron beam was then passed through a 40% polarized sample of holmium (15 kOe at 0.3°K) with the neutron polarization vector either parallel or antiparallel to that of the target nucleus. Since this corresponds essentially to the two possible channel spins for the formation of the compound nucleus with s-wave neutrons, and since a given resonance corresponds to a state of definite channel spin, it is clear that the spins of the resonances are readily determined by this method where they are well resolved. In the case of the Dubna experiments, the spins of twenty-two s-wave resonances in holmium have been reported (Alfimenkov et al., 1967).

It is also possible to determine the spins of resonances by passing an unpolarized neutron beam through a target which is either polarized or unpolarized and comparing the transmission in the two experiments. In this case if the cross sections are different for the two channel spins, the polarized target itself partially polarizes the neutron beam as it passes through it so that the transmission is no longer an exponential function of the thickness of the target. This can be expressed mathematically by writing:

$$T_{pol} = T_{unpol} \cosh \left\{ \frac{n \, \Delta\sigma \, If_1}{2I + 1} \right\}$$

where T_{pol} and T_{unpol} are the fractions of the unpolarized neutron beam transmitted through the sample when it is either polarized or unpolarized, n is the superficial density of the target, I is the target nucleus spin, f_1 is the nuclear polarization of the target and $\Delta\sigma$ is the difference between the cross sections of the target nucleus for the two possible values of the channel spin

$$\Delta\sigma = \sigma(J = I - \tfrac{1}{2}) - \sigma(J = I + \tfrac{1}{2})$$

This nonexponential behavior of the transmission can be used to determine resonance spins and the use of the unpolarized beam permits a more efficient experimental geometry and so higher neutron energy resolution. An experiment of this nature on holmium (nuclear polarization 60%) is nearing completion at Harwell as a joint experiment with the U.S. National Bureau of Standards (Marshak et al., 1969). The neutron time-of-flight resolution in this case is ∼2 nsec/meter, which is more than an order of magnitude better than that used in the Dubna experiments. It is anticipated that the spins of several hundreds of resonances will be obtained. Figure 20 shows the transmission of a 2.5-cm-thick holmium sample as a function of unpolarized cross section for an unpolarized target (curve A), compared with that expected for a 50% polarized target if the resonance has spin $I + \tfrac{1}{2}$ (curve B) or $I - \tfrac{1}{2}$ (curve C). A potential scattering cross section of 7 barns is assumed.

b. *Capture Cross Sections.* Neutron capture cross-section measurements have been made in most of the electron linac laboratories (San Diego, R.P.I., Harwell, Saclay, and Geel) using either large scintillator assemblies or detectors based on the Moxon–Rae principle (see Chapter III). One problem in these measurements, particularly with the large scintillators,

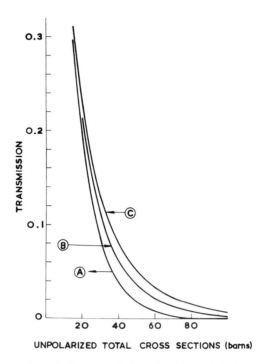

Fig. 20. The variation of the transmission of a 2.5-cm holmium sample with unpolarized cross section. Curve A refers to an unpolarized sample, curve B to a 50% polarized sample for spin $I + \frac{1}{2}$ and curve C for a resonance spin of $I - \frac{1}{2}$.

is the paralysis of the detector by the primary X-ray flash, or γ-flash from the pulsed neutron target which limits the minimum time-of-flight, or maximum neutron energy that can be used in the experiment. This problem has been solved at RPI at the expense of some loss in neutron flux, by the physical separation of the neutron producing target and the moderator slab (Fig. 21), so that the well-collimated flight path used for the capture experiments "looks at" the moderator only, the capture detector and sample being completely protected from the primary flash. This technique has permitted the use of a large scintillator at a distance of only 25 meters from the target, for neutron energies of up to about 1 MeV (Block *et al.*, 1967). The Harwell booster target is also useful in this context since the radial thickness of $\sim$10 cm of uranium acts as an excellent absorber for the primary γ-flash so that paralysis problems are minimal with this installation.

The measurement of the capture cross sections of fissile nuclei or the ratio α of capture to fission cross section, which is of great importance for reactor applications, has always presented great difficulty where the prompt

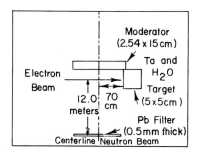

FIG. 21. RPI neutron target (Block *et al.*, 1967, p. 566).

capture γ-rays are detected because of the γ-rays and fast neutrons emitted in the competing fission process. One possible method, used successfully by a joint Oak Ridge–RPI team at RPI for ^{235}U (de Saussure *et al.*, 1967) is to remove the fission events by an anticoincidence method. Here the fissile sample is placed in an ion chamber or other fission-fragment detector of high efficiency, and the capture cross section is determined from the capture-detector yield in anticoincidence with the fission chamber after suitable corrections have been applied. This technique becomes extremely difficult in the case of more α-active materials such as ^{239}Pu, since only a rather small quantity of material may be placed in the fission chamber. A second method was used originally by Hopkins and Diven (1962) with a pulsed Van de Graaff accelerator and later by Ryabov *et al.* (1968) with the Dubna pulsed reactor. Here the liquid-scintillator tank used to detect the prompt γ-rays from capture is loaded with some material such as gadolinium which itself has a large capture cross section. The fast neutrons from a fission event are slowed down in the scintillator tank surrounding the fissile sample, and eventually captured in the gadolinium giving rise to a second pulse from the detector. This delayed pulse is used to separate fission from capture events.

A third method, also used by the ORNL–RPI team (Gwin *et al.*, 1969) makes use of the difference in the pulse-height response of their large liquid scintillator to capture and fission events occurring in a fissile sample placed at the center of the tank. About 12% of the fission events result in a pulse height which is larger than that produced by ∼99% of the capture events, so that a "high bias" placed at a suitable level permits a measure of the fission rate in the sample to be made coincidentally with a count of the total capture plus fission rate. The fission rates can then be subtracted to obtain the capture rates and so the cross section. Another subtraction technique is used at Harwell (Schomberg *et al.*, 1968) where pulse shape discrimination is used to permit the detection of knock-on protons in a liquid scintillator which is also used as a Moxon–Rae type of γ-ray detector.

Again the knowledge of the fission rate for a given neutron time-of-flight channel permits a subtraction to be made to obtain the capture cross section. All of these techniques are difficult and serious discrepancies existed between some of the earlier measurements, but the results for ^{239}Pu from the various methods described are now in reasonable agreement at least up to a neutron energy of 10 keV.

c. *Capture Gamma-Ray Spectra.* The study of the γ-ray spectra for resonance neutron capture has also received much attention in several linac laboratories in recent years due to the development of lithium-drifted germanium detectors. These studies permit the determination of partial radiation widths for transitions from the capturing state to the ground and various excited states of the compound nucleus. Interest here centers on whether the Porter–Thomas law (originally introduced to describe the distribution of reduced neutron widths) holds for the distribution of reduced, partial radiation widths. This law implies the validity of the statistical model and the absence of nuclear structure effects in resonances. The published papers from RPI, Saclay, and Harwell appear to confirm this (e.g., Rae *et al.*, 1967; Samour *et al.*, 1967, 1968; Stein *et al.*, 1970), and a full treatment of this topic is given in Chapter IV. Work done with the Brookhaven high-flux reactor and fast chopper, however, suggests the existence of a nonstatistical behavior in Tm (Beer *et al.*, 1968), and the existence of intermediate structure in neutron cross sections, particularly the fission cross sections of the heavy nuclei, suggests that structure effects must exist in the capture spectra. The search for such effects continues.

Before leaving the application of pulsed accelerator neutron sources to the study of the capture processes, we must mention the observation of the ratio of single counts to coincidence counts for two sodium iodide crystals as a function of neutron time-of-flight when detecting capture γ-rays from a sample placed at the end of a flight path. This ratio is sensitive to the multiplicity of the capture γ-ray cascade, which in turn depends on the spin of the compound nucleus. It therefore provides a powerful tool for the determination of resonance spins and was first used for this purpose by Coceva *et al.* (1965) at Ispra and has been successfully used at the Geel installation to determine the spins of large numbers of resonances (Carraro *et al.*, 1968).

d. *Neutron Scattering.* Turning now to low-energy scattering measurements, these are treated at length in Chapter III, and in the general references. The work here has been done almost entirely with electron linacs. Most of the scattering measurements have been on resonances in heavy nuclei where a scattering measurement can determine unambiguously the resonance spin, and one of the more interesting developments here has been

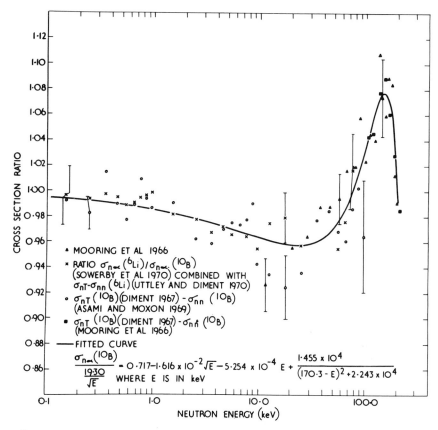

FIG. 22. Some selected experimental values of the $^{10}B(n,\alpha)$ cross section in the energy range 100 eV to 200 keV plotted relative to $19.30/\sqrt{E}$ (keV). The curve is a fit to the experimental data.

the successful use of the "bright-line" technique at Livermore (Sauter and Bowman, 1965, 1967) to measure the resonance scattering from ^{239}Pu and ^{241}Pu (see Chapter III).

More recently scattering measurements at Harwell have been extended to light elements, and a careful measurement of the scattering cross section of ^{10}B up to 100 keV (Asami and Moxon, 1969) has been combined with precise total cross-section measurements (Diment, 1967) to yield a determination of the absorption cross section of ^{10}B, an important nuclear standard. The deviation of this important cross section from $1/v$ behavior now seems well established and is shown in Fig. 22 taken from Sowerby et al. (1970). A measurement of the angular asymmetry of neutrons scattered by the weak

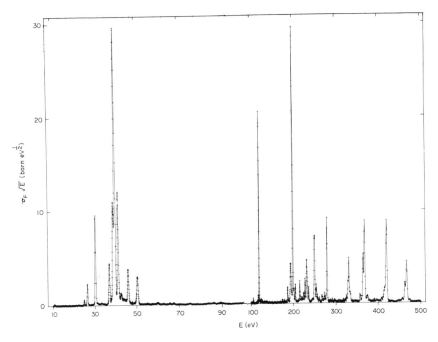

FIG. 23. Fission cross section of ^{237}Np (Michaudon, 1969, p. 485).

resonance in ^{56}Fe at 1167 eV has also been used to determine unambiguously the parity of this resonance (Asami *et al.*, 1969).

  e. Fission. The most important recent application of the linac spectrometers in the field of nuclear structure is, however, surely the study of the so-called subthreshold fission cross sections of isotopes where the excitation energy of the compound nucleus following neutron absorption is close to the fission-barrier energy. The existence of structure other than simple compound nucleus behavior was first pointed out by the Saclay group when they were unable to explain the distribution of fission widths in the resonances in ^{237}Np in terms of accepted statistical behavior (Paya *et al.*, 1967). Figures 23 and 24 show the group of resonances close to 40 eV observed in the original study, together with higher energy unresolved groups reported later (Michaudon, 1968a,b). The interesting feature of the cross section is that the resonances in which fission is observed are grouped together in clusters between which the fission cross section falls almost to zero. The spacing of the clusters is ∼100 times that of the compound-nucleus resonances. The total cross section of ^{237}Np, on the other hand, contains no such structure, showing that it is the fission widths which have large values only in certain well-defined energy regions. A similar dramatic effect

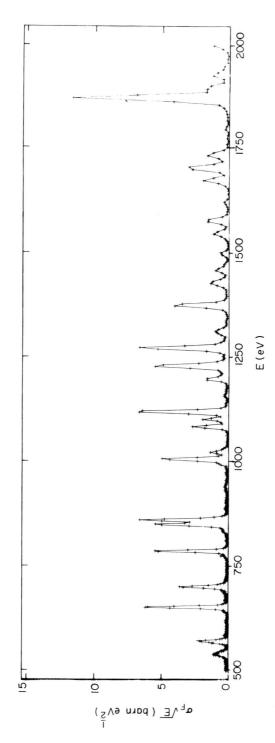

FIG. 24. Fission cross section of ^{237}Np (Michaudon, 1969, p. 485).

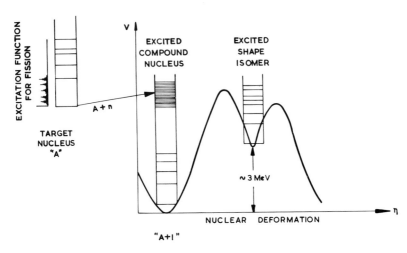

FIG. 25. Effect of second minimum in fission potential barrier.

in the subthreshold fission cross section of ^{240}Pu was observed at Geel (Migneco and Theobald, 1968) and in ^{234}U at Harwell (James and Rae, 1968), the spacing of the clusters of fission resonances being ~700 eV for ^{240}Pu and 7 keV for ^{234}U. Weigmann (1968) at Geel (and independently Lynn, 1968) explained that this structure was associated with the double-humped fission barrier postulated by Strutinsky (1967) as the result of shell effects modulating the simple liquid-drop barrier shape. Strutinsky's theory provided an explanation (Bjørnholm et al., 1967) of the sponta-neously fissioning isomeric states found at Dubna in the odd americium isotopes (Flerov and Polikhanov, 1964) which have lifetimes of the order of 1 msec, and which lie ~3 MeV above the ground state. The clusters of fissioning resonances observed in the fission cross sections of ^{237}Np, ^{240}Pu, and ^{234}U then correspond to those resonances which lie close in energy to complex states built on the isomeric "ground state" in the second minimum in the fission potential barrier (Fig. 25). The discovery of the clustering of strong fission resonances in the subthreshold region has led to a careful examination of the fission cross sections of the fissile nuclei ^{235}U and ^{239}Pu. Here, no threshold behavior is observed in the total fission cross section, but for one of the channel spins possible with s-wave neutrons the fission barrier does lie above the neutron separation energy for the compound nucleus, and there is a modulation of the fission cross section for this chan-nel spin. The effect has been demonstrated by the use of an autocorrelation analysis due to Egelstaff (1958) which was applied at Geel (Cao et al., 1968) to their high resolution measurements of the ^{235}U-fission cross section, and

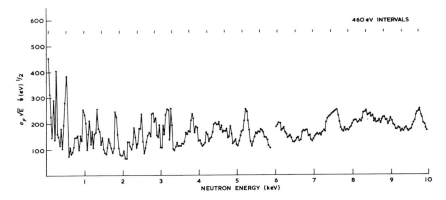

Fɪɢ. 26. Fission cross section of ²³⁹Pu averaged over 33⅓-eV intervals (James and Patrick, 1969).

at Harwell (Patrick and James, 1968) and Saclay (Blons *et al.*, 1968) to their measurements on ²³⁹Pu. The interpretation of the correlation analysis is not a simple problem, and the extraction of intermediate structure spacings by the experimenters concerned may be open to doubt (Perez *et al.*, 1969). The existence of intermediate structure in the fission cross sections is, however, well established, and is shown clearly in Fig. 26 which is a plot of the Harwell data on ²³⁹Pu averaged over 33-eV intervals in order to reduce the effect of the fine structure resonances (James and Patrick 1969).

The interpretation of the clusters of fissioning resonances has been fully developed by Lynn (1968, 1968a,b, 1969) who has also interpreted, in terms of the Strutinsky model, other forms of structure in near barrier fission, such as complex behavior of the fission cross section of ²⁴¹Am observed at Los Alamos in a nuclear-explosion experiment (Seeger *et al.*, 1967), as shown in Fig. 27.

Many aspects of the fission process other than the cross sections have also been studied with linac pulsed sources. The value of $\bar{\nu}$, the number of neutrons emitted per fission, has recently been studied at RPI for low-energy resonances in ²³⁵U and ²³⁹Pu, and in the case of ²³⁹Pu a significant difference is observed for $\bar{\nu}$ between resonances of the two possible spin states (0 and 1) (Weinstein and Block, 1969). Studies of the ratio of binary to ternary fission in the resonances of ²³⁵U have also been made at Saclay (Michaudon *et al.*, 1965) and Geel (Deruytter *et al.*, 1968). Perhaps the most sophisticated experiments on the fission process made with linac pulsed sources are those in which the angular distributions of the fission fragments are measured for a number of resonances in order to determine the effective value of K, the projection of the compound nucleus spin on the nuclear

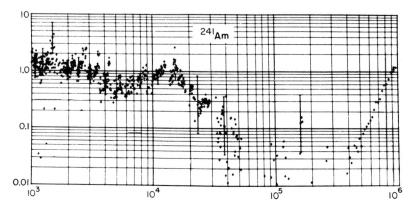

FIG. 27. Fission cross section of ²⁴¹Am (Seeger et al., 1967).

symmetry axis, for comparison with theoretical models of the barrier structure. Such an experiment carried out at Saclay with apparatus developed at
Oak Ridge by Dabbs has recently been reported (Dabbs et al., 1968). In
this experiment the anisotropy of the fission fragments is measured in the
individual resonances when fission is induced by unpolarized neutrons in
aligned ²³⁵U nuclei. Alignment is obtained through the interaction of the
electric quadrupole moment of the uranium nuclei with the crystal fields
when a crystal of $UO_2Rb(NO_3)_3$ is cooled down to 0.6°K. The 220 hours of
counting time on a flight path of 5 meters permitted the determination of
effective K values for about 30 resonances below 50 eV. Previous work
with a crystal spectrometer (Dabbs et al., 1965) has been restricted to neutron energies up to a few electron volts. A similar experiment in which a
³He - ⁴He dilution refrigerator is used to cool the $UO_2Rb(NO_3)_3$ crystals to
a temperature <0.1°K has recently been completed at Harwell. This cry-
ostat was developed by H. Postma at the Netherlands Reactor Center at
Petten and is installed on a 10-meter flight path. The greatly reduced temperature accentuates the anisotropies in the angular distributions and
simplifies analysis of the results, and the improved resolution (factor of 2)
should allow the analysis of an increased number of resonances. The counting time in this experiment was ~500 hr (Pattenden and Postma, 1969).

f. *Threshold Photoneutron Reactions.* In the preceding discussion, it was
shown that studies of the interactions of resonance neutrons with nuclei
have produced, inter alia much information on the statistical properties of
nuclear energy levels. It was first suggested by Bertozzi et al. (1965) that
a whole new range of compound nuclei could be studied using the (γ,n)
process to yield similar statistical information. They suggested exciting

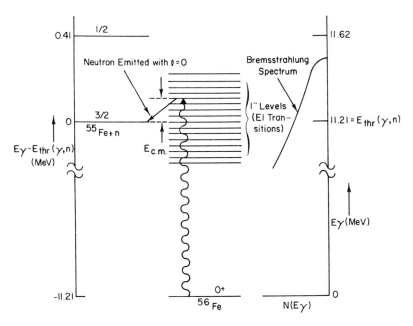

FIG. 28. Schematic representation of states involved in (γ,n) reaction (Bowman *et al.*, 1967, Fig. 1).

nuclei by bremsstrahlung to energies just above the neutron binding energy and using the time-of-flight technique to measure the neutron spectrum from lead isotopes, and obtained ground state radiative widths for several resonances in ^{208}Pb. In order to interpret the neutron spectra free from ambiguity, the maximum excitation energy must be restricted so that only neutrons leading to the ground state of the residual nucleus can be emitted. Of course, once the ground state transitions are known, it is possible to study transitions to excited states, thus increasing the amount of information obtained. Figure 28 is a schematic representation of the states studied by such a (γ,n) experiment using ^{56}Fe as an example.

The compound nuclear states studied by the (γ,n) process are the same kind of states as those investigated by the interactions of slow neutrons, but only occasionally are they identically the same. Thus, the information obtained from (γ,n) experiments, roughly speaking, doubles the number of compound nuclear levels which can be studied. The data obtained in such experiments provide information on level spacings, neutron and ground state radiation widths as well as their dependence on spin and parity, and on the multipolarity of the γ-ray producing the compound state.

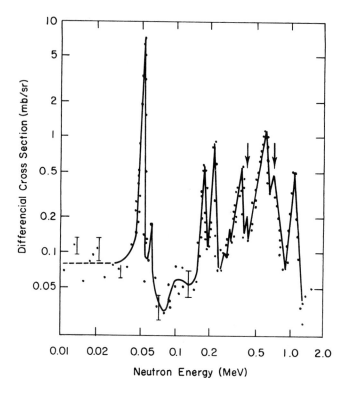

FIG. 29. Cross section for reaction ^{26}Mg(γ,n) ^{25}Mg (Berman *et al.*, 1969, Fig. 2).

The experimental requirements are short, high intensity pulses of elec-trons of well-defined energy. Bollinger (1966) in a discussion of the advan-tages of studying the (γ,n) process, suggested the use of a pulsed electron Van de Graaff accelerator, but to date the only accelerators used have been traveling-wave linacs. In the experiments, an effectively monoisotopic sample can generally be obtained by studying the isotope with the lowest neutron threshold so that natural samples can be used. Since the first measurement by Bertozzi *et al.*, the Livermore group of Bowman *et al.*, have been the most active in the field, having measured the cross sections of about sixteen nuclei (Berman *et al.*, 1966, 1967, 1969; Bowman *et al.*, 1967; Van Hemert, 1968).

Figure 29 shows their measured cross section for the reaction ^{26}Mg(γ,n) ^{25}Mg as a function of neutron energy. Several resonances are observed, and the one at 54.3 keV is particularly interesting. It appears that this state may be responsible for the primary production of neutrons in stars through the reaction ^{22}Ne(α,n) ^{25}Mg.

The (γ,n) cross section for an isolated resonance is given by

$$\sigma_{\gamma n} = \frac{\pi \lambdabar^2 g \Gamma_{\gamma 0} \Gamma_n}{(E - E_0)^2 + (\Gamma^2/4)}$$

where $\Gamma_{\gamma 0}$ is the ground state radiation width. The area under a resonance is $2\pi\lambdabar^2 g \Gamma_{\gamma 0} \Gamma_n / \Gamma$. Now the compound nucleus at the excitation energies considered here usually decays by neutron emission and so $\Gamma_n \gg \Gamma_{\gamma 0}$. Thus, shape analysis of a resonance leads to a determination of $g\Gamma_{\gamma 0}$ and Γ_n while area analysis gives $g\Gamma_{\gamma 0}$.

Just as in neutron interactions, where a measurement of the average cross section can lead to a determination of the s- or p-wave strength function, so a measurement of the average (γ,n) cross section just above threshold provides data on the γ-ray strength function $\langle\Gamma_{\gamma 0}/D\rangle$ since the average cross section is given by

$$\bar{\sigma}_{\gamma n}(E) = 2\pi\lambdabar^2 g \langle\Gamma_{\gamma 0}/D\rangle\langle\Gamma_n/\Gamma\rangle$$

The values thus obtained can be compared to the predictions of the single particle (Axel, 1962) and giant resonance (Weisskopf, 1951) models.

Other known groups making (γ,n) measurements near threshold are the photonuclear group at Harwell, using the Wantage 15-MeV linear accelerator (Patrick and Winhold, 1969) and a group in Japan using the JAERI 20-MeV linear accelerator (Asami et al., 1967). At present, no one has measured the angular distribution of the neutrons. Such a measurement could lead to the determination of the multipolarity of the γ-rays as it is not always clear that E1 photons are responsible.

At higher neutron energies, in the million electron volt region, electron linacs, notably at Harwell and RPI, have been used to study neutron spectra emitted by photons of energy corresponding to the giant E1 resonance (Firk, 1964) and above (Kaushal et al., 1968). Although not examples of resonance neutron spectroscopy, these measurements are worth mentioning here since their high-energy resolution provided some of the earliest evidence for the fine structure present in the giant resonance cross sections of the light nuclei. Later work established the validity of the quasideuteron model, as modified by Levinger (1967), in the energy region just above the giant resonance.

g. Other Applications. This brief account of some of the applications of linac pulsed sources in the field of neutron cross sections and nuclear structure will serve to indicate the scope of this work. Nor does this represent, by any means, the full extent of the use of these machines for neutron spectroscopic studies. Another field in which a great deal of effort has been expended is the study by time-of-flight techniques of neutron spectra in

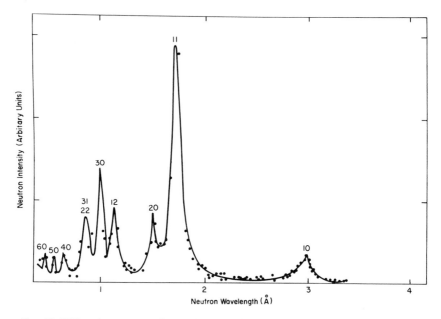

Fig. 30. Diffraction pattern of two-dimensional crystal (Sinclair *et al.*, 1969, Fig. 1).

moderators and in reactor assemblies, both thermal and fast. This work (Poole, 1965; Beyster, 1968) has received special attention at San Diego, RPI, and Harwell, where the method was first developed using the old 15-MeV accelerator (Poole, 1957). This is a field in which the programmed beam handling system on the Harwell accelerator was invaluable (it was first introduced on the 15-MeV machine) since this feature allowed the neutron spectrum work to proceed simultaneously with the cross-section work, whereas in San Diego and RPI, the two types of experiments were mutually exclusive.

Another recent application of the time-of-flight method has been to solid state physics in the field of crystallography. This work was pioneered by Buras (1966) on the Dubna pulsed reactor and has recently been undertaken with electron linacs. Diffraction patterns have been obtained for polycrystalline specimens at RPI (Moore and Kasper, 1968) and at Tohoku University (Kimura *et al.*, 1969). At Harwell, grain orientation experiments have been carried out by observing the powder diffraction patterns of a sample of bismuth telluride (Day and Sinclair, 1969). Interesting results have also been obtained at Harwell for samples of carbon fibers which show the typical diffraction pattern of a two-dimensional crystal (Fig. 30). The (*hk*) diffraction peaks are observed in the range of wavelength between

0.4 and 3 Å in a single run by the time-of-flight technique (Sinclair *et al.*, 1969).

5. THE NUCLEAR EXPLOSION AS A SOURCE OF PULSED NEUTRONS

A novel approach to neutron time-of-flight measurements involves the nuclear explosion as a pulsed neutron source. Experiments have shown that such a source can be as effective in making neutron measurements as those instruments which are the subject of the present chapter and in some cases more so. Therefore, a very brief critique of the method is in order.

As employed at Los Alamos, the source furnishes a neutron burst with a nominal duration of about 100 nsec. The flight path was typically about 185 meters so that the resolution figure was a nominal 0.5 nsec/meter. The source has the property that very small samples can be employed. The method has great potential especially for measuring fission cross sections, and many more examples of the intermediate structure in fission already mentioned can be anticipated from this work. The short duration of the single cycle time-of-flight experiment allows the use of extremely active samples which could never be studied with accelerators.

The explosion sources, however, have proved to be at a disadvantage with respect to neutron energy resolution when compared with a linac source because of the high temperature and motion of the moderator during the experiment which introduce extra resolution broadening particularly for low-energy resonances. Figure 31 (Michaudon, 1968a; Patrick, 1969) shows a part of the fission cross section of ^{239}Pu as measured at Harwell and Los Alamos (nuclear explosion) with room-temperature samples, and at Saclay with a cooled sample. The Los Alamos nuclear-explosion experiment was carried out utilizing a nominal resolution $\sim$0.5–0.6 nsec/meter. The Saclay experiment had a 50-nsec burst and a 50-meter flight path, or 1 nsec/meter, while the Harwell experiment with a 200-nsec burst and a 100-meter flight path had the worst nominal resolution at 2 nsec/meter. The danger in literally accepting these resolution figures is apparent. The resolution of the explosion experiment is clearly much worse than either of the accelerator measurements on account of the high temperature and motion of the moderator. The real resolution of the Harwell and Saclay experiments is identical, because even at room temperature the moderator time jitter at 200 eV is $\sim$100 nsec so that the effective Saclay resolution is reduced to 2 nsec/meter, while the Harwell experiment is not affected by this time jitter on account of its longer accelerator pulse. The effect of the cooled target in the Saclay measurement is seen in the sharpening of the weak

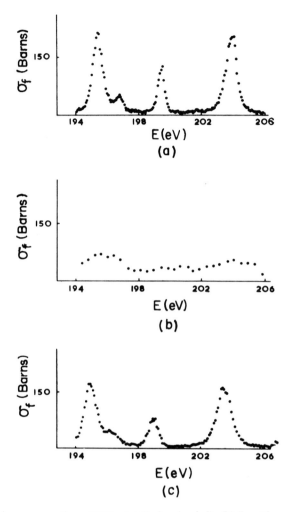

FIG. 31. Fission cross section of [239]Pu. (a) Saclay (cooled), (b) Los Alamos, (c) Harwell.

resonance at 197 eV and the higher peak cross sections. The very poor resolution of the explosion experiment at low energies is intrinsic to the method, but can be improved by careful attention to moderator design or the use of a longer flight path.

D. Comparison of Spectrometer Performances

In the Introduction to this chapter, we suggested that the two most important features of a time-of-flight spectrometer are resolution and neutron intensity. It has become apparent in the discussion that merely to take the nominal resolution, i.e., machine-pulse duration divided by the flight-path length, as a measure of the effective resolution can be very misleading. Likewise, in estimating neutron intensity at the end of a flight path, one must consider not only the neutron intensity during the pulse, and the maximum repetition rate of the accelerator but also the geometry and efficiency of the moderator and the limitation in intensity imposed by the use of overlap filters at high-repetition rates and low energies.

It is easy to see qualitatively that pulsed sources with short pulses and high-repetition rates are more useful at higher energies, while longer pulses, provided they correspond to a greater total number of neutrons emitted, can provide better operating conditions at lower energies. In order to try to be more quantitative in this comparison and to determine more exactly the energy regions of application of various types of pulsed sources, let us try to determine for a given effective resolution, the useful flux available at the detector as a function of energy for several pulsed sources.

We can write the expression for the neutron flux through 1 cm^2 distant L cm from the pulsed source in an energy interval ΔE as

$$F = I(E)\,\Delta t' f\,\Delta E/4\pi L^2 \tag{3}$$

where $I(E)$ is the energy dependent instantaneous rate of emission of neutrons from the source (including the effect of the moderator), $\Delta t'$ is the machine pulse length, and f the pulse repetition frequency. If we now keep the energy resolution $\Delta E/E$ constant, equal to R say, then we have $\Delta E/E = R = 2[(\Delta L)^2 + (v\,\Delta t)^2]^{1/2}/L$ where ΔL is the uncertainty in flight-path length, v is the neutron velocity, and Δt the full width at half-height of the total timing uncertainty. Since we wish our calculation to be independent of the particular detector and electronics used in an experiment, and since moderator timing uncertainties are included in ΔL as explained below, we shall put $\Delta t = \Delta t'$, the accelerator pulse length. We can now substitute for ΔE and L in terms of R in Eq. (3), and obtain

$$F = \frac{EI(E)f\,\Delta t'\,R^3}{16\pi[(\Delta L)^2 + (v\,\Delta t')^2]} \tag{4}$$

Now the timing uncertainty introduced by the moderation process is proportional to $1/\sqrt{E}$ or $1/v$. This is equivalent to an uncertainty in the flight-path length and its magnitude, expressed as a full width at half-height is ~2.2 cm. If we ignore length uncertainties due to the detector and moderator geometries, since these are applicable only to a particular experiment, the irreducible value of ΔL for a moderated source is thus 2.2 cm. For an unmoderated source ΔL depends on the source geometry.

The expression $I(E)$ for the energy-dependent neutron-emission rate during the pulse from a moderated source can be written

$$I(E) = \frac{\text{const} \times \text{pulse current}}{E^n \sigma_H^2(E)} \tag{5}$$

where the constant depends on the bombarding energy, the reaction used, and the geometrical efficiency of the moderator. The exponent n in the denominator is equal to unity for a moderator of infinite thickness, and in practice is less than unity. It has been measured in several cases and for the Harwell and Columbia spectrometers it has the values 0.78 and 0.90, respectively. Here, $\sigma_H(E)$ is the energy-dependent hydrogen-scattering cross section, and the inclusion of the factor $\sigma_H^2(E)$ in the denominator approximately takes care of the effect on the moderated spectrum of the dropoff in the hydrogen cross section above about 10 keV. The expression (5) has been found to give a good fit to the shape of the neutron spectra at Columbia and Harwell up to $\sim$100 keV, and should be generally applicable with a suitable choice of the exponent n.

To determine the constant in the numerator of Eq. (5) from theoretical considerations is difficult. It is easy to estimate the total peak neutron-production rate in the target from published data, but the number of those primary neutrons which interact with the moderator and finally form part of the moderated spectrum depends on the energy spectrum and angular distribution of the neutrons as they leave the thick primary target. The fraction of the primary neutrons finally radiated as a moderated spectrum is, at best, $\sim$20%, and can easily be much less for an unfavorable moderator geometry. In the case of the Columbia and Harwell spectrometers, the constant can be obtained from actual measurements of the neutron flux (Bollinger, 1960; Rainwater, 1961; Moxon, 1962; Coates et al., 1968a). These values are probably pessimistic due to collimation losses, but they serve as a guide to actual experimental moderator efficiencies, and the estimates of fluxes for other machines are based on these measurements with suitable corrections for different moderator geometries. We shall use Eqs. (4) and (5) to describe the neutron fluxes available up to about 100 keV. Above 100 keV we assume that the bare unmoderated source is used, but we retain $\Delta l = 2.2$ cm since most primary targets will have length uncertainties of at least this magnitude. An exception is the Karlsruhe cyclotron where the deuteron range is only $\sim$5 mm. Here we ignore Δl.

In order to compare different spectrometers we must take into account the effect of the overlap in time-of-flight of neutrons from successive machine cycles, since this will prevent the use of high repetition rates with a long flight path at low neutron energy. The chopper system sometimes

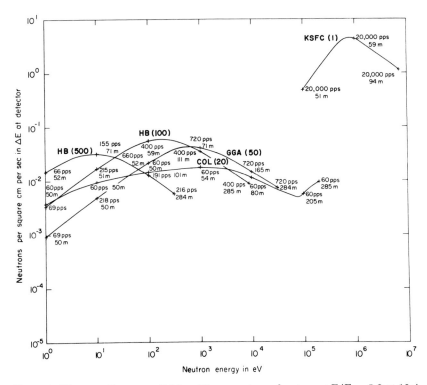

FIG. 32. Neutron fluxes available with current accelerators. $\Delta E/E = 8.8 \times 10^{-4}$. HB is Harwell Booster; COL is Columbia synchrocyclotron; SFC is Karlsruhe cyclotron; and GGA is the Gulf General Atomic linac. Numbers in brackets are machine pulse lengths in microseconds. Numbers on curves are pulse repetition frequency and flight path length in meters.

used at Columbia to prevent overlap is only feasible at very low repetition rates, so we assume that a boron filter is used which allows a 1% transmission of the unwanted neutrons at the end of a cycle. For such a filter, the transmission ϕ can easily be computed as a function of t/t_0 where t is the time-of-flight of the neutron and t_0 is the time between bursts. If $t/t_0 > 0.25$, it is better to reduce the repetition rate till $t/t_0 = 0.25$, when the product $F\phi$ is maximized. In the curves shown in Figs. 32 and 33, $F\phi$ has been maximized until as the neutron energy is increased, the maximum repetition rate of the machine has been reached.

Figure 32 shows $F\phi$, or the effective neutron flux in an energy interval RE, plotted as a function of E for $R = 8.8 \times 10^{-4}$ corresponding to a flight-path length of 50 meters for resonance neutrons. The diagram (Coates et al., 1968a) shows the fluxes obtained from the Harwell boosted linac, the San Diego linac with a tungsten target, the Columbia synchrocyclotron,

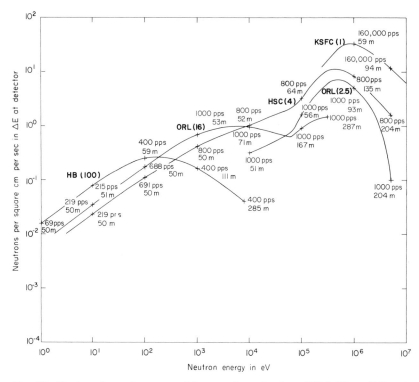

Fɪɢ. 33. Neutron fluxes for new and improved accelerators. HB is Harwell Booster; ORNL is Oak Ridge Electron Linac; HSC is Harwell Synchrocyclotron; SFC is Karlsruhe Cyclotron. Numbers in brackets are machine pulse lengths in microseconds. Numbers on curves are pulse repetition frequency and flight path lengths in meters.

and for comparison, the Karlsruhe sector focused cyclotron. Flight-path lengths have been varied to preserve the constant value of R, and flight-path lengths of up to 300 meters have been allowed. The repetition rate and flight-path length is shown on the figure for various energies for each accelerator.

We note that the Karlsruhe cyclotron with its very short pulse length is dominant in the fast neutron region, but because of its fixed high repetition rate it cannot be used below about 100 keV. The powerful unboosted linac as represented by the San Diego machine with a 50-nsec pulse gives the best performance in the kilovolt range, while the linac-booster combination, with its relatively long pulses wins out below about 1 keV. The Columbia synchrocyclotron with its intense 20-nsec pulse and fixed low prf, has a more uniform performance over the whole energy range, but its flux lies everywhere below those of the linacs.

Figure 33 shows some curves for a few new and proposed systems including ORELA, the Harwell Booster run to its heat transfer limit of 5 kW (it is currently limited for 100- and 200-nsec pulses by the capabilities of its linac injector), the Harwell synchrocyclotron with its intense 4-nsec pulse and repetition rate adjustable up to 800 pps, and once again the Karlsruhe SFC with modifications (Cierjacks et al., 1966, 1968) for comparison in the fast neutron region. The upgraded Columbia synchrocyclotron has not been included because of the lack of precise information on the improvement expected, but a conservative estimate would put it close to ORELA operating with 16-nsec pulses.

We note that the over-all flux levels have improved by about an order of magnitude over those displayed in Fig. 32, that the SFC is still dominant in the fast neutron region, and that the booster is still strong in the low-resonance region. The ORELA and the Harwell synchrocyclotron are dominant in the kiloelectron volt region.

It is possible to compare the performance of various pulsed accelerators in a number of different ways and such comparisons have been made by Harvey (1965), Bartholomew et al. (1966), and Michaudon (1966). Figure 34, taken from Michaudon's analysis, shows the neutron flux in an energy interval ΔE which is a fixed fraction of the Doppler width for a nucleus of mass 100. This type of analysis is further from the experimentally simple situation of a fixed flight-path length, but represents the desirable feature of having good resolution at all energies. We see from the diagram the same main features as in Fig. 32, namely that at low energies the existing Harwell linac-booster combination gives the highest flux, with the larger modern linac at Geel giving nearly as large a flux. The Columbia synchrocyclotron gives rather a poor flux at low energies, but competes successfully at 10 keV. When the resolution criterion is relaxed, the synchrocyclotron loses ground because with shorter flight paths, the repetition rate of the linacs can be increased, whereas that of the synchrocyclotron is fixed at 60 pps. This diagram also includes the IBR pulsed reactor with microtron injection, which is not competitive at higher energies because of its long pulse and low-repetition rate.

Figure 35 (Michaudon) shows the fluxes for the same criteria as in Fig. 34 for a number of systems which have been proposed but are as yet quite unproved. These include the most optimistic estimate of the performance of the upgraded Columbia synchrocyclotron (factor of 500 increase in neutron yield), the Canadian ING project (Bartholomew and Tunnicliffe, 1966) (canceled in 1968) and an improved PHERMEX (Motz, 1965). The ING (Intense Neutron Generator) would have utilized 1-GeV protons from a separated orbit cyclotron or proton linac, either as variable assemblies of

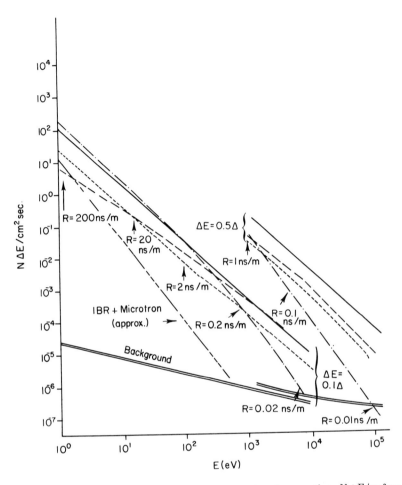

FIG. 34. Comparison of neutron velocity spectrometers in operation. $N \Delta E/cm^2$ sec = neutrons/sec cm² in the resolution width. --- , Saclay linac; —, Geel Linac (Euratom) (assuming 25 MW peak power per klystron); –·–, Harwell Linac and booster; – –, Nevis synchrocyclotron (400 MeV); —,background (¹⁰B capture γ-ray detector on electron linac 3.10^{-2} counts/cm² sec). ΔE is resolution width; Δ is doppler width (A = 100); ΔE is set equal to $k\Delta$ where $k = 0.1$ for 1 eV $< E < 10^4$ eV and where $k = 0.5$ for 10^3 eV $< E < 10^5$ eV (Michaudon, 1966, p. 796).

micropulses, or as the entire output of the accelerator collected in a storage ring and discharged as a 150-nsec pulse containing 850 A of protons, 500 times/sec. The authors point out that heat-removal problems in the neutron-producing target favor the use of high-energy protons, so that the ING storage ring would probably have represented the ultimate pulsed

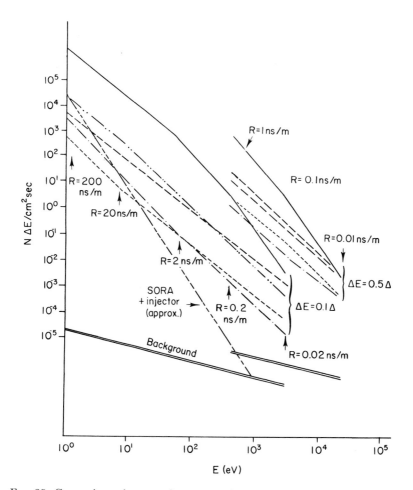

FIG. 35. Comparison of proposed neutron velocity spectrometers. N ΔE/cm² sec = neutrons/cm² sec in the resolution width. —, S.O.C. storage ring pulses (150 ns wide); —·—, S.O.C. micropulses; - - -, Electron linac (70 MeV); ——, Improved Columbia synchrocyclotron (600 MeV); —··—, PHERMEX (44 MeV, 250 A); —, background (^{10}B capture γ-ray detector on electron linac 3×10^{-2} counts/cm² sec). ΔE is resolution width. Δ is doppler width (A = 100). ΔE is set equal to $k\Delta$ where k = 0.1 for 1 eV $< E <$ 10⁴ eV (first set of curves) and where k = 0.5 for 10³ eV $< E <$ 10⁵ eV (second set of curves) (Michaudon, 1969, p. 799).

neutron source for neutron spectroscopy, and this is reflected in the fluxes shown in Fig. 35. PHERMEX was described in Section II,C,2 and the proposed design would provide 200 nsec pulses containing 70 A of electrons at 44 MeV. All of these proposed systems are compared with a good modern conventional traveling-wave electron linac capable of providing 100 kW of power in long pulses. Also shown in the diagram is a curve for the SORA pulsed reactor plus injector (Larrimore *et al.*, 1966) which is the logical development of the IBR. This diagram indicates the direction in which development is likely to go to reach the ultimate in pulsed neutron sources for neutron spectroscopy, but the question of cost effectiveness has not been considered.

Figure 36, also due to Michaudon, shows a summary of the regions of applicability of various neutron sources, including choppers on high-flux reactors for the study of very small samples of rare materials and the nuclear explosion technique developed by Diven (1966). We see that this confirms for existing systems the usefulness of the electron linac, with and without boosters, in the energy range from a few electron volts to a few hundreds of kiloelectron volts. The synchrocyclotron is also very powerful in the region above a few hundred electron volts, and if the modifications to the Columbia accelerator achieve the most optimistic estimate of an improvement of a factor of 500 in neutron flux, then this will be the most powerful source throughout the energy region. Even so, this is a very expensive accelerator, and its use as a neutron source is only practicable as a by-product of its use as a meson factory. The electron linac installations, with their multiple flight paths and flexibility of pulse widths and repetition frequencies are relatively inexpensive and are likely to continue to dominate the field of resonance neutron spectroscopy.

III. THE PULSED VAN DE GRAAFF

A. Introduction

The development of the pulsed linear positive-ion accelerator for time-of-flight application lagged considerably behind the corresponding development of the cyclotron and electron linac. This retarded development, which was essentially electronic, came as a result of the circumstance that the Van de Graaff was intrinsically suited to the study of kilovolt neutrons, whereas the cyclotrons and linacs were originally developed to study neutrons below, say, 100 eV. Thus, the pulsed Van de Graaff required a scale of time measurement which was beyond the state-of-the-art for commercially available equipment. The difference can readily be appreciated if it is recalled that a 5-keV neutron has a reciprocal velocity of about 10 nsec/cm. Depending upon the energy range of interest, the cyclotron and electron linac have traditionally employed moderators of sizeable dimensions to shift the intrinsic boiloff neutron spectrum to lower energy in order to make most efficient use of the high-energy neutrons which without modera-

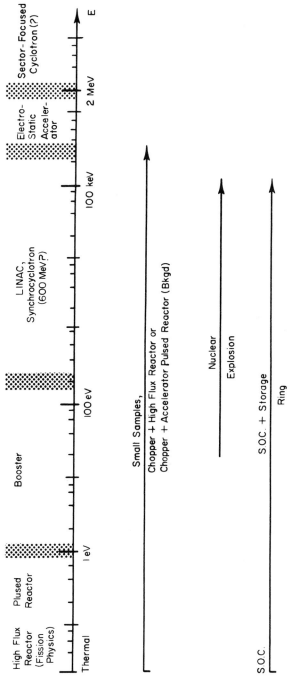

Fig. 36. Energy regions of applicability for various neutron sources (Michaudon, 1969, p. 818).

tion would normally lie outside the energy range of interest. The dimensions of the moderator were originally such as to yield neutron burst durations in the order of 100 nsec.

The pulsed Van de Graaff, as a pulsed continuous neutron source, developed (Good *et al.*, 1958) around the idea of using "low-energy" neutrons ("low" meaning 2 keV and above) as they emerge without moderation from an endothermic reaction (such as the $^7\text{Li}(p,n)^7\text{Be}$). In such cases, the neutron energy depends upon the target thickness only, through the bombarding particle energy which diminishes as it penetrates the target. Accordingly, the neutron spectrum is one which is bounded between a high- and a low-energy limit which depends upon the particle bombarding energy; clearly the mean energy (i.e., the high- and low-energy bounds) can be varied by varying the particle bombarding energy. Since no moderator is employed and the target is extremely thin, the neutron-burst duration is essentially that of the bombarding particles; this is smaller by some considerable factor than the duration of neutron bursts from the moderated cyclotron or linac as originally used. For purposes of the present discussion, the Van de Graaff neutron burst can be taken to be of the order of 1 nsec. From the considerations of a previous paragraph in which we showed that there are at least four components to the resolving time, we will see that superior time definition of a given source can be a specious advantage if not accompanied by a corresponding superiority in all other associated time uncertainties. In discussions that follow later, all time uncertainties other than that of the source are deliberately omitted, in order to compare neutron sources only. It should now be clear that the cyclotron and electron linac time-of-flight instruments were able to develop around 100 nsec electronic technology whereas the Van de Graaff was required to develop around 1 nsec technology which was almost a decade less advanced, at least commercially, than the 100 nsec counterpart.

B. Description of Pulsed Van de Graaff Neutron Spectrometer

A pulsed Van de Graaff time-of-flight spectrometer suitable for slow-neutron spectroscopy ideally consists of (a) a Van de Graaff provided at the terminal with special high-output ion source and associated pulsing optics and electronics, (b) a count-down system for selecting appropriate repetition rates, (c) a pickup for announcing the arrival time of the ion burst at the target, (d) medium-energy neutron detector capable of less than 5 nsec time response, (e) a clock to measure the flight-time intervals, and (f) a storage system for storing the frequency of measured time intervals.

Figure 37 is a schematic representation of the complete time-of-flight

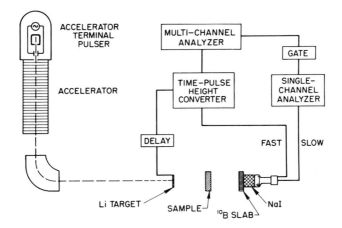

FIG. 37. Simplified schematic of pulsed Van de Graaff time-of-flight spectrometer.

system as used in the simplest application, namely, the measurement of neutron transmission. Figure 38 exhibits in detail the components of a modern system for generating positive ion bursts of 1 to 10 nsec duration.

The pulsed neutron source which supplies the periodic bursts of neutrons at accurately prescribed times is, as was stated earlier, but one essential part of a neutron time-of-flight spectrometer. Assuming a suitable detector, the remaining essential parts consist of the clock for measuring the flight times and the recording system to store the flight times in equal flight-time intervals. The principles of nanosecond-time spectrometry are now well known, and the components are available commercially. Since these principles were developed originally for the pulsed Van de Graaff, a few words of its history seem to be in order.

From the beginning of the use of the pulsed Van de Graaff for time-of-flight spectrometry, it was realized that the time intervals involved were by an order of magnitude too small to employ digital counting of time by means of scalers. The elapsed times were expected to be a few hundred to about 2500 nsec; 0.25 to 0.50 nsec time channels were thereby required, and these were further justified by the 1 nsec neutron bursts envisaged. An analog system of measurement whereby the elapsed flight times govern the amount of electric charge placed upon a condenser was an obvious solution; furthermore, it was one by which the time sorting problem was solved simultaneously with the time-measurement problem because time spectra became pulse-height spectra for which sorting analyzers already existed. The solution to the problem of time spectrometry was evolved almost simultaneously at Los Alamos (Cranberg, 1955) where time intervals up to

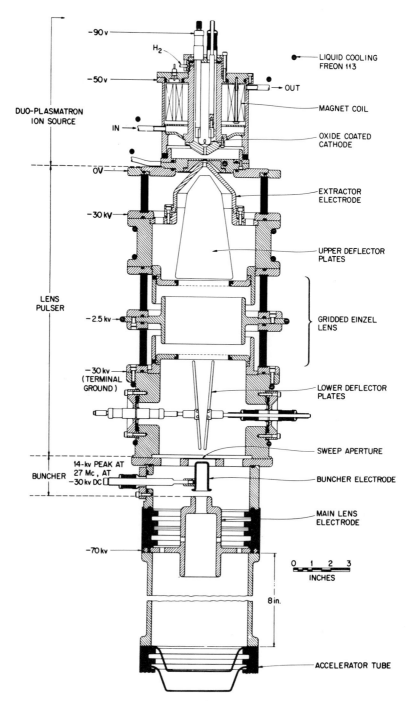

Fig. 38. Modern pulsed positive ion source capable of bursts of 1 to 10 nsec at about 10 mA or 1 mA peak current, respectively.

about 400 nsec were to be measured for application to neutron inelastic scattering applications, and at Oak Ridge (Neiler *et al.*, 1955) where time intervals up to about 2500 nsec were to be measured for application to the study of kiloelectron volt energy neutrons. The time-to-pulse height converter has been described in the literature in some considerable detail (e.g., Neiler and Good, 1960).

Because the pulsed Van de Graaff produces a neutron spectrum which is primarily free of thermal energies, it is possible to utilize a high-repetition rate, namely, 125×10^3 pps (pulse per second) to 2000×10^3 pps. This repetition rate exceeds by a factor of a hundred or more the detector count rate. In order not to subject the amplifiers and analyzers to the high-repetition rate from the target, the time-to-pulse height converter is arranged so as to accept the detector pulse first. The corresponding target pulse is made to arrive at a fixed delayed time, exceeding in magnitude the maximum elapsed flight time to be measured. Time is measured in the following way: The beam energy is reduced to below neutron threshold, and the detector, which responds in such a case to target gamma rays only, is moved to a meter or so from the target. A series of carefully calibrated delay lines in delay times ranging from 10 to about 800 nsec are inserted consecutively into the target side; the corresponding pulse heights are then observed. The result is a linear relationship between pulse height and time interval between detector and target pulses. A least-squares fit to this linear relationship is obtained. When neutrons are observed, their flight time is obviously the target delay time (which can be chosen to suit the desired number of channels, and is typically 1300 nsec) minus the time between detector and target pulses as measured by the pulse height. It is typical to make measurements at 2.5 nsec/channel when the resolution width is about 8 nsec. As has already been stated, storage of the time (pulse height) information is by means of a conventional multichannel analyzer which for greatest flexibility should have 4000 channels capable of being arranged in at least four sections of 1000 channels each.

Before concluding this brief description of the complete pulsed Van de Graaff for time-of-flight spectrometer, it is essential to mention some of the instrument's general features when so employed; it is because of these that this relatively inexpensive instrument is often able to supplement the more flexible and much more expensive electron linac.

Under satisfactory conditions and a repetition rate of 500×10^3 pps, the average current is about 4 μA, and the beam spot has a dimension of about 2 mm in diameter. With care, transmission measurements can be performed on a sample of 6 mm diameter. In the important case of separated isotopes, this ability to perform measurements upon small samples is an important feature.

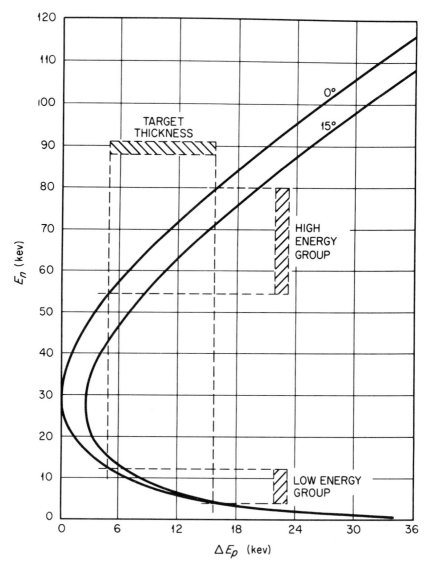

Fig. 39. Characteristic at 0° and 15° of neutron energy versus proton energy (above threshold) for the $^7Li(p,n)^7Be$ reaction.

In the Introduction to this discussion of the pulsed linear accelerators, emphasis was placed upon the intrinsic shortness of the neutron burst. However, the shortness of a neutron burst is not the most useful measure to the energy resolving capability of a time-of-flight spectrometer. It was

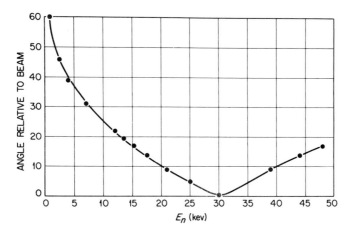

Fig. 40. Maximum angle of neutron emission versus neutron energy at 0°.

shown in I,B that the instrumental quantity $\Delta t/L$ (the shortness of burst/ flight path) is the significant quantity by which to compare different instruments. Now the cyclotron and electron linac currents are such that their neutron yields exceed by two or three orders of magnitude or more that of the Van de Graaff. And while for the Van de Graaff, Δt is smaller than that of the cyclotron and electron linac by at least an order of magnitude, the resolution $(\Delta E/E)$ under the best circumstances favors the latter instruments because of their very long flight times made possible by the very high neutron intensities. These intensity and burst-time considerations have two economic consequences. First, the total neutron yields from the Van de Graaff are so low that virtually no shielding is required to achieve resolution figures within a factor of 5 of the linac whose massive γ-radiation requires tons upon tons of shielding. Second, it is not possible to say precisely, but it seems probable that the cyclotron and the electron linac will always require ten times longer flight paths than the Van de Graaff for a given energy resolution. At the present time, the pulsed Van de Graaff uses 1 meter to 4 meters for transmission studies.

C. The ^{7}Li(p,n)^{7}Be Neutron Source

If a reaction is endothermic, then there exists an energy below which the reaction cannot occur and precisely at which the reaction products have zero energy in the center-of-mass system. Thus just at threshold they move in the laboratory with center-of-mass velocity and in its direction. So it is that at zero degrees to the beam, the reaction neutrons of the ^{7}Li(p,n)^{7}Be reaction have 30 keV and are kinematically "focused" in the beam direction. As the energy is increased above threshold, the velocities of the reaction products in the center-of-mass increase and a corresponding spread in both energy and angle occurs in the laboratory. The properties of the $^7Li(p,n)^7Be$ reaction which have just been described qualitatively are shown quantitatively in Figs. 39 and 40. They show, respectively, the neutron energetic dependence on neutron emergence angle and proton energy, and the kinematic "focusing" near threshold. Because of this focusing, the "intensity" of neutrons increases as 30 keV energy is approached along the lower energy branches of the curves in Fig. 39. As a result, neutrons in the region 2–30 keV have the property that as the neutron energy is increased approaching 30 keV (the primary proton energy decreased), the neutron intensity per energy interval increases and the flight path can be increased accordingly.

The logarithmic energy resolution dE/E can be maintained about constant in this important energy region by increasing the flight path to maintain constant flight time since by Eq. (2a), $dE/E = 2\,dt/t$. From the foregoing considerations, it is seen that the $^7Li(p,n)^7Be$ reaction which with a thin target is the source of neutrons of precisely known energy, becomes with a thick target a continuous source with especially suitable properties for time-of-flight measurements.

From the cross section of the $^7Li(p,n)^7Be$ reaction and its kinematical properties, it is possible to estimate the counting rate per timing channel. If N is neutrons per second, and dt is the incremental time at the detector, then

$$\frac{dN}{dt} = \frac{dN}{dE_n}\frac{dE_n}{dt} = \frac{dN}{dE_n}\,2\frac{E_n}{t} \tag{6}$$

Thus for an assumed constant dN/dE (not of course strictly true) dN/dt remains constant if E/t remains constant. It has been shown (Good, 1961) that if ΔE_p is the proton bombarding energy relative to threshold, and E_p is the proton bombarding energy, then an estimate of dN/dE for E_p not too close to $\Delta E = 0$ is:

$$\frac{dN}{dE_n} = 4.5 \times 10^6 \times \frac{\sigma(E)}{0.2} \times \left(\frac{a}{r}\right)^2 \frac{[\frac{1}{7}(E_p/\Delta E_p)^{1/2} + 7(\Delta E_p/E_p)^{1/2} - 2]}{[50(\Delta E_p/E_p)^{1/2} - 7]} \times I \tag{7}$$

where a is the detector radius, r the detector distance, and the constants are such that dE_n is in kiloelectron volts, N is neutrons per seconds, I is current in microamperes. From this expression an estimate of about 0.5 neutron/μA $\times$ sec $\times$ eV is obtained for the case of 5 keV neutrons incident upon a 10-cm diameter detector at 1 meter from the ^7Li target.

D. Applications

This chapter is primarily concerned with the intrinsic properties of pulsed neutron sources. These sources differ so drastically in their basic features that it is of interest to compare the features of a given experiment as performed with the several types of instruments. For this purpose it suffices to give bare outlines of the experimental situations; the reader is referred to other chapters and the original papers for details.

1. TRANSMISSION OR TOTAL CROSS-SECTION MEASUREMENTS

The simplest measurements to perform are those of the well-known transmission type. For several years, this type of measurement assumed special interest because of the theoretical importance of the quantity $\langle \Gamma_n^0 \rangle / \langle D \rangle$ that is deduced from the resonances which are observed.

The method for measuring total cross sections is well known. In one particular geometrical arrangement, for example, the samples are positioned in such a way as to overshadow the neutron detector by about 50% depending upon the geometrical shape of the sample. The shadowing is conveniently accomplished by substituting a lamp of filament dimensions comparable to those of the neutron source. The transmission measurements are carried out in the usual manner whereby the sample and open beam positions are alternated (automatically) at intervals of approximately ten minutes to reduce effects of drift during prolonged measurements. Lengths of flight paths must be such that flight times do not exceed the time interval limit of the time to pulse-height converter. The measurements down to lowest energies of about 2 keV are therefore carried out with present instruments at a distance of about a 1-meter flight path. In general, a measurement commences at say, 2 keV on the shortest desired flight path, the detector distance is then increased 20% and the bombarding energy is changed so that the mean energy of the source neutrons is increased 40%. Below 30 keV the source neutrons are those in the lower of the two energy groups, Fig. 39, and for these the mean energy increases with decreasing primary proton energy. This procedure is then repeated, the principle being that the total energy range under observation is covered stepwise in energy steps that are 40% of each preceding minimum energy, each spectrum occupying the same portion of the time spectrometer because the flight times are maintained as constant as possible as we explained earlier.

The detector can be elaborated on at considerable length. For the energy region from 2 keV to about 100 keV which is approximately the energy range appropriate to this present chapter, the $^{10}B(n;\alpha,\gamma)^7Li$ reaction probably remains the most satisfactory. This detector utilizes the ^{10}B to convert the neutrons to 0.5 MeV γ-rays which can be detected with relatively high efficiency and relatively good signal-to-noise by means of a NaI spectrometer. This is a standard slow-neutron detector and does not need discussion here excepting to point out the two ways in which the NaI can be employed. One way is generally used by the electron linacs in which the NaI spectrometer has several NaI crystals arranged radially around the ^{10}B slab with the crystals shielded from the direct beam. An ingenious, but less satisfactory, approach employed at Oak Ridge is to place the single large NaI crystal directly behind the ^{10}B slab. As the slab becomes increasingly transparent, the NaI becomes a secondary detector of poor resolution on account of its length which contributes corresponding timing uncertainty. The detector consisting of ^{10}B + NaI will then produce pulses of two categories: pulses attributable to ^{10}B$(n;\alpha,\gamma)$ and pulses attributable to NaI(n,γ), the total number being the sum of the two. The latter number can be monitored continuously and simultaneously on account of the continuous nature of the NaI(n,γ) spectrum [in contrast to the 0.5-MeV line spectrum from ^{10}B $(n;\alpha,\gamma)$] by a gate set to be triggered by γ-rays of energy greater than 0.5 MeV. Since the poor resolution component behaves like a time-dependent background, it can be subtracted from the observed spectrum to obtain the real or desired result.

The instrument as described has been used in the measurement of many total cross sections, with special emphasis on separated isotopes. Typical results are shown for the cases of ^{65}Cu and ^{90}Zr in Figs. 41 and 42, respectively.

2. Radiative-Capture Cross Sections

The fundamental objective in the measurement of neutron radiative capture is the study of partial radiation widths Γ_{γ_i} although the capture cross sections and the total radiation widths Γ_γ are important by themselves. As is well known, the radiation widths of the primary γ-ray in a given cascade, $\Gamma_{\gamma i}$, can be obtained from Γ_γ, if the spectrum of γ-rays is sufficiently well known. In the low-energy region (<100 eV), neutron interactions are, except for very special cases, limited to elastic scattering and radiative capture with radiative capture predominating. In the kiloelectron volt region, elastic scattering is the dominating process, however, and the instrumental problems involved are entirely different from those at low energy where radiative capture dominates. The essential problem in the kiloelectron volt energy region is to recognize the γ-rays from capture in the presence of the gamma spectrometer's own capture gamma rays arising from those neutrons scattered into it by the target under investigation.

For example, in the case of kiloelectron volt neutrons, there may be from

10^2 to 10^3 times as many neutrons scattered from a sample as gamma rays emitted; hence the problem of differentiating the sample's gamma rays from those of the detector itself is severe. A generally applicable solution to this problem is to interpose between scatterer and spectrometer a combination of moderator and capturing nuclide (as for example ^{10}B or ^6Li) in such a way as to remove differentially the neutrons relative to gamma rays.

Two types of detectors for measuring total radiative-capture cross sections are well known, viz., the large scintillator tank and the Moxon–Rae detector. The large scintillator, in principle, because of its size converts each capture event, independent of the number of deexciting gamma rays into a summation pulse whose height depends only upon neutron binding energy with an efficiency which does not depend strongly on what the binding energy is. Of course this ideal is not achieved and the result is a pulse spectrum which indeed depends upon the mode of deexcitation. The Moxon–Rae detector, on the other hand, employs the principle of the gamma-ray Geiger detectors in an almost opposite approach to that of the large-tank scintillator. The Moxon–Rae is of small dimensions and its efficiency depends in first approximation only on the binding energy, independent of mode of decay. In the case of the large liquid scintillator practically all of the neutrons that are captured become so only after having been moderated to essentially thermal energy in the tank. By introducing a nuclide like ^{10}B in the liquid of the tank, the pulses resulting from these captured neutrons are rendered so small as to be almost eliminated by a bias which at the same time produces almost negligible loss in detection efficiency. In contrast to the large scintillator, the Moxon–Rae because of its small size is found to be very insensitive to neutrons relative to gamma rays!

The purpose of the brief account just given of the large scintillator and the Moxon–Rae gamma-ray detectors is to emphasize those features of radiative neutron capture which must be taken into account when the neutron energy exceeds about 1 keV. In the context of the present discussion, it should be noted that both types of detector have been successfully employed with different time-of-flight neutron sources. The first studies into radiative-capture cross sections as a continuous function of energy in the range of 10 to 100 keV were made with the pulsed Van de Graaff, however.

Since the pulsed Van de Graaff is capable of about 1 nsec resolving time, the energy resolution attainable with it as a neutron source for radiative-capture cross section studies depends upon the resolving time of the large scintillator tank and the Moxon–Rae detectors. For the large scintillator, this figure is 15–25 nsec (Gibbons *et al.*, 1961) which corresponds, with the

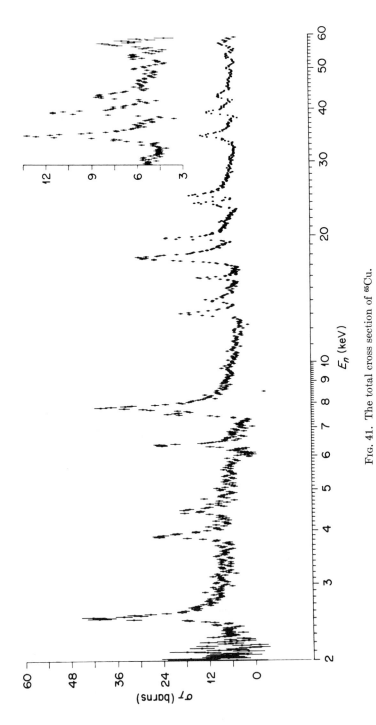

Fig. 41. The total cross section of ^{65}Cu.

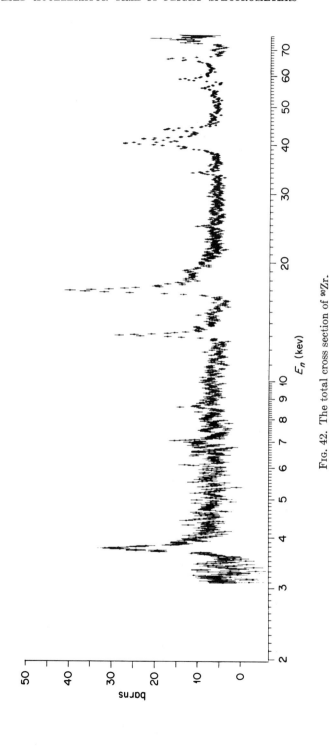

Fɪɢ. 42. The total cross section of ^{90}Zr.

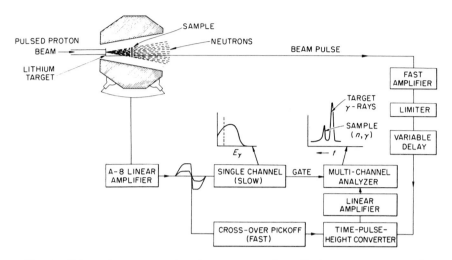

Fig. 43. Schematic representation of arrangements for radiative capture cross section measurements by means of the large scintillator tank.

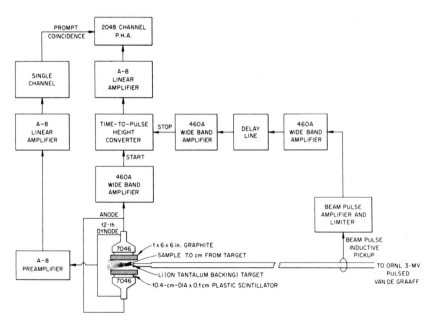

Fig. 44. Schematic representation of arrangements for radiative capture cross section measurements by means of the Moxon–Rae detector.

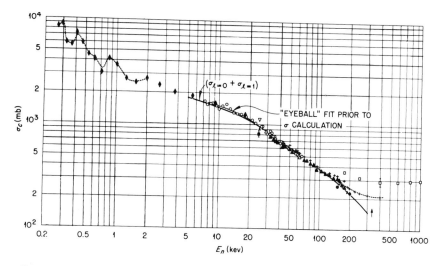

FIG. 45. The radiative capture cross section of In as obtained with the pulsed Van de Graaff time of flight spectrometer $\langle \Gamma_n^0/D \rangle = 0.31 \times 10^{-4}$; $\frac{1}{3}\langle \Gamma_n'/D \rangle = 4.0 \times 10^{-4}$; $\Gamma_\gamma = 60 \times 10^{-3}$ eV; $D_{OBS} = 6.7$ eV. $+$, ORNL In versus ^{235}U, then Li target (1.21×10^{22} atoms/cm²); ▲, ORNL T(p,n) runs (1.21×10^{22} atoms/cm²); ○, ORNL Li(p,n) runs (1.21×10^{22} atoms/cm²); □, Diven et al., 1900; A, ORNL activation $\times 1.15$; ▽, UCRL activation $\times 1.15$; ●, WISC. activation $\times 1.15$; ■, ORNL spherical siter; △, ORNL threshold; ◆, ORNL fast chopper 1.95×10^{21} atoms/cm².

pulsed Van de Graaff as a neutron source, to an over-all figure of merit (i.e., $\Delta t/L$) of 15 to 20 nsec/meter. It is a surprising and tremendously important fact that the Moxon–Rae detector has a time resolution of 3 nsec (Macklin et al., 1963) so that with the pulsed Van de Graaff as a neutron source it is possible to obtain 15–25 nsec/meter at a flight path of only 15 to 20 cm thereby permitting the use of small samples.

Block-circuit diagrams and schematic geometry for the large scintillator tank and the Moxon–Rae detectors as employed by the pulsed Van de Graaff are shown in Figs. 43 and 44. Figure 45 from Gibbons et al. (1961), shows an example of the type of result obtained with the large scintillator tank. In this latter example, interest centered upon average cross sections. The mass region shown is not a proper one for observing individual resonances. Figure 46 shows a result obtained with the Moxon–Rae detector for sulfur. This nuclide is in the mass region for which there exists resolvable resonances in the 2–100 keV region of energy. The particular resonance shown represents a situation common in the kiloelectron volt energy region, namely: This level was found in capture because of its normal Γ_γ. The level had escaped previous detection in transmission studies which measured Γ_n. This particular resonance is given as an example because it represents a case

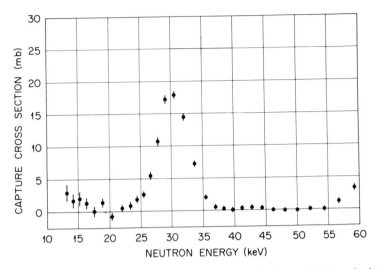

Fɪɢ. 46. The radiative capture cross section of sulfur obtained with the pulsed Van de Graaff time-of-flight spectrometer.

in which Γ_n, though small, is large enough to be measured. Figure 47 shows the total cross section of S as measured with a sample whose thickness was chosen to afford optimum opportunity for observing levels of small Γ_n.

It is appropriate in the present discussion of radiative-capture measurements to mention those features of the pulsed Van de Graaff and $^7Li(p,n)^7Be$ neutron source which have special utility. Of course, there are the general characteristics of short flight paths for a given precision of energy measurement and low levels of extraneous radiation. However, there are features of special utility to the measurement of capture cross sections. One of these concerns the case of large tank measurements in which it is essential to have information about the pulse spectra from the tank in order to properly interpret the capture cross-section measurements which are normally obtained by integrating all pulses above some arbitrary lower cutoff bias. Since the background for such measurements is rather large, its pulse spectrum yield must be known with precision also. Figure 48 shows how it was possible to obtain this background spectrum yield coincident with the spectral yield from the sample by gating an analyzer to monitor continuously the pulse spectrum at a time after each target pulse and just prior to the following neutron capturing events, since in such experiments the background arises predominantly from thermal capture. The focusing property of the $^7Li(p,n)$ reaction near threshold gives to the pulsed Van de Graff when used with this target the possibility of especially high precision of

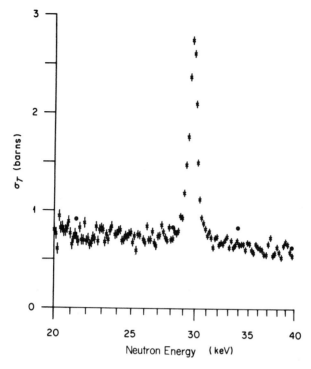

FIG. 47. The total cross section of sulfur (4.26 barns/atom).

radiative-capture cross-section measurement at threshold neutron energy because of the favorable geometry afforded by the intrinsic neutron collimation. Thus many average capture cross sections have been measured in this way for neutrons of about 30 keV energy. The ³H(p,n)³He reaction has the same characteristic features as the ⁷Li(p,n)⁷Be, but it is not so suitable as a source with the pulsed Van de Graaff. However, it has been used alternatively with the ⁷Li(p,n) source to provide intrinsically collimated neutrons of 65 keV (which is their energy at threshold) for measurements of radiative capture (Gibbons et al., 1961).

In some regions of nuclear mass, e.g., mass region for the filling of the $2s$-$1d$ shell and the mass region of lead, the resonant level spacings are such that the resonances are resolvable with 15 to 25 nsec/meter energy resolution. It is then possible to measure the total radiation widths. In such cases it is also generally possible to measure the resonant γ-ray spectra and from the two measurements to derive partial radiation widths Γ_{γ_i} of the initial γ-ray transitions. The first successful γ-ray spectroscopy at kiloelectron volt neutron energies was performed with a NaI-spectrometer.

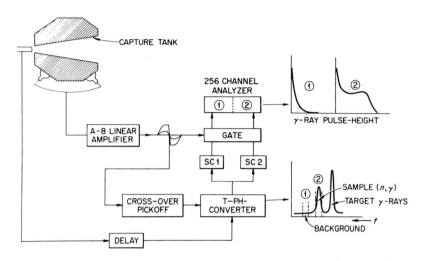

Fig. 48. Block diagram to indicate the means for determining the background spectrum, simultaneously with the radiative capture spectrum.

The configuration, electronically as well as physically, is shown in Fig. 49. In this application it is necessary to take into account that the background pulse spectra will be time dependent on the neutron flight-time scale of times and that this background spectrum must be subtracted from the observed one. These time-dependent background spectra arise because of scattered neutrons that capture in the detectors, although a special shield composed of moderators and absorbers like ^{10}B or ^{6}Li is used differentially to absorb neutrons relative to gamma rays.

The challenge of neutron radiative-capture spectroscopy is its multidimensional aspect which is fully capable of being mastered by modern instrumental electronic pulse techniques. Radiative-capture γ-ray spectra obtained simultaneously as a function of neutron energy are shown for ^{206}Pb in Fig. 50 against the three parameters, namely, relative counts, neutron flight-time, gamma-ray pulse height. The reason ^{206}Pb was chosen as an example for resonance neutron gamma-ray spectra is that it represents one of the first cases observed in which the neutron widths are so small that the resonances escaped detection in the first total cross-section measurements (Biggerstaff et al., 1967); their presence was revealed by their radiative capture, and they were only subsequently found in transmission measurements. A fitting conclusion to this discussion of the use of the pulsed Van de Graaff for studies of resonant neutron gamma-ray spectra is the observation that in the kiloelectron volt energy region the germanium gamma-ray spectrometer opens a challenging and interesting area for future investigation.

FIG. 49. Diagram to describe the geometry and the circuitry for the two parameter measurement of radiative capture spectra versus neutron time-of-flight.

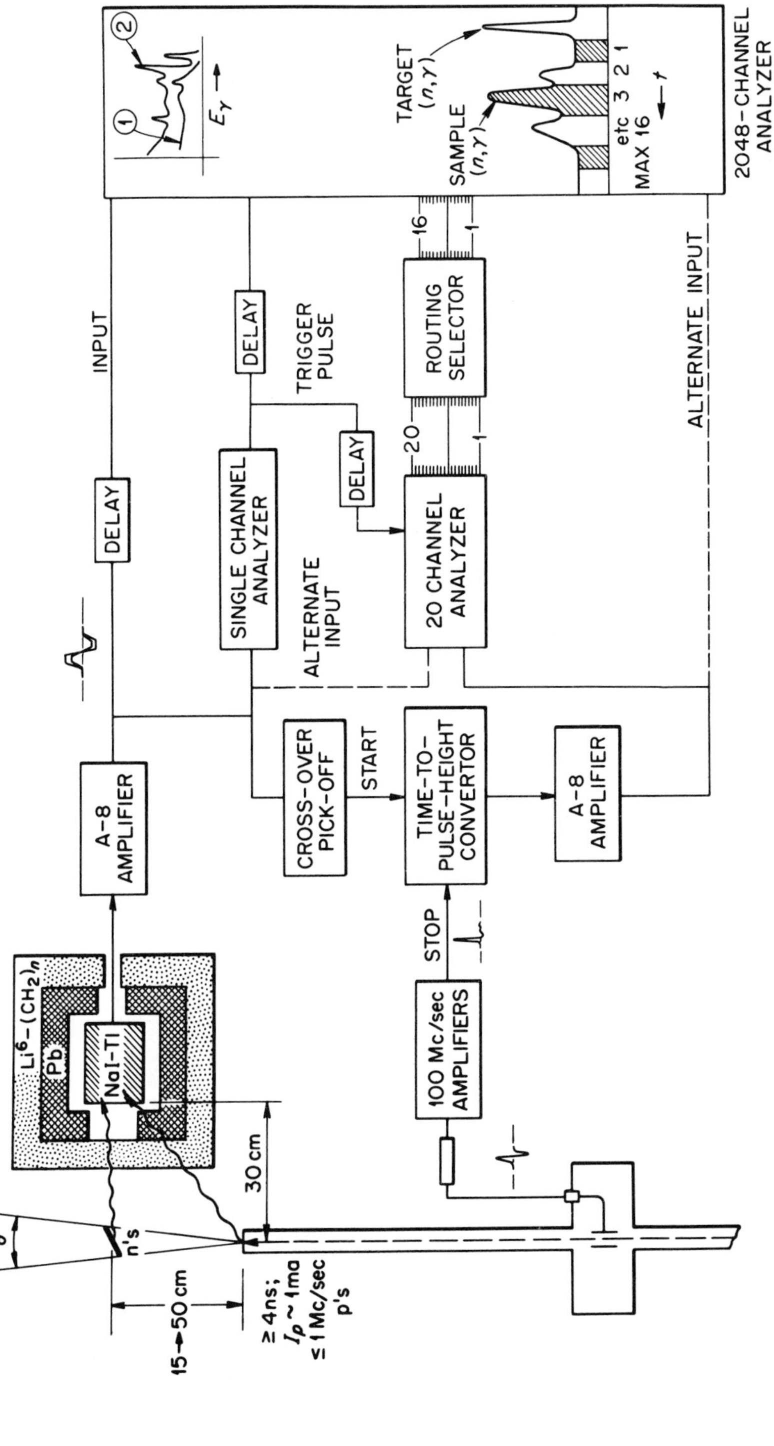

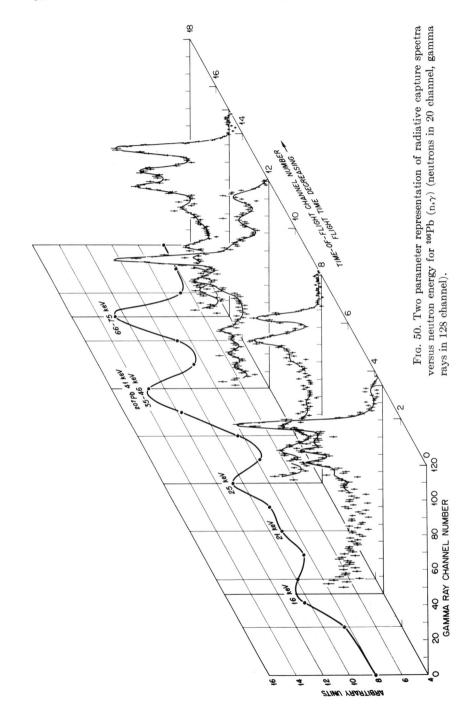

FIG. 50. Two parameter representation of radiative capture spectra versus neutron energy for ^{206}Pb (n,γ) (neutrons in 20 channel, gamma rays in 128 channel).

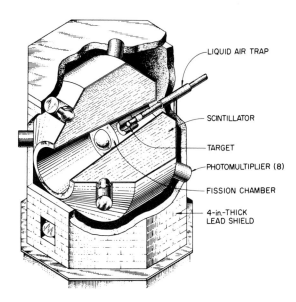

  LIQUID AIR TRAP

  SCINTILLATOR

  TARGET

  PHOTOMULTIPLIER (8)

  FISSION CHAMBER

  4-in.-THICK
  LEAD SHIELD

FIG. 51. Schematic representation of the large scintillator tank as used for the measurement of the capture to fission ratios.

3. FISSION

A review of the pulsed Van de Graaff would not be complete without at least mentioning its application to fission. An example is the measurement of the quantity α which is the ratio of capture to fission cross sections. The problem is essentially one of γ-ray detection, hence an appropriate detector is the large tank shown previously, Fig. 51. In this case, however, the tank serves a double purpose not only of detecting simple radiative capture but also in simultaneously establishing whether or not a certain sequence of events characteristic of fission occurs. If the neutron target is a fissionable material, then absorption of a neutron can be followed either by radiative capture in which only prompt gamma rays are emitted, or it can be followed by fission in which case prompt gamma rays are first emitted, then fission takes place. If the tank is then properly designed, each of the several prompt neutrons characteristic of fission will independently be moderated and captured, and if the capture material is suitably chosen, these terminal events can be separately detected. It is the sequence of these thermal capture events and knowledge of their number which establishes that the *initial* prompt gamma ray was associated with a fission absorption event and not a radiative-capture absorption event. A special requirement of the experiment is a very low-repetition rate because it is necessary to allow of the order of 20 μsec between neutron bursts.

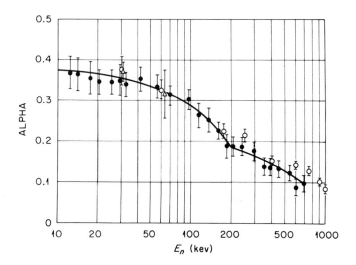

F IG. 52. Ratio of capture to fission for ^{235}U as measured by the pulsed Van de Graaff.
● △, present experiment; ○, Hopkins and Diven (1962).

The details of this experiment are similar to, but more complicated than, that of simple radiative capture. Certainly for the case of ^{235}U, only energy average values of α can be obtained because of the small level spacing. The result of an experiment to measure α for the case of ^{235}U is shown in Fig. 52.

E. Theory and Critique of the Pulsed Van de Graaff

The pulsed Van de Graaff as presently used, exploits less of its potential than other types of pulsed neutron sources do. Part of the disparity in resolution between the pulsed Van de Graaff and other time-of-flight instruments is a consequence of the large peak currents that these other instruments employ. The best present-day pulsed Van de Graaff's employ peak currents before bunching of 1 mA or less, in spite of the fact that techniques for 100 mA peak current before bunching are employed in other applications. Indeed, pulsed Van de Graaffs in present use can hardly be said to have advanced beyond the first or, at most, the second generation of instrument.

In order to understand what consequences would follow if the higher peak current resources available to the Van de Graaff were to be utilized, it is necessary to review, at least briefly, the theory of ion-burst production. No attempt will be made to outline the actual physical problems that would require solution for such further development. However, it must be anticipated that the complexity and cost of such a new instrument would be considerable compared with those of the present instruments.

It is possible, in principle, to devise many types of electromagnetic configurations for producing ion bursts. In practice, it has proved most convenient to employ a two-step process which consists of a "chopping" operation to produce a relatively long burst, followed by a bunching operation. The reason for this will be given. In an analytical discussion of ion-burst production it is convenient to consider the same two-step process.

An elementary "chopper" to carry out the first step of the process of ion-burst production consists of a pair of deflector plates and a defining-slit beyond (Fowler and Good, 1960). In principle, a pulse gate on the deflector system could be used to achieve the first step of burst production. Such a method is much more difficult than the one more commonly used in which the beam of ions is swept, oscilloscope fashion, across a screen with an aperture and a burst appears beyond as the beam crosses the aperture. Only this method of chopping will be considered. In this case the burst time Δt_{burst} is related to a combined aperture size and beam diameter Δy by the relation

$$\Delta t_{burst} = \Delta y / \text{writing speed} \tag{8}$$

The writing speed obviously has the dimensions of a velocity which, in turn, can be derived from the product of acceleration and a time. By definition the writing speed, which will be designated ws, is obtained by evaluating $\Delta y / \Delta t_{burst}$ at the aperture which defines the burst. The result generally has the form

$$\text{ws} = K(eV_0/md)\tau \tag{9}$$

where V_0 is the peak voltage across the deflection plates, d is the distance between these plates, m is the ion mass, τ is the deflector plate to aperture transit time, and K is a constant which depends in a dimensionless way upon the transit time between the deflector plates. Thus for a step-function-applied potential

$$\text{ws} = 2(eV_0/md)\tau_2 \tag{10}$$

If $V(t)$ is a sinusoidal function of angular velocity ω and $\tau_2 > \tau_1$, then

$$\text{ws} = \omega\tau_1(eV_0/md)\tau_2 \tag{11}$$

In these expressions, τ_1 is the deflector plate transit time and τ_2 is the transit time to the aperture beyond, that defines the burst duration. The acceleration, eV_0/md, has a value of about $1/50$ cm/nsec^2 at 20 keV/cm. For $\tau_2 \approx 50$ nsec it would appear that a writing speed ≈ 1 cm/nsec is achievable. Thus, the production of 1-nsec burst appears feasible by applying a sinusoidal sweep voltage.

The expressions just given apply only to a highly idealized beam of particles. Certain limitations must be expected when the physical properties of the beam are taken into account. Without going into the details which are given elsewhere (Fowler and Good, 1960), it is possible to see the origin of these limitations. Consider a system consisting of a pair of deflection plates with a defining slit or aperture beyond. When the aperture is located on the axis of the deflector plates and $\tau_2 \gg \tau_1$, then the beam enters the deflector plates at a time $\approx -\frac{1}{2}\tau_1$, and leaves the deflector plates at a time $\approx +\frac{1}{2}\tau_1$ with respect to the time $t = 0$ measured by the sinusoidal sweep voltage. If the mean beam diameter in the plates is W, then there appears, across the beam on entering deflector plates spaced a distance d apart, an energy spread of about $\frac{1}{2}\tau_1 e V_0 w/e$. In a static situation, this energy spread is removed on leaving the deflector plates. However, with sweep voltage applied, as described, the change in phase in transit through the plates causes an amount to be added on leaving the deflector plates equal to that contributed on entering them. Thus a total energy spread $\approx (w/d) \times \tau_1 e V_0$ is introduced by the sweeper, and this energy spread gives rise to a *time spread* as discussed in I.B. Its amount is $\Delta t(\Delta E) = \frac{1}{2} \Delta E/E$. Now from $\Delta t = \Delta y/\text{ws.}$, we can estimate the *burst* duration from (9). If the effective aperture Δy is, say, w' then

$$\Delta t_{\text{burst}} = \frac{\text{width of beam}}{\text{writing speed}} = \frac{w'}{K(eV/md)\tau_2} \tag{12}$$

If w is the beam width, then $(w/d)KeV_0$ is the energy spread from the sweeping process which we will call ΔE. Then we can write

$$\Delta t_{\text{burst}} = w'wm/\Delta E\tau_2 \tag{13}$$

This result can be manipulated to give the relation

$$\Delta t_{\text{burst}} = \frac{ww'}{v^2 \frac{1}{2} \Delta(E/E)\tau_2} = \frac{ww'}{v^2 \, \Delta t(\Delta E)} \tag{14}$$

where v is the ion velocity. Thus the burst duration shortens in inverse proportion to ΔE, the energy spread across the beam. A minimum duration is obtained in the case in which beam sweeping is used to generate ion bursts when

$$\Delta t(\Delta E) = \Delta t(\text{writing speed}) \approx (ww')^{1/2}/v \tag{15}$$

From Eq. (13) it is also easy to derive the estimate

$$\Delta E \, \Delta t \approx \Delta p_y \, \Delta y \tag{16}$$

in which ΔE is the energy spread introduced by the sweeping process, and Δt is the burst duration at the target. The quantity $\Delta p_y \, \Delta y$ is the well-known differential invariant which measures the quality of a beam of given size; it is a quantity that is determined by the ion source and which remains constant in accelerators of steady beams. Eq. (16) states (a) that when ion-burst production is achieved by beam sweeping, the product $\Delta E \, \Delta t$ is a quantity determined by the ion source, and (b) that ΔE must increase as Δt decreases. In practically every application of the Van de Graaff, the energy spread of the beam is a quantity that must and can be controlled to meet certain specifications. The pulsed beam applications are, in general, no exceptions. In the present chapter the $^7\text{Li}(p,n)^7\text{Be}$ reaction has served as the neutron source with the target thickness chosen so as to provide the neutrons with upper and lower energy bounds to suit the particular experiment. It is not profitable to utilize a beam with energy spread ΔE exceeding the prescribed target thickness. It follows for a pulsed neutron source, that the product $\Delta E \, \Delta t$ is a quantity that is fixed by the specification of neutron energy range and the time resolution. From the equation $\Delta E \, \Delta t = \Delta p_y \, \Delta y$ the volume element $\Delta p_y \, \Delta y$ is likewise determined, thus fixing the yield of the ion source.

Clearly, beam sweeping is not a very efficient means of producing short ion bursts because of the unprofitable manner in which extra energy is introduced into the beam. It is thus suggested that a better procedure would be to start with a relatively long burst with negligible extra energy from the beam sweeping and apply a modulating voltage to this burst so as to achieve klystron bunching at the target. The bunching conditions are well known. If a burst of initial duration Δt_1 and energy spread ΔE_1 is uniformly modulated to a maximum energy ΔE_2 (the higher energy particles leaving the modulating gap last), then in a transit time $\tau = L/v$, where L is the gap to target distance, the last ions to leave the gap will overtake the first ones if

$$(\Delta E_2/E)\tau = 2 \, \Delta t_1 \tag{17}$$

Assume that an initial burst of duration Δt_1 is shortened and that the duration of the burst at the target is Δt_2, then the burst duration will have been shortened *and*, in addition, the peak current will be increased in the ratio $\Delta t_2/\Delta t_1$. The bunching condition (17) was derived assuming exact time focusing as though the beam were ideally monoenergetic. The consequence of the initial beam energy spread is easily taken into account. Consider two ions at the lead of the burst which initially have an energy spread ΔE_1, characteristic of the beam. In the bunching flight time, $\tau = L/v$, these two ions will arrive separated in time by an interval

$$\Delta t_2 = \tfrac{1}{2}(\Delta E_1/E)\tau \qquad\qquad (17a)$$

Eliminating τ/E between Eqs. (17) and (17a) we see the relation:

$$\Delta t_2/\Delta t_1 = \Delta E_1/\Delta E_2; \qquad \Delta E_1\,\Delta t_1 = \Delta E_2\,\Delta t_2 \qquad (18)$$

The advantage of ion-beam bunching as a means for producing short ion bursts is clearly set forth in this relationship. The peak current (leaving physical dimensions of the beam at the target out of account) is inversely proportional to the duration of the burst, and this is in turn inversely proportional to the inclusive energy spread in each burst. Thus, bunching as a means of beam-burst production makes possible a significant improvement in the peak currents and resolving time simultaneously.

It will be recalled that bursts which are produced by "beam sweeping" are limited in peak current through restrictions imposed on ΔE and Δt by experimental requirements; the limitation comes about from the property of "beam sweeping" whereby $\Delta E\,\Delta t \approx \Delta p_y\,\Delta y$. The process of klystron bunching appears to impose no such limitation, but some consequences of large pulse peak must be expected. These can be estimated from a bilinear invariant which is an extension to pulsed beams of the one which characterizes the steady beam under focal conditions. The pulsed beam relationship is (Fowler and Good, 1960)

$$[\Delta E\,\Delta t + \Delta p_r\,\Delta r]_{\text{source}} = [\Delta E\,\Delta t + \Delta p_r\,\Delta r]_{\text{target}} \qquad (19)$$

It is seen to be an extension of the well-known Abbey Law of static optics that

$$[\Delta p_r\,\Delta r]_{\text{objcct}} = [\Delta p_r\,\Delta r]_{\text{image}} \qquad (20)$$

The differential bilinear invariant for pulsed beams (19) shows that so far as differential elements of a beam are concerned, time compression (always accompanied by peak current multiplication) can be achieved at the expense of energy spread or spatial-directional definition, thus establishing an equivalence for various methods for ion-burst production. Thus also, beyond increasing the ion-source performance, i.e., the density of ions in phase space, higher peak currents can only be achieved ultimately at the expense of beam quality, i.e., energy spread, beam definition at target, and so forth.

It is now possible to make some general remarks about the present state-of-the-art of the pulsed positive-ion linear accelerator. These can quite logically begin with some observations which pertain to the "quality" of ion source. With the adaptation of the Duoplasmatron (Moak $et\ al.$, 1959) to Van de Graaff use, a source of nearly 95% ionization efficiency became available for the pulsed beam application. Unfortunately, however, as adapted to the Van de Graaff the mass 1 component is, at best, about 60% compared to the 95% figure of the older rf type sources. Nevertheless, the Duoplasmatron is almost the best that can be expected and some means could conceivably be found to improve upon its mass 1 relative yield. One such means exists, but it could only be applied to a completely new generation of pulsed positive-ion accelerators. It is an established fact that when the parameters of the Duoplasmatron are chosen for ion-current yields in the order of 100 mA or more, conditions in the ion source favor mass hydrogen ions. These higher yield ion sources achieve their greater performance primarily from their total volume (in phase space) of ions, but some increase in density is also afforded because of the improvement in the mass 1 content. Thus in scaling up the output of an ion source by increasing its power and gas consumption, some intrinsic improvement is also achieved because of the peculiarities of the Duoplasmatron ion source. Bunching was first conceived as a means of obtaining higher peak currents without incurring the demands of a larger ion source. The bilinear invariant (19) shows, however, that bunching has a value of more intrinsic significance: Since ΔE of the ion source is reasonably small (compared with the 5 keV or so target thickness), the duration of the burst can be increased without necessarily increasing $\Delta p_r \, \Delta r$.

The quantity $\Delta E \, \Delta t$ has at the target an upper limit of about 5000 eV-nsec which is set by the Δt limitations of the other instrumental components exclusive of the pulsed beam itself and the thickness of the target which ΔE cannot profitably exceed. Having reached this limit, larger currents require increasing $\Delta p_r \, \Delta r$ exclusively, either by coupling with $\Delta E \, \Delta t$ or by larger initial ion-source yield.

From the preceding arguments there appears the not very surprising result that beyond a certain figure for peak current (probably 10 mA or less) the neutron time-of-flight energy resolution can, indeed, be improved, but the energy resolution per unit of area cannot be improved very much. Thus there will be a distinct tendency for better energy resolution to require larger neutron target specimens.

In a previous section of this chapter, various species of time-of-flight instruments were compared, all of which utilize continuous energy-spectrum sources. Table I compares the pulsed Van de Graaff with other sources. From the estimate in Section III,C, of $\Delta N/\Delta E_n \approx 0.5$ neutrons/μA $\times$ sec $\times$ eV and from

$$\frac{\Delta N}{\Delta t} = \frac{\Delta N}{\Delta E_N} \frac{\Delta E_N}{\Delta t} = \frac{\Delta N}{\Delta E_N} 2 \frac{E_N}{t}$$

there results $\Delta N/\Delta t \approx 5$ neutrons/μA $\times$ sec $\times$ nsec. In the case of a 10-cm diameter detector (Good, 1961) and a 3-μA average current, there results about 1 neutron/sec $\times$ cm^2 for $\Delta E/E \approx 0.01$. This performance of the pulsed Van de Graaff can roughly be compared with other time-of-flight instruments by referring to Fig. 32. The present electron linacs and proton cyclotrons have an order of magnitude or so in their favor regarding the relative number of neutrons per unit area at the detector if account is taken of the higher energy resolution quoted in the figures. For the newer installations reviewed in Fig. 33 the flux advantage is 2 to 3 orders of magnitude over the Van de Graaff.

It is clear that justification for considering the pulsed Van de Graaff in the context of "resonance neutron physics" must rely on factors not contained in Tables I and III, and Figs. 32 and 33. Two factors of some importance to the experiments of interest to resonance neutron physics have already been mentioned, but they need to be emphasized at this point.

They are (a) the ability to produce a narrow energy band of nonmonoenergetic neutrons as a property of the Van de Graaff alone, and (b) the requirement of very small amounts of sample. In Tables I and III, it would have been appropriate to include not only neutron production rate, but also "neutron production rate per unit area of source," this also being equivalent to "per unit area of sample." Such a figure of merit would be high for the Van de Graaff whose neutron source size is a few tenths of a square centimeter.

It is the unrealized potential of the pulsed Van de Graaff type of instrument and the special properties of the source of neutrons that it provides that make a place for this instrument in the study of "resonance" neutrons.

REFERENCES

ADCOCK, D. R., BEYSTER, J. R., and COLE, J. L. (1967). *IEEE Trans. Nucl. Sci.* **NS-14,** 721.

ALFIMENKOV, V. P., LUSHCHIKOV, V. I., NIKOLENKO, V. G., TARAN, Yu.V., and SHAPIRO, F. L. (1967). *Joint Inst. for Nucl. Res., Dubna.* Preprint No. P3-3208.

ALLARD, C., and SALOME, J. M. (1966). Progr. Rept. on Nucl. Data Res. in the Euratom Community Jan. 1st–Dec. 31st 1965. EANDC(E)66 U, p. 134.

ALVAREZ, L. W. (1938). *Phys. Rev.* **54,** 609.

ASAMI, A., and MOXON, M. C. (1969). UKAEA-Rept. AERE-R5980 [to be published in *J. Nucl. Energy* (1970)].

ASAMI, A., FUKETA, T., KAWARASAKI, Y., NAKAJIMA, Y., OKUBO, M., SAKUTA, T., TAKAHASHI, K., and TAKEKOSHI, H. (1967). Japan At. Energy Rept. JAERI-1138.

ASAMI, A., MOXON, M. C., and STEIN, W. E. (1969). *Phys. Letters* **28B,** 656.

AUSTIN, N. A., and FULTZ, S. C. (1959). *Rev. Sci. Instr.* **30,** 284.

AXEL, P. (1962). *Phys. Rev.* **126,** 671.

BAKER, C. P., and BACHER, R. F. (1941). *Phys. Rev.* **59,** 332.

BAREFORD, C. F., and KELLIHER, M. G. (1953). *Phillips Tech. Rev.* **15,** 1.

BARTHOLOMEW, G. A., and TUNNICLIFFE, P. R. (1966). "Intense Neutron Generator" AECL-2600.

BARTHOLOMEW, G. A., KATZ, L., and McNEILL, K. G. (1966). "Intense Neutron Generator" AECL-2600, Section VII, p. 128.

BEER, M., LONE, M. A., CHRIEN, R. E., WASSON, O. A., BHAT, M. R., and MUETHER, H. R. (1968). *Phys. Rev. Letters* **20**, 340.

BERGERE, R. (1961). "Neutron Time of Flight Methods," p. 329, Euratom, Brussels.

BERMAN, B. L., SIDHU, G. S., and BOWMAN, C. D. (1966). *Phys. Rev. Letters* **17**, 761.

BERMAN, B. L., VAN HEMERT, R. L., and BOWMAN, C. D. (1967). *Phys. Rev.* **163**, 958.

BERMAN, B. L., VAN HEMERT, R. L., and BOWMAN, C. D. (1969). *Phys. Rev. Letters* **23**, 386.

BERTOZZI, W., SARGENT, C. P., and TURCHINETZ, W. (1965). *Phys. Letters* **6**, 108.

BETZ, N. A., REYNOLDS, J. W., and SLAUGHTER, G. G. (1969). (to be completed by J.A.H.). *Proc. Skytop Conf. Computer Systems Nucl. Phys.* CONF-690301, EANDC (U.S.) 121U, p. 218.

BEYSTER, J. R. (1968). "Neutron Thermalization and Reactor Spectra," Vol. II, p. 3. IAEA, Vienna.

BEYSTER, J. R., and RUSSELL, J. L. (1966). "Pulsed High Intensity Fission Neutron Sources." USAEC CONF-650217, p. 36.

BIGGERSTAFF, J. A., BIRD, J. R., GIBBONS, J. H., and GOOD, W. M. (1967). *Phys. Rev.* **154**, 1136.

BJØRNHOLM, S., BORGGREN, J., WESTGAARD, L., and KARNAUKHOV, V. A. (1967). *Nucl. Phys.* **A95**, 420.

BLOCK, R. C., HOCKENBURY, R. W., BARTOLOME, Z. M., and FULLWOOD, R. R. (1967). "Nuclear Data for Reactors," Vol. I, p. 565. IAEA, Vienna.

BLOKHIN, G. E. *et al.* (19 authors) (1962). "Physics of Fast and Intermediate Reactors," Vol. III, p. 399. IAEA, Vienna.

BLONS, J., EGGERMANN, C., and MICHAUDON, A. (1968). *Compt. Rend. Acad. Sci. Paris* **267**, 901.

BOLLINGER, L. M. (1960). "Nuclear Spectroscopy" (Fay Ajzenberg-Selove, ed.), Part A, p. 344. Academic Press, New York and London.

BOLLINGER, L. M. (1966). USAEC CONF-660303, Vol. 2, p. 1064.

BOWMAN, C. D., SIDHU, G. S., and BERMAN, B. L. (1967). *Phys. Rev.* **163**, 951.

BOYD, T. J., ROGERS, B. T., TESCHE, F. R., and VENABLE, D. (1965). *Rev. Sci. Instr.* **36**, 1401.

BURAS, B. (1966). "Intense Neutron Sources." USAEC CONF-660925, p. 677.

CAO, M. G., MIGNECO, E., and THEOBALD, J. P. (1968). *Phys. Letters* **27B**, 409.

CARRARO, G., COCEVA, C., CORVI, F., and GIACOBBE, P. (1968). Progr. Rept. on Nucl. Data Res. in the Euratom Community, Jan. 1st to Dec. 31st 1967. (EANDC(E)89U), p. 146.

CIERJACKS, S., FORTI, P., KROPP, L., and UNSELD, H. (1966). "Intense Neutron Sources." USAEC CONF-660925, p. 589.

CIERJACKS, S., DUELLI, B., FORTI, P., KOPSCH, D., KROPP, L., LOSEL, M., NEBE, J., SCHWEICKERT, H., and UNSELD, H. (1968). *Rev. Sci. Instr.* **39**, 1279.

COATES, M. S., ENDACOTT, D. A. J., GAYTHER, D. B., JAMES, G. D., and LANGSFORD, A. (1968). UKAEA Rept. AERE-PR/NP13, p. 59.

COATES, M. S., DIMENT, K. M., and RAE, E. R. (1968a). Unpublished rept. EANDC-(UK)97 AL.

COCEVA, C., CORVI, F., GIACOBBE, P., and STEFANO, M. (1965). *Phys. Letters* **16**, 159.

COCKCROFT, J. D. (1949). *Nature* **163**, 869.

CRANBERG, L. (1955). *Proc. Intern. Conf. Peaceful Uses At. Energy, Geneva* **4**, 40.

DABBS, J. W. T., WALTER, F. J., and PARKER, G. W. (1965). "Physics and Chemistry of Fission," Vol. 1, p. 39. IAEA, Vienna.

DABBS, J. W. T., EGGERMANN, C., CAUVIN, B., MICHAUDON, A., and SANCHE, M. (1968). *Bull A.P.S.*, Ser. II, 13, 1407.

DAY, D., and SINCLAIR, R. N. (1969). *J. Phys.* (*C*) **2**, 870.

DERUYTTER, A. J., WAGEMANS, C., and FURETTA, C. (1968). Progr. Rept. on Nucl. Data in the Euratom Community, Jan. 1st to Dec. 31st 1967. (EANDC(E)89 U, p. 146.

DE SAUSSURE, G., WESTON, L. W., GWIN, R., INGLE, R. W., TODD, J. H., HOCKENBURY, R. W., FULLWOOD, R. R., and LOTTIN, A. (1967). "Nuclear Data for Reactors," Vol. 2, p. 233. IAEA, Vienna.

DIMENT, K. M. (1967). UKAEA Rept. AERE-R 5224.

DIVEN, B. C. (1966). "Intense Neutron Sources." USAEC CONF-660925, p. 539.

DUPEN, W. D. (1966). Stanford Linear Accelerator Centre Report SLAC-62.

EGELSTAFF, P. A. (1958). *J. Nucl. Energy* **7**, 35.

FERTEL, G. E. F., GIBBS, D. F., MOON, P. B., THOMPSON, G. P., and WYNN-WILLIAMS, C. E. (1940). *Proc. Roy. Soc.* **175**, 316.

FIRK, F. W. K. (1964). *Nucl. Phys.* **52**, 437.

FIRK, F. W. K., REID, G. W., and GALLAGHER, J. F. (1958). *Nucl. Instr.* **3**, 309.

FIRK, F. W. K., LYNN, J. E., and MOXON, M. C. (1960). *Proc. Kingston Intern. Conf. Nucl. Structure* (*Toronto*) p. 757.

FLEROV, G. N., and POLIKHANOV, S. M. (1964). *Compt. Rend. Cong. Intern. Phys. Nucl.* **1**, 407.

FOWLER, T. K., and GOOD, W. M. (1960). *Nucl. Instr. Methods* **7**, 245.

FRY, D. W., HARVIE, R. B., MULLETT, L. B., and WALKINSHAW, W. (1948). *Nature* **162**, 859.

FULTZ, S. C. (1969). Private communication.

GAERTTNER, E. R., YEATER, M. L., and FULLWOOD, R. R. (1962). "Neutron Physics," p. 263. Academic Press, London and New York.

GARG, J. B., RAINWATER, L. J., and HAVENS, W. W. Jr. (1965). *Phys. Rev.* **137B**, 547.

GIBBONS, J. H., MACKLIN, R. L., MILLER, P. D., and NEILER, J. H. (1961). *Phys. Rev.* **122**, 182.

GOOD, W. M. (1961). "Neutron Time of flight Methods," p. 309. Euratom, Brussels.

GOOD, W. M., NEILER, J. H., and GIBBONS, J. H. (1958). *Phys. Rev.* **109**, 926.

GWIN, R., WESTON, L. W., DE SAUSSURE, G., INGLE, R. W., TODD, J. H., GILLESPIE, F. E., HOCKENBURY, R. W., and BLOCK, R. C. (1969). ORNL Report ORNL-TM-2598.

HAHN, J., DHAWAN, S., MAYER, R., and GILLMAN, C. (1968). Pegram Nucl. Phys. Lab. Progr. Rept. to USAEC, Jan. 1967 to Jan. 1968, p. 104.

HARVEY, J. A. (1963). Unpublished EANDC paper EANDC (US) 46 U.

HARVEY, J. A. (1965). "Pulsed High Intensity Fission Neutron Sources," USAEC CONF-650217, p. 124.

HARVEY, J. A. (1969). Private communication. See also ORNL-4230, p. 125; ORNL-4280, p. 33.

HAVENS, W. W., JR. (1955). *Proc. Intern. Conf. Peaceful Uses At. Energy Geneva* **4**, 74.

HAVENS, W. W., JR. (1966). "Intense Neutron Sources," USAEC CONF-660925, p. 565.

HAVENS, W. W., Jr., and RAINWATER, L. J. (1951). *Phys. Rev.* **83**, 1123.

HAVENS, W. W., JR., MELKONIAN, E., RAINWATER, L. J., and ROSEN, J. L. (1959). *Phys. Rev.* **116**, 1538.

HOPKINS, J. C., and DIVEN, B. C. (1962). *Nucl. Sci. Eng.* **12**, 169.

JAMES, G. D., and PA RICK, B. H. (1969). "Physics and Chemistry of Fission," p. 391. IAEA, Vienna.

JAMES, G. D., and RAE, E. R. (1968). *Nucl. Phys.* **A118**, 313.

KAUSHAL, N. N., WINHOLD, E. J., YERGIN, P. F., MEDICUS, H. A., and AUGUSTSON, R. H. (1968). *Phys. Rev.* **175**, 1330.

KIMURA, M., SUGAWARA, M., OYAMADA, M., YAMADA, Y., TOMIYOSHI, S., SUZUKI, T., WATANABE, N., and TAKEDA, S. (1969). *Nucl. Instr. Methods* **71**, 102.

LARRIMORE, J. A., HAAS, R., GIEGERICH, K., RAIEVSKI, V., and KLEY, W. (1966). "Intense Neutron Sources," USAEC CONF-660925, p. 373.

LEBOUTET, H., PICARD, E., and VASTEL, M. (1957). *L'Onde Electrique* **37**, 28.

LEISS, J. E. (1966). Intense Neutron Sources. USAEC CONF-660925, p. 605.

LE PIPEC, C., MICHAUDON, A., RIBON, P., and OLIVIER, E. (1961). "Neutron Time of Flight Methods," p. 105. Euratom, Brussels.

LEVINGER, J. S. (1967). *Proc. Intern. Conf. Low Intermediate Energy Electromagnetic Interactions* (Acad. of Sci., USSR, Moscow) **3**, 411.

LIOU, H., WYNCHANK, S., SLAGOWITZ, M., CEULEMANS, H., RAINWATER, L. J., and HAVENS, W. W., JR. (1968). Pegram Nuclear Physics Laboratory Progress Report to USAEC, Jan. 1967 to Jan. 1968, p. 1.

LYNN, J. E. (1968a). UKAEA Rept. AERE-R 5891.

LYNN, J. E. (1968b). "Nuclear Structure," p. 463. IAEA, Vienna.

LYNN, J. E. (1968). "The Theory of Neutron Resonance Reactions," p. 459. Oxford Univ. Press (Clarendon), London and New York.

LYNN, J. E. (1969). "Physics and Chemistry of Fission," p. 249, IAEA, Vienna.

MACKLIN, R. L., GIBBONS, J. H., and INADA, T. (1963). *Nucl. Phys.* **43**, 353.

MALYSHEV, A. V. (1966). "Nuclear Structure with Neutrons," p. 236. North-Holland, Amsterdam.

MARSHAK, H., POSTMA, H., SAILOR, V. L., SHORE, F. J., and REYNOLDS, C. A. (1962). *Phys. Rev.* **128**, 1287.

MARSHAK, H., UTTLEY, C. A., and LYNN, J. E. (1969). Private communication.

McNEILL, K. G. (1967). Progr. Rept. on the Univ. of Toronto Electron Linear Accelerator, Univ. of Toronto, Dept. of Phys.

MEHTA, G. K., and MELKONIAN, E. (1966). *Proc. Symp. Nucl. Solid State Phys.* Bombay, India, USAEC CONF-660221, p. 33.

MEHTA, G. K., LEBOWITZ, J. M., and MELKONIAN, E. (1967). *Proc. Symp. Nucl. Solid State Phys.* Kanpur, India.

MELKONIAN, E., and MEHTA, G. K. (1965). "Physics and Chemistry of Fission," Vol. 2, p. 355. IAEA, Vienna.

MERRISON, A. W., and WIBLIN, E. R. (1952). *Proc. Roy. Soc.* **A215**, 278.

MICHAUDON, A. (1964). Saclay Rept. CEA-R 2552.

MICHAUDON, A. (1966). "Intense Neutron Sources," USAEC CONF-660925, p. 789.

MICHAUDON, A. (1968a). "Neutron Cross Sections and Technology." NBS Special Publication 299, p. 427.

MICHAUDON, A. (1968b). "Nuclear Structure," p. 483. IAEA, Vienna.

MICHAUDON, A. (1969). Private communication.

MICHAUDON, A., LOTTIN, A., PAYA, D., and TROCHON, J. (1965). *Nucl. Phys.* **69**, 573.

MIGNECO, E., and THEOBALD, J. P. (1968). *Nucl. Phys.* **A112**, 603.

MOAK, C. D., BANTA, H. E., THURSTON, J. N., JOHNSON, J. N., and KING, R. F. (1959). *Rev. Sci. Inst.* **30**, 697.

MOORE, M. J., and KASPER, J. S. (1968). *Nature* **219**, 848.

MOORING, F. P., MONAHAN, J. E., and HUDDLESTON, C. M. (1966). *Nucl. Phys.* **82**, 16.

MOTZ, H. T. (1965). "Pulsed High Intensity Fission Neutron Sources." USAEC CONF-650217, p. 77.

MOXON, M. C. (1962). Private communication.

NEAL, R. B. (1968). "The Stanford Two-Mile Accelerator," Benjamin, New York.

NEILER, J. H., and GOOD, W. M. (1960). "Fast Neutron Physics" (J. B. Marion and J. L. Fowler, eds.). Wiley (Interscience), New York.

NEILER, J. H., KELLEY, G. G., BELL, P. R., and BANTA, H. E. (1955). Oak Ridge Natl. Lab., Phys. Div. Semiannual Progr. Rept. ORNL-1975.

PATRICK, B. H., and JAMES, G. D. (1968). *Phys. Letters* **28B**, 259.

PATRICK, B. H., and WINHOLD, E. J. (1969). Private communication.

PATRICK, B. H., SCHOMBERG, M. G., SOWERBY, M. G., and JOLLY, J. E. (1967). "Nuclear Data for Reactors," Vol. II, p. 117. IAEA, Vienna.

PATTENDEN, N. J., and POSTMA, H. (1969). "Physics and Chemistry of Fission," p. 330. IAEA, Vienna.

PAYA, D., DERRIEN, H., FUBINI, A., MICHAUDON, A., and RIBON, P. (1967). "Nuclear Data for Reactors," Vol. 2, p. 128. IAEA, Vienna.

PEREZ, R. B., DE SAUSSURE, G., and MOORE, M. N. (1969). "Physics and Chemistry of Fission," p. 283. IAEA, Vienna.

PEVZNER, M. I. (1968). Private communication.

POOLE, M. J. (1957). *J. Nucl. Energy* **5**, 325.

POOLE, M. J., and WIBLIN, E. R. (1958). *Proc. Intern. Conf. Peaceful Uses At. Energy 2nd Geneva* **14**, 266.

POOLE, M. J. (1965). "Pulsed Neutron Research," Vol. 1, p. 425. IAEA, Vienna.

POOLE, M. J. (1966). "Intense Neutron Sources." USAEC CONF-660925, p. 451.

POOLE, M. J., and ROBINSON, A. H. (1964). UKAEA Rept. AERE PR/NP 7, p. 47.

RAE, E. R. (1967). "Fundamentals in Nuclear Theory," Chapter 14, p. 831. IAEA, Vienna.

RAE, E. R., and BOWEY, E. M. (1953). *Proc. Phys. Soc.* **A66**, 1073.

RAE, E. R., MOYER, W. R., FULLWOOD, R. R., and ANDREWS, J. L. (1967). *Phys. Rev.* **155**, 1301.

RAINWATER, L. J. (1961). "Neutron Time of Flight Methods," p. 321. Euratom, Brussels.

RAINWATER, L. J. (1966). "Neutron Cross Section Technology." USAEC CONF-660303, Vol. 1, p. 381.

RAINWATER, L. J., and HAVENS, W. W., JR. (1946). *Phys. Rev.* **70**, 136.

RAINWATER, L. J., HAVENS, W. W., JR., DUNNING, J. R., and WU, C. S. (1947). *Phys. Rev.* **71**, 65.

RAINWATER, L. J., HAVENS, W. W., JR., DESJARDINS, J. S., and ROSEN, J. L. (1960). *Rev. Sci. Instr.* **31**, 481.

RIBON, P., and MICHAUDON, A. (1961). "Neutron Time of Flight Methods," p. 357. Euratom, Brussels.

ROLAND, S. *et al.* (1967). *Bull. Informations Sci. Tech.* **113**, 35.

RYABOV, YU. V., DON SIK SO., CHIKOV, N., and JANEVA, N. (1968). *At. Energiya* **24**, 351.

SAMOUR, C., JACKSON, H. E., CHEVILLON, P. L., JULIEN, J., and MORGENSTERN, J. (1967). "Nuclear Data for Reactors" Vol. 1, p. 175. IAEA, Vienna.

SAMOUR, C., ALVES, R. N., JACKSON, H. E., JULIEN, J., and MORGENSTERN, J. (1968).

"Neutron Cross Sections and Technology," NBS Special Publication 299. Vol. 2, p. 669.

SAUTER, G. D., and BOWMAN, C. D. (1965). *Phys. Rev. Letters* **15**, 761.

SAUTER, G. D., and BOWMAN, C. D. (1967). *Nucl. Instr. Methods* **55**, 141.

SCHOMBERG, M. G., SOWERBY, M. G., and EVANS, F. W. (1968). "Fast Reactor Physics," p. 289. IAEA, Vienna.

SCHULTZ, H. L., and WADEY, W. G. (1951). *Rev. Sci. Instr.* **22**, 383.

SCHULTZ, H. L., PIEPER, G. F., and ROSLER, L. (1956). *Rev. Sci. Instr.* **27**, 437.

SCHWARTZ, R. B., SCHRACK, R. A., and HEATON, H. T. (1968). "Neutron Cross Sections and Technology," NBS Special Publication 299. Vol. 2, p. 721.

SEEGER, P. A., HEMMENDINGER, A., and DIVEN, B. C. (1967). *Nucl. Phys.* **A96**, 605.

SHAPIRO, F. L. (1966). "Nuclear Structure Study with Neutrons," p. 223. North-Holland, Amsterdam.

SHAPIRO, F. L. (1968). Private communication.

SINCLAIR, R. N., WEDGWOOD, A., HARRIS, D. H. C., and EGELSTAFF, P. A. (1969). UKAEA Report AERE-R6052.

SOWERBY, M. G., PATRICK, B. H., UTTLEY, C. A., and DIMENT, K. M. (1970). AERE-R6316.

STEIN, W. E., THOMAS, B. W., and RAE, E. R. (1970). *Phys. Rev.* **C1**, 1468.

STOLOVY, A. (1960). *Phys. Rev.* **118**, 211.

STRUTINSKY, V. M. (1967). *Nucl. Phys.* **A95**, 420.

SUND, R. E., WALTON, R. B., NORRIS, N. J., and MACGREGOR, M. H. (1964). *Nucl. Instr. Methods* **27**, 109.

TAKEKOSHI, H. (1965). Report to EANDC on Nuclear Data Measuring Facilities in Japan (Japanese Nuclear Data Committee), EANDC(J)2 U, p. 11.

TATARCZUK, J. R., and BLOCK, R. C. (1967). RPI Rept. RPI-328-100, p. 191.

TAYLOR, A. E. (1969). Private communication.

UTTLEY, C. A., and DIMENT, K. M. (1970). Private communication.

VAN HEMERT, R. L. (1968). Thesis entitled Threshold-Photoneutron Cross Sections for Light Nuclei, issued as rept. UCRL-50501.

VENABLE, D. (1964). *Phys. Today* December 1964, p. 19.

VLADIMIRSKY, V. V., and SOKOLOVSKY, V. V. (1958). *Proc. Intern. Conf. Peaceful Uses At. Energy 2nd Geneva* **14**, 283.

VORONKOV, R. M., PEVZNER, M. I., FLEREV, N. N., AREFIEF, A. V., KOROLEV, V. M., MOSKALEV, S. S., BASALAEV, M. I., and OSIFOV, V. P. (1962). *At. Energy (USSR)* **13**, 327.

WEIGMANN, H. (1968). *Z. Phys.* **214**, 7.

WEINSTEIN, S., and BLOCK, R. C. (1969). *Phys. Rev. Letters* **22**, 195.

WEISSKOPF, V. F. (1951). *Phys. Rev.* **83**, 1073.

WESTON, L. W. (1964). *Nucl. Sci. Eng.* **20**, 80.

WIBLIN, E. R. (1955). *Proc. Intern. Conf. Peaceful Uses At. Energy Geneva* **4**, 35.

WYNCHANK, S., GARG, J. B., HAVENS, W. W., JR., and RAINWATER, L. J. (1968). *Phys. Rev.* **166**, 1234.

YEATER, M. L., GAERTTNER, E. R., and BALDWIN, G. C. (1957). *Rev. Sci. Instr.* **28**, 514.

II

TOTAL NEUTRON CROSS SECTION MEASUREMENTS

F . W . K . F I R K
YALE UNIVERSITY
NEW HAVEN, CONNECTICUT

E . M E L K O N I A N
COLUMBIA UNIVERSITY
NEW YORK, NEW YORK

I. INTRODUCTION

A. General Considerations

During the past twenty years, numerous papers have reported measurements of total neutron cross sections for incident neutron energies up to 50 keV [see, for example, the compilations of data by Hughes and Schwartz (1958) and Stehn *et al.* (1965)]. The reasons for the popularity of this branch of low energy nuclear physics are twofold, namely: the relative simplicity of the experiments discussed in this section and the significance of the results in both reactor physics and theoretical nuclear physics, discussed in Section V.

A total neutron cross section is defined as

$$\frac{\text{number of events of all types per unit time per nucleus}}{\text{number of incident neutrons per unit area per unit time}}$$

or, qualitatively, as the effective area which a target nucleus presents to an incoming neutron. The outstanding feature of low energy total neutron cross sections, measured as a function of incident neutron energy, is the appearance of sharp resonances which correspond to discrete energy levels in the compound state formed by the target nucleus plus neutron. (We are concerned here only with nuclear effects and not with atomic, molecular, magnetic, or crystalline effects observed at very low neutron energies.)

A total neutron cross section $\sigma_T(E)$ is determined by measuring the transmission $T(E)$ of monoergic neutrons of energy E through an element of uniform thickness. This is related to the cross section by the equation

$$T(E) = \exp\{-n\sigma_T(E)\} \tag{1}$$

where n is the number of nuclei per square centimeter of the element normal to the incident neutron beam. The detector of the transmitted neutrons must subtend a small solid angle at the sample so that elastically scattered neutrons are not detected. Furthermore, the incident neutron beam must be suitably collimated at the sample position in order that the number of nuclei per square centimeter of sample, normal to the beam, is well defined. The experiment involves only the determination of the ratio of counting rate in a neutron detector with and without an absorber in the neutron beam and a measurement of the neutron energy. No absolute measurement of the incident neutron flux is required. Corrections for multiple scattering of neutrons within the sample, or for the attenuation of reaction γ-rays by the sample (as required in partial cross-section measurements) are unnecessary since a transmission measurement is not concerned with the details of a nuclear reaction; it is simply concerned with the fact that a nuclear encounter has taken place and has thereby prevented the neutron from reaching the detector.

Although fundamentally straightforward, there are several difficulties associated with measuring the true total neutron cross section as given by Eq. (1). First, the neutron spectrometer used to measure the neutron energy E' has a finite resolving power so that the observed transmission $T_{obs}(E)$ is given by

$$T_{obs}(E) = \int_{E_1}^{E_2} R(E' - E)T(E') \, dE' \qquad (2)$$

where $R(E' - E)$ is the instrumental resolution function such that $R(E' - E) = 0$ for $E_1 > E' > E_2$. Second, the target nucleus is not at rest in the laboratory system but has a distribution of velocities which corresponds to the thermal motion of atoms in the lattice of the sample material. This so-called Doppler effect can drastically distort the shapes of narrow resonances which appear in the total neutron cross section. The above two effects are treated in detail in Section III,A.

B. Theoretical Considerations

The form of the cross section for a reaction proceeding through an isolated resonance was first given by Breit and Wigner (1936). The basis for their initial work was not well founded, however, because time-dependent perturbation theory had been used to describe strong nuclear interactions. Nevertheless, the shape of the cross section agreed well with observation due to the fact that the shape only depends upon the formation of a long-lived intermediate state. Such a compound state is a consequence of the strength of nuclear forces.

A rigorous theory of nuclear reactions was introduced by Kapur and Peierls (1938) which does not depend upon the approximations of the perturbation method and, furthermore, does not depend upon particular reaction mechanisms such as compound nucleus formation. An important feature of their theory (and many later developments) is the occurrence of a complete set of formal states, defined in a volume of nuclear dimensions, which result from imposing certain boundary conditions at the surface of the volume. The theory is particularly well suited to a description of compound nucleus states which can be identified with the formal states. The Kapur–Peierls boundary conditions are energy-dependent and complex so that most of the parameters which occur are also energy-dependent. Their theory, therefore, has not become fashionable in the interpretation of low energy nuclear reactions. Wigner and Eisenbud (1947) introduced different boundary conditions that lead to parameters that have a clearer physical significance.

Breit (1940, 1946) attempted to remove from the theory those parameters, such as interaction radii, whose values cannot affect the cross sections. More recently, Humblet and Rosenfeld (1961) have introduced a theory that places emphasis on the behavior of the wave functions in the external region and on a correct description of the cross sections themselves. Again, it is not straightforward to interpret measured resonance parameters due to the appearance of complex phase shifts in the "widths" of the decaying states. We shall therefore use the Wigner–Eisenbud (1947) theory bearing in mind, however, that all the theories since Kapur and Peierls are formally correct. The following outline shows how the significant quantities which occur in discussions of slow neutron interactions appear in the formal theory [for a detailed discussion see Lane and Thomas (1958)].

Consider the problem of the elastic scattering of an s-wave neutron (assumed spinless) by a square potential well (Vogt, 1962). The radial part of the Schrödinger equation is then

$$-(\hbar^2/2m)(d^2\phi/dr^2) + V\phi = E\phi \tag{3}$$

where

$$V = V_1 \quad \text{for} \quad r \lesssim a; \quad V = 0 \quad \text{for} \quad r > a$$

The mass of the particle is m and the radius of the well is a.

The solutions of Eq. (3) are

$$\phi = A \sin Kr \quad \text{for} \quad r \lesssim a \tag{4}$$

and

$$\phi = B(e^{-ikr} - Ue^{ikr}) \quad \text{for} \quad r > a \tag{5}$$

where A is an arbitrary constant, K and k are the wave numbers inside and outside the well, respectively, and B is a normalizing factor chosen so that the incoming wave has unit flux

$$B = (4\pi v)^{-1/2} \tag{6}$$

where v is the neutron velocity. Here, U is termed the collision function which may also be defined in terms of a phase shift

$$U = e^{2i\delta} \tag{7}$$

The scattering cross section σ_{nn} can then be written

$$\sigma_{nn} = (\pi/k^2)|1 - U|^2 = (4\pi/k^2) \sin^2\delta \tag{8}$$

The wave function ϕ which describes the scattering resonance at low energies is almost a standing wave. The term "almost" applies because the neutron is unbound and eventually leaves the well after a time interval $\sim 10^{-14}$ sec.

The formal theory is concerned with constructing a complete set of standing (stationary) waves X_λ of which one closely resembles the true wave function ϕ.

In expanding the wave function ϕ in terms of the standing waves X_λ, one term may dominate the expansion and this term then corresponds to the standing wave associated with the resonance.

The stationary states X_λ are solutions of the equation

$$-(\hbar^2/2m)(d^2X_\lambda/dr^2) + VX_\lambda = E_\lambda X_\lambda \tag{9}$$

subject to the boundary condition at the radius a

$$[r(dX_\lambda/dr) = bX_\lambda]_{r=a} \tag{10}$$

where b is an arbitrary number. The X_λ form a complete orthogonal set so that the actual wave function may be expanded, thus,

$$\phi = \sum_\lambda A_\lambda X_\lambda \tag{11}$$

where

$$A_\lambda = \int_0^a X_\lambda \phi \, dr \tag{12}$$

Multiplying Eq. (3) by X_λ and Eq. (9) by ϕ, subtracting and integrating by parts twice, and using Eq. (10), we find

$$(\hbar^2/2ma)\{a\phi'(a) - b\phi(a)\}X_\lambda(a) = \int_0^a (E_\lambda - E)X_\lambda \phi \, dr \tag{13}$$

where

$$\phi' = d\phi/dr$$

Using Eqs. (12) and (13) gives

$$A_\lambda = (E_\lambda - E)^{-1}(\hbar^2/2ma)X_\lambda(a)\{a\phi'(a) - b\phi(a)\} \tag{14}$$

so that

$$\phi = \sum_\lambda (E_\lambda - E)^{-1}(\hbar^2/2ma)X_\lambda^2(a)\{a\phi'(a) - b\phi(a)\} \tag{15}$$

$$= R\{a\phi'(a) - b\phi(a)\} \tag{16}$$

where

$$R = \sum_\lambda \gamma_\lambda^2/(E_\lambda - E) \tag{17}$$

and

$$\gamma_\lambda^2 = (\hbar^2/2ma)X_\lambda^2(a) \tag{18}$$

The R-function, as defined by Eq. (17), is one of the principal quantities appearing in the formal theory.

The collision function U can be obtained in terms of R by considering the logarithmic derivative ϕ'/ϕ at $r = a$, giving

$$\phi'(a)/\phi(a) = (1 + bR)/aR \tag{19}$$

Equation (5) is then rewritten as follows:

$$\phi = I - UO \tag{20}$$

where

$$I = (4\pi v)^{-1/2}e^{-ikr} \qquad \text{and} \qquad O = I^*$$

the incoming and outgoing waves, respectively. Substituting these values of $\phi(a)$ and $\phi'(a)$ in Eq. (19), the collision function becomes

$$U = e^{-2ika}\left(\frac{1 + bR + ikaR}{1 + bR - ikaR}\right) \tag{21}$$

If there is only a single resonance centered at an energy E_0, say, the R-function is dominated by one term

$$R_0 \approx \gamma_0^2/(E_0 - E) \tag{22}$$

and the cross section close to the resonance becomes

$$\sigma_{nn}^0 = \frac{\pi}{k^2}\left| 2\sin kae^{ika} - \frac{\Gamma_0}{(E_0 - E + \Delta_0) - \frac{1}{2}\Gamma_0} \right|^2 \tag{23}$$

where $\Gamma_0 = 2ka\gamma_0^2$ is the width of the level and $\Delta_0 = b\gamma_0^2$ is the level shift which shifts the maximum of the resonance from E_0 to $E_0 + \Delta_0$.

The simplest boundary condition is such that $b = 0$; the real resonance states have zero logarithmic derivative at $r = a$ so that when $b = 0$, the standing wave resembles the true scattering state as closely as possible. Furthermore, when $b = 0$, the level shift is zero so that the eigenvalue of the standing wave problem coincides with the energy of the peak cross section.

The above theory may be extended to the general case of nuclear reactions without changing the basic principles involved in generating the R-function. In practice, however, there are frequently too many details to handle in a satisfactory way. These difficulties are due to the fact that, in general, an excited state of a nucleus can lose energy in many different ways; for example, by emitting neutrons, protons, γ-radiation, etc. These different modes of decay are referred to as channels. The simple R-function obtained above for the case of elastic scattering must be replaced by an R-matrix in which the value of the wave function for a channel, at the nuclear surface, is related to the value of the derivative of the wave function for all the channels, also at the surface. The rows and columns of the R-matrix refer to the different channels, and the expressions for I, O, U, and b also become matrices. Equation (21) still applies when written in a matrix notation.

In its most general form, the total neutron cross section $\sigma_{n,T}$ (J) may be written [see, for example, Vogt (1959)]

$$\sigma_{n,T}(J) = (2\pi/k^2)g(J) \sum_{s=|I-i|}^{I+i} \sum_{l=|J-s|}^{J+s} [1 - \text{Re } U_{nsl,nsl}(J)] \qquad (24)$$

where k is the wave number of the relative motion of the incident neutron and the target nucleus and $g(J)$ is a statistical weighting factor where

$$g(J) = \frac{2J + 1}{(2i + 1)(2I + 1)}$$

I is the ground state spin of the target nucleus;
i is the intrinsic spin of the incident neutron $(i = \frac{1}{2})$;
J is the total angular momentum of the compound state $(\mathbf{J} = \mathbf{l} + \mathbf{s})$;
s is the channel spin $(\mathbf{s} = \mathbf{I} + \mathbf{i})$;
l is the orbital angular momentum of the relative motion of the incident neutron and target nucleus; and
Re $U_{nsl,nsl}(J)$ is the real part of the collision function diagonal element $U_{nsl,nsl}$ defined as in Eq. (21) when written in an appropriate matrix (many-channel) form.

For neutron energies up to 50 keV, the most probable processes that occur are elastic scattering of the incident neutron and radiative capture. The case in which the compound state undergoes fission is treated in Chapter V and will not be considered here.

The major problem in the practical application of the formal theory is to develop a suitable form for the collision function U. Thomas (1955) obtained particularly useful expressions for U and his approach will be used here.

At neutron energies below 50 keV nuclear reactions are predominantly s-wave $(l = 0)$ since the penetration factors for incident neutrons of higher orbital angular momenta are small. Exceptions occur for nuclei with mass numbers $90 \gtrsim A \gtrsim 110$ and $A \gtrsim 230$. In these cases, p-wave $(l = 1)$ interactions have been observed at neutron energies below 1 keV.

The following formulas are developed for s-wave interactions: the extension of the theory to include higher l-values is tedious but straightforward [see Lane and Thomas (1958)].

If the expression for the total cross section is restricted to include a single component of total angular momentum J, Eq. (24) becomes

$$\sigma_{n,T} = (2\pi/k^2)g(1 - \text{Re } U_{nn}) \qquad (l = 0) \qquad (25)$$

Thomas (1955) showed that for elastic scattering and radiative capture of s-wave neutrons, the diagonal elements U_{nn} of the collision function can be written

$$U_{nn} = \exp\{-2ika\}(1 + ikaR_{nn})/(1 - ikaR_{nn}) \qquad (26)$$

where the reduced R-function R_{nn} is

$$R_{nn} = \sum_{\lambda} \gamma_{\lambda n}^2/(E_\lambda - E - \tfrac{1}{2}i\Gamma_{\lambda\gamma}) \qquad (27)$$

in which

E_λ is the resonance energy of level λ;
$\gamma_{\lambda n}^2$ is the reduced neutron width $(\gamma_{\lambda n}^2 = \Gamma_{\lambda n}/2ka)$;
$\Gamma_{\lambda n}$ is the neutron width of level λ;
$\Gamma_{\lambda\gamma}$ is the total radiation width of level λ; a is the nuclear radius; and
k is the wave number of the incident neutron.*
The level shift is assumed to be zero.

Equation (27) is a good approximation provided that all the partial radiation widths [the eliminated channels in Thomas (1955)] are much less than the level spacing and, furthermore, that their amplitudes are random in sign. The collision function U_{nn} given by Eq. (26) is valid irrespective of whether the total widths Γ_λ are greater or smaller than the level spacings.

If there is only one level in the sum over λ of Eq. (27), then the total cross section becomes

$$\sigma_{n,T} = \frac{\sigma_0}{1 + x^2} + \sigma_0 \tan(2ka) \cdot \frac{x}{1 + x^2} + \frac{4\pi}{k^2} g \sin^2(ka) \qquad (28)$$

where

$$\sigma_0 = (4\pi/k^2)g(\Gamma_n/\Gamma) \cos(2ka)$$

is the peak total cross section

* In terms of the laboratory energy E_L of the incident neutron, the wave number k is

$$k = \frac{2\pi}{\lambda} = \left(\frac{M_r}{M_r + 1}\right)[2.1968 \times 10^9 \times (E_L(\text{eV}))^{1/2}] \quad \text{cm}^{-1}$$

where λ is the de Broglie wavelength, M_r the mass of the target nucleus.

$$x = (2/\Gamma)(E - E_\lambda), \quad \text{and} \quad \Gamma = \Gamma_n + \Gamma_\gamma$$

Equation (28) is the Breit–Wigner single level form for the total cross section in the neighborhood of a resonance of total angular momentum J (Breit and Wigner, 1936). It consists of three parts:

(i) a resonance term: $\quad\quad \sigma_{\text{res}} = \sigma_0/(1 + x^2)$
(ii) an interference term: $\quad \sigma_{\text{int}} = \sigma_0 \tan(2ka) \cdot [x/(1 + x^2)]$
(iii) a term representing part of the potential or hard sphere scattering.

For nuclei with $I \neq 0$, resonances of both spin states $(I \pm \frac{1}{2})$ are present, in which case, the total cross section is obtained by adding the contributions as indicated in Eq. (24).

If the contribution from the opposite spin state is added to the last term of Eq. (28), then the potential scattering cross section σ_{pot} is obtained

$$\sigma_{\text{pot}} = (4\pi/k^2) \sin^2(ka) \approx 4\pi a^2 \quad \text{for} \quad ka \ll 1 \tag{29}$$

This is the cross section of an impenetrable sphere of radius a.

The cumulative effect of distant levels on the cross section may be found by writing Eq. (27) as follows:

$$R_{nn} = \frac{\gamma_{\lambda n}^2}{E_\lambda - E - \frac{1}{2}i\Gamma_{\lambda\gamma}} + R_{nn}^\infty \tag{30}$$

Here, R_{nn}^∞ is the contribution from all other levels of the same spin and parity as the isolated level under consideration.

If Eq. (30) is written in the form

$$R_{nn} = A + iB$$

then the total cross section becomes

$$\sigma_{n,T} = \frac{2\pi}{k^2} g \left\{ \frac{\rho^2(A^2 + B^2)(1 + \cos\phi) + 2\rho(B - A\sin\phi) + 1 - \cos\phi}{1 + 2\rho B + \rho^2(A^2 + B^2)} \right\} \tag{31}$$

where $\rho = ka$ and $\phi = 2ka$.

If R_{nn}^∞ is assumed to be a real constant (no absorption effects included), then the effect of adding this contribution from distant levels may be clearly seen by considering the approximations

$$2(E_\lambda - E) \gg \Gamma \quad \text{and} \quad ka \ll 1$$

in which case, we obtain

$$\sigma_{n,T} \approx \frac{\sigma_0 \Gamma^2}{4(E_\lambda - E)^2} - \frac{\sigma_0 \Gamma ka(1 - R_{nn}^\infty)}{(E_\lambda - E)} + 4\pi g[a(1 - R_{nn}^\infty)]^2 \tag{32}$$

In this low energy approximation, the effect of distant levels is to modify the nuclear radius. Using the notation of Feshbach et al. (1954), we have

$$R' = a(1 - R_{nn}^{\infty})$$ (33)

More general results are given by Lane and Thomas (1958) in which modifications to the level shift and partial widths are discussed.

The importance of including effects of R_{nn}^{∞} is demonstrated in Sections III,A and III,C.

II. EXPERIMENTAL TECHNIQUES

A. Typical "Good-Geometry" Arrangements

In the previous section, it was pointed out that to measure total cross sections by the transmission method, the neutron detector must not detect neutrons elastically scattered from the sample. Also, the incident neutron beam must be well collimated at the sample position in order that the number of nuclei per square centimeter of sample is accurately defined. A "good geometry" arrangement is therefore used, whereby the sample subtends a small solid angle at the source and at the detector.

The four principal sources of neutrons used for measurements of total cross sections in the energy range of interest are: (i) neutron choppers, (ii) pulsed electron linear accelerators and pulsed cyclotrons, (iii) crystal spectrometers, and (iv) pulsed and dc Van de Graaff machines. Each of these devices has characteristics (treated in detail in other chapters) which lead to different experimental arrangements in transmission measurements.

The accuracy of total cross section measurements is frequently limited by uncertainties in the magnitude of the background. The following treatment gives a typical method for determining the background in a time-of-flight spectrometer. In such spectrometers, the background can generally be resolved into a time-constant component T and an energy (or time-of-flight) dependent component $B(t)$. At time τ, the observed neutron spectrum is simply the sum of T, $B(\tau)$, and $C(\tau)$ where $C(\tau)$ is the true spectrum. The energy-dependent background $B(\tau)$ arises from neutrons with an apparent flight time τ which originate from source neutrons with high energies of average flight time $\bar{\tau}$ ($< \tau$). These high energy neutrons are scattered into the neutron detector and its shielding, and are finally detected at a later time τ after being moderated in the surrounding materials.

If a sample, that has a strong resonance at time τ, is placed in the beam, then all the true spectrum neutrons $C(\tau)$ can be prevented from reaching the detector thus leaving only the background components. The observed sample-in spectrum $IN_{obs}(\tau)$ is then given by

$$IN_{obs}(\tau) = T + B(\tau) \exp\{-n\sigma_{eff}(\bar{\tau})\} \tag{34}$$

Thus, $B(\tau)$ is determined provided $\sigma_{eff}(\bar{\tau})$ is known. The effective cross section $\sigma_{eff}(\bar{\tau})$ for the removal of fast neutrons can be obtained by making measurements with successively "blacker" samples. For the Harwell 40-MeV electron linear accelerator, $\sigma_{eff}(\bar{\tau})$ corresponds to the total cross section of a particular sample averaged over the energy range 500–1000 keV.

The "time-constant" background T may fluctuate during a measurement due to changes in the general background level (caused, for example, by switching on or off a nearby reactor or accelerator). It is then usual to monitor T by means of a timing channel (background gate) which opens at the end of a machine cycle. In most spectrometers it is found that the energy-dependent background $B(t)$ is negligible at very long flight times so that T may be readily determined.

Another important point in measuring total cross sections is the need for good stability in the efficiency of the neutron detection system and beam monitoring sytsem. Measurements often take many days to complete so that stability is of prime importance in relating the sample-in and sample-out results. The effects of drifts in the efficiency of the system may be greatly reduced by using "sample changers" which automatically change from the sample-in to the sample-out condition at short intervals of about 10 min duration. The use of two or more independent beam monitors is also advisable when cross section data with accuracies of better than 5% are required.

Typical examples of high quality total cross section data currently available are shown in Figs. 1–3. "High quality" is used here in the sense of good energy resolution coupled with good statistical accuracy. Some measurements have the additional advantage of the use of small quantities of samples (in particular separated isotopes).

The major installations carrying out total cross section measurements have already been referred to in Chapter I.

B. Neutron Detectors

A variety of neutron detectors have been used for transmission measurements from ~1 eV to 50 keV. The factors which are desirable in any particular detector may be listed as follows:

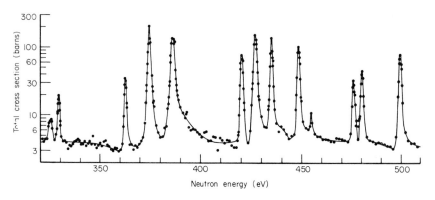

FIG. 1

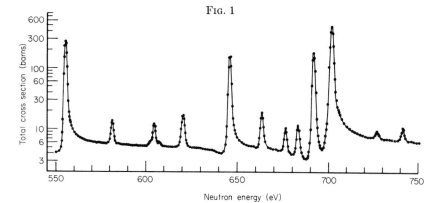

FIG. 2

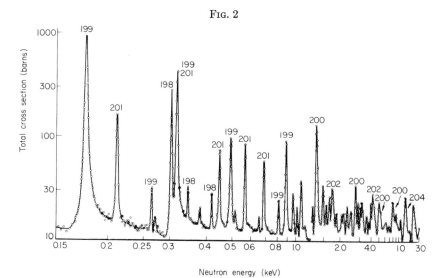

FIG. 3

(i) High detection efficiency which is a smooth function of neutron energy.

(ii) Timing resolution compatible with the duration of the neutron burst for time-of-flight applications.

(iii) Low sensitivity to fast neutron and γ-ray backgrounds.

(iv) Ease of production in various shapes and sizes.

(v) Simplicity of the associated electronics.

In most cases, the reactions $^{10}B(n,\alpha\gamma)^7Li + 2.3$ MeV or $^6Li(n,\alpha)T + 4.8$ MeV are used. The thermal capture cross sections for these reactions are large: $\sim$4000 b and $\sim$1000 b, respectively. They exhibit a smooth $1/V$ dependence in the energy range of interest. A number of other reactions have been used, e.g., $Sm(n, \gamma)$ and $Ag(n, \gamma)$, but these have not found widespread application due to low average capture cross sections and pronounced resonance structure.

Although the $^{10}B(n,\alpha\gamma)$ and $^6Li(n,\alpha)$ reactions are common to most slow neutron detectors, the detection mechanism varies considerably. For example, the ions created in a $^{10}BF_3$ gas counter due to the passage of an α-particle, may be collected and a measurable current obtained. Alternatively, the 480 keV γ-ray which results more than 90% of the time when ^{10}B captures a slow neutron, may be observed with a NaI(Tl) scintillation counter (Rae and Bowey, 1953). Recently, the scintillations produced by the heavily ionizing reaction products in boron- and lithium-loaded glass scintillators have been successfully observed (Voitovetskii et al., 1959; Bollinger et al., 1959; Firk et al., 1961; and Bollinger et al., 1962).

The more important characteristics of each type of detector are listed in Table I. These data are mostly taken from a compilation by Brooks (1961). It is evident from Table I that a number of detectors combine reasonable efficiencies with the demanding requirement of fast time resolution.

A frequent problem is that of the sensitivity of neutron detectors to background γ-rays. This has been discussed by Brooks (1961) who made

---

FIG. 1. The observed total neutron cross section of iodine (Garg et al., 1965). This measurement was made using the Columbia University pulsed synchrocyclotron with a time-of-flight resolution of 0.5 nsec m^{-1}.

FIG. 2. The observed total neutron cross section of rhodium (Ribon et al., 1961). This measurement was made using the Saclay 25 MeV linear electron accelerator.

FIG. 3. The observed total neutron cross section of mercury (Carpenter and Bollinger, 1960). This measurement was made using the ANL fast chopper with a time-of-flight resolution of 12 nsec m^{-1}.

TABLE I

CHARACTERISTICS OF SOME POPULAR NEUTRON DETECTORS[a]

Detector	Typical thickness (cm)	Efficiency at 1 keV (%)	Practical[b] timing resln. (nsec)	Neutron peak	
				Full width at ½ max (%)	Equivalent electron energy (MeV)
$^{10}BF_3$ counter (150 cm of Hg)	10	1	500	5	2.3
^{10}B–NaI(Tl)	2 (^{10}B) 3 (NaI)	25	$\gtrsim 3$	10	0.48
^{10}B-loaded liquid scint.	1	30	400	60	0.1
^{10}B-loaded glass scint.	1	30	not quoted	50	0.2
^{10}B–ZnS(Ag)	0.05	1	100	no peak	0.2
^6Li-loaded glass scint.	2.5	15–20	2	10–25	1.6
^6LiI(Eu) crystal	2.5	20	not quoted	12	4.1

[a] Brooks (1961).

[b] These figures represent the best performances so far achieved. From the known decay times and scintillation efficiencies of both the ^{10}B–NaI(Tl) and ^6Li-loaded glass, it is theoretically possible to achieve timing resolutions ~ 1 nsec.

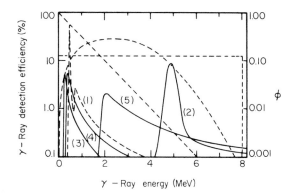

FIG. 4. Efficiency for detecting γ-rays of some neutron detectors listed in Table I (Brooks, 1961). (1) ¹⁰B + NaI (Tl); (2) ⁶LiI (Eu); (3) ¹⁰B-loaded liquid; (4) ¹⁰B-loaded glass; (5) ⁶Li-loaded glass. The broken curves represent the three different assumed shapes for the γ-ray spectra.

quantitative estimates of the γ-ray sensitivity of a number of the detectors listed in Table I. Brooks assumed three different γ-ray spectra incident on the detectors: (i) a "flat" spectrum; (ii) a typical fission γ-ray spectrum; and (iii) a typical capture γ-ray spectrum from a medium or heavy nucleus. The γ-ray detection efficiency E_γ versus γ-ray energy for five different detectors is shown in Fig. 4. The ¹⁰BF₃ counter is the most insensitive γ-ray detector and is normally used where low energy neutrons ($\gtrsim 100$ eV) are to be detected in the presence of severe γ-ray backgrounds.

The choice of detector naturally depends upon the particular requirements of an experiment. However, the trend in recent years has been away from ¹⁰BF₃ counters and toward scintillation detectors. This is due to the need for high efficiencies in the kilovolt region coupled with the need for improved time resolutions of $\gtrsim 0.1$ μsec.

The increasing interest in nanosecond neutron time-of-flight experiments has resulted in the development of a fast NaI–¹⁰B system by Good et al. (1958) and the ⁶Li-loaded glass scintillator by Firk et al. (1961). The time resolutions reported for these two detectors are ∼5 and ∼2 nsec, respectively. These resolutions are obtained when observing γ-rays and must, of course, be increased appropriately to include the neutron capture time in the detector (Bollinger et al., 1962).

In those applications which do not require fast time resolution, the boron-loaded liquid scintillator offers exceptionally high efficiency in the energy region of interest (Muehlhause and Thomas, 1953, and subsequent developments by the Argonne group).

III. ANALYSIS OF NEUTRON SPECTROMETER TRANSMISSION DATA

In the case of a single isotope* with nonzero nuclear spin, there are three classes of effects which must be considered in attempting an analysis:

(i) the intrinsic effects of (a) interference between resonance and potential scattering for s-wave levels and (b) the interference between levels of the same angular momentum and parity;

(ii) the mixed intrinsic-experimental Doppler effect (Section III,A,2) which can be modified to a limited extent by changing the temperature of the sample; and

(iii) the purely experimental effects of (a) lack of perfect resolution (i.e., the inability, within practical limits, of reducing the energy spread of the neutron beam to such a small extent that no significant variation of cross section occurs within the interval), and (b) lack of infinite statistical accuracy on each transmission value.

In addition, serious experimental difficulties often arise from the presence of various types of background which occur because of the detection of γ-rays and of neutrons with energies different from those under consideration. These background effects are associated with each particular experimental apparatus and cannot be discussed in general. They will therefore be omitted in our further discussion, and it will be assumed that transmission data have been derived for which the proper background corrections have been made.

The analysis of a given set of transmission data in the most general form is a formidable task and is not usually attempted. (However, with the steady increase in both the availability and the storage capacity of digital computers, this picture is slowly changing, and machine calculations are being made in which increasingly greater complications are considered simultaneously.) Situations are sought in which only a limited number of factors are present, the remainder being either negligible or included as small corrections. The most useful division is in terms of resolution. Where the resolution width is less than the level width, the shape of the observed resonance is significant and "shape analysis" (described in Section A) can be attempted. This involves a match of the experimental data, point by point, with the theoretical curves containing suitably chosen parameters. For most of the observed levels, the resolution width is larger than the level width, and the observed width is then no longer an indication of the

---

* In the general case of a mixture of isotopes, isotopic assignments of the levels must first be made before a complete analysis can be attempted; the problem then reduces to that under consideration. Nevertheless, it is frequently possible to get partial results.

level width. In general, the resolution function is not known with sufficient accuracy to disentangle its effects from the observed width. Clearly, it is difficult to deduce resonance parameters from such a resolution-broadened curve. "Area analysis," described in Section B, is then the most common form of analysis and transmission measurements on samples of several thicknesses are useful in the determination of resonance parameters.

A. Shape Analysis

1. Low Neutron Energies ($\stackrel{\sim}{<}1$ keV)

Except for the case of the fissionable isotopes, which is treated separately in Chapter V, the level widths below 1 keV neutron energy are generally small compared with level spacings so that level–level interference is small and can usually be neglected. This is fortunate, since the inclusion of level–level interference terms demands a knowledge of the spins J of all the levels, and such information is available only in a few favorable cases.

The only other intrinsic effect left is that of the interference between resonance and potential scattering, and this is readily included in the analysis.

Before describing some of the techniques of shape analysis, it is necessary to discuss the Doppler broadening of resonances previously mentioned in Section I,A. The following outlines the theory of this effect.

2. Doppler Broadening of the Resonance Form

In order to take into account the thermal motions of the target nuclei, the recoil of the compound nucleus must be considered. The simplest case is that of a target nucleus which is initially at rest and free to move: the resonance term given by Eq. (28) then becomes

$$\sigma_T(E_L) = \frac{\sigma_0(\Gamma/2)^2}{[(E_L - R) - E_\lambda]^2 + (\Gamma/2)^2} \tag{35}$$

where E_L is the laboratory energy of the incident neutron, and R is the recoil energy of the compound nucleus such that

$$R = mE_L/(m + M) \tag{36}$$

where m is the mass of the neutron, and M the mass of the target nucleus. Note that Eq. (35) is usually written with $R = 0$ (i.e., the system in which the compound nucleus is at rest).

In practice, the target nuclei are not at rest but have a distribution of velocities characteristic of their environment and its temperature. Bethe and Placzek (1937) modified Eq. (35) for the case in which the target nuclei are treated as a classical gas. It is necessary to include in the energy dependence of Eq. (35) the relative motion of the target nucleus and incident neutron and then to average over the Maxwellian distribution of velocities of the atoms. Their result is obtained as follows.

In general, the "Doppler-broadened" cross section $\sigma_\Delta(E_L)$ may be written in terms of an energy transfer function $S(E_t) = S(E_L - E)$ which is convoluted with $\sigma(E)$ to give*

$$\sigma_\Delta(E_L) = \int S(E_L - E) \cdot \sigma(E) \, d(E_L - E) \tag{37}$$

The form of the function $S(E_L - E)$ depends on the model which is assumed for the medium in which the target nuclei are bound. For a classical gas, the energy transfer $(E_L - E)$ is

$$(E_L - E) \approx R + mvw_x \tag{38}$$

where v is the laboratory energy of the neutron, and w_x is the velocity of the target nucleus in the direction of the incident neutron beam. The distribution of velocities w_x is given by the Maxwell–Boltzmann formula

$$p(w_x) \, dw_x = M/(2\pi MkT)^{1/2} \exp\{-Mw_x{}^2/2kT\} \, dw_x \tag{39}$$

where T is the gas temperature, and k is the Boltzmann constant. The function $S(E_L - E)$ then becomes

$$S(E_L - E) = (1/\Delta \sqrt{\pi}) \exp\{-(E_L - E - R)^2/\Delta^2\} \tag{40}$$

where

$$\Delta = 2(RkT)^{1/2}$$

is the "Doppler width." Jackson and Lynn (1962) point out that the same result is obtained for a classical solid in which the nuclei are considered as linear harmonic oscillators whose energies are described by Boltzmann statistics.

Lamb (1939) calculated the shape of a resonance for target nuclei bound in a quantum mechanical crystal. He discussed the oscillation of a target nucleus bound in a harmonic oscillator well (HOW). The energy of the oscillation is then one of the eigenvalues of the HOW, and the expectation value of this energy is equal to the recoil energy R: There is a finite probability of the oscillator remaining in its initial state [cf. the Mössbauer effect (1958)].

---

* Note that the Doppler broadening acts on the cross section, whereas the broadening arising from lack of perfect resolution acts upon the transmission.

For nuclei bound in a Debye crystal, Lamb calculated the modes of oscillation and the resulting resonance shape using dispersion theory. The conclusions were:

(i) for a cold crystal (relative to the Debye temperature θ_D) and for a narrow resonance width and low recoil energy (both relative to $k\theta_D$), the "recoilless" peak is expected. Some structure may also appear at higher energies due to the creation or annihilation of one or more phonons in the crystal lattice.

(ii) at high crystal temperature (i.e., weak binding such that $\Delta + \Gamma \gg 2k\theta_D$), the result is the same as that of the classical gas broadening outlined above except that the mean energy per degree of freedom ($\frac{1}{2}kT$) must be replaced by the quantum-mechanical mean energy

$$\bar{\epsilon} = \tfrac{1}{2} \int_0^{\nu_m} \coth(h\nu/2kT)g(\nu)h\nu\,d\nu \qquad (41)$$

where $g(\nu)$ is the frequency spectrum of the lattice, and ν_m is the maximum frequency in the spectrum. (Generally, $\bar{\epsilon} > \frac{1}{2}kT$ but approaches it at high temperature.)

Landon (1954) verified the classical gas approximation by studying the resonance profile of the 1.26 eV resonance in rhodium as a function of temperature. Jackson and Lynn (1962) have carried out measurements on ^{189}Os, ^{238}U, and U_3O_8 for a range of sample temperatures down to 4°K. For the metal samples, they obtained good agreement with their data using Lamb's theory and a simple Einstein frequency spectrum for the lattice.

The Doppler broadened cross section corresponding to Eq. (28) is usually written

$$\sigma_T(\beta,x) = \sigma_0\Psi(\beta,x) + \sigma_0 \tan(2a/\lambda)\Phi(\beta,x) + \sigma_{\text{pot}} \qquad (42)$$

where

$$\Psi(\beta,x) = (\beta\sqrt{\pi})^{-1} \int_{-\infty}^{\infty} (1 + y^2)^{-1} \exp\{-(x - y)^2/\beta^2\}\,dy \qquad (43)$$

and

$$\Phi(\beta,x) = (\beta\sqrt{\pi})^{-1} \int_{-\infty}^{\infty} [y/(1 + y^2)] \exp\{-(x - y)^2/\beta^2\}\,dy \qquad (44)$$

Here,

$$\beta = 2\Delta/\Gamma, \qquad x = (2/\Gamma)(E - E_\lambda) \qquad \text{and} \qquad y = (2/\Gamma)(E_{\text{CM}} - E_\lambda)$$

(E_{CM} is the center-of-mass energy of the neutron).

Other dimensionless parameters used in the literature in place of β are

$$\xi \equiv \Gamma/\Delta = 2/\beta: \qquad t = (\Delta/\Gamma)^2 = \tfrac{1}{4}\beta^2$$

Extensive tables of the function $\psi(\beta,x)$ have been published (Rose *et al.*, 1953). However, since most analysis is currently performed using computers, these functions are computed as needed.

3. SOME TECHNIQUES OF SHAPE ANALYSIS

The "wings" of a resonance are usually unaffected by resolution and Doppler broadening since the cross section is only varying slowly with energy in this region. The Breit–Wigner formula Eq. (28) may therefore be applied directly. A least-squares analysis frequently gives useful combinations of resonance parameters. Specifically, the cross section on the wings of a resonance is

$$\sigma_{n,T,\text{wings}} = \frac{\sigma_0 \Gamma^2}{4(E - E_\lambda)^2} + \frac{\sigma_0 \Gamma(a/\lambda)}{E - E_\lambda} + \sigma_{\text{pot}} \tag{45}$$

Assuming that the resonance energy E_λ can be obtained adequately by inspection, a curve fit to the data gives σ_{pot}, $\sigma_0 \Gamma^2$ (proportional to $g\Gamma_n\Gamma$), and $\sigma_0 \Gamma a$ (where σ_0 is proportional to $g\Gamma_n$). The main requirement here is that the wings do not include significant contributions from other levels.

The central region of a resonance, on the other hand, is usually affected by both types of broadening. All the methods of shape analysis used for the central region have been a variation of the "trial and error" method. Here, a preliminary analysis or inspection is used to give approximate values of the parameters. The Doppler-broadened cross section, transmission and resolution-broadened transmission are then computed successively and are compared with the experimental transmission values. An inspection shows the changes that are necessary to the input values. A new set of corrected transmission values are then calculated, and comparison with observations again made. The whole process is repeated until a "satisfactory" fit is obtained. Early applications of this method were made by Havens and Rainwater (1946), McDaniel (1946), and Meyer (1949).

Sailor (1953) introduced an improved method for finding the corrections to be applied to the initial choices of the parameters. The transmission function is expanded in a Taylor series about the values determined by the initial parameters, and terms are kept only to the first order in the corrections for σ_0 and Γ. The method of least-squares is then used to find those values of the corrections which reduce to a minimum, the sum of the squares of the differences between observed and calculated transmission values. This process can be repeated until it is shown that the higher terms in the Taylor expansion are indeed negligible. Seidl *et al.* (1954) have developed a similar method of analysis.

Until recently, the above methods could be applied in a few special cases in which only one low-lying level was adequately resolved in a given isotope. With the recent increase in spectrometer resolution, there are now frequent cases in which many levels in a given isotope are sufficiently well resolved for a shape analysis to be applied. Desk calculation methods become inadequate and digital computers must be used. Harvey and Atta (1961) have set up schemes for the IBM 704 and IBM 7090 computers which can shape-analyze a transmission curve for as many as six levels at once. The method is similar in principle to the Sailor procedure in that transmission values for assumed input parameters are calculated, and least-squares solutions made to calculate corrections to the input parameters. However, the program is of far larger scope: up to six resonances may be analyzed together and iteration is made to a much higher degree of precision. Certain other refinements, to be discussed below, are also included.

In the Harvey–Atta method, the cross section is represented as a sum of single-level Breit–Wigner resonances including interference between resonance and potential scattering, but not including interference between resonances. Doppler and resolution broadening are both represented by Gaussian functions. The resolution width can be of the form $R_0 + R_1 i$ (i = channel number) to take into account an energy-dependent part of the resolution such as effects of burst and channel width, and an energy dependent part, which might arise from a finite detector depth or variation in neutron moderation time in the source. Provision is made for a super-imposed transmission variation of the form

$$P(E) = K_0 + (K_1/\sqrt{E}) + (K_2/E) \tag{46}$$

which may arise from contributions of resonance levels outside the energy region being analyzed. This variation can also take into account a variation of the neutron flux with energy when the program is used to analyze partial cross section data (such as capture or scattering). The least-squares fit then gives values of the resonance energy, the full width, and the product $fg\Gamma_n^0$ where f is the fractional abundance of the isotope to which the resonance belongs; g is the statistical weight factor; and Γ_n^0 is the reduced neutron width. The program includes estimates of the errors in the finally-selected values of the parameters.

An accompanying program performs area analyses (Section III,B) on as many as thirty resonances at once. The parameters deduced from the area analysis may then be used as initial estimates for the shape analysis program.

An additional advantage of the Harvey–Atta procedure is that a large amount of overlap of levels can be tolerated. This is due to the summation

of the contributions of all the levels in the energy interval under consideration. Such a procedure is valid provided that level–level interference is
negligible.

Several other specialized approaches to the shape analysis problem are
discussed by Melkonian (1956) and Michaudon (1961).

4. ENERGY $\gtrsim 1$ keV, NO DOPPLER BROADENING AND GOOD RESOLUTION

The improved resolution of neutron spectrometers during recent years
has resulted in an increasing number of total cross section measurements
above 1 keV. In these cases, levels with large neutron scattering widths
compared with the level spacings are frequently encountered so that interference effects between levels of the same spins and parities become important. It is then necessary to use the general multilevel form for the cross
section as given in Eqs. (25) and (26).

Measurements of this type were first reported by Bollinger et al. (1955)
who determined the total cross section of manganese from thermal energies
to ~10 keV. A theoretical analysis of these measurements was made by
Krotkov (1955).

More recently, the total cross section of vanadium has been measured
up to 25 keV by Firk et al. (1963b) and the data analyzed using the Thomas
approximation given in Eq. (26). The observed total cross section is shown
in Fig. 5. The energies, total widths and spins [see Eq. (80)] of each level
are determined from a shape analysis of the observed total cross section.
The shape analysis is possible since the effects of resolution and Doppler
broadening are negligible at almost all points of the measurement. The
parameters giving the best fit to the data (obtained from a least-squares
analysis programmed for an IBM 704 computer) are shown in Table II.

TABLE II

BEST VALUES OF RESONANCE PARAMETERS FOR THE
NEUTRON CROSS SECTION OF $^{51}V^a$

E (keV)	J	$\gamma^2_{\lambda n}$ (keV)	Γ_λ (keV)
4.169 ± 0.005	4	3.304 ± 0.035	0.508 ± 0.006
6.886 ± 0.010	3	6.475 ± 0.070	1.28 ± 0.014
11.810 ± 0.004	3	21.25 ± 0.18	5.50 ± 0.05
16.6 ± 0.1	4	1.13 ± 0.03	0.35 ± 0.10
17.4 ± 0.1	4	0.3 ± 0.1	0.09 ± 0.03
22.32 ± 0.03	3	2.68 ± 0.11	0.95 ± 0.04

^{a} Firk et al. (1963b).

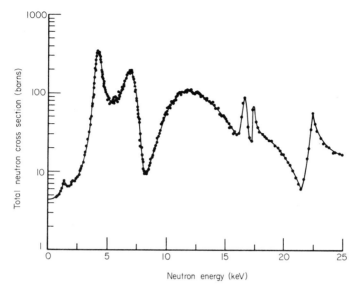

FIG. 5. The observed total neutron cross section of vanadium up to an energy of 25 keV (Firk *et al.*, 1963b).

A feature of the analysis of the vanadium total cross section is the inclusion of the effect arising from distant levels. The following method is used to determine this effect. In the low energy approximation of Eq. (32) the thermal scattering length a_J for spin J may be written

$$a_J = a(1 - R_{nnJ}^{\infty}) \tag{47}$$

$$= a \left(1 - \sum_{\lambda} (\gamma_{\lambda nJ})^2 / E_{\lambda J} - R_{nnJ}^{\infty\prime} \right) \tag{48}$$

where a is the nuclear radius, $\sum_{\lambda}(\gamma_{\lambda nJ})^2/E_{\lambda J}$ is the contribution at thermal energy ($E \approx 0$) due to levels of spin J in the energy range 0–25 keV and $R_{nnJ}^{\infty\prime}$ is the cumulative effect of levels of spin J outside the range 0–25 keV. From the thermal scattering data of Peterson and Levy (1952), two pairs of values of a_J can be obtained, and therefore two pairs of values of $R_{nnJ}^{\infty\prime}$. The best fit to the data is obtained for both $R_{nnJ=3}^{\infty\prime}$ and $R_{nnJ=4}^{\infty\prime}$ negative. The sensitivity of the fit to the correct choice of $R_{nnJ}^{\infty\prime}$ is clearly shown in Fig. 6 in which $R_{nnJ}^{\infty\prime}$ is considered a constant. In a final analysis of the data, $R_{nnJ}^{\infty\prime}$ is resolved into an energy dependent part (due to levels outside the range 0–25 keV which are, nevertheless, close enough to exert a mild energy dependence) plus a constant. The theoretical implications of the

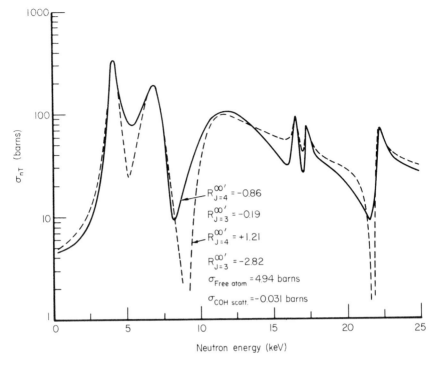

INITIAL PARAMETERS

E_λ (eV)	Γ_λ (eV)	J	E_λ (eV)	Γ_λ (eV)	J
4,150	540	4	16,600	190	4
6,800	1200	3	17,400	200	4
11,700	4400	3	22,300	950	3

Fig. 6. A multilevel fit to the vanadium total cross section using constant values of $R_{nnJ}^{\infty\prime}$. The pairs of values of $R_{nnJ}^{\infty\prime}$ are consistent with the thermal scattering cross section data. The resonance energies, total widths, and spins of the six levels were obtained initially by inspection. Final values of the resonance parameters deduced from a least-squares analysis of the cross section are given in Table II (Firk *et al.*, 1963b).

vanadium results are discussed in the paper of Firk *et al.* (1963b) and in Section V.

B. Area Analysis

The most widely used method of analyzing neutron transmission data to determine resonance parameters is the "area-method" developed by

Havens and Rainwater (1946, 1951). The usefulness of this method arises from the fact that the total area under an isolated transmission dip (plotted as a function of energy, say), is independent of the resolution function of the neutron spectrometer. A quantitative treatment of the validity of the area method of analysis has been given by Lynn and Rae (1957).

The expression for the area under a transmission dip in terms of the resonance parameters and sample thickness may be obtained using the single-level Breit–Wigner form for the total cross section given by Eq. (28). This form is suitable in most cases of analysis met in the energy range below 50 keV. Multilevel interference effects have been considered by Lynn (1958) and these may be included where necessary.

Using Eq. (28) the transmission is

$$T(x) = \exp\{-n\sigma_T(x)\} = \exp\left\{-\frac{n\sigma_0}{1+x^2} - \frac{n\sigma_0 x}{1+x^2}\tan(2a/\lambda) - 4\pi na^2\right\}$$
(49)

The slowly varying contribution to the transmission from nearby and distant levels may be included in the term $\exp(-4\pi na^2)$ where a now becomes an effective scattering length [cf. R' of Eq. (33)].

Writing $T_P = \exp(-4\pi na^2)$ the area under a transmission dip (in energy units) is defined as

$$A_E(\infty) = \int_{-\infty}^{\infty} [1 - T(E)/T_P]\, dE$$
(50)

or, in the x-notation

$$A_x(\infty) = \int_{-\infty}^{\infty} [1 - T(x)/T_P]\, dx = (2/\Gamma)A_E$$
(51)

1. No Doppler Broadening and No Interference Term

If the interference term in Eq. (28) is neglected (valid for $n\sigma_0 \gtrsim 50$) and Doppler broadening is negligible, then the area function A_x' (say) is given by

$$A_x'(\infty) = \int_{-\infty}^{\infty}\left[1 - \exp\left(\frac{-n\sigma_0}{1+x^2}\right)\right] dx$$
(52)

Substituting $x = \tan(\phi/2)$ this becomes

$$A_x'(\infty) = (n\sigma_0/2)\exp(-n\sigma_0/2)\left\{\int_{-\pi}^{\pi}\exp(-n\sigma_0\cos\phi/2)[1 - \cos\phi]\, d\phi\right\}$$
(53)

which is readily shown to be

$$A_x'(\infty) = \pi n\sigma_0 \exp(-n\sigma_0/2)[I_0(n\sigma_0/2) + I_1(n\sigma_0/2)]$$
(54)

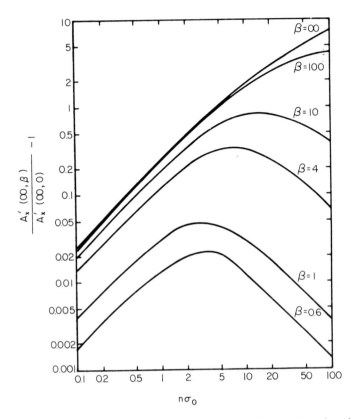

F$_{\text{IG}}$. 7. A curve of the relative Doppler broadening of the area function $A_x'(\infty,\beta)$ as a function of $n\sigma_0$ (Melkonian *et al.*, 1953).

where I_0 and I_1 are Bessel functions of imaginary argument of order 0 and 1, respectively. It is interesting to note that the analytical expression for $A_x'(\infty)$ was first given by Ladenburg and Reiche (1913), in connection with the absorption by the earth's atmosphere, of spectral lines from the sun.

The asymptotic values for Eq. (54) are, in the energy notation

$$\lim_{n\sigma_0 \to 0} A_E'(\infty) = \pi n\sigma_0 \Gamma/2 \tag{55}$$

and

$$\lim_{n\sigma_0 \to \infty} A_E'(\infty) = (\pi n\sigma_0)^{1/2}\Gamma \tag{56}$$

Equations (55) and (56) represent the "thin" and "thick" sample approximations, respectively, and have been widely used to determine σ_0 and Γ by careful choice of sample thickness. Melkonian *et al.* (1953) have discussed

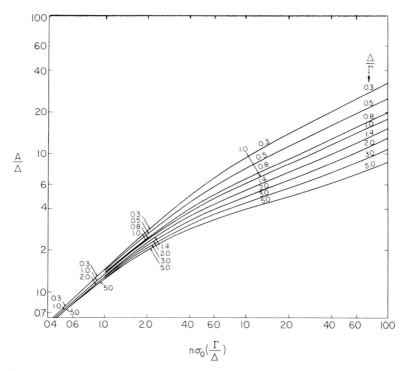

$$n\sigma_0\left(\frac{\Gamma}{\Delta}\right)$$

Fig. 8. A set of curves of $A_E'(\infty,\beta)/\Delta$ as a function of $n\sigma_0(\Gamma/\Delta)$ for a range of values of Δ/Γ. These curves are used in the Hughes' method of area analysis (Hughes, 1955a).

the problem for sample thicknesses intermediate in value between the limits imposed in Eqs. (55) and (56). These authors show that for sample thicknesses such that $1 \gtrsim n\sigma_0 \gtrsim 10$ the area is given by

$$A_E'(\infty) \propto \sigma_0\Gamma^P \qquad \text{where} \quad 2 > P > 1$$

2. Doppler Effect but No Interference Term

The effect of Doppler broadening on the total cross section in the neighborhood of a resonance and the consequent change in the area under a transmission dip was first considered by Melkonian et al. (1953). In their treatment, the interference term was neglected.

On averaging over the thermal velocity distribution of the atoms in the sample the resonance term $\propto (1 + x^2)^{-1}$ becomes

$$(1 + x^2)^{-1} \to \Psi(\beta,x) = (\beta\sqrt{\pi})^{-1} \int_{-\infty}^{\infty} (1 + y^2)^{-1} \exp\{-(x - y)^2/\beta^2\}\, dy$$

where the terms are defined in Section III,A,2.

The Doppler broadened area function is now written

$$A_x'(\infty,\beta) = \int_{-\infty}^{\infty} \{1 - \exp[-n\sigma_0\Psi(\beta,x)]\}\, dx \qquad (57)$$

The integral in Eq. (57) has been numerically integrated (Melkonian et al., 1953) and a curve of the relative Doppler broadening of the area as a function of $n\sigma_0$ presented. Their curve is shown in Fig. 7. It is seen that the effect of Doppler broadening is to increase the range of validity of the thick sample approximation. In this treatment, it is also possible to correct simultaneously for the "wing areas" excluded by the practical need to limit the measured area to a finite interval about the resonance energy.

A useful method of area analysis due to Hughes (1955a) enables corrections for Doppler broadening and finite sample thickness to be made simultaneously. Corrections for the excluded wing areas must, however, be made by successive approximations. Figure 8 gives a set of curves of $A_E'(\infty,\beta)/\Delta$ as a function of $n\sigma_0(\Gamma/\Delta)$ for a wide range of (Δ/Γ). A typical example readily demonstrates the Hughes' method. The analysis of the data given by Firk (1958) for the 23.9 eV resonance in tantalum proceeds as follows.

The Doppler width $\Delta = 0.116$ eV

Sample thicknesses (atoms/b)	Observed area $A_E'(\infty,\beta)$ (eV)
0.000304	0.142 ± 0.010
0.00076	0.279 ± 0.011
0.00954	0.915 ± 0.027

For each sample thickness $A_E'(\infty,\beta)/\Delta$ is a constant. Using Hughes' curve shown in Fig. 8, a table is constructed for sample thickness $n = 0.000304$ atom/b (here, $A_E'(\infty,\beta)/\Delta = 1.22$). The details are given in Table III.

This procedure is repeated for the different sample thicknesses, and the

TABLE III

DETERMINATION OF THE FUNCTIONAL RELATIONSHIP $\sigma_0 = f(\Gamma)$ FOR THE 23.9 eV
RESONANCE IN Ta USING HUGHES' METHOD[a]

Δ/Γ	Γ (eV)	$n\sigma_0(\Gamma/\Delta)$	$n\sigma_0$	σ_0 (b)
0.2	0.58	0.81	0.162	533
0.4	0.28	0.84	0.336	1105
0.7	0.166	0.87	0.609	2003
1.0	0.116	0.90	0.900	2960
2.0	0.058	0.94	1.88	6184

[a] Firk (1958).

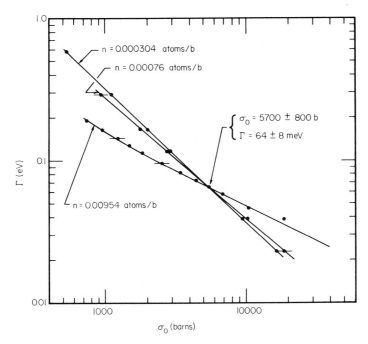

FIG. 9. The results of an area analysis of the 23.9 eV resonance in Ta (Firk, 1958) Three different sample thicknesses were used in order to determine σ_0 and Γ.

values of σ_0 and Γ are then determined from a plot of log σ_0 versus log Γ as shown in Fig. 9. The results are

$$\sigma_0 = 5700 \pm 800 \quad \text{b}; \qquad \Gamma = 64 \pm 8 \quad \text{meV}$$

3. Inclusion of the Interference Term

a. No Doppler Effect. The asymmetry due to interference between resonance and potential scattering is clearly demonstrated in the transmission curves shown in Fig. 10 (Lynn *et al.*, 1958b). As the sample thickness increases the contribution to the area above the line, T_P becomes more and more pronounced. The sign of the area [as defined in Eq. (50)] above the line $T/T_P = 1$ (positive) is opposite to that of the area below $T/T_P = 1$. The result is that the area is decreased by the interference effect.

The method of area analysis including the interference effect was pioneered by Lynn (1958) whose method is outlined below.

The area function including the interference term may be written

$$A_x(\infty, \beta) = A_x'(\infty, \beta) - \sum_{m=1}^{\infty} C_m(\beta) \tag{58}$$

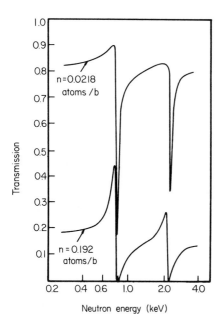

Fig. 10. An example of interference between resonance and potential scattering for different values of sample thickness. The resonances are those observed in the cross section of bismuth (Lynn *et al.*, 1958).

where the positive quantity $C(\beta) = \Sigma_m C_m(\beta)$ takes into account the net decrease in the total area. In the case of no Doppler broadening ($\beta = 0$), Lynn gives the following expression for $C_m(0)$:

$$C_m(0) = \pi n\sigma_0 \exp\left(\frac{-n\sigma_0}{2}\right) \frac{(2m-3)!!}{(2m)!} \left(\frac{2n\sigma_0 a^2}{\lambda^2}\right)^m \left[I_{m-1}\left(\frac{n\sigma_0}{2}\right) + I_m\left(\frac{n\sigma_0}{2}\right) \right] \quad (59)$$

For energies above ~ 10 keV* terms involving a/λ explicitly should be replaced by $\frac{1}{2}\tan(2a/\lambda)$.

Lynn (1960a) has made extensive calculations of a natural area function defined by

$$A_N(\infty, 0) = \frac{1}{4}\tan(2a/\lambda)A_x(\infty, 0) \quad (60)$$

These calculations are available (Lynn, 1960b) for $n\sigma_0$ in the range 1–1024 and $\frac{1}{2}n\sigma_0 \tan^2(2a/\lambda)$ in the range 0.02–2.7.

---

* Seth (1959), using analytical methods similar to Lynn, has discussed a method of area analysis, including the interference term but excluding Doppler broadening, for neutron energies in the range 10–1000 keV ($a/\lambda > 0.1$). This energy range is, however, mostly outside the scope of the present article.

b. *Doppler Effect Included.* The expression for the area function $A_x(\infty,\beta)$ including Doppler broadening of both the resonance and interference terms is

$$A_x(\infty,\beta) = \int_{-\infty}^{\infty} \{1 - \exp[-n\sigma_0\Psi(\beta,x)]\}\ dx$$

$$- \sum_{m=1}^{\infty} \frac{(n\sigma_0\tan(2a/\lambda))^{2m}}{(2m)!} \int_{-\infty}^{\infty} \Phi^{2m}(\beta,x)$$

$$\times \exp[-n\sigma_0\Psi(\beta,x)]\ dx \qquad (61)$$

where

$$\Phi(\beta,x) = (\beta\sqrt{\pi})^{-1} \int_{-\infty}^{\infty} y/(1+y^2) \exp\{-(x-y)^2/\beta^2\}\ dy$$

Lynn (1960b) has calculated $A_x(\infty,\beta)$ by numerical integration methods and has presented tables of the area function:

$$A_N(\infty,\beta) = \tfrac{1}{4}\tan(2a/\lambda)A_x(\infty,\beta) \qquad (62)$$

for wide ranges of $\beta(1(\sqrt{2})64)$ and $n\sigma_0\tan^2(2a/\lambda)$ $(0.014865(\sqrt{2})6.4)$.

c. *Determination of Resonance Parameters Using Lynn's General Area Function.* Before comparing an experimentally obtained area with a theoretical area, in order to determine the resonance parameters, the following corrections are necessary:

(i) A wing correction to the area ($\Delta A_E(\epsilon)$, say), due to finite energy limits.
(ii) A wing correction to the area ($\delta A_E(\epsilon)$, say), due to asymmetry of the transmission curve.
(iii) An evaluation of the effective potential transmission T_p.
(iv) An evaluation of the true resonance energy E_λ.

These corrections have been discussed in detail (Lynn et al., 1958b; Lynn, 1960a). An important feature emerges, namely, that the above four correction terms can be expressed as functions of the parameters $n\sigma_0\tan^2(2a/\lambda)$ and $\Gamma/\tan(2a/\lambda)$. The theoretical area function is also dependent upon these two quantities. The two parameters $n\sigma_0\tan^2(2a/\lambda)$ and $\Gamma/\tan(2a/\lambda)$ are, therefore, regarded as the resonance parameters to be determined from the experimental data. (Note that there are three unknowns: σ_0, Γ, and a.) The method of analysis is therefore to assume a set of values of $n\sigma_0\tan^2(2a/\lambda)$ and $\Gamma/\tan(2a/\lambda)$ which cover a large area of the $n\sigma_0\tan^2(2a/\lambda)$, $\Gamma/\tan(2a/\lambda)$ plane in the region where the true values of these parameters are expected to lie. The above four correction terms are cal-

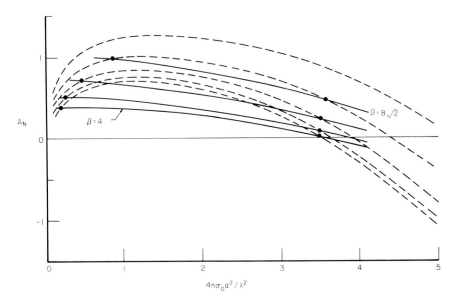

Fɪɢ. 11. An example of Lynn's general area method of analysis. A comparison of experimental and theoretical area functions for the 348 eV resonance in the cross section of ^{238}U (Firk et al., 1963a). Note the negative value of the area for large values of $n\sigma_0$. (- - -, theoretical; —, experimental data; $n = 0.0605$ atoms/b.)

culated for each pair of values in the set. Using these correction terms the following pseudocorrected area is calculated from the experimental data:

$$A_E^{\text{exp}} = \left[\sum_i \Delta E_i - \sum_i T_i \, \Delta E_i/T_p \right]_{E_1}^{E_2} + \Delta A_E(\epsilon) - \delta A_E(\epsilon) \qquad (63)$$

where ΔE_i is the width of an energy channel, and T_i the observed transmission associated with a point i located between the energies E_1 and E_2.

The pseudocorrected experimental areas A_E^{exp} are now compared with the theoretical areas obtained using the same set of assumed parameters. If the correct resonance parameters have been chosen for the wing corrections, potential transmission, and true resonance energy, then A_E^{exp} should agree with the theoretical area function A_E, computed from these parameters.

Agreement is obtained not only for the true parameters but also for other values of the resonance parameters which correspond to a functional relation

$$\tfrac{1}{4}\sigma_0 \tan^2 (2a/\lambda) = f(2\Gamma/\tan (2a/\lambda)) \qquad (64)$$

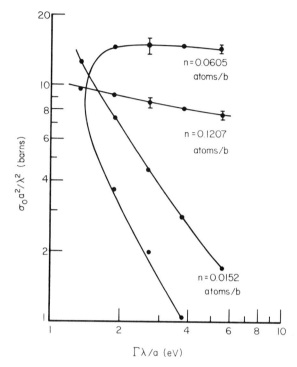

FIG. 12. A determination of the resonance parameters (σ_0, Γ, and a) for the 348 eV resonance in ^{238}U $+$ n using three "thick" samples ($n\sigma_0 \gg 10$) (Firk *et al.*, 1963a).

This gives rise to a self-consistent curve for a given resonance and sample thickness.

By using transmission data obtained from a range of sample thicknesses a set of self-consistent curves may be obtained. These curves will intersect at the correct values of the resonance parameters within the experimental errors.

As an example of the above method of analysis, consider the data obtained for the 348 eV resonance in ^{238}U (Firk *et al.*, 1963a). A comparison between the pseudocorrected experimental data and the corresponding theoretical data is shown in Fig. 11. Samples containing $n = 0.01524$, 0.06045, and 0.12072×10^{24} atoms/cm^2 were used. The self-consistent curves from Fig. 11 are then plotted in Fig. 12, and the resonance parameters ($\sigma_0 a^2$ and Γ/a) determined from the intersection of the three curves (using a least-squares analysis).

d. Determination of Resonance Parameters Using the Harvey–Atta Method of Area Analysis. Recently, there has been a considerable increase in the number of resonances resolved below an energy of about 1 keV. The analysis of such data is no longer practicable using hand calculations which treat each resonance separately. In addition to the shape analysis method, Harvey and Atta have produced a computer program which will analyze up to thirty resonances simultaneously using an area method (Harvey and Atta, 1961). Their method has not been developed in such a general way as that due to Lynn: for instance, there are no trigonometrical factors included in the expressions for the transmission and the inclusion of level–level interference would require considerable changes in the program. Nevertheless, the method is valuable for analyzing transmission data for those nuclides in which the average level spacing is large compared with the average width.

The computer calculates the following Doppler broadened cross section and resolution broadened transmission

$$\sigma_\Delta(E') = (\Delta\sqrt{\pi})^{-1} \int_0^\infty \sigma(E'') \exp - \{(E' - E'')/\Delta\}^2 \, dE'' \tag{65}$$

and

$$T(E_i) = (R\sqrt{\pi})^{-1} \int_0^\infty \exp - \{n\sigma_\Delta(E')\} \exp - \{(E_i - E')/R\}^2 \, dE' \tag{66}$$

where all the terms in Eqs. (65) and (66), except R, have been defined previously. Here, R is the full width at half maximum of the resolution function which is assumed to be a Gaussian.

Estimates of the resonance parameters are fed into the analysis program and the theoretical areas under all the transmission dips are then calculated. Iterations are made until the computed experimental areas and theoretical areas agree within a specified accuracy. The "optimum" values of the resonance parameters are thus determined. As shown in Section III,B,2, the use of samples of different thicknesses yields values of $g\Gamma_n$ and Γ for each resonance.

Two important features of the method are:

(i) The ability to analyze many resonances simultaneously by including the cumulative effects of neighboring resonances on each other.

(ii) The inclusion of resolution-broadening in the expression for the transmission. This procedure is necessary in cases in which the wings of resonances are still affected by resolution and neighboring resonances are closely spaced. An example of the simultaneous area

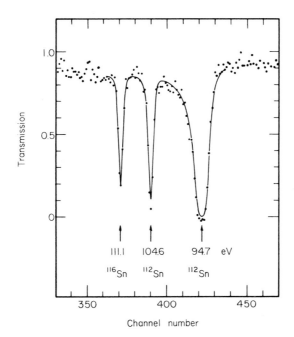

FIG. 13. An example of the simultaneous analysis of three resonances in tin using the Harvey–Atta area analysis method (Harvey and Atta, 1961; Khan and Harvey, 1962).

analysis of resonances in tin is shown in Fig. 13 (Khan and Harvey, 1962).

C. Average Cross Sections

Although the resolution of neutron spectrometers has been greatly improved during recent years, it is still not possible to resolve individual resonances in heavy elements for energies greater than a few kilovolts. Under these circumstances the average total cross section or transmission may be used to obtain the average properties of the resonances, i.e., the s- and p-wave strength functions.

Early measurements of s-wave strength functions using the average transmission technique (Hughes and Zimmerman, 1956; and Gayther and Nicholson, 1957) were limited in accuracy due to the use of thin samples (transmission close to unity). This procedure was necessary in order to avoid large corrections due to preferential attenuation of neutrons near resonances thereby making it difficult to relate the average transmission to the true average cross section.

The computations by Lynn (1960b) of Doppler broadened area functions for wide ranges of resonance parameters and sample thickness enable the average transmission method to be used for thick samples. In addition, important effects due to interference between resonance and potential scattering are also included. Recently, Uttley and Jones (1961) have used both the average transmission and average cross section methods to determine the s- and p-wave strength functions of $(^{238}U + n)$ and $(^{232}Th + n)$. These authors point out that their analysis is inadequate as it is necessary to weight the area functions A_z with a Porter–Thomas distribution of reduced neutron widths (Porter and Thomas, 1956) when averaging over an energy interval which contains many resonances. Such a weighting factor is included by Lynn (1963a) in the analysis of the average $(^{238}U + n)$ and $(^{232}Th + n)$ data. The expression for the total cross section in the neighborhood of a single resonance (resonance energy E_0) may be obtained using the Thomas approximation given in Section I. It may be written (Lynn, 1963a):

$$\sigma_T(E) = \sigma_{sl}(E) + \sigma_{pot} + \sigma_{rr} + \sigma_{slw} \tag{67}$$

| single level term | potential scattering term | resonance–resonance interference term | single level wings from other resonance |

where

$$\sigma_{sl}(x) = \sigma_0/(1 + x^2) + 2ka\sigma_0[x/(1 + x^2)]$$
$$\sigma_{pot} = 4\pi a^2(1 - R^{\infty\prime})^2$$
$$\sigma_{rr} \approx -4\pi a^2\pi^2 S_0^2 \quad \text{(see Thomas, 1955)}$$
$$\sigma_{slw} \approx -8\pi a^2(1 - R^{\infty\prime})\pi^2 S_0(\epsilon/3D)$$
$$+ 4\pi\lambda a S_0(\Gamma/D)\{(\pi^2/6) + (\pi^4/360) + \ldots\}$$

and

$$\epsilon = E - E_0$$

A uniform level approximation is assumed in which the spacing is D (eV), and the equal reduced widths are $\gamma_{\lambda n}^2$ so that $S_0 = \gamma_{\lambda n}^2/D$. The term σ_{rr} is essentially energy independent in the region of interest. The energy independent part of σ_{slw} may then be included with σ_{pot} and σ_{rr} to give an effective constant cross section σ_c, say. The neutron transmission is then given by

$$T(E) = \exp\{-n[\sigma_{sl}(E) + \sigma_c - 8\pi a^2(1 - R^{\infty\prime})\pi^2 S_0(\epsilon/3D)]\} \tag{68}$$

For values of sample thickness n, used in practice, the final term in Eq. (68) may be expanded using the binominal theorem so that

$$T(E) \approx \exp\{-n[\sigma_{s1}(E) + \sigma_c]\}(1 + r\epsilon) \qquad (69)$$

where

$$r = 8\pi n a^2(1 - R^{\infty\prime})\pi^2(S_0/3D)$$

The average transmission $\bar{T}$ in the uniform level approximation is

$$\bar{T} = D^{-1} \int_{-D/2}^{+D/2} T(E)\, d\epsilon$$
$$= \exp(-n\sigma_c)\{D - \int_{-D/2}^{+D/2} [1 - \exp\{-n\sigma_{s1}(E)\}(1 + r\epsilon)]\, d\epsilon\} \qquad (70)$$

where the integral in Eq. (70) is $\hat{A}_E(D/2)$ used by Lynn (1958).

The first order approximation to Eq. (70) is

$$\bar{T} = \exp\{-n\sigma_c\}[1 - (\Gamma/2D)A_x] \qquad (71)$$

Lynn discusses this approximation in detail with special reference to the magnitude of the many level effects. He concludes that the higher order terms due to these effects are negligible in the cases of (^{238}U + n) and (^{232}Th + n) for the typical range of sample thicknesses used in experiments.

In order to obtain the mean transmission $\langle T \rangle$ of neutrons through many resonances, the average transmission $\bar{T}$ must be averaged over a Porter–Thomas distribution of neutron widths. We then have

$$\langle T \rangle = \exp\{-n\sigma_c\}[1 - G] \qquad (72)$$

where

$$2DG = \int_0^\infty (\Gamma_\gamma + \Gamma_n)A_x P(\Gamma_n)\, d\Gamma_n$$

and

$$P(\Gamma_n)\, d\Gamma_n = (2\pi\Gamma_n\bar{\Gamma}_n)^{-1/2} \exp\{-\Gamma_n/2\bar{\Gamma}_n\}\, d\Gamma_n$$

Here, A_x is the Doppler broadened area function defined in Eq. (61).

If the self-screening effect of p-wave resonances is small (which is true in ^{239}U and ^{233}Th), the transmission of p-wave neutrons is given by exp $\{-n\bar{\sigma}_p\}$ where $\bar{\sigma}_p$ is the average p-wave cross section. The average transmission for s- and p-wave neutrons is then

$$\langle T \rangle = \exp\{-n(\sigma_c + \bar{\sigma}_p)\}(1 - G) \qquad (73)$$

The Thomas approach outlined in Section I is then used in its p-wave form to obtain $\bar{\sigma}_p$. Lynn gives

$$\bar{\sigma}_p = 6\pi\lambda^2(1 - \text{Re } \bar{U}_1) \qquad (74)$$

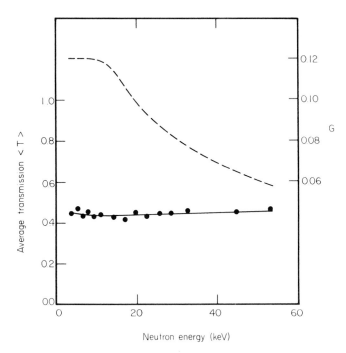

FIG. 14. An analysis of the average transmission data of ^{238}U in order to extract the p-wave strength function (Lynn, 1963a).

where the averaged collision function for p-waves is

$$\bar{U}_1 = \exp\{-2i\phi_1\}(1 + iP_1\bar{R}_1)/(1 - iP_1\bar{R}_1) \tag{75}$$

in which

$$\bar{R}_1 = R_1^\infty + i\pi S_1; \qquad S_1 = (\gamma_{\lambda n}^2/D)_{l=1}$$
$$P_1 = (ka)^3/[1 + (ka)^2]; \qquad \phi_1 = \tan^{-1}\{-j_1(ka)/n_1(ka)\}$$

Here, R_1^∞ is the cumulative effect of p-wave levels, and $j_1(ka)$ and $n_1(ka)$ are spherical Bessel and Neumann functions, respectively.

A least-squares fit to Eq. (73) is made to the data of (^{238}U $+ n$) and (^{232}Th $+ n$) using the s-wave potential scattering cross section (σ_{pot}) and the p-wave strength function (S_1) as variables. Lynn's results for ^{238}U are shown in Fig. 14. The accuracy obtained for the p-wave strength functions show clearly the power of this method compared with the determination of individual resonance parameters. (It is to be noted that more than 100 p-wave resonances would need to be identified and analyzed to obtain an error of $\pm 15\%$ on the p-wave strength function.)

IV. THE SELF-INDICATION METHOD

A. Experimental Technique

In the above sections, we have been discussing transmission measurements made with the use of a "flat" detector, i.e., one whose efficiency for detecting neutrons varies slowly with energy and is therefore constant over an energy interval in which the resonance effect of a single level is significant. Alternatively, it is possible to place the element under study at the detector position and observe, as a function of neutron energy, the γ-rays emitted in the process of radiative capture of the neutrons. There are two main motivations for this: (a) the desire to measure capture cross sections as a function of neutron energy in their own right, a topic discussed in detail in Chapter III; and (b) the fact that the response of the γ-ray detection system is very fast and the thickness of the primary detector, i.e., the sample, in the direction of the neutron flight path is negligible. Full advantage can therefore be taken of the resolution capability of the time-of-flight spectrometer system.

The second motivation was (until recently) the primary consideration in the operation of the Columbia University (Nevis) synchrocyclotron spectrometer system. The burst width and timing gates were both 0.1 μsec. In the detection system actually used, the sample under study was suspended normal to the beam direction, and the γ-rays emitted were detected by shielded plastic scintillation detectors placed above and below the sample.

Two distinct types of measurements can be made: one with various thicknesses of sample at the detector position only, and the other with the sample material at both the detector position and the transmission position (the "self-indication" method).

B. Analysis of the Data

1. SAMPLE AT DETECTOR POSITION ONLY

a. *Data Are Taken for Various Thicknesses of Sample.* If ϵ is the (unknown) efficiency for detecting γ-rays, a single resonance will then give a response in excess of background of $\epsilon(\Gamma_\gamma/\Gamma)(1 - T)$ (neutron flux). If a "thick" sample is included among the various samples, such that the transmission T is zero for several points in the neighborhood of the resonance energy, $\epsilon(\Gamma_\gamma/\Gamma)$ (neutron flux) is determined and can be used to convert the data for the various sample thicknesses to "$(1 - T)$ data." These data are then comparable with the transmission dips obtained with flat detectors and may be handled exactly as in Section III.

b. It Is Sometimes Observed That ϵ Is Constant for All the Levels of a Given Element. If the value of Γ_γ/Γ is known for one or more levels (obtained from other methods of measurements say), then a thick sample with $T = 0$ can be used to determine ϵ. (Occasionally, a simultaneous analysis of many levels can yield ϵ without recourse to external information.) For other levels one obtains $(\Gamma_\gamma/\Gamma)(1 - T)$. The area of the resonance can then be used as above. If the factor Γ_γ/Γ is not too close to unity, the areas can lead to a determination of the spin weighting factor g when these data are combined with conventional transmission measurements.

2. The Self-Indication Method: Samples at Both the Detector and Transmission Positions

Here, the element under investigation, together with the γ-ray detection system, is considered as a detector, and various thicknesses of the same element are used for making transmission measurements. Where the resolution warrants, channel-by-channel transmission values can be computed exactly as for flat detectors and the results treated similarly. Except at the lower neutron energies, this is seldom done because the resolution also effects the sample-out spectrum. The subsequent treatment of self-indication is based on area analysis. We define a quantity T_{si}, the self-indication transmission, as

$$
\begin{aligned}
T_{\text{si}} &= \frac{\text{area with transmission sample}}{\text{area without transmission sample}} \\
&= \frac{\int_{-\infty}^{\infty} \{1 - \exp[-n_D\sigma_0\Psi(\beta,x)]\} \exp[-n_T\sigma_0\Psi(\beta,x)] \, dx}{\int_{-\infty}^{\infty} \{1 - \exp[-n_D\sigma_0\Psi(\beta,x)]\} \, dx} \tag{76}
\end{aligned}
$$

$$
= \frac{\int_{-\infty}^{\infty} \{1 - \exp[-(n_D + n_T)\sigma_0\Psi(\beta,x)]\} \, dx - \int_{-\infty}^{\infty} \{1 - \exp[-n_T\sigma_0\Psi(\beta,x)]\} \, dx}{\int_{-\infty}^{\infty} \{1 - \exp[-n_D\sigma_0\Psi(\beta,x)]\} \, dx} \tag{77}
$$

where n_D and n_T refer, respectively, to the number of nuclei per square centimeter in the detector sample and in the transmission sample. The three integrals in the last expression for T_{si} are those referred to above as $A_x'(\infty,\beta)$. Hence, T_{si} can be calculated from previously evaluated expressions. Here, T_{si} is a function of $n_T\sigma_D$, $n_D\sigma_D$, and β, or any suitable combination of these. Graphs of these relationships have been prepared. As an example, Fig. 15 shows T_{si} versus $n\sigma_0\Gamma/\Delta$ for various values of Δ/Γ in the special case of $n_D = n_T$.

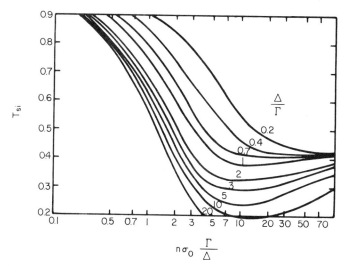

FIG 15. A set of curves of T_{si} versus $n\sigma_0\Gamma/\Delta$ for various values of Δ/Γ in the special case of $n_D = n_T$.

Each measurement of T_{si} with given values of n_D and n_T then gives a relationship between σ_0 and Γ (since n_D and n_T are known), and two measurements will then allow determinations of σ_0 and Γ. This, then, is the strict application of self-indication. Frequently, however, the curves of σ_0 versus Γ (or equivalent) for various measurements intersect at such small angles that small experimental errors result in a large uncertainty in the parameters. In this case, self-indication data may be combined with other measurements to deduce the parameters. The g value cannot, of course, be determined by self-indication alone.

Data on ^{238}U (Rosen *et al.*, 1960) and on Ag, Au, and Ta (Desjardins *et al.*, 1960) have been analyzed by a combination of methods, including self-indication. Figure 16 shows an example (Rosen *et al.*, 1960) where self-indication on one transmission sample thickness is combined with an area measurement under the "D only" (capture yield) curve normalized by an internal determination of ϵ. Figure 17 (Desjardins *et al.*, 1960) shows an example for Ag but with two different transmission sample thicknesses. Also, since $I = \frac{1}{2}$ for Ag, g is either $\frac{1}{4}$ or $\frac{3}{4}$, and the area under the "D only" curve gives a relationship for each value of g. Here, $J = 1$ ($g = \frac{3}{4}$) is clearly selected as the value for the spin J. Figure 18 shows an example for Au (Desjardins *et al.*, 1960), with additional relationships deduced from flat detector measurements. These are examples of favorable cases in which

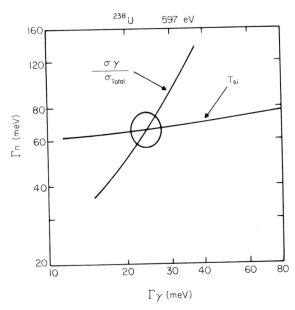

Fig. 16. Determination of resonance parameters using a combination of self-indication (transmission) data and capture yield data (Rosen *et al.*, 1960). ($\Gamma_n = 66 \pm 10$ meV; $\Gamma_\gamma = 24.0 \pm 3.0$ meV.)

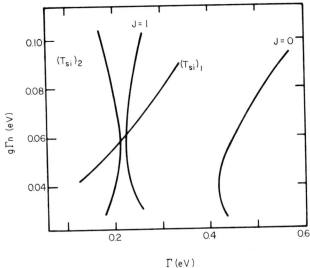

Fig. 17. Determination of resonance parameters using self-indication data with two different transmission sample thicknesses (Desjardins *et al.*, 1960). [Ag (134 eV).]

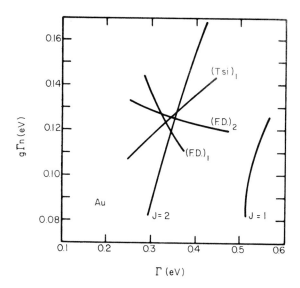

FIG. 18. Determination of resonance parameters using a combination of self-indication and conventional transmission data (Desjardins *et al.*, 1960). [Au (606 eV).]

complete evaluation of the parameters can be made. In many cases, only partial results can be obtained.

The measurements made by detection of γ-rays have some disadvantages when compared with transmission measurements, particularly for thicker samples: (i) the self-absorption of the capture γ-rays in the sample itself produces complicated variations in ϵ; (ii) scattering of neutrons of energy slightly higher than a resonance energy, but which now have a large probability of capture because their energy is reduced to the resonance energy; and (iii) the natural radioactivity of some samples. Corrections for multiple scattering of neutrons in the sample have been made recently by Lynn (1963b) using a Monte Carlo type calculation: such corrections greatly extend the scope of the self-indication method.

V. DATA OBTAINED FROM TOTAL NEUTRON CROSS SECTION MEASUREMENTS

Our present knowledge of neutron resonance parameters and the statistical properties of these parameters is based largely on the results of total cross section measurements. Although such measurements only give values of the resonance energy E_λ, the total width Γ_λ and the neutron width multiplied by the spin weighting factor $g\Gamma_\lambda$, a number of assumptions can frequently be made which enable the determination of other resonance parameters such as the radiation width Γ_γ or the total spin J. For example, a useful assumption is the near constancy of the total radiation width Γ_γ for a particular isotope. In the cases of medium and heavy elements ($A >$ 100), this assumption is theoretically well founded since there are many states which can be reached by γ-rays in the decay of a resonance to the ground state. Experimentally, this assumption has been established within an accuracy of better than 10% in a number of nuclei. A detailed discussion of both resonance parameters and their statistical properties is given by Lynn (1968). The following is an outline of the information which has been obtained from total cross section measurements.

A. Resonance Parameters

1. Measurement of the Total Radiation Width

At low neutron energies ($\gtrsim 100$ eV, say) radiative capture is the predominant process so that a measurement of the total width can yield accurate values of Γ_γ.

Cases of interest are:

(i) The neutron width extremely small so that $\Gamma_\gamma \approx \Gamma$. This straightforward case provided many of the early (and accurate) values of radiation widths (Sailor, 1953; Landon and Sailor, 1955).

(ii) The neutron width comparable with the radiation width ($\Gamma_n \approx \Gamma_\gamma$). In such cases, the radiation width is simply obtained from

$$\Gamma_\gamma = \Gamma - \Gamma_n \tag{78}$$

$$= \Gamma - \sigma_0 \Gamma / 4\pi \lambda^2 g \tag{79}$$

in which both σ_0 and Γ are known from accurate total cross section data. The spin gactor g presents no difficulty if the target spin I is zero ($g = 1$), or if I is large so that $g \approx \frac{1}{2}$. This method has been widely used to determine radiation widths.

2. Measurement of the Neutron Width

As mentioned above, the factor $g\Gamma_n$ is determined directly from total cross section measurements. For those elements in which $g = 1$ or $g \approx \frac{1}{2}$ a large number of neutron widths have been obtained (Harvey et al., 1955). In the case of $^{238}U + n$, 100 resonances have been analyzed and the values of Γ_n found (Firk et al., 1963a).

Other results with good statistical accuracy have been reported for the reaction $^{127}I + n$ (Garg et al., 1965) and the data found to agree with the Porter–Thomas distribution (see Section V,B,2).

3. Measurement of the Resonance Spins

For resonances in the kilovolt region, the predominant decay process is by elastic neutron scattering in which case it is sometimes possible to determine the spin J of the resonances since

$$\sigma_0 \approx 4\pi\lambda^2 g(\Gamma_n/\Gamma) \approx 4\pi\lambda^2 g \qquad \text{for} \quad \Gamma_n \gg \Gamma_\gamma \qquad (80)$$

An accurate measurement of the peak total cross section and the resonance energy determines the statistical weight factor g, and hence the spin J. This method has been used on a number of occasions when the resolution of the spectrometer has been sufficient to measure the peak cross section with precision (Bollinger et al., 1955; Lynn et al., 1958a; Good et al., 1958; Firk et al., 1963b; Coté et al., 1964, and Morgenstern et al., 1965).

B. Statistical Properties of Resonances

1. The Distribution of Level Spacings

The simplest distribution obtainable from slow neutron resonance data is the distribution of level spacings. Even in this case, however, a number of difficulties are soon encountered. For instance, the rapid decrease in resolving power of neutron spectrometers as the neutron energy increases results in an increasing number of unobserved resonances at high energy. In general, therefore, only a limited number of resonances at low energy are available for inclusion in the distributions. At the present time, the largest number of resonances clearly resolved in a single nuclide is about 200 (Garg et al., 1964a).

Another frequent difficulty is due to lack of knowledge of the spins of resonances for those nuclei in which $I \neq 0$; for high spin values ($J > 3$, say), data are essentially nonexistent.

In spite of these limitations, the main features of the level spacing distributions are now known, however. It is clear from a number of experiments (Harvey and Hughes, 1958; Rosen et al., 1960; Firk et al., 1963a; and Garg et al., 1964a) that the distribution is well described by that proposed by Wigner (1957), namely:

$$P(D)\, dD = D/2\bar{D}^2 \exp\{-D^2/4\bar{D}^2\}\, dD \qquad (81)$$

where D is the level spacing and $\bar{D}$ is the mean level spacing. Equation (81) predicts zero probability of finding levels with zero spacing. This "level repulsion" effect had been noted by Gurevich and Pevzner (1956) and Lane et al. (1956).

The excellent agreement between the observations of Garg et al. (1964a) and the distribution given by Eq. (81) is demonstrated in Fig. 19.

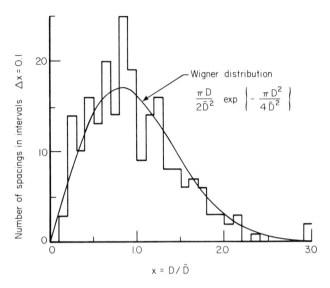

FIG. 19. The observed distribution of level spacings in $^{238}U + n$ compared with a Wigner distribution (Garg et al., 1964a).

2. The Distribution of Reduced Neutron Widths

Using the data obtained by the Brookhaven fast chopper group prior to 1955, Hughes and Harvey (1955b) determined, empirically, the distribution of reduced neutron widths $[\Gamma_{\lambda n}^0 = \Gamma_{\lambda n}/(E_\lambda)^{1/2}]$. They concluded that the distribution has the form

$$P(x)\, dx = (2\sqrt{\pi})^{-1}(x/2)^{-1/2} \exp(-x/2)\, dx \qquad (82)$$

where

$$x = \Gamma_n^0/\langle \Gamma_n^0 \rangle_{\mathrm{av}}$$

in which $\langle \Gamma_n^0 \rangle_{\mathrm{av}}$ is the average reduced neutron width.

This distribution was predicted theoretically by Brink (1955) who argued that the reduced width amplitudes [see Eq. (18)] should have a Gaussian distribution with zero mean, thus

$$P(\gamma_{\lambda n})\, d\gamma_{\lambda n} = (2\pi \langle \gamma_{\lambda n}^2 \rangle_{\mathrm{av}})^{-1/2} \exp(-\gamma_{\lambda n}^2/2\langle \gamma_{\lambda n}^2 \rangle_{\mathrm{av}})\, d\gamma_{\lambda n} \qquad (83)$$

so that the distribution of reduced widths becomes

$$P(\gamma_{\lambda n}^2)\, d\gamma_{\lambda n}^2 = (2\pi \gamma_{\lambda n}^2 \langle \gamma_{\lambda n}^2 \rangle_{\mathrm{av}})^{-1/2} \exp(-\gamma_{\lambda n}^2/2\langle \gamma_{\lambda n}^2 \rangle_{\mathrm{av}})\, d\gamma_{n\lambda}^2 \qquad (84)$$

Porter and Thomas (1956) derived Eq. (84), independently, using arguments based upon the physical properties of the overlap integral in the

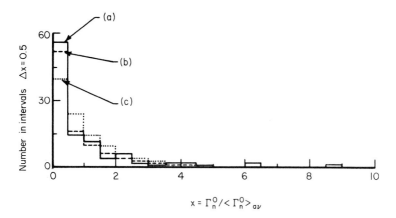

FIG. 20. The observed distribution of reduced neutron widths for 100 levels in ²³⁸U + n compared with Porter–Thomas and exponential distributions (Firk et al., 1963a). (a) Experimental data; (b) theoretical fit: a Porter–Thomas distribution; (c) theoretical fit: an exponential distribution.

definition of the reduced width amplitude [see Eq. (18)]. They concluded that positive and negative values of the integrand are equally likely which is equivalent to Brink's conjecture.

The Porter–Thomas–Brink frequency function is a special case $(\nu = 1)$ of the chi-squared frequency function with ν degrees of freedom:

$$\rho_\nu(x)\,dx = \bar{\Gamma}^{-1}(\nu/2)(\nu/2\bar{x})^{\nu/2}x^{(\nu-2)/2}e^{-\nu x/2\bar{x}}\,dx \tag{85}$$

where $\bar{\Gamma}$ is the incomplete gamma function.

An exponential frequency function is obtained from Eq. (85) for a value of $\nu = 2$.

In order to find the frequency function which best fits the data, it is necessary to determine ν and the error associated with it. The maximum likelihood method is used (Kendall, 1946). For the chi-squared family the most likely value of ν is given by

$$n^{-1}\sum_{i=1}^{n}\ln x_i = \ln\bar{x} = \phi(\nu/2) - \ln(\nu/2) \tag{86}$$

where x_i $(i = 1$ to $n)$ are the observed values, $\bar{x}$ is the mean value, and $\phi(\nu/2)$ is the logarithmic derivative of $\bar{\Gamma}(\nu/2)$.

The most recent results based upon 416 reduced neutron widths (Desjardins et al., 1960; Firk et al., 1963a; Garg et al., 1964a; and Garg et al., 1965) have been analyzed by Garrison (1964), and a value of $\nu = 1.04 \pm 0.10$ obtained which is in excellent agreement with the distribution given in Eq. (84). A typical example is shown in Fig. 20 in which the observed distribution of 100 reduced neutron widths in $^{238}U + n$ (Firk et al., 1963a) is shown together with the theoretical distributions for $\nu = 1$ and $\nu = 2$.

C. Neutron Strength Functions

One of the most significant quantities obtained from the results of slow neutron total cross section studies is the s-wave neutron strength function and its variation with mass number. This quantity is directly related to the average compound nucleus cross section (Feshbach et al., 1954). For example, on integrating over a Breit–Wigner line shape (ignoring interference and resonant self-absorption effects) and then summing over an appropriate energy interval E', we find

$$\sigma_{\mathrm{res}} \times E' = (4\pi^2/k^2)\rho \sum_\lambda \gamma_{\lambda n}^2 \tag{87}$$

where $\rho = ka$, and k is assumed to be constant throughout E'. If the average spacing between levels is $\bar{D}$, and there are λ levels in E', and $\langle \gamma_n^2 \rangle_{\mathrm{av}}$ is the average reduced neutron width, then

$$\bar{\sigma}_{res} = (4\pi^2/k^2)\rho S_0 \tag{88}$$

where

$$S_0 = \langle \gamma_n^2 \rangle_{\mathrm{av}}/\bar{D}$$

is the s-wave neutron strength function.

In the theoretical development of the complex potential model of nuclear reactions, the R-function defined in Eq. (17) for elastic scattering becomes modified

$$R_{\mathrm{op}} = \sum_P \gamma_P^2/(E_P - E - iW) \tag{89}$$

where a complex potential $V = -(V_0 + iW)$ has been used in the original Schrödinger Eq. (3), and E_P is the energy of a single particle resonance in the real potential V_0.

The strength function S is related to the imaginary part of R_{op} (Lane and Thomas, 1958), thus

$$R_{\mathrm{op}} = R^\infty + i\pi S \tag{90}$$

so that

$$S = \frac{W}{\pi} \sum_P \frac{\gamma_P^2}{(E_P - E)^2 + (W)^2} \tag{91}$$

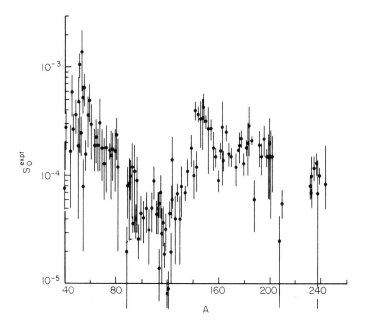

FIG. 21. The observed s-wave neutron strength function S_0^{expt} as a function of mass number (Lynn, 1968).

It is not practicable to study the location of these single particle resonances in a particular nucleus because many l-waves participate at higher energies so that the underlying simple structure becomes ill defined. An alternative approach is to study the variations of the strength function, normalized to an energy of 1 eV, as a function of mass number A (equivalent to a variation in nuclear radius). The advantage of this method is due to the dominance of s-wave interactions at such a low energy, thus enabling the determination of accurate values of neutron widths for many resonances whose spins are known. It is therefore customary to define the experimental neutron strength function for s-wave as

$$S_0^{\text{expt}} = \langle \Gamma_{\lambda n}^0 \rangle_{av} / \bar{D} = 2k_1 a S_0 \qquad (92)$$

where

$$\Gamma_{\lambda n}^0 = \Gamma_{\lambda n} / (E_\lambda / 1)^{1/2}$$

and k_1 is the neutron wave number at 1 eV.

The most recent compilation of data giving S_0^{expt} as a function of A is due to Lynn (1968) and is shown in Fig. 21.

The general form of the curve agrees well with the predictions of the original complex potential model of Feshbach *et al.* (1954) who used a square complex potential with a radius constant $r_0 = 1.45 \times 10^{-13}$ cm and a depth for the real part of the well $V_0 = -42$ MeV. Maxima occur in the strength function at $A \sim 60$ (3s-state) and at $A \sim 155$ (4s-state). The early experimental work of Carter *et al.* (1954) resulted in a best fit to the data using a value of $W_0 = -3.4$ MeV for the depth of the imaginary part of the potential.

Many refinements to the theory have since been reported in order to obtain improved agreement with experiment. The most important features of these refinements are:

(i) the use of a potential well with a diffuse edge (e.g., Woods–Saxon form);

(ii) the use of a surface peaked imaginary part of the potential;

(iii) the use of a nonspherical complex potential (static); and

(iv) inclusion of rotational and vibrational modes of the target nucleus induced by the incident nucleon.

Effects such as the possible coupling of the first 2^+ excited state to the two-phonon triplet (0^+, 2^+, 4^+) have been considered by Furuoya and Sugie (1963), but the predicted third peak associated with the 3s-state is not evident in the present results.

Possible effects due to spin-orbit splitting, particularly for the d-state which is important in studying the coupling of rotations and quadrupole vibrations, are likewise not evident. However, the accuracy of many strength function measurements is not sufficiently high to rule out the possibility of these more subtle effects.

Since the existence of spin exchange forces in the nucleon-nucleon interaction is established, it is not unreasonable to postulate the existence of a spin-spin term in the optical potential which couples the target spin I to the incident particle spin i in a manner analogous to the isospin coupling term. Such a term would result in an angular momentum dependence of the s-wave neutron strength function (Firk *et al.*, 1963b). Recent evidence of such effects has been reported in ^{77}Se $+ n$ (Julien *et al.*, 1962), in ^{75}As $+ n$ (Julien *et al.*, 1964; Garg *et al.*, 1964b) in ^{59}Co $+ n$ (Morgenstern *et al.*, 1965) and in ^{197}Au $+ n$ by Julien *et al.* (1966).

Perhaps a clearer demonstration of spin–spin effects is obtained from the observed spin dependence of the scattering length a_J [see Eq. (47)] deduced in the case of ^{51}V $+ n$ by Firk *et al.* (1963b). Lynn (1968) estimates that the difference in scattering lengths in ^{51}V $+ n$ leads to a value of the coupling strength $V_2 \sim 10$ W where

$$\Delta V_{i \cdot I} = (V_2/A)(\boldsymbol{i} \cdot \boldsymbol{I}) \tag{93}$$

In addition to the s-wave strength function, several measurements of the p-wave strength function have been reported. The two main methods used to deduce the p-wave strength function are:

(i) deviations (for small reduced neutron widths) from the s-wave Porter–Thomas distribution (Rosen et al., 1960; Desjardins et al., 1960; Michaudon and Ribon, 1962; Garg et al., 1965; Le Poittevin et al., 1965); and

(ii) analysis of average transmissions in the kilovolt region (Uttley and Jones, 1961; Lynn, 1963a; and Newstead, 1967).

Although both these methods have many difficulties associated with them, it is possible to obtain p-wave strength functions with errors as low as $\pm 15\%$ in favorable cases.

At the present time, there is no conclusive evidence to support the contention that the p-wave strength function exhibits a spin–orbit splitting in the region $A \sim 100$.

It is clear from the above results that total cross section measurements form the backbone of slow neutron spectroscopy, and that they will continue to do so in the future. The performance of modern electron and proton accelerators indicates that time-of-flight resolutions of less than 0.01 nsec m^{-1} will shortly be achieved in the energy range above several kilovolts. The majority of resonances in all nuclei will then be resolved up to an energy of about 10 keV. The analysis of the data from total, capture, and scattering cross section measurements (Rae et al., 1958; Evans et al., 1959; Iliescu et al., 1965; and Asghar et al., 1966) will give greatly improved values of both the individual and the statistical properties of neutron resonances, thereby adding to our general knowledge of nuclear structure.

REFERENCES

Asghar, M., Chaffey, C. M., Moxon, M. C., Pattenden, N. J., Rae, E. R., and Uttley, C. A. (1966). Nucl. Phys. **76.**

Bethe, H. A., and Placzek, G. (1937). Phys. Rev. **51,** 462.

Bollinger, L. M., Dahlberg, D. A., Palmer, R. R., and Thomas, G. E. (1955). Phys. Rev. **100,** 126.

Bollinger, L. M., Thomas, G. E., and Ginther, R. J. (1959). Rev. Sci. Instr. **30,** L.1135.

Bollinger, L. M., Thomas, G. E., and Ginther, R. J. (1962). Nucl. Instr. Methods **17,** 97.

Breit, G. (1940). Phys. Rev. **58,** 506.

Breit, G. (1946). Phys. Rev. **69,** 472.

Breit, G., and Wigner, E. P. (1936). Phys. Rev. **49,** 519, 642.

Brink, D. M. (1955). Thesis, Univ. of Oxford (unpublished).

BROOKS, F. D. (1961). *In* "Neutron Time-of-Flight Methods" (J. Spaepen, ed.), p. 389. E.A.E.C. (Euratom), Brussels.

CARTER, R. S., HARVEY, J. A., HUGHES, D. J., and PILCHER, V. E. (1954). *Phys. Rev.* **96**, 113.

CARPENTER, R. T., and BOLLINGER, L. M. (1960). *Nucl. Phys.* **21**, 66.

CORGE, C., HUYNH, V-D., JULIEN, J., MORGENSTERN, J., and NETTER, F. (1961). *In* "Neutron Time-of-Flight Methods" (J. Spaepen, ed.), p. 545. E.A.E.C. (Euratom), Brussels.

CoTÉ, R. E., BOLLINGER, L. M., and THOMAS, G. E. (1964). *Phys. Rev.* **134**, B1047.

DESJARDINS, J. S., ROSEN, J. L., HAVENS, W. W., JR., and RAINWATER, J. (1960). *Phys. Rev.* **120**, 2214.

EVANS, J. E., KINSEY, B. B., WATERS, J. R., and WILLIAMS, G. H. (1959). *Nucl. Phys.* **9**, 205.

FESHBACH, H., PORTER, C. E., and WEISSKOPF, V. F. (1954). *Phys. Rev.* **96**, 448.

FIRK, F. W. K. (1958). *Nucl. Phys.* **9**, 198.

FIRK, F. W. K., SLAUGHTER, G. G., and GINTHER, R. J. (1961). *Nucl. Instr. Methods.* **13**, 313.

FIRK, F. W. K., LYNN, J. E., and MOXON, M. C. (1963a). *Nucl. Phys.* **41**, 614.

FIRK, F. W. K., LYNN, J. E., and MOXON, M. C. (1963b). *Proc. Phys. Soc.* **82**, 201.

FURUOYA, I., and SUGIE, A. (1963). *Nucl. Phys.* **44**, 44.

GARG, J. B., RAINWATER, J., PETERSON, J. S., and HAVENS, W. W., JR. (1964a). *Phys. Rev.* **134**, B985.

GARG, J. B., HAVENS, W. W., JR., and RAINWATER, J. (1964b). *Phys. Rev.* **136**, B177.

GARG, J. B., RAINWATER, J., and HAVENS, W. W., JR. (1965). *Phys. Rev.* **137**, B547.

GARRISON, J. D. (1964). *Ann. Phys.* **30**, 269.

GAYTHER, D. B., and NICHOLSON, K. P. (1957). *Proc. Phys. Soc.* **A 70**, 51.

GOOD, W. M., NEILER, J. H., and GIBBONS, J. H. (1958). *Phys. Rev.* **109**, 926.

GUREVICH, I. I., and PEVZNER, M. I. (1956). *Proc. Intern. Conf. Nucl. Reactions, Amsterdam. Physica* **XXII**, 1132.

HARVEY, J. A., and HUGHES, D. J. (1958). *Phys. Rev.* **109**, 471.

HARVEY, J. A., and ATTA, S. E. (1961). *In* "Neutron Time-of-Flight Methods" (J. Spaepen, ed.), p. 55. E.A.E.C. (Euratom), Brussels.

HARVEY, J. A., HUGHES, D. J., CARTER, R. E., and PILCHER, V. E. (1955). *Phys. Rev.* **99**, 10.

HAVENS, W. W., JR., and RAINWATER, L. J. (1946). *Phys. Rev.* **70**, 154.

HAVENS, W. W., JR., and RAINWATER, L. J. (1951). *Phys. Rev.* **83**, 1123.

HUGHES, D. J. (1955a). *J. Nucl. Energy* **1**, 237.

HUGHES, D. J., and HARVEY, J. A. (1955b). *Phys. Rev.* **99**, 1032.

HUGHES, D. J., and ZIMMERMAN, R. L. (1956). *In* "Nuclear Reactions" (P. M. Endt and P. B. Smith, eds.), Vol. 1, p. 380. North Holland Publ., Amsterdam.

HUGHES, D. J., and SCHWARTZ, R. B. (1958). Brookhaven National Laboratory Rept., BNL 325, 2nd ed.

HUMBLET, J., and ROSENFELD, L. (1961). *Nucl. Phys.* **26**, 529.

ILIESCU, N., SAN, KIM, H., PIKELNER, L. B., SHARAPOV, E. I., and SIRAZHET, H. (1965). *Nucl. Phys.* **72**, 298.

JACKSON, H. E., and LYNN, J. E. (1962). *Phys. Rev.* **127**, 461.

JULIEN, J., CORGE, C., HUYNH, V. D., MORGENSTERN, J., and NETTER, F. (1962). *Phys. Letters* **3**, 69.

JULIEN, J., BIANCHI, G., CORGE, C., HUYNH, V. D., LE POITTEVIN, G., MORGENSTERN, J., NETTER, F., and SAMOUR, C., (1964). *Phys. Letters* **10**, 86.

JULIEN, J., DE BARROS, S., BIANCHI, G., CORGE, C., HUYNH, V. D., LE POITTEVIN, G., MORGENSTERN, J., NETTER, F., SAMOUR, C., and VASTEL, M. (1966). *Nucl. Phys.* **76**, 391.

KAPUR, P. L., and PEIERLS, R. (1938). *Proc. Roy. Soc. (London)* **A166**, 277.

KENDALL, M. G. (1946). "The Advanced Theory of Statistics," Vol. II, Chapter 17. Griffin, London.

KHAN, F. A., and HARVEY, J. A. (1962). *Bull. Am. Phys. Soc.* **7**, 289.

KROTKOV, R. (1955). *Can. J. Phys.* **33**, 622.

LADENBERG, R., and REICHE, F. (1913). *Ann. Phys.* **42**, 181.

LAMB, W. E. (1939). *Phys. Rev.* **55**, 190.

LANDON, H. H. (1954). *Phys. Rev.* **94**, 1215.

LANDON, H. H., and SAILOR, V. L. (1955). *Phys. Rev.* **98**, 1267.

LANE, A. M., and THOMAS, R. G. (1958). *Rev. Mod. Phys.* **30**, 257.

LANE, A. M., LYNN, J. E., and STORY, J. S. (1956). Atomic Energy Research Establishment, Harwell, England, Rept. No. T/M 137.

LE POITTEVIN, G., DE BARROS, S., HUYNH, V. D., JULIEN, J., MORGENSTERN, J., NETTER F., and SAMOUR, C. (1965). *Nucl. Phys.* **70**, 497.

LYNN, J. E. (1958), *Nucl. Phys.* **7**, 599.

LYNN, J. E. (1960a). *Nucl. Instr. Methods* **9**, 315.

LYNN, J. E. (1960b). Atomic Energy Research Establishment, Harwell, England, Rept. Nos. R-3353 and R-3354.

LYNN, J. E. (1963a). *Proc. Phys. Soc.* **82**, 903.

LYNN, J. E. (1963b). Private communication.

LYNN, J. E. (1968). "The Theory of Neutron Resonance Reactions." Univ. Press, Oxford.

LYNN, J. E., and RAE, E. R. (1957). *J. Nucl. Energy* **4**, 418.

LYNN, J. E., FIRK, F. W. K., and MOXON, M. C. (1958a). *Nucl. Phys.* **5**, 603.

LYNN, J. E., MOXON, M. C., and FIRK, F. W. K. (1958b). *Nucl. Phys.* **7**, 613.

McDANIEL, B. D. (1946). *Phys. Rev.* **70**, 832.

MELKONIAN, E. (1956). *Proc. Intern. Conf. Peaceful Uses At. Energy, Geneva* **4**.

MELKONIAN, E., HAVENS, W. W., JR., and RAINWATER, J. (1953). *Phys. Rev.* **92**, 702.

MEYER, R. R. (1949). *Phys. Rev.* **75**, 773.

MICHAUDON, A. (1961). *In* "Neutron Time-of-Flight Methods" (J. Spaepen, ed.), p. 531. E.A.E.C. (Euratom), Brussels.

MICHAUDON, A., and RIBON, P. (1962). Private communication.

MORGENSTERN, J., BIANCHI, G., CORGE, C., HUYNH, V. D., JULIEN, J., NETTER, F., LE POITTEVIN, G., and VASTEL, R. (1965). *Nucl. Phys.* **62**, 529.

MÖSSBAUER, R. L. (1958). *Z. Physik.* **151**, 124.

MUEHLHAUSE, C. O., and THOMAS, G. E. (1953). *Nucleonics* **11**, 44.

NEWSTEAD, C. M. (1967). Thesis, Univ. of Oxford (unpublished).

PETERSON, S. W., and LEVY, H. A. (1952). *Phys. Rev.* **87**, 462.

PORTER, C. E., and THOMAS, R. G. (1956). *Phys. Rev.* **104**, 483.

RAE, E. R., and BOWEY, E. M. (1953). *Proc. Phys. Soc.* **A66**, 1073.

RAE, E. R., COLLINS, E. R., KINSEY, B. B., LYNN, J. E., and WIBLIN, E. R. (1958). *Nucl. Phys.* **5**, 89.

RIBON, P., DIMITRIJEVICK, Z., MICHAUDON, A., and WAGNER, P. (1961). *J. Phys. Radium* **22**, 708.

ROSE, M. E., MIRANKER, W., LEAK, P., and RABINOWITZ, G. (1953). Brookhaven Natl. Lab. Rept, BNL. 257.

ROSEN, J. L., DESJARDINS, J. S., RAINWATER, J., and HAVENS, W. W. JR., (1960). *Phys. Rev.* **118**, 687.

SAILOR, V. L. (1953). *Phys. Rev.* **91**, 53.

SEIDL, F. G. P., HUGHES, D. J., PALEVSKY, H., LEVIN, J. S., KATO, W. Y., and SJÖSTRAND, N. G. (1954). *Phys. Rev.* **95**, 476.

SETH, K. K. (1959). *Ann. Phy.* **8**, 223.

STEHN, J. R., GOLDBERG, M. D., WIENER-CHASMAN, RENATE, MUGHABGHAB, S. F., MAGURNO, B. A., and MAY, V. M. (1965). Brookhaven Natl. Lab. Rept., BNL 325, 2nd ed., supplement 2.

THOMAS, R. G. (1955). *Phys. Rev.* **97**, 224.

UTTLEY, C. A., and JONES, R. H. (1961). *In* "Neutron Time-of-Flight Methods" (J. Spaepen, ed.), p. 109. E.A.E.C. (Euratom), Brussels.

UTTLEY, C. A. and JONES, R. H. (1962). Private communication.

VOGT, E. (1959). *In* "Nuclear Reactions" (P. M. Endt and M. Demeur, eds.), Vol. 1, p. 215. North-Holland Publ., Amsterdam.

VOGT, E. (1962). *Rev. Mod. Phys.* **34**, 723.

VOITOVETSKII, V. K., TOLMACHEVA, N. S., and ARSAEV, M. I. (1959). *At. Eneng.* **6**, 321, 472.

WIGNER, E. P. and EISENBUD, L. (1947). *Phys. Rev.* **72**, 29.

WIGNER, E. P. (1957). *Proc. Gatlinburg Conf. Neutron Time-of-Flight Methods*, Oak Ridge Natl. Lab. Rept., O.R.N.L. 2309.

III

NEUTRON SCATTERING AND CAPTURE
CROSS-SECTION MEASUREMENTS

ERNEST R. RAE
ATOMIC ENERGY RESEARCH ESTABLISHMENT
HARWELL, GREAT BRITAIN

ROBERT C. BLOCK
DEPARTMENT OF NUCLEAR SCIENCE
RENSSELAER POLYTECHNIC INSTITUTE, TROY, NEW YORK

I. INTRODUCTION

A. Measurements in the Resonance Region; Determination of Partial Widths and Spins

The measurement of neutron total cross sections is a most powerful tool in the study of neutron interactions with nuclei in the resonance region. Most of the information presently available on the systematics of neutron resonances, notably neutron widths, level spacings, and strength functions has been derived from such measurements. In favorable cases radiation widths, spins, and (with appropriate assumptions) even fission widths have been determined from total cross-section measurements; but, in general, the extraction of these further parameters requires the measurement of the appropriate partial cross sections. To understand the limitations of the total cross-section measurements and the importance of a knowledge of the partial cross sections in the determination of level parameters, it is necessary to consider the relationships existing between the resonance parameters and the various measured quantities.

In principle, it is necessary to use a multilevel formalism to describe the cross sections observed in the resonance region. In practice, however, the multilevel description is only necessary when the levels are so wide that overlapping occurs. This can happen in regions of high s-wave strength function and wide level spacing when strong interference between levels is observed in the elastic scattering cross section. It can also be observed in the fission cross section of heavy nuclei where the large fission widths cause overlapping of resonances even at very low neutron energies. We shall ignore these difficulties, which occur in special regions of the periodic table, and consider the majority of cases in which the single level Breit–Wigner formula gives an adequate description of the cross sections.

If we exclude the case of certain light elements with large (n, α) cross sections, we can write the total cross section for neutrons below the threshold for inelastic scattering, incident with energy E on a nonfissile nucleus at rest, as the sum of the partial cross sections for elastic scattering and radiative capture:

$$\sigma_t(E) = \sigma_n(E) + \sigma_\gamma(E) \tag{1}$$

where σ_t, σ_n, and σ_γ are, respectively, the total, elastic scattering, and radiative capture cross sections. The partial cross sections close to a resonance (with spin J) at resonance energy E_0 may be expressed in terms of the Breit–Wigner single-level formulas for low energy s-wave interactions with nuclei having spin I:

$$\sigma_n(E) = \pi \lambda_0^2 g \frac{\Gamma_n^2}{(E - E_0)^2 + (\Gamma/2)^2} + 4\pi \lambda_0 g \frac{\Gamma_n a_J (E - E_0)}{(E - E_0)^2 + (\Gamma/2)^2} + \sigma_p \quad (2)$$

$$\sigma_\gamma(E) = \pi \lambda \lambda_0 g \frac{\Gamma_n \Gamma_\gamma}{(E - E_0)^2 + (\Gamma/2)^2} \quad (3)$$

$$g = \frac{2J + 1}{2(2I + 1)} \qquad (J = I \pm \tfrac{1}{2}) \quad (4)$$

In the above, λ is the de Broglie wavelength, divided by 2π, of the relative motion; λ_0 is the de Broglie wavelength at the resonance energy E_0; Γ_n is the neutron width at resonance*; Γ_γ is the radiation width, and

$$\Gamma = (\lambda_0/\lambda)\Gamma_n + \Gamma_\gamma \approx \Gamma_n + \Gamma_\gamma$$

since, in the cases to be discussed, the variation of (λ_0/λ) is small over the energy region where the resonance cross section is significantly large. Finally,

$$\sigma_p = 4\pi \sum_J g a_J^2$$

is the effective potential scattering cross section, where a_J is the scattering length for states of total angular momentum J.

It is convenient to write σ_0, $\sigma_{0\gamma}$, and σ_{0n} for the total, capture and scattering cross sections, respectively, at exact resonance. These quantities are defined by

$$\sigma_0 = 4\pi \lambda_0^2 g \Gamma_n / \Gamma = \sigma_t(E_0) - \sigma_p \quad (5)$$

$$\sigma_{0\gamma} = \sigma_0 \Gamma_\gamma / \Gamma = \sigma_\gamma(E_0) \quad (6)$$

$$\sigma_{0n} = \sigma_0 \Gamma_n / \Gamma = \sigma_n(E_0) - \sigma_p \quad (7)$$

In practice, due to the thermal motion of the target nuclei and the finite resolution of the spectrometer, the "true" cross sections at exact resonance are never observed. The effect of thermal motion (Doppler effect), however, can be taken into account (Bethe, 1937; Lamb, 1939). When the spectrometer resolution is sufficiently good and the effect of thermal motion does not significantly broaden the width of the resonance, it is possible to obtain σ_0 and Γ directly from the shape and magnitude of the observed total cross section. For the special case in which $\Gamma_n \gg \Gamma_\gamma$ (of frequent occurrence in the kilovolt region), it is possible to use Eq. (5) to determine directly the spin weighting factor g. With very good resolution, it is clear from Eqs. (6) and (7) that a measurement of the peak scattering or capture

---

* Here, Γ_n is a constant in this notation, and it is equal to the energy-dependent neutron width at the resonance energy E_0.

cross section taken along with a measurement of the peak total cross section, allows us in principle to determine g, Γ_n, and Γ_γ. In practice, however, the spectrometer resolution is not negligible and is usually imperfectly known. In this case the shapes of the resonances are distorted and the analysis of the resonances in the cross sections is usually made by the method of area analysis. This method determines the area above the transmission curve due to a single resonance. This area, which we designate A_t, is thus given by

$$A_t = \iint R(E - E')[1 - T(E')/T_p]\, dE'\, dE$$

where $R(E - E')$ is the normalized resolution function centered about E with a variation in E', $T(E')$ is the transmission of the sample at energy E', and T_p is the transmission that would have been observed if the resonance were not there. Generally, T_p is equal to $\exp(-N\sigma_p)$, where σ_p is the potential scattering cross section, although T_p can also be interpreted as the net transmission due to potential scattering and the cross section contributions from distant resonances. Here $T(E')$ is divided by T_p so that the area A_t represents the net resonance contribution of neutrons removed from the beam. If the resolution width is small compared to the level spacing, the limits of integration can effectively be set at $-\infty$ and $+\infty$, and then the area reduces to

$$A_t = \int_{-\infty}^{\infty} [1 - T(E')/T_p]\, dE' \int_{-\infty}^{\infty} R(E - E')\, dE$$

$$= \int_{-\infty}^{\infty} [1 - T(E')/T_p]\, dE'$$

since $\int_{-\infty}^{+\infty} R(E - E')\, dE = 1$ for a normalized resolution function. The area above the dip in a transmission curve due to a single level, ignoring resonance—potential interference [the middle term on the right-hand side of Eq. (2)] has been calculated, tabulated and presented graphically for a wide range of level parameters and sample thickness (see, for example, Hughes, 1955). In the limit of very thin, or very thick samples, the thermal motion of the atoms has no first order effect on the resonance areas and the following asymptotic forms apply:

$$A_t = \pi N \sigma_0 \Gamma/2 \qquad (N\sigma_0 \ll 1) \tag{8}$$

and

$$A_t^2 = \pi N \sigma_0 \Gamma^2 \qquad (N\sigma_0 \gg 1) \tag{9}$$

where A_t is the transmission area above the resonance dip, in energy units, N is the number of atoms per unit area of sample, and σ_0 and Γ have the usual meanings. From a transmission measurement on a very thin sample then, we obtain the quantity

$$(A_t)_{\text{thin}}/2\pi^2 N\lambda_0^2 = g\Gamma_n \tag{10}$$

and from a thick sample:

$$(A_t)_{\text{thick}}^2/4\pi^2 N\lambda_0^2 = g\Gamma_n\Gamma \tag{11}$$

It is clear from Eqs. (10) and (11) that thick and thin sample transmission measurements yield values of Γ and $g\Gamma_n$. In the special case in which $\Gamma_n \gg \Gamma_\gamma$, then $\Gamma_n \approx \Gamma$ and g and Γ_n are, therefore, determined from the area measurement. Alternatively, in the limit $\Gamma_n \ll \Gamma$, then $\Gamma_\gamma \approx \Gamma$ and Γ_γ is determined, together with the product $g\Gamma_n$.

Let us now consider what is obtained from area measurements of partial cross sections, in particular from scattering and capture measurements. Let us assume that we have an experimental arrangement whereby we can measure the fraction of neutrons of a given energy, present in the neutron beam, which is either scattered or captured. We designate this fraction of scatters or captures as the "yield" Y_x, where x denotes the type of partial interaction. Thus, the yield is equal to the number of detected partial events divided by the product of the detector efficiency times the number of neutrons incident upon the sample. It is convenient to subdivide further this yield into a primary yield Y_{xp}, and a multiple yield* Y_{xm}, where the former refers to partial interactions which take place at the first collision, and the latter takes into account all higher order, or multiple, collisions of neutrons which are initially scattered in the sample. Thus, $Y_x = Y_{xp} + Y_{xm}$. For the case of scattering measurements, the effect of multiple interactions is to absorb scattered neutrons which otherwise would escape from the sample. Thus in scattering, Y_{nm} expresses the loss of neutrons and is a negative quantity. For a detector of less than 4π solid angle, however, Y_{nm} must also include neutrons which are scattered into the detector by a second or later interaction (inscattering).

The primary yields can be determined analytically, but the multiple yields are exceedingly complicated and are usually computed by Monte Carlo methods. It is thus convenient to think of the multiple yield as a "correction" to the primary yield although in some cases, it can even exceed the primary yield.

For neutrons of energy E incident upon a sample, the number of primary interactions per incident neutron at a depth n and in an element of thickness dn is equal to

---

* The multiple yield is also referred to as multiple scattering.

$$d\,Y_{xp}(E)$$

$$= \exp[-n \int_{-\infty}^{+\infty} \sigma_t(E')S(E - E')\,dE'] \int_{-\infty}^{+\infty} dn\,\sigma_x(E')S(E - E')\,dE' \tag{12}$$

where the first term is the transmission of neutrons to a depth n, and the second term is the number of interactions of type x which take place in the thickness dn. The thermal motion of the target nuclei is explicitly taken into account by integration over $S(E - E')\,dE$; where $S(E - E')\,dE$ is the probability of the interaction energy being E' when the incident neutron energy (relative to a stationary nucleus) is E. For the gas model of thermal motions (Lamb, 1939) $S(E - E')$ is given by:

$$S(E - E') = \exp[(E - E')/(4E_0kT/A)^{1/2}]^2/(4E_0kT/A)^{1/2}\sqrt{\pi} \tag{13}$$

where E_0 is taken to be the energy at the resonance, k is the Boltzmann constant, T is the effective temperature, and A is the atomic weight of the target nucleus.

If we define the Doppler broadened cross sections as

$$\sigma_{\Delta t}(E) = \int_{-\infty}^{+\infty} \sigma_t(E')S(E - E')\,dE' \tag{14}$$

$$\sigma_{\Delta x}(E) = \int_{-\infty}^{+\infty} \sigma_x(E')S(E - E')\,dE' \tag{15}$$

then (12) reduces to

$$dY_{xp}(E) = \exp\{-n\sigma_{\Delta t}(E)\}\sigma_{\Delta x}(E)\,dn \tag{16}$$

The primary yield is then equal to the integral over the sample thickness N, or

$$Y_{xp}(E) = [1 - \exp\{-N\sigma_{\Delta t}(E)\}]\sigma_{\Delta x}(E)/\sigma_{\Delta t}(E) \tag{17}$$

In an actual experiment, it is rare that the instrumental resolution is good enough to measure the yield as a function of energy over the resonance. Rather, the area under the resonance is determined from the experiment, and for an isolated resonance this is equal to

$$\int_{-\infty}^{+\infty} \int_{-\infty}^{+\infty} Y_x(E)R(E - E')\,dE'\,dE = \int_{-\infty}^{+\infty} R(E - E')\,dE' \int_{-\infty}^{+\infty} Y_x(E)\,dE$$

$$= \int_{-\infty}^{+\infty} Y_x(E)\,dE \equiv A_x \tag{18}$$

where the resolution function $R(E - E')$ integrates to unity when the integral is taken over all energies (or in practice, over the energy region in the vicinity of the resonance where the product of the yield times resolution function is significant). This integrated yield (in energy units) is called the "partial area" under the resonance, and it is designated A_n, A_γ, or A_f, respectively, for scattering, capture, or fission. For the case of scattering, the integral in (18) from $-\infty$ to $+\infty$ is infinite because of the potential scattering contribution to the cross section. In practice, however, scattering experiments upon isolated resonances are usually carried out with thin samples so that the potential scattering is small compared to the resonance scattering in the vicinity of the resonance. In this case the scattering area can be approximately evaluated as the area above the potential scattering that would have been observed in the absence of the resonance. For thick samples this approximation is not valid, and (18) would have to be evaluated with finite limits of integration. If the resolution function $R(E - E')$ is not too wide, the limits of integration of E' in (18) can effectively be set at several resolution widths on either side of the resonance, and $R(E - E')$ can still be integrated out of the equation.

The actual computation of the integrated yield in (18), with either finite or infinite limits of integration, is quite complicated, and generally requires a mixture of analytical integration and Monte Carlo methods.

However, for the case of thin samples ($N\sigma_0 \ll 1$), multiple effects can be neglected and (18) reduces to

$$
\begin{aligned}
A_x &= \int_{-\infty}^{+\infty} Y_x(E)\, dE \approx \int_{-\infty}^{+\infty} Y_{xp}(E)\, dE \\
&\approx \int_{-\infty}^{+\infty} [1 - 1 + N\sigma_{\Delta t}(E) - \cdots]\, \sigma_{\Delta x}(E)\, dE/\sigma_{\Delta t}(E) \\
&\approx \int_{-\infty}^{+\infty} N\sigma_{\Delta x}(E)\, dE \\
&\approx N \int_{-\infty}^{+\infty} dE \int_{-\infty}^{+\infty} \sigma_x(E')S(E - E')\, dE' \\
&\approx N \int_{-\infty}^{+\infty} S(E - E')\, dE \int_{-\infty}^{+\infty} \sigma_x(E')\, dE' \\
&\approx N \int_{-\infty}^{+\infty} \sigma_x(E')\, dE'
\end{aligned}
\tag{19}
$$

where $\int_{-\infty}^{+\infty} S(E - E')\, dE$ integrates out to unity. Thus for thin samples, the area under a partial yield curve is just equal to the integrated partial cross section times the sample thickness.

Let us now consider the case of an isolated Breit–Wigner resonance and a thin sample. One can substitute the value of the scattering cross section or capture cross section in Eq. (2) or (3), respectively, into (19) and determine the integral. If we neglect the potential cross section and assume that the width of the level Γ is much smaller than the resonance energy E_0, then the thin sample partial areas become

$$A_n \approx \pi N \sigma_0 \Gamma_n/2 \tag{20}$$

$$A_\gamma \approx \pi N \sigma_0 \Gamma_\gamma/2 \tag{21}$$

From Eqs. (20) and (21), we then obtain the quantities $\sigma_0 \Gamma_n$ and $\sigma_0 \Gamma_\gamma$, and can write

$$A_n/2\pi^2 N \lambda_0^2 = g\Gamma_n^2/\Gamma \tag{22}$$

and

$$A_\gamma/2\pi^2 N \lambda_0^2 = g\Gamma_n\Gamma_\gamma/\Gamma \tag{23}$$

In principle, either (22) or (23) taken in conjunction with (10) and (11) will yield g, Γ_n, and Γ. In practice, if $\Gamma_n \ll \Gamma_\gamma$, (23) reduces to (10), and the scattering measurement (22) is necessary to achieve a solution. If, on the other hand the opposite is true, (22) reduces to (10), making the capture measurement (23) the important factor. Thus, in addition to the transmission measurements, it is generally necessary to measure the partial area corresponding to the *smaller* partial width to extract all the resonance parameters. If the nucleus being studied is fissile, a further parameter is added to the set, but provided the levels being studied are well separated, so that the single level approximation is valid, a further datum is obtained from the fission resonance area A_f, and we can write

$$A_f/2\pi^2 N \lambda_0^2 = g\Gamma_n\Gamma_f/\Gamma \tag{24}$$

Again, in principle this extra equation, together with thick and thin sample transmission measurements, and either scattering or capture yield measurements, permits a determination of all the level parameters. In this case Γ_n is normally much less than Γ_γ or Γ_f, and the set of equations used would be (10), (11), (22), and (24).

In practice, several difficulties arise in applying this simple thin-sample method of solving simultaneous equations. First, in the case of transmission measurements, the total areas A_t for $N\sigma_0 \ll 1$ are extremely small and are difficult to measure with adequate statistical accuracy, while in the case $N\sigma_0 \gg 1$, the neglect of resonance-potential interference cannot, in general, be justified. Transmission measurements are therefore usually made for a range of sample thicknesses which do not satisfy either asymptotic limit, and consequently no simple analytic expressions as (8) or (9) exist for the resonance area A_t. From the graphs of Hughes (1955), however, it is possible to construct a curve of Γ_n versus Γ for each measured resonance area and sample thickness.

In the case of partial cross-section measurements, the problem of multiple interactions of neutrons initially scattered in the sample severely complicates the analysis. One method to avoid rigorous multiple interaction corrections is to make partial cross-section measurements with a range of sample thicknesses, and then asymptotically obtain the limiting value for $N \rightarrow 0$ of A_x/N (where A_x represents A_n, A_γ, or A_f). For the smaller sample thicknesses first-order analytic multiple interaction corrections can be readily applied. From the limiting value of A_x/N the functional relationship between Γ_n and Γ determined from Eqs. (22)–(24) is obtained. This method has been used, for example, by Rae *et al.* (1958) in determining the g values of the resonances in Ag. In general, however, the statistical accuracy is better with the thicker samples (which are more affected by multiple interactions in the sample), but the limiting value of A_x/N depends critically on the thinner sample data. In addition, the recent kilovolt capture cross-section measurements of Moxon (1966) and Block *et al.* (1967) critically show the need for multiple interaction corrections to be applied to the partial cross-section data. Modern high-speed computers, however, now make it practical to make multiple interaction corrections by the Monte Carlo method. Various codes have been written (Lynn, 1963; Sullivan and Werner, 1964; Friedes, 1964; Fröhner, 1966) which compute the total interaction of a beam of neutrons incident upon uniform thickness samples over the energy region of a Breit–Wigner resonance. All these codes use the Doppler-broadened total cross section to compute the neutron mean free path in the slab, and typical computation times of 1 to 5 min per resonance have been obtained. Lynn (1963) and Friedes (1964) have studied the effect of either including the motion of the target nuclei in the scattering kinematics or considering the target nucleus to be at rest when the neutron does scatter. For most low energy resonances the target nucleus can be taken as stationary (but, of course, the Doppler-motion must be taken into account to determine the neutron mean free path in the sample). For the case of kilovolt energy partial measurements, the use of thick samples does not permit one to ignore the potential cross section or the interference terms in (2), and a proper calculation must be made including the effects of thermal motion and multiple interactions.

The result of any partial measurement is still a functional relationship between the parameters g, Γ_n, Γ_γ, and Γ_f for each sample thickness N, where for thin samples Eqs. (22)–(24) apply, and for thick samples the relationship is obtained through Monte Carlo calculations. In a typical partial measurement the number of sample thicknesses used exceeds the number of parameters to be obtained, and a set of "best values" can be obtained for the parameters. For example, Rae *et al.* (1958) have developed a graphical solution. Figure 1, taken from their paper, illustrates the method as applied to the 5.2 eV level in silver. Two sets of curves were constructed from the experimental data with the aid of Eqs. (10), (11), (22),

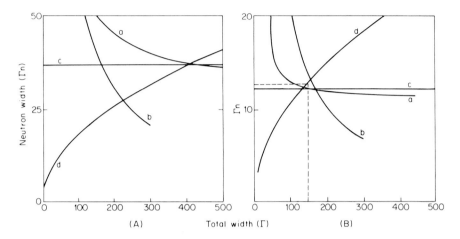

FIG. 1. Combination of the experimental data to determine Γ_n and Γ for the 5.2 eV resonance in silver and the appropriate statistical weight factor g. Ordinate and abscissa scales are in millielectron volts. Curves show the variation of Γ_n with Γ as determined by: a, radiative capture in thin sample; b, thick-sample transmission; c, thin-sample transmission; d, elastic scattering [(A) $g = \frac{1}{4}$, $\chi^2_{min} = 75$; (B) $g = \frac{3}{4}$, $\chi^2_{min} = 3.2$]. (Rae et al., 1958.)

and (23) and the curves of Hughes (1955). The two sets correspond to the two possible values of g; and if the data were perfect, the curves would intersect in a common point for the correct choice of g. In this example $g = \frac{3}{4}$ is clearly the more likely value, and a method of least-squares analysis developed by the same authors provides a preferred value of g, and most probable values of Γ_n and Γ_γ together with standard errors. This method has since been programmed for use with a digital computer (Evans et al., 1958). Graphical methods have also been developed by Rosen et al. (1960), Desjardins et al. (1960), Iliescu et al. (1965), Fröhner and Haddad (1965), and Fröhner (1966) to obtain values of partial widths and spins from a combination of capture, self-indication, and transmission measurements. Fröhner et al. (1966) have compared the relative sensitivity of transmission, scattering, capture, and self-indication measurements; and their results are summarized in Fig. 2. Theoretical analysis curves are plotted in Fig. 2 for the four types of measurements for thick and thin samples; strong, intermediate, and weak resonances; and both the "correct" and "incorrect" spin assignments. These curves serve as a guide to the experimenter in determining which particular set of resonance measurements will be required to determine in the most efficient way all the resonance parameters.

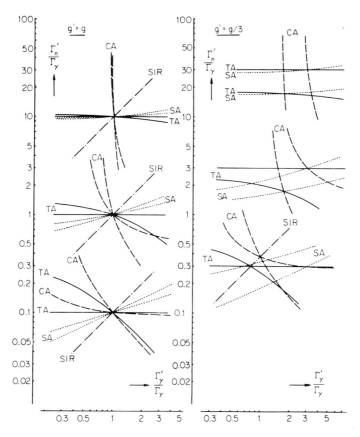

FIG. 2. Theoretical analysis derived with the assumption that the samples are very thin or very thick for the transmission areas (TA), capture areas (CA), and scattering areas (SA), and that the self-indication ratios (SIR) were measured with a thin first and a thick second sample, or vice-versa. The abscissas and ordinates are the calculated values Γ_γ' and Γ_n' in units of the true radiation width Γ_γ, on a log-log scale. Three cases are shown: a strong, an intermediate, and a weak resonance ($\Gamma_n : \Gamma_\gamma = 10, 1, 0.1$, respectively). The curves on the left-hand side are derived with the correct spin factor, the curves on the right-hand side show what is obtained when the spin factor is taken three times too small (which for s-wave neutrons corresponds to an incorrect compound spin of $J = 0$, and a correct compound spin of $J = 1$). (The correct curves are independent of J) (Fröhner et al., 1966).

This, then, is an outline of the method by which area analysis of elastic scattering and capture data can be used in conjunction with transmission measurements to provide values of g, Γ_n, and Γ_γ for low energy neutron resonances.

B. Measurements of Average Cross Sections

The yield $Y_x(E)$ is defined in Section A as the fraction of neutrons of a given energy E, present in the neutron beam, which is either scattered or captured (or absorbed to undergo fission). When measurements are carried out in the unresolved or continuum region, the average yield $\overline{Y_x(E)}$ is determined, where $\overline{Y_x(E)}$ is equal to the yield averaged over the experimental resolution function. It is also convenient to divide the average yield into a primary average yield and a multiple average yield, where the primary and multiple components have the same meaning as in Section A. Partial cross section measurements, in particular capture cross section measurements, are frequently carried out with samples sufficiently thin such that multiple and self-protection effects can be neglected. Then Eq. (17) can be directly applied to average yield measurements as follows:

$$\overline{Y_x(E)} \approx \overline{Y_{xp}(E)} \approx \overline{[1 - \exp\{-N\sigma_{\Delta t}(E)\}]\sigma_{\Delta x}(E)/\sigma_{\Delta t}(E)} \approx \overline{N\sigma_{\Delta x}(E)}$$

where the exponential term has been expanded. Thus, for thin samples the average partial cross section is just equal to the average partial yield divided by the sample thickness and no information is required for the total or other partial cross sections. For the case of thick samples the data must be corrected for multiple and self-protection effects in order to determine partial cross sections. Thus, in order to convert thick-sample yields to partial cross sections, a knowledge of the total cross section is necessary. If the measurements are carried out in the region where the total and partial cross sections are not rapidly varying over the energy range spanned by the incident neutron beam and subsequently scattered neutrons, these corrections are quite straightforward. The methods developed for infinite slab geometry by Case et al. (1953) and for disk geometry by Schmitt (1960) are quite useful here. On the other hand, if the measurements are carried out in the unresolved region where the cross sections do fluctuate rapidly over the energy span of the incident and scattered neutrons, statistical methods based on the distributions of neutron widths and level spacings must be applied to the data to extract the partial cross sections. Methods based on average resonance parameters have been reported for this unresolved region by Dresner (1962) and Bogart and Semler (1966). An alternate technique that appears promising is to construct a "ladder," or statistical sample, of resonances by selecting resonances at random from a statistical distribution of level spacings and widths. Then the yield data can be corrected for multiple and self-protection effects by applying discrete-resonance multiple-correction techniques to the ladder of resonances. Of course this ladder method is useful only if the experimental resolution is sufficiently large to average over many resonances; otherwise large fluctuations will be encountered because of the inherently poor statistics resulting from a few levels.

There is another distinct difference between partial cross section measurements in the resolved and in the average regions. For the resonance case the yield data resemble a resonance line shape with a peak at the resonance energy and with "wings" which rapidly decrease to a background value. An exact determination of the background is not necessary here since the partial area A_x, as defined by Eq. (18), can readily be determined as the integral of the yield curve above the background value of the yield evaluated at the wings of the resonance. For the average partial cross section time-of-flight measurements there is no built-in value of the background yield, and this background must be separately determined. This is most frequently determined by placing a resonance absorber or scatterer in the incident neutron beam to remove essentially all neutrons in the vicinity of the resonance. Typically samples of aluminum are placed in the beam to determine the background near 35 keV, sodium near 2.8 keV, manganese near 330 eV, cobalt near 132 eV, and tungsten near 18 eV.

II. NEUTRON SCATTERING MEASUREMENTS

A. General Considerations

Traditionally, neutron scattering measurements have been made in one of two basic ways. The first, or conventional method, Fig. 3a, consists in placing a sample of the material under study in a well-collimated beam of neutrons, and observing the fraction of this beam, the scattering yield, which is scattered by the sample. All the neutrons in the beam may lie within a narrow energy interval, having been diffracted or passed through a rotating monochromator; or alternatively, the source of neutrons may be pulsed by a chopper or pulsed accelerator, and the energy of the neutrons determined by a time-of-flight measurement. In either case, the energy of the neutron is determined *before* scattering, and the object of the experiment is to observe the fraction of neutrons scattered by the sample. An ideal detector would completely surround the sample (4π-detector) and have an efficiency independent of neutron energy, so that all scattered neutrons would be detected with the same efficiency whatever their angular distribution or energy loss after scattering. If used in a time-of-flight experiment, this ideal detector would also be both vanishingly thin and very fast in its operation.

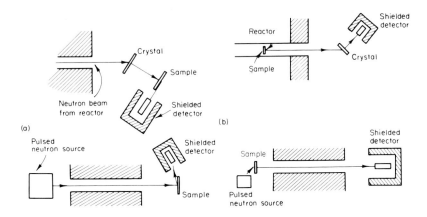

F<small>IG</small>. 3. Basic geometry of neutron scattering experiments in which (a) the energy is determined before scattering, and (b) the energy is determined after scattering ("bright-line"). The upper two figures illustrate a reactor beam with a crystal spectrometer, and the lower two figures, a pulsed neutron source and a time-of-flight spectrometer.

The second basic method of measuring neutron scattering yields has been called the "bright-line" method (Borst, 1953) and is illustrated in Fig. 3b. In this method the scattering sample is placed close to, or in the center of, the neutron source, and the scattered neutrons travel down a well-collimated drift tube to the neutron spectrometer where the energy of the neutrons *after* scattering is determined. The neutron energy may be determined by means of a crystal spectrometer (as in the case of Borst) or equally well by time-of-flight (as in the case of Sauter and Bowman, 1965).

The main advantage of the bright-line method over the conventional method is that the troublesome gamma rays due to capture in the sample or the troublesome fast neutrons from fission events are prevented by the neutron spectrometer from reaching the detector along with the scattered slow neutrons. Against these advantages must be placed disadvantages of a more complex experimental arrangement, of sample cooling problems when a reactor source is used (as in the case of Borst), and of a lack of exact knowledge of the incident neutron energy because of the variation of target nucleus recoil momentum with the angle of the scattered neutron.

Until recently, the great majority of resonance scattering measurements have been made using the arrangement in Fig. 3a. We shall, therefore, begin our study of actual scattering measurements by considering this method.

B. Experimental Measurements

1. EXPERIMENTS IN WHICH THE NEUTRON ENERGY IS DETERMINED BEFORE SCATTERING

Perhaps the earliest measurements of resonance elastic scattering made use of the Columbia Velocity Selector (Rainwater et al., 1947) and were made on the 4.9 eV resonance in gold (Tittman et al., 1950; Tittman and Sheer, 1951) and 5.2 and 16 eV resonances in silver (Sheer and Moore, 1955). Thick samples were used with BF_3 detectors, and only first-order scattering corrections were made.

Brockhouse (Brockhouse et al., 1951; Brockhouse, 1953) at Chalk River also studied scattering from thick samples using an annulus of BF_3 detectors in conjunction with the Chalk River crystal spectrometer (Hurst et al., 1950). The lowest energy resonances in cadmium and samarium were studied. Foote (1958) and Oleksa (1958) at Brookhaven used the technique of placing a thin sample at a glancing angle to the neutron beam in order to obtain useful counting rates without the difficulties associated with thick samples. This method was used to measure the scattering cross section of ^{235}U and ^{233}U up to 8 and 3 eV, respectively. The observed cross sections were, however, essentially flat and no resonance data were obtained.

Also in 1958, the Harwell team produced the first of a series of papers on resonance scattering (Rae et al., 1958; Evans et al., 1958; Waters et al., 1959; Fraser and Schwartz, 1962) in which resonance scattering measurements were made on levels in silver, tantalum, tungsten, and ^{239}Pu. For this work the scattering chamber designed by Collins (1954) was used in conjunction with the 15-MeV Harwell electron linac (and later the 25-MeV linac) time-of-flight system.

More recent papers include a fast chopper measurement of the ^{233}U scattering cross section up to 20 eV by Moore and Simpson (1962) at the MTR site at Idaho; low energy measurements on resonances in ^{115}In, ^{177}Hf, ^{147}Sm, ^{149}Sm, ^{157}Gd, ^{193}Ir, Re, and ^{235}U using a crystal spectrometer at Mol (Ceulemans and Poortmans, 1961; Ceulemans et al., 1962; Poortmans and Ceulemans, 1963; Poortmans and Ceulemans, 1966; Poortmans et al., 1967); scattering measurements up to a few hundred electron volts on resonances in Zn, Mo, Ru, Rb, Rh, and Ba, at the USSR pulsed reactor facility at Dubna (Pikelmer et al., 1963; Wang et al., 1963; Iliescu et al., 1965; San et al., 1966); and scattering measurements on three low energy resonances in ^{191}Ir at the BNL-Chalk River fast chopper (Bhat, 1966). The Idaho work gives a cross-section curve which is fitted well by the theoretical multilevel curve based on the parameters of Moore and Reich (1960), and the Mol and Chalk River experiments determined the spins of the levels studied. Scattering measurements on the resonances in Rh up to ~700 eV have been carried out at the Saclay electron linac (Ribon *et al.*, 1966), using essentially the technique developed at Harwell.

The BF_3 detectors at Harwell have now been replaced with ^6Li glass detectors and higher resolution measurements have been carried out on ^{107}Ag, ^{109}Ag, ^{159}Tb, ^{165}Ho, ^{197}Au, ^{232}Th, ^{238}U, ^{239}Pu, and ^{240}Pu (Asghar and Brooks, 1966; Asghar *et al.*, 1966a,b,c, 1967; Asghar, 1967). At Rensselaer a new technique has been developed (King *et al.*, 1967a,b; King and Block, 1968; King, 1969) for measuring scattering in the presence of intense capture and fission; measurements have been reported for Rh, Cd, W, and ^{239}Pu.

Other laboratories where neutron resonance scattering studies have been undertaken, but where no published results are yet available, include Oak Ridge (Harvey, 1959), Chalk River (Singh, 1963), and Rensselaer (Simpson *et al.*, 1966).

In all neutron resonance measurements the primary limitation on the energy range, and therefore on the number of levels studied, is one of resolution. This is especially true in scattering measurements because as the neutron energy increases, the ratio Γ_n/Γ increases; also the increased energy loss on collision leads to less severe multiple interaction problems. Thus apart from the deterioration in resolution, experimental conditions become rather easier with increasing neutron energy. Since crystal spectrometers "run out" of resolution at a few tens of electron volts, all the measurements involving the study of a number of resonances have been carried out with time-of-flight velocity selectors, either choppers or pulsed accelerators, with the pulsed accelerators providing the highest fluxes together with good neutron energy resolution. With any time-of-flight system, resolution can always be improved at the expense of intensity by increasing the length of the flight path, so in a sense, a logical way to improve the energy range studied is to use even more powerful pulsed sources. There is, however, also the possibility of improving resolution by improving the neutron detectors used. This improvement may be one of efficiency, thus enabling the flight path length to be increased, or it may be due to a reduction in the size or time jitter of the detector, thus giving a direct improvement in resolution. Thus it is perhaps more significant to categorize these scattering experiments by the type of detector employed rather than the particular spectrometer.

The BF_3 proportional counter was one of the first detectors to be used in neutron scattering measurements. This detector has the desirable

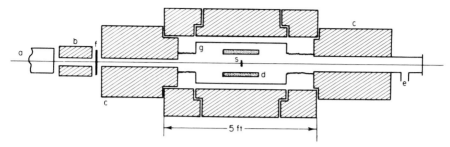

FIG. 4. General arrangement of scattering apparatus. a, evacuated flight path; b, final collimation; c, boric oxide shielding; d, active volume of neutron detector; e, diffusion pump; f, boron or cadmium filter; g, vacuum chamber; s, scattering sample. (Collins, 1954.)

characteristics of good efficiency at low neutron energies and a very low response to gamma rays. This detector has, however, the disadvantage of large size and poor timing resolution. Since many of the published measurements at the Harwell laboratory have been carried out using this detector, we shall consider these experiments in some detail.

The neutron source was in each case, an electron linac which produced a pulsed, "white" spectrum of neutrons. These traversed a collimated flight path of length approximately 12 m and were scattered by a circular thin foil of the material under study into an annulus of BF_3 counters, the output from which was fed to a multichannel time analyzer. The number of neutrons scattered by the foil under study was compared with that scattered by standard lead foils which had a known constant scattering cross section. In this way the measurement of absolute neutron fluxes was avoided.

The detector (Collins, 1954; Rae et al., 1958) consisted of eighteen boron trifluoride counters containing approximately 45 g of ^{10}B. These had the advantage of giving nearly 100% discrimination between scattered neutrons and capture gamma rays. These counters had a time jitter of ~2 μsec, so that at 12 m the best resolution was ~160 nsec/m. They were arranged to form an annular detecting volume effectively 38 cm long, 30 cm outside diameter and 13 cm inside diameter. The detector assembly was mounted in an aluminum chamber which, together with the neutron collimators, was evacuated by a diffusion pump. The axis of the detector was accurately aligned with the axis of the collimated neutron beam, and the samples, in the form of flat foils normal to the neutron beam, were placed at the center of the detector. The complete assembly (Fig. 4) was heavily shielded with 1.5 tons of boric oxide. The background, measured with no sample in the chamber, was closely proportional to the total machine output during a run and corresponded to the detection of less than 0.5% of the total flux from the collimated beam incident on the scattering sample.

The overall detector efficiency for an isotropic point source of 1 eV neutrons at the center of the detector was calculated to be approximately $12\frac{1}{2}\%$ and for neutron energies greater than 1 eV the efficiency varied approximately as $E^{-1/2}$. A neutron incident on the sample with energy E will enter the detector with an energy $E' < E$ after scattering. The consequent increase in detector efficiency, after averaging over all angles of scattering, is just $(A + 1)/A$ per collision, where A is the mass number of the scattering nucleus. This effect is sufficiently small to be neglected when comparing the scattering from heavy targets with that from lead.

The counts recorded in the time-of-flight channels were normalized to the neutron output of the acclerator by comparison with the counts recorded from beam monitors. Efficient monitoring of this neutron flux was most important, particularly since many runs took several days to complete. Three independent monitoring circuits were used, and the overall stability of the electronics was checked by observing the constancy from run to run of the ratios of the counts observed in each of the three monitoring systems.

Typical portions of scattering counts from single foil samples of lead and silver are shown in Fig. 5. From the known scattering cross section of lead, the silver scattering counts can be converted to scattering yield curves. Quantitative analysis was based on the theoretical yield of a very thin sample in which self-absorption is negligible, viz. Eq. (20) of Section I.

$$\sigma_0 \Gamma_n = 2A_n/\pi N \qquad (24a)$$

In practice, however, the samples used were seldom sufficiently thin for this equation to hold although the thinnest possible samples were used. The analysis was therefore based on the limiting relationship

$$\sigma_0 \Gamma_n = (2/\pi) \underset{N \to 0}{\mathcal{L}} (A_n/N) \qquad (25)$$

For samples of finite thickness, the partial scattering area is

$$A_n = \int_{E_1}^{E_2} dE \int_{E_3}^{E_4} R(E - E') Y_n(E') \, dE' \qquad (26)$$

where $R(E - E')$ is the resolution function, and $Y_n(E')$ is the yield defined in Section I. For the case of resonances in which the level spacing is much greater than the width of the resolution function, it can readily be shown

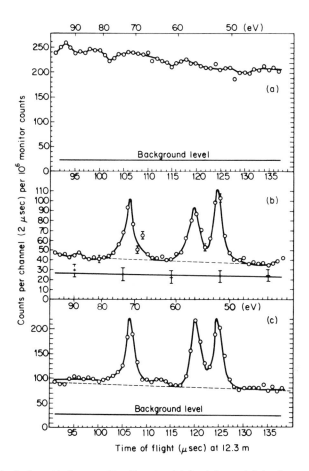

FIG. 5. Typical scattering results. Counts obtained from (a) lead sample of 7.815 $\times 10^{22}$ atoms/cm² (b) natural silver sample of 8.02 $\times 10^{22}$ atoms/cm²; (c) natural silver sample of 2.99 $\times 10^{20}$ atoms/cm². The solid lines sketched through the experimental points have no quantitative significance. The statistical S.D. on the experimental points in (a) and (c) is approximately equal to the diameter of the circles as plotted. (Rae et al., 1958.)

that the resolution function can be integrated out to unity, and then A_n reduces to

$$A_n = \int_{E_1}^{E_2} Y_n(E')\, dE' \tag{27}$$

where E_1 and E_2 are chosen to include all the experimentally observed scattering yield (and hence approximate the theoretical limits $-\infty$ and

$+\infty$). We can separate the primary and multiple contributions to the partial areas, and A_n reduces to

$$A_n = \int_{E_1}^{E_2} [Y_{np}(E') + Y_{nm}(E')]\, dE' = \int_{E_1}^{E_2} Y_{np}(E')\, dE' + \int_{E_1}^{E_2} Y_{nm}(E')\, dE'$$

(28)

where $Y_{np}(E')$ and $Y_{nm}(E')$ are the primary and multiple yields defined in Section I. In principle, A_n can then be determined directly by analytically determining the primary partial area and by Monte Carlo, or other, techniques determining the multiple partial area. It is at this point, however, that the nature of the scattering problem greatly simplifies the analysis. As pointed out in Section I, the effect of multiple interactions is to absorb neutrons which are scattered in the primary collision. Thus the multiple yield $Y_{nm}(E')$ is a negative number expressing the self-screening of the sample for scattered neutrons. Also, as emphasized in Section I, resonance scattering measurements are usually carried out when $\Gamma_n < \Gamma_\gamma$, therefore the multiple yield $Y_{nm}(E')$ can be analytically described by a rapidly converging series of second, third, etc., interactions. In many cases the straightforward second interaction, or self-screening of the sample for primary scattered neutrons, is adequate to determine $Y_{nm}(E')$, and hence $\int_{E_1}^{E_2} Y_{nm}(E')\, dE'$.

If the primary scattering area is factored out, A_n reduces to

$$A_n = \int_{E_1}^{E_2} Y_{np}(E')\, dE' \left[1 + \frac{\int_{E_1}^{E_2} Y_{nm}(E')\, dE'}{\int_{E_1}^{E_2} Y_{np}(E')\, dE'} \right] = \phi(N) \int_{E_1}^{E_2} Y_{np}(E')\, dE'$$

(29)

where $\phi(N)$ is the self-absorption correction factor which, for a given resonance, depends only on the sample thickness N. Substituting expression (17) of Section I, for the primary yield the scattering area in (29) becomes

$$A_n = \phi(N) \int_{E_1}^{E_2} [1 - \exp(-N\sigma_{\Delta t}(E'))][\sigma_{\Delta n}(E')/\sigma_{\Delta t}(E')]\, dE' \quad (30)$$

This was the relation used by the Harwell group in analyzing their data. Values of A_n/N were calculated by both first order analytic and Monte Carlo methods for the 5.2 eV level in silver in the range $0 < N\sigma_0 < 2$ (Rae et al., 1958; Morton, 1957) with good agreement between the two methods, and the function was shown to converge smoothly to its limit for $N \to 0$. On the basis of these calculations $\mathcal{L}_{N \to 0}(A_n/N)$ was determined graphically for each resonance, and Fig. 6 shows these extrapolations to zero thickness for four levels. (Note that the resonance areas have been

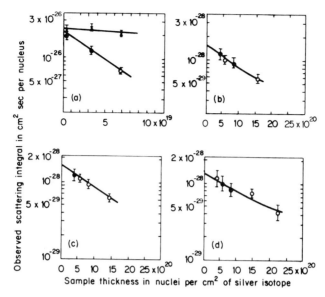

FIG. 6. Typical extrapolations, to zero thickness, of the resonance scattering integral. The semilog plot is used for convenience in displaying results. The points, shown as crosses, in diagram (a) show the result of applying the corrections, calculated by the Monte Carlo method, for finite sample thickness [(a) 5.2 eV; (b) 51.8 eV; (c) 56 eV; (d) 71.5 eV] (Rae *et al.*, 1958).

measured on the observed time-of-flight curves, and so are expressed in the units square centimeter seconds/nucleus.) Values of $\sigma_0 \Gamma_n$ were obtained by this method for ten levels in silver.

The method of measurement used in the experiment described above on silver was used essentially unchanged in the work on tantalum and tungsten (Evans *et al.*, 1958; Waters *et al.*, 1959); Rh (Ribon *et al.*, 1966), Xe, ^{155}Gd, and ^{157}Gd (Ribon *et al.*, 1967a), and in the work on Ir carried out at the BNL–AECL fast chopper installation at Chalk River (Bhat, 1966). For the Ir data, the Monte Carlo calculation of Friedes (1964) was used to treat the scattering areas.

The scattering measurements upon fissile nuclei have to contend with a much more difficult experimental situation. In the first place, the widths of resonances in the fissile nuclei are larger (due to competition from capture and fission), and hence Γ_n/Γ for the low energy resonances is quite small. Since the energy loss after a primary collision is very small for the heavy fissile nuclei, the probability of escape of a neutron after a resonance scattering in a foil of moderate thickness is quite small. For example, in the ^{239}Pu scattering measurements at Harwell (Fraser and Schwartz, 1962) only one thin sample of PuO_2 baked on an aluminum foil was used. The value of $N\sigma_{\Delta 0}$ (where N is the number of atoms per unit area, and $\sigma_{\Delta 0}$ the Doppler broadened peak cross section) varied from 0.05 to 0.11 for the 8 resonances measured, and $\mathfrak{L}_{N\to 0}(A_n/N)$ was calculated for each resonance from the observed scattering areas for the single sample. The measurement, using the same counter system as in the earlier work on Ag, Ta, and W, with a flight path of 15 m, was only made possible with such a thin sample because of the introduction of the much more powerful 25-MeV electron linac with its boosted neutron source (see Chapter I) which provided an increase in neutron flux of at least a factor of 100 over that used in the earlier experiments. Even so, the extremely low fraction of the neutron beam scattered in the peaks of the resonances ($\sim$0.1%) made the observed resolution-broadened resonance yields very difficult to determine against the chamber background ($\sim$1%).

A second major difficulty in the study of a fissile nucleus is the large resonance yield of fast neutrons, on average $\sim$100 fast neutrons per scattered slow neutron in the energy range studied. Fortunately, the $1/v$ BF_3 counter system used in this early work is relatively insensitive to the fast neutrons. Even so, with such a large fast neutron yield, the observed counting rate due to the fast neutrons was of the same order as that due to the resonance neutrons. Consequently, the true resonance scattering was determined by making measurements of the observed counting rate as a function of time-of-flight with and without a sleeve of B_4C placed between the scattering sample and the detector, a method also employed by Moore and Simpson (1962) in their chopper measurement on ^{233}U. The observed yields with and without the absorbing sleeve in position are shown in Figs. 7a and 7b. The resonance areas were found for both curves by comparison with a standard lead scatterer, and the observed scattering areas determined by subtraction, due allowance being made for the transmission of the filter. These observed areas were then corrected for self-screening and self-absorption (the correction varying from 4 to 12%), and the resulting values of $\sigma_0\Gamma_n$ for the eight levels analyzed were used in conjunction with the total and fission cross-section measurements of Bollinger et al. (1958) to deduce values of the spin weighting factor g.

Using essentially the same technique as Fraser and Schwartz (1962), a ^{239}Pu scattering measurement has been carried out at Rensselaer (Simpson et al., 1966). Their results are shown in Figs. 8a and 8b for the energy range from 2 to 55 eV, and they generally confirm the spin assignments of Sauter and Bowman, (1965), for the resonances below 30 eV. Poortmans et al., (1967) used an annulus of ^3He proportional counters with the Mol crystal spectrometer to measure resonance scattering from the 8 and 12 eV resonances in ^{235}U. By taking advantage of the differences in the pulse

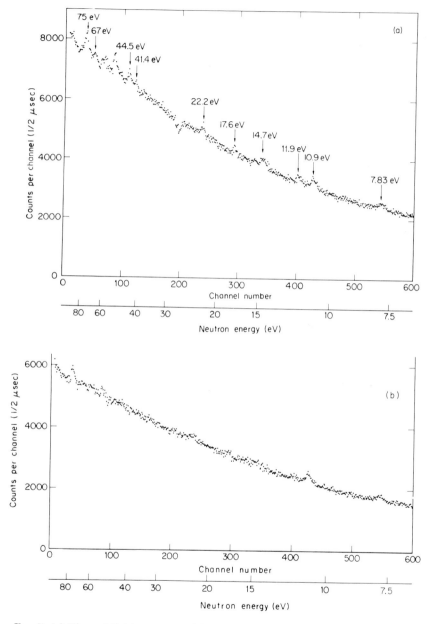

Fig. 7. (a) Time-of-flight spectrum with ^{239}Pu sample of 4.66×10^{19} atoms/cm^2 with a Cd filter of 0.7 g/cm^2. (b) Time-of-flight spectrum with ^{239}Pu sample inside a tube of B$_4$C acting as an absorber for the resonant scattered neutrons. A Cd filter of 0.7 g/cm^2 was placed in beam. (Fraser and Schwartz, 1962.)

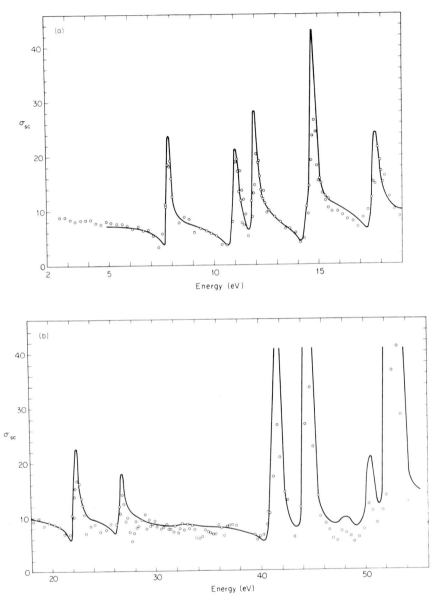

FIG. 8. Neutron scattering cross section of ^{239}Pu from 2 to 55 eV. (Simpson *et al.*, 1966.)

height in the ³He counter between a resonance energy neutron and a fission energy neutron, they were able to discriminate against fission neutrons by a factor of about 1000.

The BF₃ and ³He proportional counters are very suitable for scattering measurements because of their good discrimination against the detection of gamma rays and fission neutrons. They have, however, a low efficiency for detecting neutrons; they are physically large; and they have poor time resolution. The BF₃ annulus at Harwell, for example, had an efficiency of only ~0.4% for neutrons of 1 keV, was 0.4 m long, and had a time jitter of ~2 μsec. Since high resolution measurements can only be obtained at time-of-flight spectrometers by either increasing the flight path or reducing the time jitter, these detectors are generally unsatisfactory.

The group at Dubna (Pikelner *et al.*, 1963) have developed a rather unusual detector for the IBR pulsed reactor (Blokhim *et al.*, 1961). This reactor develops a relatively long pulse of 36 μsec duration, so they developed a neutron detector that has a constant efficiency of about 10% below 300 eV, but which has a timing uncertainty of ~15 μsec. The detector is massive in size (0.5 m long by ~0.4 m diameter) and consists of a lami-

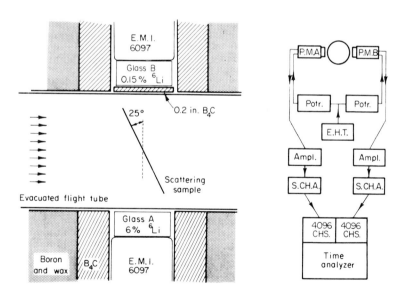

Fig. 9. Schematic of the Harwell lithium-glass scattering detector. Glass A contains 6% ⁶Li and detects scattered neutrons and capture gamma rays, while glass B contains only 0.15% ⁶Li and is shielded by B₄C and hence primarily detects capture gamma rays. Signals from both detectors are simultaneously recorded in 4096-channel time-of-flight analyzers (Asghar and Brooks, 1966).

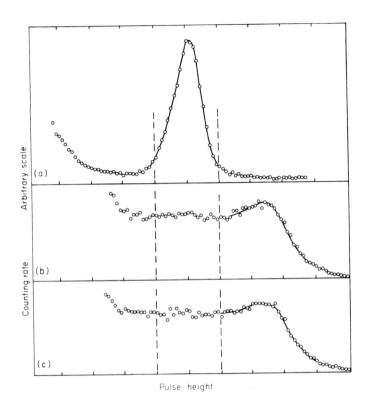

FIG. 10. Pulse height distributions in lithium-loaded glass. Glass A contains 6% ^{6}Li and glass B contains 0.15% ^{6}Li; (a) thermal neutrons on glass A (the peak is due to ^{6}Li(n,α) ^{3}H events); (b) and (c) ThC″ gamma rays on glasses A and B, respectively. (Asghar and Brooks, 1966.)

nated annulus of Plexiglass and ZnS(Ag) viewed by photomultipliers. They placed this detector at the end of a 500 m flight path and have reported scattering measurements on Rh up to ~300 eV with a resolution of ~80 nsec/m. This detector, however, is of no use to the higher resolution accelerator spectrometers which use short-burst widths and modest flight paths to achieve resolution.

In order to increase the energy range of the scattering measurements at Harwell, Asghar and Brooks (1966) have replaced the BF$_3$ counters by ^{6}Li-loaded glass scintillators. The ^{6}Li glass detector (Firk et al., 1961) is both fast and efficient; it has a timing uncertainty for detecting the ^{6}Li (n,α) ^{3}H reaction of $\lesssim 5$ nsec and has a neutron efficiency of 7%/cm thickness at 1 keV. This detector however, is sensitive to gamma rays, and scattering experiments must be carefully designed either to measure or to

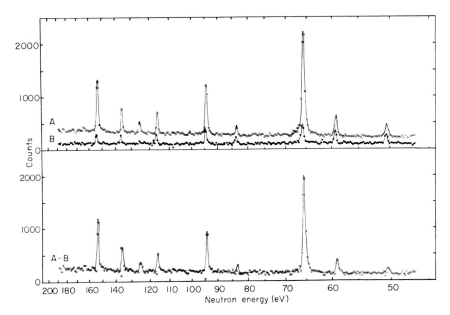

FIG. 11. The number of counts per timing channel plotted against neutron energy for a ^{169}Tm sample. Curve A is the data from the ^{6}Li enriched glass detector ($n = 3.21 \times 10^{-4}$ atoms/b; resolution = 20 r sec/m); curve B is the capture background determined from the low-enrichment glass detector; and curve (A–B) is the difference between the two and represents the net scattering (Asghar and Brooks, 1966).

reject the gamma-ray background. Asghar and Brooks, using the Harwell 25-MeV electron linac pulsed neutron source, initially set up two lithium-loaded glass detectors arranged symmetrically with respect to the scattering sample, (see Fig. 9) at a distance of 50 m from the neutron source. The two detectors ($1\frac{3}{4}$ in. in diameter by $\frac{3}{4}$ in. thick) were identical except that one glass was loaded with lithium enriched to 96% in ^{6}Li while the other contained naturally abundant lithium. Both scintillator detectors had approximately the same efficiency for detecting capture gamma rays, while the one containing the enriched lithium had a neutron detection efficiency 40 times that of the other at high neutron energy, its efficiency at 1 keV being approximately 10% for neutrons passing through it. Scattering measurements were made by recording simultaneously on a digital magnetic tape analyzer the time-of-flight yield of both detectors, each equipped with a single channel pulse height analyzer to select only pulses which occurred in the region of the ^{6}Li (n,α) ^{3}H peak. The pulse height response of this detector to neutrons and gamma rays is plotted in Fig. 10. A typical pair of time-of-flight yield curves for thulium is shown in Fig. 11. The resonance

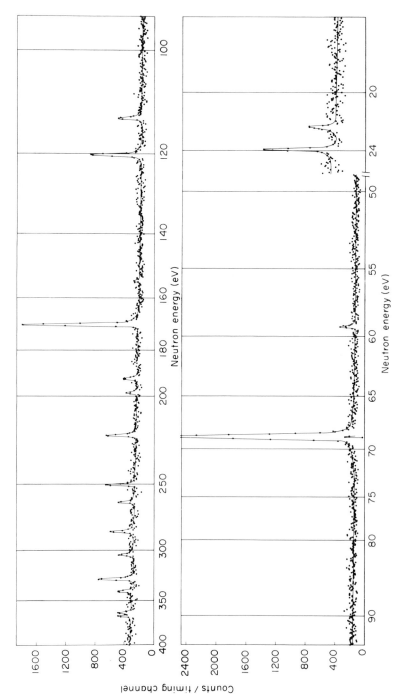

Fig. 12. Neutron scattering counts per timing channel for a 1.317×10^{-4} atom/barn thick sample of ^{232}Th. The Harwell five-glass scattering detector was used with a resolution of 5 nsec/m (Asghar et al., 1966a).

scattering areas are obtained by taking differences between the areas obtained from the two curves, flux normalization being, as usual, to the lead cross section. Asghar and Brooks further improved the efficiency of their scattering detector by incorporating five scintillating glasses (three enriched in ^{6}Li and two with naturally abundant Li) into one well-shielded assembly at the Harwell linac's 50 m flight station. A typical set of data for scattering from a thin sample of Th is shown in Fig. 12. These data show a considerable improvement in both signal-to-background ratio and neutron energy resolution compared to the older scattering experiments. With this apparatus resonance scattering measurements can be carried out well into the kilovolt energy range. These Th data have been combined with transmission and capture measurements to determine the resonance parameters and resonance integral of Th (Asghar et al., 1966a). Scattering measurements, in conjunction with capture measurements, have been reported on Ag, Au, Tb, and Ho (Asghar et al., 1966b) and on ^{240}Pu (Asghar et al., 1967) using this detector. Asghar (1967) recently incorporated a stilbene crystal in the 5-glass detector to measure fast neutrons produced in fission. With this modification he measured scattering from ^{239}Pu and was able adequately to subtract the gamma-ray and fission neutron background from the ^{6}Li glass counts. In this measurement he was able to determine the spins of 47 resonances in ^{239}Pu.

Now the overall efficiency of the lithium glass detector at 1 keV including the solid angle subtended by the detector at the sample is only $\sim$0.5%, which is similar to that of the BF$_3$ annulus (Rae et al., 1958). The much smaller size of the glass detectors, however, means that the uncertainty in flight path length due to detector geometry has been reduced from $\sim$20 to $\sim$7 cm, thus giving an improvement of about a factor of $2\frac{1}{2}$ in resolution at low energies where the length uncertainty predominates. At high energies, where the time uncertainty is dominant, the improvement is even greater, the time jitter being reduced from $\sim$2 μsec to $\leq$5 nsec.

The time uncertainty for the glass thus reduces to the neutron pulse length, this being $\sim$0.1 μsec for the Harwell accelerator. The effective improvement in resolution at high energies is thus a factor of 20. Figure 13 shows the energy resolution as a function of neutron energy for both the BF$_3$ annular and the 5-glass-assembly detector over the energy range 1 eV to 1 keV, each at 50 m from the pulsed source with an accelerator pulse duration of 0.25 μsec.

Another technique to reduce the gamma-ray sensitivity in these neutron scintillation detectors is to employ pulse shape discrimination. Coceva (1963) achieved pulse shape discrimination in lithium-loaded glass detectors, and his data suggest that a factor of 10 discrimination against gamma

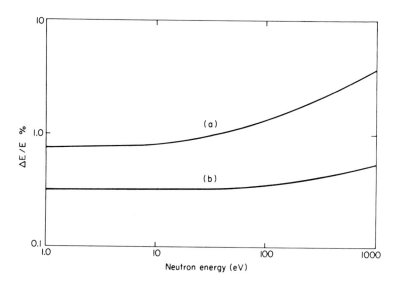

FIG. 13. A comparison of percent neutron energy resolution between (a) the Harwell BF₃ annular scattering detector (Collins, 1954) and (b) the assembly of five lithium-loaded glass scintillators (Asghar *et al.*, 1966a). The comparison is made with both detectors located at 50 m from the neutron source and with an accelerator pulse duration of 0.25 μsec.

radiation can be obtained at the expense of about a factor of 2 in neutron detection efficiency. Brooks (1961) has shown that pulse shape discrimination against gamma-ray events can be achieved with boron-loaded liquids, thus offering a high detector efficiency with a moderate time resolution of ∼0.4 μsec. At that time, however, it was extremely difficult to use this technique because of the very small pulses produced by the ^{10}B (n,α) process. Jackson and Thomas (1965) have developed a boron-loaded scintillator utilizing pulse-shape discrimination which promises to be extremely useful for scattering measurements. By suitable choice of scintillating solution and electronics, they have reduced the gamma-ray sensitivity by a factor of 500 at the expense of only a 5% loss in neutron efficiency. A detector of this type has been used by the Geel group (Cao *et al.*, 1968) to measure scattering from ^{240}Pu resonances below 300 eV, and by the Saclay group (Asghar, private communication) to measure scattering in Rh.

At Rensselaer King *et al.* (1967b) have combined the ^{6}Li glass scintillator with a large liquid scintillator to suppress the background in the ^{6}Li glass detector due to gamma rays and, for fissile nuclei, fast neutrons. Their apparatus, shown in Fig. 14, consists of a 1.25 m-diameter liquid scintillator

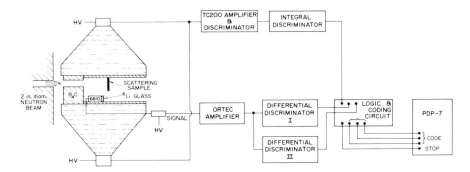

Fig. 14. Neutron scattering detector utilizing a lithium glass neutron detector surrounded by a large-diameter (1.25 m) liquid-scintillator capture detector at a 25 m flight station. The capture detector is operated in anticoincidence with the neutron detector to suppress the capture background in the glass (King *et al.*, 1967a).

detector (Block *et al.*, 1961) through which passes a collimated neutron beam from the Rensselaer linac neutron target. The scattering sample is placed in the beam at the center of the large scintillator, and a ⁶Li glass neutron detector (2 in. diameter by 0.5 in. thick) is placed at approximately 120° to the scattered beam. The ⁶Li glass detector is operated in anticoincidence with the liquid scintillator. Since the liquid scintillator has an efficiency of $\gtrsim 90\%$ for detecting neutron capture or fission, the anticoincidence circuit rejects $\gtrsim 90\%$ of the capture or fission background pulses in the glass. This technique is thus useful in resonances where the scattering is very small compared to capture (or fission). Data taken with this technique for a scattering sample of rhodium are plotted in Fig. 15. Four fields of time-of-flight data were simultaneously recorded during the experiment. The upper two plots were taken with the ⁶Li glass detector's single channel analyzer set on the neutron peak, and the lower two plots with the window set in the valley below the neutron peak (see Fig. 10). The ⁶Li glass pulses were recorded both in coincidence and in anticoincidence with the large liquid scintillator. The upper anticoincidence curve represents mostly neutron scattering counts with a small addition of capture background counts. The second curve, in coincidence with the large scintillator represents those capture events in Rh which lead to a pulse-height in the glass detector that falls within the neutron window. If all the pulses from the glass (which fall within the neutron window) were recorded, the sum of the upper two curves would result; this sum is what is recorded in the experiment of Asghar and Brooks (1966). Although the liquid scintillator rejects $\sim 90\%$ of the capture events, it is still necessary to correct for this remaining $\sim 10\%$ background. To do this the lower two curves, which cor-

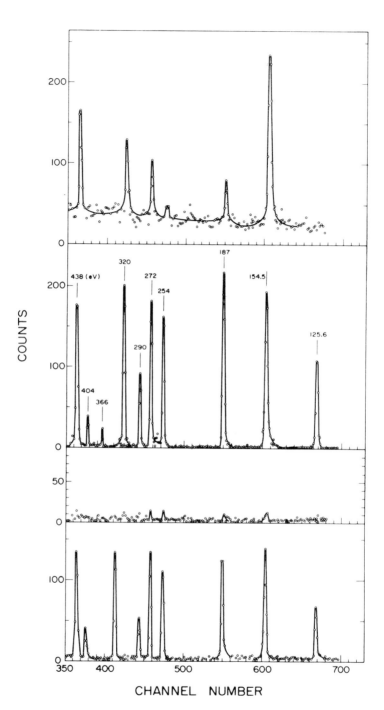

Fig. 15

respond mostly to capture events in the glass, can be combined with the upper two curves to determine the background. Recently, this technique was improved by placing a second hole through the large scintillator detector, transverse to the neutron beam hole. A pair of 5-in. diameter by 0.5-in. thick ^{6}Li glass detectors were placed in the transverse hole at $\sim$90° to the scattering sample, and a larger diameter neutron beam was allowed to fall on the sample. The counting rate was increased by over one order of magnitude, but the rejection of capture and fission background in the glass detector was still about 90%. Measurements have been reported on Rh, Cd, and ^{239}Pu with the improved system (King and Block, 1968; King, 1969).

2. Experiments in which the Neutron Energy Is Determined after Scattering (Bright-Line)

It has been pointed out by Bollinger that the main usefulness of the bright-line technique is likely to be in the study of the fissile nuclides, where the background due to the fast fission neutrons, together with capture gamma rays, could be completely eliminated either by the crystal spectrometer or by the time-of-flight system where these unwanted radiations would reach the detector long before the slow neutrons under study.

Early published measurements in which this method has been used in the resonance region are those of Borst (1953) and Wood (1956). These authors used the Brookhaven graphite reactor as a neutron source, the scattering sample being placed in the center of an evacuated pipe passing through the reactor. Neutrons scattered in the foil traveled a distance of 20 ft to the diffraction spectrometer, passing through a 35-hole collimator mounted in the reactor shield which prevented neutrons originating within the reactor or scattered within the region of appreciable neutron density from reaching the crystal. Apart from the scattering sample itself, the crystal could see only a neutron trap mounted within the reactor shield at the far end of the pipe. This trap consisted of a thin steel shell filled with boron carbide, and no neutron contribution could be observed from this source. In the earlier work Borst used a sodium chloride crystal, while in

---

Fig. 15. Lithium-loaded glass counts versus neutron time-of-flight for a 0.010″ thick Rh scattering sample. The upper two plots (a) and (b), were taken with the lithium glass detector's single channel analyzer set on the neutron peak, and the lower two plots, (c) and (d), with the single-channel analyzer set in the valley below the neutron peak. Lithium glass pulses were recorded both in coincidence (b) and (d) and in anticoincidence (a) and (c) with the large liquid scintillator (King, 1969).

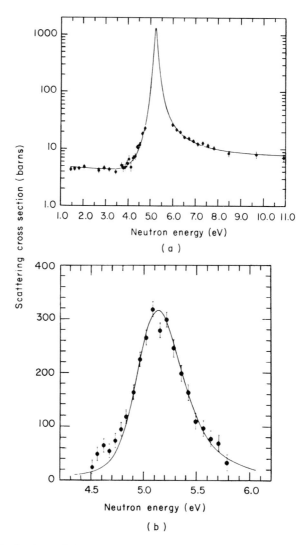

FIG. 16. (a) The theoretical scattering cross section of the 5.2 eV silver resonance. The points shown have been corrected for the effect of sample thickness, and these corrected points were used for the scattering wing analysis. (b) The comparison of the observed silver scattering cross section with the computed curve for $\sigma_0 = 15,500$ b and $g = \frac{3}{4}$. The spin assignment ($J = 1$) is verified by this curve since the other spin ($J = 0$) would give an expected cross section approximately three times as large as that shown. The counting statistics were as indicated (Wood, 1956).

Wood's work a beryllium crystal was used, giving an energy resolution width $\Delta E = 0.0273\ E^{3/2}$ eV. Scattering from the sample under study was compared with that from standard samples of graphite and lead, thus eliminating absolute neutron counting. Scattering measurements were made on several samples of indium, silver, gold, and tantalum, the observed yield curves being converted to cross-section curves after the application of corrections for self-absorption and neutron energy shift. Theoretical cross-section curves were then fitted to the observed points using a least-squares method, Doppler and resolution broadening corrections being made in the region of the resonance peak. The best fits to the data on the 5.2 eV level in silver are shown in Figs. 16a and 16b. The parameters for this level giving the best fit to the scattering data, and also to the transmission measurements of Wood (1956), are $g = \frac{3}{4}$, $\Gamma_n = 13.4 \pm 0.6$ meV, $\Gamma_\gamma = 136 \pm 6$ meV. These figures can be compared with the results of the Harwell analysis of this level (Rae *et al.*, 1958): $g = \frac{3}{4}$, $\Gamma_n = 12.7 \pm 0.23$ meV, $\Gamma_\gamma = 136 \pm 8$ meV. The agreement, despite the very different techniques used, is excellent.

It should be noted that all of the Brookhaven bright-line work was carried out by fitting theoretical curves to the corrected experimental yield points. The fact that curve fitting was used, together with relatively thick samples for off-resonance study, meant that all possible information was extracted from the experimental data, including values of the potential scattering cross section for both silver isotopes.

The limitation of the Brookhaven bright-line measurements was one of resolution. Their study of silver, for example, was limited to the first level at 5.2 eV. They failed completely to resolve the scattering peak due to the next level at 16 eV. It should be noted that the energy spread in the scattered neutrons due to variations in the angle of scattering introduces a limit to the energy resolution obtainable with this type of experiment. In the Brookhaven experiment where neutrons were incident on the sample from all angles, this limit is $\sim 0.02\ E$ for a medium weight nucleus. With very powerful pulsed accelerator sources, however, the sample could be placed at a distance from the source, thus limiting the angular spread in the scattering, and so reducing the limiting energy spread. Bright-line experiments of this latter type have in the past few years been reported at the Rensselaer, Harwell, and Kurchatov linear accelerator laboratories, but to date no results have been published. A bright-line experiment has been developed at the Livermore Linac Laboratory (Sauter and Bowman, 1965) which has demonstrated the high neutron intensity, low neutron background, and low sensitivity to radioactivity in the scattering sample that

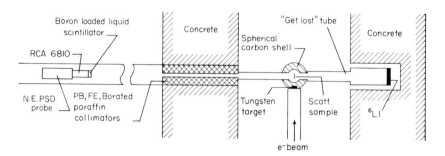

Fig. 17. The Livermore "bright-line" scattering experimental arrangement. Photo-neutrons produced in a tungsten target are moderated in a hollow spherical shell of graphite. The scattering sample is placed inside an evacuated tube passing through the carbon shell, and scattered neutrons are detected at the end of an evacuated flight path with a boron-loaded liquid-scintillator detector (Sauter and Bowman, 1967).

is inherently present in the bright-line technique. In this experiment (Fig. 17), the photoneutrons produced in a tungsten target are moderated in a hollow spherical shell of graphite. The scattering sample is placed inside the graphite shell, and this "4π"-moderator geometry produces an extremely high intensity of neutrons incident upon the scatterer. An evacuated flight tube passes through the center of the graphite shell and encloses the scattering sample. The neutron dector is placed at one end of this flight tube; a ^{6}Li-neutron absorber to reduce the number of backscattered neutrons reaching the detector is located at the other end.

The Livermore group has measured scattering from ^{239}Pu (Sauter and Bowman, 1965) and have assigned spins to 15 resonances from 7.9 to 75.2 eV. Their spin assignments are in agreement with the subsequent results of Asghar (1967) and Simpson et al. (1966), but they do disagree with three of the earlier spin assignments of Fraser and Schwartz (1962). Recently, this group (Sauter and Bowman, 1967) has measured the scattering cross section of ^{241}Pu (13 yr half-life). Their preliminary results, plotted in Fig. 18 for the energy range from 2 to 12 eV strikingly illustrates the power of this technique for measuring scattering from highly radioactive samples. These data have not been corrected for Doppler or resolution broadening and the effects of ^{242}Pu and ^{241}Pu and ^{241}Am contaminants show up, respectively, in the energy region below 3 eV and the peak at 5.4 eV. More recent measurements have also been reported for ^{233}U, ^{235}U, and ^{241}Pu (Sauter and Bowman, 1968). Their technique is, however, limited to rather low neutron energies by the intrinsic poor resolution of the bright-line method.

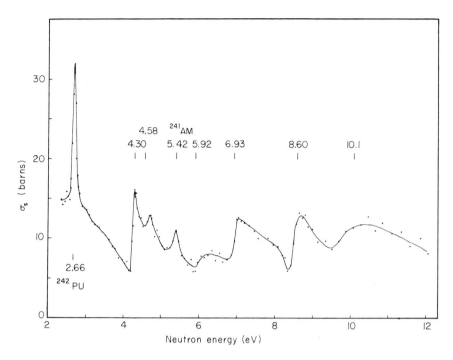

FIG. 18. Scattering cross section of 241Pu from 2 to 12 eV using the "bright-line" method. The data are not corrected for Doppler or resolution broadening, and the effects of 242Pu and 241Am contaminants show up in the energy region below 3 eV and the peak at 5.4 eV (Sauter and Bowman, 1967).

III. NEUTRON CAPTURE MEASUREMENTS

A. General Considerations

Neutron capture cross sections can be measured in a variety of ways, depending upon the nucleus, the energy range of the neutrons, and the degree of energy resolution required. Frequently, only the capture cross section averaged over a reactor energy spectrum is desired, and this can readily be obtained by irradiating a sample in a reactor and then measuring either the induced radioactivity or resultant stable nuclear masses in a mass spectrometer. Another technique is to measure the change in re-activity of a reactor when the sample is inserted (or oscillated) in the reactor. When a monochromatic source of neutrons is available, e.g., 24-keV neutrons from an Sb–Be source or 30-keV neutrons from a $Li(p,n)$

threshold reaction, the shell transmission technique can be employed to measure neutron capture at a few discrete energies. Another method is to measure the activity induced by monochromatic neutrons produced in a charged particle reaction. All of these techniques, however, generally suffer from poor resolution and are primarily used to measure average capture cross sections. These techniques have been reviewed by Cameron *et al.* (1963); and since these methods have not changed significantly, they will not be reviewed here.

For high resolution studies of the capture of neutrons into resonance levels, the only practical method having general application is the observation of the prompt gamma rays emitted when the compound nucleus de-excites to its ground state. This technique is highly suitable for use with time-of-flight spectrometers over a wide range of neutron energies up to (and in some cases beyond) the threshold for inelastic scattering. Gamma-ray detectors can be used here which have very good time resolution, and thin foil samples normally employed introduce negligible geometrical uncertainty into the flight path length. The two main problems are: (i) to find a detector with low neutron sensitivity whose efficiency for detecting gamma rays is independent of fluctuations in the spectrum and multiplicity of the prompt capture gamma-ray cascades, and (ii) to interpret the observed yield curves in terms of the resonance parameters or capture cross sections. Two types of detectors approach the ideal characteristics of low neutron sensitivity, good time resolution, and an efficiency independent of the capture gamma-ray spectrum and multiplicity. The first is the large liquid scintillation tank (Diven *et al.*, 1960; Gibbons *et al.*, 1961; Block *et al.*, 1961; Haddad *et al.*, 1964; Kompe, 1967). This detector achieves near independence of the capture spectrum and multiplicity by being so large that even the hardest gamma rays have less than a 50% probability of escape without detection from a sample placed at the center of an axial tube through the tank. The efficiency of the large scintillator detector therefore approaches 100% for typical capture gamma-ray cascades. The time resolution is good since liquid scintillators have a fast decay for the light produced, and hence the time resolution depends mainly on the path length of the light before reaching a phototube. This time is typically of the order of several nanoseconds. Low sensitivity to neutrons is achieved by loading with boron and biasing out the small pulses caused by capture of the moderated neutrons in the boron and also by lining the inside of the detector with a neutron absorber like ^{6}LiH.

The second type of detector to approach the ideal characteristics was originally developed at the Harwell Laboratory and is referred to as the "Moxon–Rae" detector (Moxon and Rae, 1961; Macklin *et al.*, 1963b).

The simplest version of this device consists of a sleeve of graphite of wall thickness about 1.25 in. thick surrounding the capture sample. Secondary electrons produced by gamma rays in the graphite sleeve are detected by a ring of phototubes each equipped with a sheet of plastic phosphor of thickness ∼0.020 in. This detector, like the thick-walled Geiger–Müller counter, has an efficiency almost directly proportional to the photon energy, up to some maximum energy determined by the thickness of the graphite, in this case ∼10 MeV. Since the efficiency of this detector is low (∼0.3%/MeV of gamma-ray energy) the overall efficiency for detection of a capture cascade is the sum of the efficiencies for the individual photons emitted and so depends only on the total energy released; that is, it depends only on the binding energy of the captured neutrons. The timing resolution of this detector is extremely good, depending essentially on the phototubes used, and Macklin et al. (1963b) report a resolution of 3 nsec for one of these detectors used in conjunction with a pulsed Van de Graaff neutron source at Oak Ridge. Sensitivity to neutrons is intrinsically low, provided their energy is not high enough to produce knock-on proton counts in the plastic phosphor. The onset of this process currently restricts the useful upper limit of this detector to around 100 keV when used with pulsed sources giving a continuous spectrum of neutrons.

Both of these "ideal" detectors will be discussed in more detail in the next section of this chapter. A significant fraction of the published data, however, has been obtained with the aid of detectors having far from ideal characteristics. Small organic scintillators were used in the early work of Meservey at Columbia (1952, 1954), and by Albert and Gaerttner (1954) and Gaerttner et al. (1954) at the Knolls Atomic Power Laboratory. Rae and Bowey (1957) at Harwell used sodium iodide crystals, and the Columbia synchrocyclotron group at Nevis (Rosen et al., 1960; Desjardins et al., 1960) used plastic phosphors, subtending a rather small solid angle at the sample to detect the capture-gamma rays. For the lead slowing down spectrometer in the USSR (Bergmann et al., 1955), both proportional counters and CaF$_2$ scintillators have been used to detect the gamma rays. The detectors used in all of these measurements were undoubtedly sensitive to fluctuations in the capture spectra; but, nevertheless, valuable data have been obtained from the work because, with few exceptions, the multiplicity and general form of the gamma-ray spectra from resonance capture in medium and heavy nuclides do not fluctuate strongly from resonance to resonance.

Apart from the choice of a detector, the other problem in the measurement of capture cross sections is the determination of capture yields. The yield is defined in Section I as the fraction of neutrons at a given energy, present in the neutron beam, which is either scattered or captured (or absorbed to undergo fission). This yield is experimentally determined as the ratio of the number of detected partial events to the product of the detector efficiency times the number of neutrons incident upon the sample. Thus

$$Y_x(E) = C_x(E)/\eta_x\phi(E) \tag{31}$$

where $Y_x(E)$ is the yield at incident neutron energy E; x represents either scattering, capture, or fission; $C_x(E)$ is the number of events (corrected for background) detected by the partial detector; η_x is the efficiency of the detector; and $\phi(E)$ is the number of neutrons of energy E that were incident upon the sample. The problem in determining capture yields from Eq. (31) is slightly different from that encountered in the scattering measurements. In the latter case, measurements of absolute efficiency η_x and absolute flux $\phi(E)$ were avoided by comparing the scattering from the nuclide under investigation with that from lead, or some other nucleus having a well-known energy-independent scattering cross section. The calibration of a capture experiment is not quite so simple since nature does not provide energy-independent capture cross sections. Under favorable circumstances absolute flux and efficiency measurements can be avoided by measurements of the saturation yield from low energy resonances. In this case if the resolution is good, if $\Gamma_\gamma \gg \Gamma_n$, and if the sample is thick enough to ensure the condition $N\sigma_0 \gg 1$, then near the peak of the resonance 100% of the incident neutron beam is captured and Y_x in Eq. (31) is equal to unity. Thus the counts near the peak, C_x, are equal to $\eta_x\phi$, and it only remains to determine the energy dependence of the neutron flux in order to be able to convert observed counting rate curves to capture yields. When time-of-flight measurements are made with Van de Graaff pulsed sources, however, the lowest practical neutron energy is ~5 keV and the saturated resonance method cannot be applied. In this case, the measured relative capture cross sections are normalized either to absolute spherical shell measurements or to some known capture or fission cross section. Beyond this point the analysis of the data and the extraction of parameters from resonance areas follows methods rather similar to those developed for the analysis of scattering yields.

B. Experimental Measurements with Prompt Gamma-Ray Detectors

1. Work with Spectrum-Sensitive Detectors

Among the first resonance capture cross-section measurements were those of Meservey (1954). This author used two stilbene scintillator detectors to record the capture radiation from 2×3 in. rectangular samples placed 6 m from the Columbia cyclotron pulsed neutron source. The whole detecting system was shielded with lead and lithium to reduce background and to avoid sensitivity to slow neutrons. Measurements were reported on cadmium, barium, strontium, and silver, one level each being studied in strontium and silver, three in cadmium, and three in barium. Values of $\sigma_{0\gamma}\Gamma$ or $\sigma_{0\gamma}\Gamma^2$ based on resonance area measurements were reported for each of the levels observed, and $\sigma_0\Gamma$ and Γ were deduced for the level at 0.177 eV in cadmium by a theoretical fit to the observed capture cross section. Apart from the data on the first levels in cadmium and silver, where a saturation count rate was determined, the other measured capture integrals were only very rough values based on assuming the detector efficiency for the barium and strontium radiation to be similar to that for cadmium.

Another early measurement was that described by Albert and Gaerttner (1954) and Gaerttner *et al.* (1954). These authors used two annular liquid-scintillator detectors 8-in. OD, 6-in. long, and with a 4-in. central hole, placed close together with the neutron beam passing through both, and with the capture samples inserted in the narrow gap between the detectors. This detecting system, suitably shielded, was placed 6 m from the pulsed neutron target of the General Electric betatron. Operation of a single detector gave a very high background counting rate, so the system was operated with the two annuli in coincidence. Under these conditions the detector efficiency was $\sim5\%$ and the background was $\sim3\%$ of the count rate observed with a thick samarium sample inserted in the beam. Such a coincidence detector has an average efficiency $\langle \epsilon \rangle$ for detecting capture gamma-ray cascades of the form

$$\langle \epsilon \rangle \approx \langle \eta^2 \nu(\nu - 1) \rangle$$

Since the efficiency η of each detector for a single photon decreases with increasing photon energy, and since increasing the mean photon energy corresponds to reducing the value of the gamma-ray multiplicity ν, it is clear that the overall efficiency of the coincidence detector is extremely sensitive to fluctuations in ν. A measurement of the ratio of coincidence to single counts for several resonances in tantalum indicated that $\langle \epsilon \rangle$ was constant to within the errors of the measurement, that is, to $\pm20\%$. On the basis of this measurement, it was assumed in analyzing the data that $\langle \epsilon \rangle$ was constant for any particular nuclide though this assumption introduced a large uncertainty into the results. The basis of the analysis was to assume that the samples were thin enough to ignore potential scattering and multiple scattering effects and write (assuming good resolution) the expression for the integrated capture yield as:

$$A_\gamma = \int_{\text{res}} Y_\gamma(E) \, dE \approx \int_{\text{res}} [1 - T(E/T_p)](\sigma_{\Delta\gamma}/\sigma_{\Delta t}) \, dE \approx A_t(\Gamma_\gamma/\Gamma) \quad (32)$$

where $A_t = \int_{\text{res}} [1 - T(E)/T_p] \, dE$ is the transmission area above the resonance. In the limit of thick samples

$$A_t^2 = \pi N \sigma_0 \Gamma^2 \tag{9}$$

and so it follows that

$$A_\gamma^2 \approx \pi N \sigma_0 \Gamma_\gamma^2$$

Measurements were made on two sample thicknesses of tantalum, and the quantity $\sigma_0 \Gamma_\gamma^2$ was determined for seven levels in this nucleus.

Rae and Bowey (1957) made measurements of the capture cross section of silver using as detectors four sodium iodide crystals 1.75-in. diameter by 2-in. long. These were arranged symmetrically around a 3-in. diameter capturing sample placed in an evacuated tube passing through the apparatus. The crystals were separated from the sample by a filter of B_4C and water of thickness greater than 1 in. to prevent low energy scattered neutrons from reaching the detectors. The whole assembly was enclosed in a massive lead shield surrounded by boxes containing B_2O_3 to shield against slow neutrons from the outside. The detecting system was placed 11.5 m from the neutron target of the Harwell 15-MeV electron linac. The four detectors could either be connected in parallel or could be arranged as two pairs, a coincidence being demanded between the two pairs. Various preliminary experiments were carried out in order to estimate the fluctuations in the detector efficiency from level to level, including a measurement of the ratio of single counts to coincidences, a study of the pulse height spectra from various levels, and a comparison of observed peak heights with minimum transmission in a transmission experiment. As a result of these experiments, it was estimated that the detector efficiency was constant for the two isotopes of silver and for all the levels observed, to an accuracy of $\pm 5\%$. This uncertainty was combined with the other statistical errors in the measurement in arriving at final integrated resonance yields. The final measurements were made with the detector bias set at 2 MeV, which gave a capture detection efficiency $\sim 3\%$ and a background of $\sim 2\%$ of the saturation count from the 5.2 eV level. The 5.2 eV level was used for calibration purposes, five samples being studied, the thicker ones absorbing 99% of the neutrons at resonance energy. The energy dependence of the neutron spectrum was determined with the aid of a BF_3 counter. Three sample thicknesses (0.001, 0.002, and 0.020 in.) were used to study the levels up to 100 eV. The integrated yield was determined for all those cases where $N\sigma_0 \ll 1$ in which case Eq. (21) applies, and

$$A_\gamma \approx (\pi/2)N\sigma_0\Gamma_\gamma \qquad (21)$$

The equality was made more exact by applying self-screening corrections for the small, but finite, thickness of the samples. The values of $\sigma_0\Gamma_\gamma$ obtained were combined with scattering and total cross-section measurements as described in Section I,B to obtain the parameters g, Γ_n, and Γ_γ for all the levels observed (Rae et al., 1958). The same equipment and methods were used for capture area measurements on tantalum (Evans et al., 1958) and tungsten (Waters et al., 1959), the energy range studied being in each case approximately 0 to 100 eV.

Using essentially the same techniques as the Harwell group, capture measurements have been made on samples of ^{111}Cd, ^{112}Cd, ^{113}Cd, and ^{114}Cd (Adamchuk et al., 1966) and ^{162}Dy and ^{163}Dy (Danelyan et al., 1964) at the electron linear accelerator of the Kurchatov Institute in Moscow. The samples were placed 15 m from the pulsed neutron target, and the detector consisted of four NaI crystals imbedded in B_4C to reduce neutron capture in the iodine (Danelyan and Efimov, 1963).

The group at the Columbia University's Nevis Laboratory have made capture measurements on U, Ag, Ta, and Au (Rosen et al., 1960; Desjardins et al., 1960). These authors used plastic scintillation detectors to observe the capture gamma-ray yield from metallic foils placed normal to the neutron beam from the synchrocyclotron at a distance of 35 m. The capture gamma-ray detector with a suitable foil in position was used primarily by the Columbia group as a self-indication transmission detector (see Chapter I), but the resonance capture yields themselves were also employed in conjunction with the self-indication data to deduce radiation widths and spins. The resolution of the Columbia system (5–10 nsec/m) was very much better than had been used in previous capture measurements, the Harwell measurements, for example, having a resolution of $\sim$170 nsec/m, but the Columbia measurements did suffer from poor signal-to-background ratios because of the intense background of gamma rays produced by the synchrocyclotron. Because of the improved resolution, however, the number of levels studied was greatly increased, being about 80 in the case of silver, and the energy range covered extended to about 1 keV in the cases of uranium and gold. With the high resolution available, normalization of the capture yields was established from the saturation yields of several levels, and the authors stated that efficiency of the detector appeared to be relatively insensitive to any fluctuations in the gamma-ray spectrum from level to level. From the integrated capture yields, taken in conjunction with the self-indication transmission data, the Columbia group were able to determine Γ_γ for 28 levels above 100 eV in uranium, 16 levels above

100 eV in silver, 28 levels above 40 eV in gold, and 13 levels above 65 eV in tantalum.

A considerable number of capture measurements have been carried out with the lead slowing down spectrometer. This technique was developed in the USSR, and was first reported at the 1955 Geneva Conference (Bergmann et al., 1955). The spectrometer consists of a large assembly of lead about 2 m on a side, and a pulse of fast neutrons is produced in the center of this assembly by a Cockcroft–Walton accelerator through the $T(d,n)$ reaction. The neutrons slow down almost uniformly as a function of time, and near the surface of the assembly the neutron population is centered about a mean energy which varies slowly with time. The energy spread is essentially a Gaussian of $\sim 30\%$ full width at half maximum. Either a CaF_2 scintillator or a proportional counter gamma-ray detector is located in a hole near the surface of the lead, and a sample of the nuclide of interest is wrapped about the capture-gamma detector. With this instrument capture measurements have been carried out from thermal energies to $\lesssim 100$ keV on samples of Cl, Fe, Ag, and Au (Isakov et al., 1961; Kashukeev et al., 1961) Ni, Cu, and Mo (Kapchigaskev and Popov, 1963); separated Tl isotopes (Konks and Shapiro, 1964); Mo, Fe, and Ag (Mitzel and Plendl, 1964 at the Karlsruhe Lab); ^{50}Cr, ^{52}Cr, ^{53}Cr, (Kapchigaskev and Popov, 1964); V, Zr, ^{90}Zr, ^{91}Zr, and ^{94}Zr (Kapchigaskev, 1965); La, Pr, Ta, and Au (Konks et al., 1964); Cu, Fe, Pb, Ni, Br, Rb, ^{85}Rb, Nb, Mo, ^{98}Mo, ^{100}Mo, Sb, Rh, Ag, In, I, Cs, W, Ir, and Au (Popov, 1964 and Popov and Shapiro, 1962); V, Ga, Y, and separated isotopes of Zr, Mo, Te, and W (Bergmann et al., 1966); ^{45}Sc and ^{35}Cl (Romanov and Shapiro, 1965); and Ho and Lu (Konks and Fenin, 1966). The result of a typical capture measurement using this technique is shown in Fig. 19 for samples of ^{203}Tl and ^{205}Tl (Konks and Shapiro, 1964). The data are impressive in the energy range covered, but the intrinsic limitation of $\sim 30\%$ energy resolution severely restricts resonance measurements to, at best, the first few resonances. The efficiency of the capture detector is calibrated against either a known thermal capture cross section or capture in a resonance of known parameters. Although the CaF_2 detector is spectrum sensitive, the (thick walled) proportional counter has an efficiency proportional to gamma-ray energy up to ~ 4 MeV (Kashukeev et al., 1961) and hence an essentially constant efficiency for capture events in a given isotope.

A new type of spectrum-sensitive detector has been developed recently at Oak Ridge for kilovolt-capture measurements (Macklin and Gibbons, 1967). This detector, named the "total energy detector," consists of two plastic scintillators, each approximately 10 cm in diameter by 8 cm long viewed by a photomultiplier. By applying a weighting function to the pulse

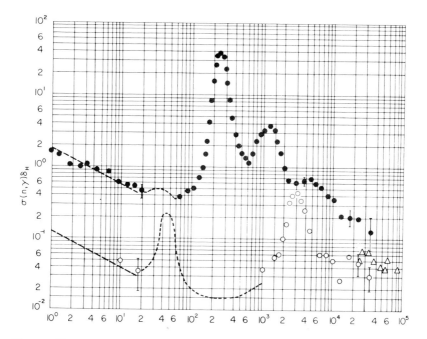

FIG. 19. The capture cross section of ²⁰³Tl (●) and ²⁰⁵Tl (○) from 1 to 100,000 eV. Measured with the lead slowing-down spectrometer (Konks and Shapiro, 1964).

height obtained in the scintillator, the authors report that the efficiency dependence of this detector upon gamma-ray cascade is removed, and that the detector efficiency depends only upon the neutron binding energy. This detector has an overall efficiency of ~20%, can be used with as little as 3 g of capture sample at the Oak Ridge pulsed Van de Graaff accelerator, and has a timing jitter of ~2 nsec. Measurements from 30 to 220 keV have been reported for V, Fe, Ni, ⁸⁷Sr, ⁸⁶Sr, Y, Nb, Rh, Ag, ¹²²Te, ¹²³Te, ¹²⁴Te, ¹²⁵Te, ¹²⁶Te, ¹²⁸Te, ¹³⁰Te, I, Eu, ¹⁷⁵Lu, ¹⁷⁶Lu, W, Au, ²⁰⁴Pb, and ²⁰⁸Pb.

2. LARGE LIQUID SCINTILLATOR DETECTORS

In building a large liquid-scintillator detector, the object is to produce a detector with a 4π geometry whose efficiency is 100%. In such an ideal detector all the gamma-ray energy is absorbed so that the capture of a neutron in a sample placed at the center of an axial hole of small diameter will give rise to a pulse whose height is proportional to the binding energy of the neutron (plus its incident energy). Unfortunately, this ideal detector is impossible to achieve with any practical size of detector since a sphere

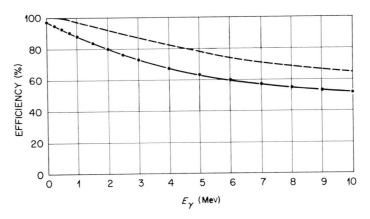

Fig. 20. Calculated tank γ-ray intrinsic (or "zero-bias") efficiency for the Oak Ridge 1.2-m diameter liquid scintillator (·——·——·). The dashed curve represents the efficiency if there were no neutron flight tube passing through the detector. (Gibbons *et al.*, 1961.)

of ~3 m diameter would be required to detect an 8 MeV photon with an efficiency of ~90%. Fortunately, however, capture-gamma reactions rarely take place without three or four gamma rays emitted in a cascade, and then a scintillator of approximately 1 m in diameter is essentially 100% efficient in detecting capture if the detector could be operated at zero bias. Figure 20 shows the calculated "zero bias" efficiency of just such a detector (Gibbons *et al.*, 1961) for monochromatic gamma rays. According to this curve, if a typical capture cascade proceeds by the emission of three 2.5 MeV gamma rays, the probability of detecting none of the 2.5 MeV gamma rays is $(1 - 0.75)^3 \approx 1.5\%$. Thus, the efficiency is $\approx 98.5\%$ for detecting such a capture cascade if the bias could be set at zero. In practice, of course, this cannot be done, and background considerations make it necessary to apply a bias corresponding to about 1 MeV. Under these conditions the mean efficiency for detecting capture gamma-ray cascades is ~90% for typical large scintillators, and for target nuclides with a high level density the fluctuations in efficiency should be quite small.

The first capture measurements using this technique were carried out at Los Alamos (Diven *et al.*, 1960) to measure capture from 175 keV to 1 MeV in twenty-eight elements. They first constructed a 0.48 m diameter detector, and then replaced it with a 1-m diameter detector viewed by twenty-eight 5-in. diameter phototubes. These authors observed that the 0.48-m diameter detector did not exhibit a peak in the pulse height spectrum for neutron capture, corresponding to the binding energy of the captured neutron, but to the contrary produced a monotonically falling pulse height

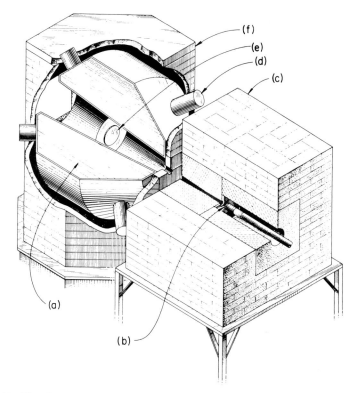

FIG. 21. The Oak Ridge 1.2-m diameter liquid scintillator capture detector. This detector is designed for use with $^7\mathrm{Li}(p,n)$ targets and has a conical neutron flight tube passing through its center (Gibbons *et al.*, 1961). [(a) scintillator; (b) target; (c) lithium-loaded paraffin; (d) photomultiplier (8); (e) sample; (f) 4-in. thick lead shield.]

distribution. The 1-m detector, however, did produce a pulse height distribution which peaked at, or near, the binding energy. If one were forced to operate at a gamma-ray bias of $\sim$3 MeV, then the smaller detector would miss $\sim$50% of the captures while the larger detector would only miss $\sim$25% of the captures. Therefore, as one goes to a larger detector the operating bias can be raised to discriminate against the background without sacrificing efficiency, and thus it is possible to go to rather large ($\lesssim$2-m diameter) detectors without getting saturated with background counts. Another advantage of a high bias is the ability of the large scintillator to bias out low-energy gamma rays caused by neutron inelastic scattering and hence enable capture measurements to make up to a few million electron volts.

Figure 21 shows the tank developed at Oak Ridge for high-energy

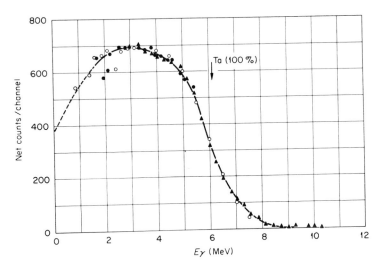

Fig. 22. Pulse height spectrum from the Oak Ridge 1.2-m diameter capture detector for 65 keV neutron capture in tantalum (Gibbons *et al.*, 1961).

average capture cross-section measurements (Gibbons *et al.*, 1961; Macklin *et al.*, 1963a). It is a good approximation to a sphere 1.2 m in diameter, and high reflectivity of the inner walls is achieved by coating them with α-Al_2O_3 in a binder of sodium silicate. The tank is filled with double-distilled xylene containing 4 to 5 g/liter of p-terphenyl and 0.02 g/liter 2,5-di-(biphenyl)-oxazole. Five liters of tri-methyl borate prepared from 97% ^{10}B is added to this mixture to ensure capture of moderated neutrons in boron rather than in hydrogen. The scintillator is viewed by eight 5-in. diameter phototubes (in contrast with the twenty-eight phototubes used in the Los Alamos detector) in direct contact with the liquid, and the whole detector is shielded by 4 in. of lead. The detector is used with the Oak Ridge 3-MV pulsed Van de Graaff. Figure 22 shows the pulse height spectrum observed with this detector due to neutron capture in a tantalum target. The dashed part of the pulse height curve shows the type of extrapolation to zero pulse height used by these authors to estimate the counts lost below the bias setting.

A smaller tank 0.70 m in diameter viewed by four 5-in. diameter photo-tubes was used by Block *et al.* (1961) in conjunction with the Oak Ridge fast chopper to study, by time of flight, the capture cross sections of 25 elements in the atomic weight range $90 < A < 200$ and the energy range 200–8000 eV. The time resolution of the system was $\sim$170 nsec/m so that although resonance structure was visible at the lower energies, measure-

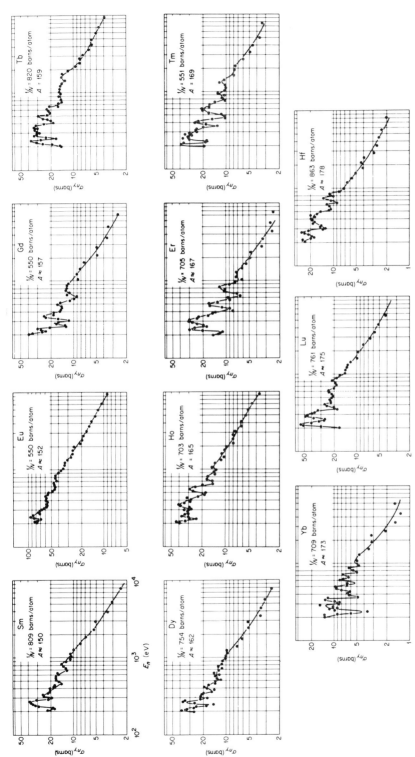

Fig. 23. Radiative capture cross sections of the rare earth elements measured with a 0.70-m diameter liquid-scintillator capture detector and the Oak Ridge fast chopper (Block *et al.*, 1961).

ments of the parameters of individual levels was not feasible. The samples used were thin enough for multiple scattering effects to be negligible in the energy range studied, so the observed capture counts were corrected for the known neutron energy spectrum (as determined by a BF_3 proportional counter) and displayed as observed capture cross section curves, normalization being to the higher energy work of Gibbons et al. (1961). The data obtained in the rare earth mass region are shown in Fig. 23.

Another large scintillator detector developed at Oak Ridge (Block et al., 1961) was installed in 1962 at the 25-m flight station of the Rensselaer Polytechnic Institute Electron Linac Laboratory. This detector, shown in Fig. 24, is 1.25 m in diameter and is viewed by only two 16-in. diameter phototubes. This 1.25-m, two-phototube detector has essentially the same pulse height response as the 1.2-m, eight-phototube detector at the Oak Ridge Van de Graaff, but it does have a timing jitter of ~25 nsec due to the large phototubes.

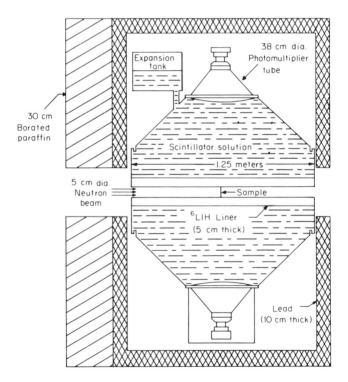

FIG. 24. The 1.25-m diameter liquid-scintillator capture detector. This detector was designed and built at Oak Ridge and installed at the Rensselaer Linac Laboratory. Only two 16-in.-diameter photomultipliers view the scintillating solution.

This detector has recently been modified to permit scattering measurements and higher resolution capture measurements to be carried out. Eight 58AVP photomultipliers have been added to the detector to reduce the timing jitter to less than 6 nsec (Hockenbury and Moyer, 1969), and a transverse tube now passes through the center of the detector to accommodate ^{6}Li-glass neutron detectors. The scintillator solution is similar to that used by Gibbons *et al.*, 1961, but elemental boron was added to the liquid (in the form of methyl borate) instead of highly enriched ^{10}B; a boron-to-hydrogen atom ratio of 0.0056 is used here. In addition a 2-cm thick hollow cylinder of ^{6}LiH surrounds the capture sample to reduce the scattered neutrons entering the scintillator. A prompt neutron sensitivity of $\lesssim 10^{-5}$ has been reported for this detector for scattered 85-keV neutrons (Hockenbury *et al.*, 1969), so that measurements of resonances in the kilovolt energy range, where $\Gamma_n \gg \Gamma_\gamma$ can readily be carried out without being limited by scattered neutrons. Pulse height spectra taken with this detector for capture in Au and Sn resonances are plotted in Figs. 25a and 25b, respectively. The Au spectra are characterized by a high energy peak due to a hard component of gamma rays in the capture cascade. (It is interesting that this "hardness" does not vary from resonance to resonance although the capture takes place in resonances of two different spins.) On the other hand, the spectra for capture in the 46 eV resonance of ^{118}Sn differs considerably from capture in the 360 eV resonance of ^{118}Sn. Therefore, it is reasonable to assume a constant response, and hence efficiency, for capture in heavy nuclei, but great care must be taken in determining the efficiency for resonance capture in light and medium weight nuclei. In favorable cases, it is possible to make isotopic identification of the resonance by correlating the peaks in the capture spectra with the neutron binding energy. Capture measurements have been carried out at Rensselaer on samples of ^{182}W, ^{183}W, ^{184}W, and ^{186}W from approximately 1 eV to 10 keV, spin assignments being made in some cases (Block *et al.*, 1964a, 1965), and on samples of Na, Mg, Al, S, F, ^{56}Fe, ^{57}Fe, ^{58}Fe, Ni, ^{60}Ni, ^{90}Zr, ^{91}Zr, ^{92}Zr, ^{94}Zr, ^{182}W, ^{183}W, ^{184}W, ^{186}W, Pb, ^{204}Pb, and ^{207}Pb from 1 to 200 keV (Block *et al.*, 1967; Bartolome *et al.*, 1968, 1969; Hockenbury *et al.*, 1969). Isotopic assignments have been made with this instrument for resonances in ^{176}Hf, ^{177}Hf, ^{178}Hf, ^{179}Hf, ^{180}Hf from 1 to 300 eV (Russell *et al.*, 1963) and in ^{120}Sn, ^{122}Sn, and ^{138}La (Block *et al.*, 1964b; Fuketa *et al.*, 1964). The capture yield divided by sample thickness* for a 0.25-in. thick sample of elemental Fe is plotted in Fig. 26; a resolution of $\sim$1.3 nsec/m was achieved during the high-energy measurements. The neutron flux

---

* The yield divided by sample thickness is, for thin samples, equal to the capture cross section [see Eq. (17) Section I,A]. For thin samples, the yield reduces to: $Y_{xp} \approx N\sigma_{\Delta\gamma}$.

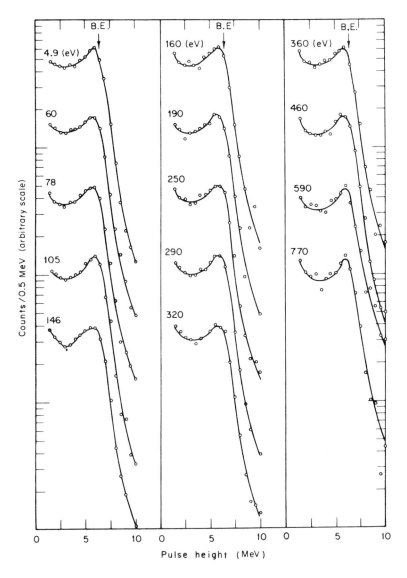

was determined by a ^{10}B–NaI detector, with multiple scattering corrections applied to determine the relative efficiency of the detector, and normalization was made at low energies by the saturated resonance method and a Ag sample. By placing a fission chamber inside the large capture

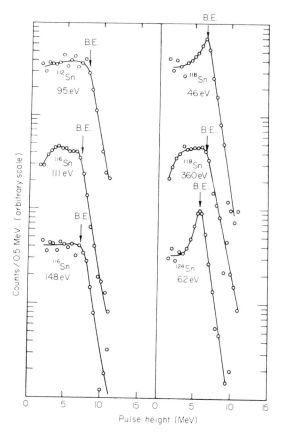

FIG. 25b

FIG. 25. Pulse height spectra for neutron capture in a 1.25-m diameter detector for isolated resonances in (a) Au and (b) Sn. The pulse height corresponding to the binding energy is indicated.

detector, capture measurements have been made on ^{233}U, ^{235}U, and ^{239}Pu, using the capture detector in anticoincidence with the fission detector (de Saussure et al., 1965, 1966; Gwin et al., 1969). A similar measurement in the kilovolt region was also carried out with the 1.2-m scintillator at the Oak Ridge 3-MV Van de Graaff laboratory (R. Gwin, private communication).

A 4000-liter capture detector has been constructed at the General Atomic Laboratory (Haddad et al., 1964) which is composed of a cluster of forty-four, 9-in. diameter, hollow Plexiglass cylinders filled with decaline-base solution and viewed by pairs of photomultipliers (see Fig. 27). The volume of this detector is equivalent to that of a spherical detector 1.9 m in diame-

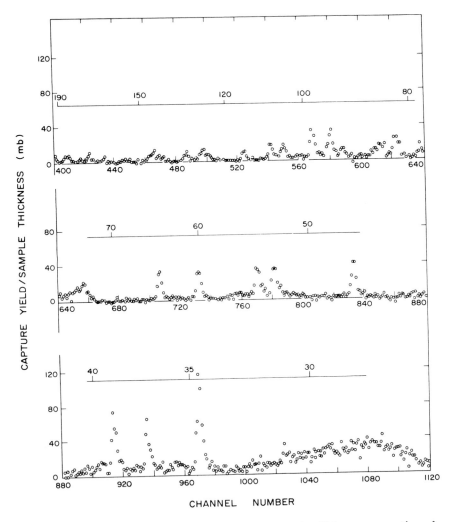

FIG. 26. The neutron capture yield divided by sample thickness versus time-of-flight for a 0.25-in. thick sample of iron; a resolution of ~1.3 nsec/m was achieved at the higher neutron energies (keV). (Hockenbury *et al.*, 1969.)

ter, and it has an 80% "zero bias" efficiency for 8-MeV photons. This is the largest liquid-scintillator capture detector in existence, and at a bias of 3 MeV it has an efficiency of approximately 80% for detecting capture in typical heavy nuclei (as compared to an efficiency of ~65% at a 3 MeV bias for the 1.25-m diameter scintillator at Rensselaer). Capture measure-

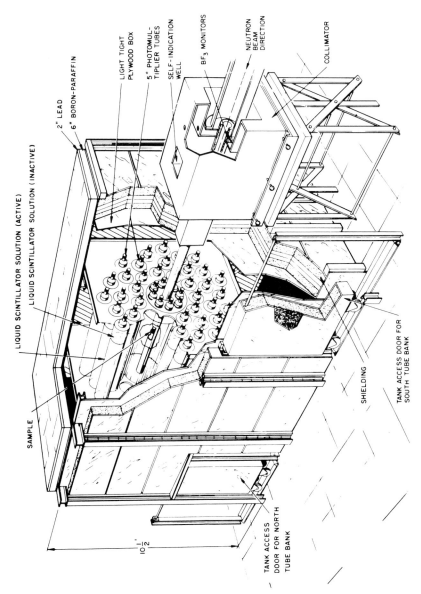

FIG. 27. The 4000-liter liquid-scintillator capture detector at the General Atomic laboratory. This detector consists of forty-four, 9-in. diameter hollow Plexiglass cylinders filled with scintillating solution and each viewed by pairs of photomultipliers (Haddad *et al.*, 1964).

ments have been carried out on samples of Er and Au from 10 eV to ~40 keV (Haddad *et al.*, 1964) Zr in the kilovolt region (Lopez *et al.*, 1968) and Na at the 2.85 keV resonance (Friesenhahn *et al.*, 1968); a combination of capture and self-indication measurements on Th from 20 to 220 eV has determined the radiation widths of eleven Th resonances (Haddad *et al.*, 1965); and capture and self-indication measurements have been made on Nb from 30 to 1400 eV (Lopez *et al.*, 1967). The relative neutron flux is determined by a BF_3 proportional counter, and normalization is again made with the saturated resonance method.

A 1.1-m diameter scintillator detector has been installed at the IBIS Van de Graaff accelerator at Karlsruhe (Kompe, 1967). This detector uses fast photomultipliers to achieve a timing resolution of ~3 nsec. The average capture cross sections of twelve medium to heavy elements have been measured from 10 to 150 keV with a resolution of ~7 nsec/m although Kompe states that an ultimate resolution of ~2 nsec/m is possible with this apparatus. More recently elliptical-shaped large-volume liquid scintillator detectors have been constructed at Oak Ridge (E. Silver, private communication) and the Tokai-Mura laboratory of J.A.E.R.I. (T. Fuketa, private communication). These detectors are elongated along the neutron beam axis so that small angle capture-gamma rays pass through the same amount of scintillator fluid as those emitted at angles near 90°.

At the Dubna IBR pulsed reactor laboratory (Blokhim *et al.*, 1961) a 400-liter liquid scintillator has been constructed which is operated as a pair of 200-liter detectors in coincidence (Pikelmer *et al.*, 1963). The coincidence technique reduces the background considerably, but since they achieve only a 30% capture detection efficiency, their results are quite sensitive to variations in the gamma-ray cascades as was the case for the early measurements at the General Electric betatron (see Section III,B,1). The scintillator is located at the 1000-m flight station at Dubna and a resolution of ~40 nsec/m is obtained. Capture measurements have been reported for Rh up to ~1 keV (Wang *et al.*, 1963); Nb and Rb from 400 to 1300 eV (Iliescu *et al.*, 1965); Ba, Zn, Rb, Nb, Mo, and Ru to 400 eV (San *et al.*, 1966); and Yb, [171]Yb, [172]Yb, [173]Yb, [174]Yb, and [176]Yb to 150 eV (Wang *et al.*, 1966).

A rather small $0.5 \times 0.5 \times 0.5 \ m^3$ scintillator has been utilized at the Obninsk laboratory in conjunction with a pulsed Cockcroft–Walton accelerator. Capture measurements have been made from 30 to 170 keV with a resolution of ~20 nsec/m on Ta, W, and Re, and isotopes of W (Dovbenko *et al.*, 1966), and on [107]Ag, [109]Ag, [161]Dy, [162]Dy, [164]Dy, [182]W, [184]W, [186]W, Ta, W, and Re (Kononov *et al.*, 1967).

An excellent suggestion which might at the same time reduce the size of the scintillator and improve its efficiency for higher energy gamma rays has been made by Macklin (1961). The suggestion is that use should be made of napthalene-based scintillators, either liquid or plastic, since these can have high refractive indices, which would permit loading with small particles of lead glass. A lead-glass-loaded naphthalene-based scintillator would yield perhaps only 20% of the light obtained from a pure scintillator, but for capture-gamma-ray studies with 5–10 MeV of photon energy available, there should still be adequate light output. The density of the glass-scintillator mixture would be close to 2.4 g/cc which would permit a reduction in linear dimensions of the tank of the order of a factor 2 to 3 together with improved characteristics for high energy photons. Unfortunately, no one has been able to produce the powdered glass in the correct size without introducing either chemical impurities or serious changes in refractive index so that this detector has not been successful to date.

3. DETECTORS WITH EFFICIENCY DIRECTLY PROPORTIONAL TO PHOTON ENERGY (THE MOXON–RAE DETECTOR)

Three difficulties are inherent in the use of large scintillator detectors. One is the decrease in efficiency with increasing photon energy due to the predominantly Compton interactions with the low-Z materials in the scintillator. Another is the sheer size of the detectors which makes effective shielding both difficult and expensive, and consequently these large detectors, in general, have a high background. The third difficulty used to be their rather poor timing resolution of ~20 nsec, but the recent improvement demonstrated at Karlsruhe (Kompe, 1967) and at Rensselaer (Hockenbury and Moyer, 1969) have reduced this to ≲6 nsec.

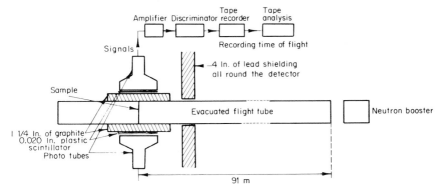

FIG. 28. The Moxon–Rae neutron capture cross-section detector. Four capture assemblies are used, each assembly consisting of a 1.25-in. thick graphite converter, a 0.020-in. thick plastic scintillator, and a 5-in. diameter photomultiplier.

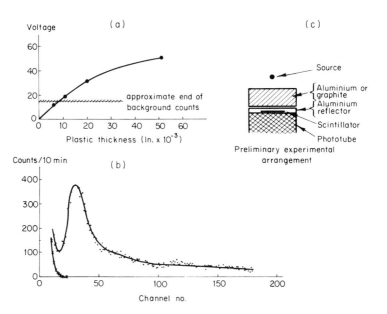

FIG. 29. (a) Pulse height corresponding to the peak in the distribution versus plastic thickness; (b) pulse height distribution in the 0.020-in. thick plastic scintillator of a Moxon–Rae detector with an aluminum radiator and a ^{60}Co gamma-ray source ($\cdots\cdots$, counts from source plus background; $\times$, background counts); (c) experimental arrangement.

The detector developed by Moxon and Rae (1961, 1963) is a fast scintillation detector analogue of the thick-walled gas counter as used in the lead slowing down time spectrometer (Kashukeev $et\ al.$, 1961). It provides an improved performance over the large scintillation detector in all the three respects mentioned above at the expense of a large reduction in overall detection efficiency, which for this detector is only 2–4%. Where the detector is used with nanosecond pulsed sources, this reduction in efficiency can be offset by a reduction in flight path length due to the better time resolution of $\sim$3 nsec (Macklin $et\ al.$, 1962, 1963b,c). The original detector (Moxon and Rae, 1963) is shown in Fig. 28. Gamma rays following the capture of a neutron in the sample eject secondary electrons from the four slabs of graphite arranged symmetrically around the evacuated flight pipe, and the electrons emerging from the surface of the graphite are detected in thin (0.020-in. thick) sheets of plastic phosphor attached to four 5-in. diameter photomultipliers. Figure 29 shows the pulse height spectrum obtained with a ^{60}Co source, and illustrates the clean separation of the peak (corresponding to 50 to 100 keV of electron energy dissipated in the plastic) from the background. The higher the energy of the photon interacting with the graphite,

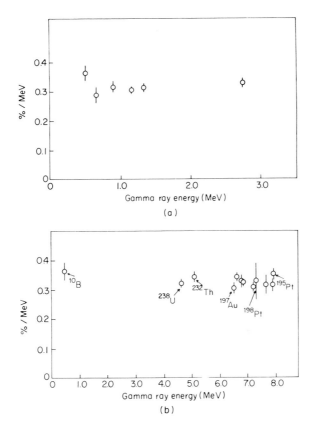

FIG. 30. Efficiency per million electron volts versus photon energy for the Moxon–Rae detector with a 1.25-in. thick graphite converter. (a) Efficiency per million electron volts for single gamma rays; (b) efficiency per million electron volts for neutron capture, normalized to 0.36% per million electron volts for the 0.48 MeV gamma ray from ^{10}B + n.

the greater the range of the secondary electrons, and the greater the thickness from which electrons can be ejected; thus one would expect the efficiency of the detector to increase with photon energy. In fact, for the thickness of graphite used in the detector (1.25 in.) the measured efficiency is very nearly directly proportional to the photon energy, except below 1 MeV. Figure 30a shows the efficiency per million electron volts of photon energy measured for individual photons by a beta-gamma coincidence method (except for ^{137}Cs where a calibrated source was used) while Fig. 30b shows the efficiency per million electron volts for capture gamma-ray cascades from a variety of nuclides. The points in Fig. 30b were obtained from the saturation yields of low-energy resonances and from known thicknesses of

^{10}B. They were normalized to the individual photon measurements by set-ting the efficiency for the 0.48-MeV gamma ray from ^{10}B $+$ n equal to that for positron annihilation as obtained in Fig. 30a. Apart from the high point at $\sim$0.5 MeV in both figures, the flatness of the efficiency curves is impres-sive and demonstrates the suitability of the detector for counting capture events since the efficiency for detecting a cascade is approximately equal to the sum of the efficiencies for the component photons, and so is a constant for constant total energy release.

This dependence of efficiency on binding energy does present a problem when measurements are made on a sample of mixed isotopes, and hence different binding energies and detector efficiencies. Isotopic assignments to the resonances must be made to interpret the resonance capture, and some model of relative capture cross section per isotope must be assumed to interpret the average capture cross section in the unresolved region.

Macklin et $al.$ (1963b) have made measurements of the efficiency per million electron volts of photon energy of a similar detector, but with a thinner (1-in. thick) graphite converter, for single photons up to 6 MeV, and their measurements are shown in Fig. 31a together with a calculated efficiency curve. It is to be noted that their experimental points also show a rise at about 0.5 MeV together with a high point at 3 MeV. The theoretical curve shows a steady decrease in efficiency per million electron volts with increasing energy from about 1.5 MeV upward although the drop-off is no worse than that calculated for the large tank. The experi-mental points are in both cases better than the calculated curve, and in any case the Oak Ridge group (Macklin et $al.$, 1963b) have shown that the calculated efficiency curve can be greatly improved by the addition of a small amount of high-Z material to the graphite, so as to give an efficiency curve linear to a few percent from 1 to 10 MeV. They have also made measurements on such a mixture, and the excellent agreement obtained between theory and experiment is shown in Fig. 31b.

Measurements of the sensitivity of the detector to scattered neutrons have been carried out both at Oak Ridge and at Harwell using pulsed Van de Graaff accelerators to separate the effect of scattered neutrons from that of capture gamma rays by time-of-flight. In each case the estimate of the ratio of the efficiency for detecting scattered neutrons to that for capture cascades was $\lesssim 3 \times 10^{-4}$. Thus only when $\Gamma_n/\Gamma_\gamma \gtrsim 1000$ does the effect of the scattered neutrons become measurable. This means that measurement of capture cross sections and radiation widths is possible using this detector over a very wide range of Z.

The original detector was used in conjunction with the Harwell 25-MeV electron linac to make measurements on Fe, Rh, Ag, In, Au, Th, and U with a nominal resolution of 10 nsec/m over the energy range between a

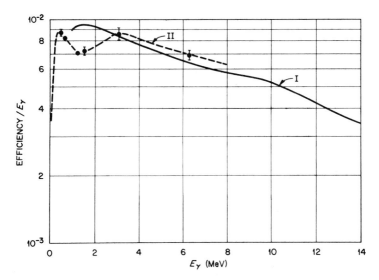

FIG. 31a. (I) Calculated curve for zero bias compton and pair first interactions and (II) empirical calibrations of the efficiency per million electron volts of the Moxon–Rae detector with a 2.54-cm thick graphite converter, source distance: 5.08 cm, electron density: 5.1×10^{23} cm^{-3} (Macklin *et al.*, 1963a).

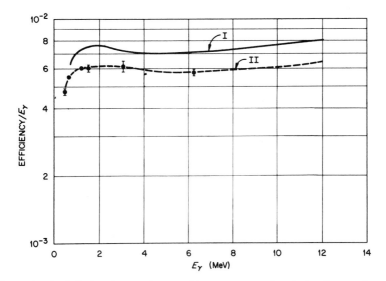

FIG. 31b. (I) Calculated curve for zero-bias compton and pair first interactions and (II) empirical calibrations of the efficiency per million electron volts of the Moxon–Rae detector with a 2.38-cm thick (graphite + Bi$_2$O$_3$) graphite converter, source distance: 5.08 cm, electron density: 8×10^{23} cm^{-3}, composition: C 78.5 at.%, Bi 8.6 at.%, and O 12.9 at.% (Macklin *et al.*, 1963a).

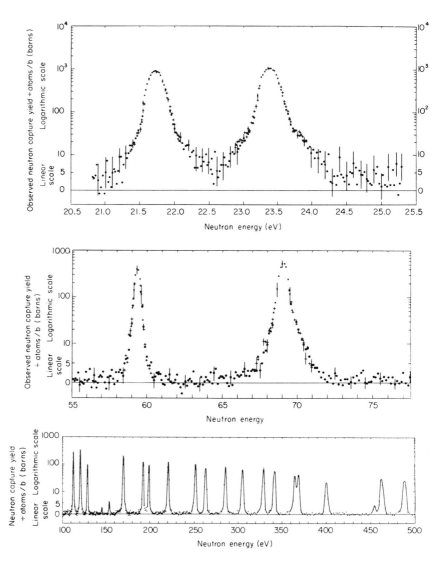

FIG. 32. The capture yield (divided by sample thickness) versus neutron energy for Th (•, $n = 2.776 \times 10^{-3}$ atom/b; +, $n = 1.418 \times 10^{-3}$ atom/b; ✗, $n = 0.7132 \times 10^{-3}$ atom/b). (Asghar et al., 1966a.)

few electron volts and 100 keV (Moxon and Mycock, 1962; Moxon and Chaffey, 1963; Moxon and Rae, 1963). The detector was calibrated by observing the saturation yield from 32 resonances below 100 eV in various elements, and the shape of the neutron spectrum was determined as a

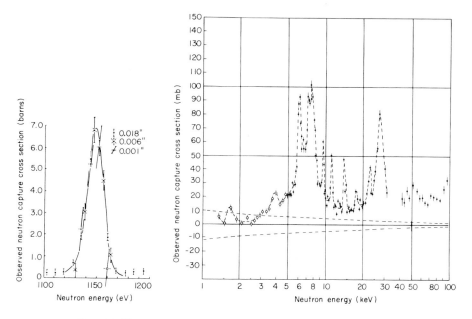

FIG. 33. The neutron capture cross section of Fe (Moxon, 1966).

function of energy by observing the 480-keV gamma-ray yield from samples of ^{10}B of various thicknesses. In the resonance region the background observed in the study of thin samples was less than 1% of the saturation yields, and conventional resonance area analysis, including first-order corrections for multiple scattering, was employed to determine the parameters in levels in thorium and uranium. These data were then combined with transmission and scattering data (Corvi, 1962; Uttley and Jones, 1963; Firk et al., 1963) to determine the radiation widths of (about 30) levels. More recently, further measurements have been made up to ~1 keV using the improved Harwell 45-MeV electron linac. These have been combined with transmission and scattering measurements to yield resonance parameters for ^{232}Th (Asghar et al., 1966a) where the radiation widths of 23 resonances were determined, for ^{107}Ag, ^{109}Ag, ^{169}Tm, ^{197}Au, ^{159}Tb, and ^{165}Ho (Asghar et al., 1966b), for ^{240}Pu (Asghar et al., 1967), and for ^{238}U (Asghar et al., 1966c). This detector has also been used to obtain capture data up to 50 keV in Fe and Co (Moxon, 1966) and Pb (Block and Moxon, 1963) from which resonance parameters were obtained. Data were also obtained for Na which were used in combination with transmission measurements (Moxon and Pattenden, 1967) to yield resonance parameters. A Monte Carlo code written by Lynn (1963) is used to interpret the capture yields

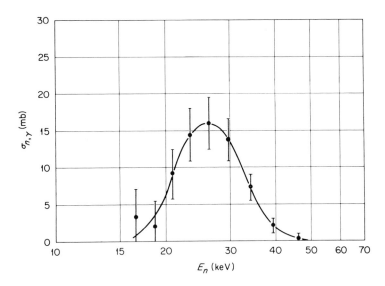

FIG. 34. The neutron capture cross section of sulfur in the vicinity of a resonance near 27 keV (^{32}S: 2.8×10^{22} atoms/cm^2). (Macklin *et al.*, 1963b.)

from the thicker samples. Typical data (in the form of capture yields divided by sample thickness) are shown in Fig. 32 for Th, and in Fig. 33 for Fe.

Macklin *et al.* (1962, 1963b,c, 1965) used this type of detector at a 7-cm flight path of the Oak Ridge 3 MV pulsed Van de Graaff accelerator to measure 30 keV capture in Mo, Cd, Sn, Sn isotopes, Ta, W, Pt, and Au, and also resonance capture in S and F from 10 to 50 keV. Even with a 7-cm flight path the nominal resolution was 40 nsec/m, and they resolved resonances in fluorine and sulfur. Their results for the cross section of sulfur are shown in Fig. 34.

The Moxon–Rae type of detector has been adapted for use with the nuclear explosion time-of-flight experiments by the Los Alamos group (Glass *et al.*, 1966, 1968; Byers *et al.*, 1966, Harlow *et al.*, 1968). They have replaced the scintillator and phototubes shown in Fig. 28 with a solid state detector so that they can measure the resultant conversion-electron current. The neutron intensity achieved in these experiments is so high that very thin capture samples can be used, and hence the thin-sample analysis [see Eq. (21)] can be applied to the data. A typical capture cross section obtained by this technique is shown for capture in ^{238}U in Fig. 35.

A high efficiency version of the Moxon–Rae detector has been developed at the CBNM laboratory at Geel (Weigmann *et al.*, 1967). This detector

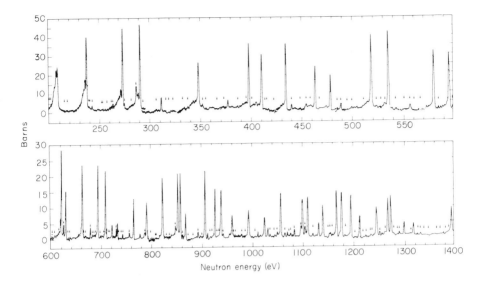

Fig. 35. The capture yield (divided by sample thickness) versus neutron energy for ^{238}U ($1/n = 200$b). These data were obtained during the Petrel nuclear explosion. The arrows indicate weak resonances observed in either these data or in data obtained with thicker samples (Glass et al., 1968).

consists of six 3.5-cm thick plastic scintillator slabs which are alternately viewed by two photomultipliers. By demanding a coincidence between the photomultipliers, one essentially has the same property as the conventional Moxon–Rae detector, namely, that the conversion electron travels from one plastic scintillator slab into an adjacent one. Furthermore, the coincidence requirement can be used to eliminate the sensitivity to recoil protons, thereby extending the useful neutron energy range. This Geel detector is shown in Fig. 36, and it is constructed out of 1.5-cm thick sheets of scintillating plastic. Boron plastic sheets 0.5-cm thick have been inserted between five of the 1.5-cm thick pairs of sheets to reduce hydrogen capture in the plastic. Comparative measurements with this detector and both the conventional Moxon–Rae and Bi-loaded Moxon–Rae detectors indicate that this detector has an absolute efficiency of 10% at 8 MeV, and that the efficiency is linear with photon energy. Capture measurements have been carried out with this detector at the Geel Linac on Mo from 10 eV to 25 keV and $g\Gamma_n$ or Γ_γ has been determined for 18 resonances (Weigmann and Schmid, 1967); measurements have also been reported for Cu and ^{240}Pu (Weigmann et al., 1968).

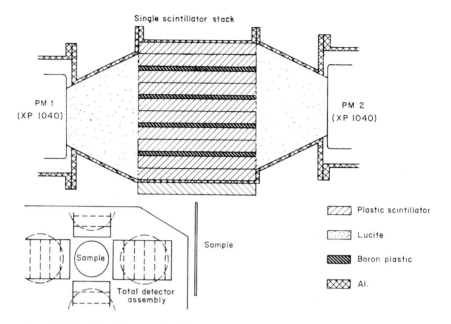

FIG. 36. The Geel improved-efficiency Moxon–Rae detector. Conversion electrons passing from one plastic scintillator sheet to an adjacent plastic scintillator sheet produce a coincidence in the two photomultipliers. This detector has effectively five pairs of converter plates in series which adds to the overall capture detection efficiency. (Weigmann *et al.*, 1967.)

C. Shell Transmission and Activation Measurements

These two techniques are of limited general application to low-energy neutron spectroscopy. The shell transmission method requires a monochromatic neutron source, preferably isotropic, and a rather thick hollow sphere of the sample to be measured.

The lowest-energy neutron source applicable to shell transmission measurements is the 24-keV Sb-Be source, and measurements have been carried out on medium to heavy weight nuclei where the energy spread of the Sb-Be source averages over several levels in the target nucleus (Macklin *et al.*, 1957; Vervier, 1958; Booth *et al.*, 1958; Kononov *et al.*, 1958; Belanova, 1958, 1960; Belanova *et al.*, 1965, 1967; Schmitt and Cook, 1960; Chaubey and Sehgal, 1965; Robertson, 1965; and Ryves *et al.*, 1966). These measurements are capable of an accuracy of $\sim 5\%$ (Belanova *et al.*, 1967), and they serve as normalization for time-of-flight data or as an independent check upon other methods.

The activation technique requires a radioactive product nucleus of reasonable half-life and a monochromatic neutron source to irradiate the sample. Measurements have been carried out at distinct energies with radioactive sources like Sb-Be, and also as a function of neutron energy by taking advantage in a nuclear reaction of the variation of neutron energy with charged particle energy. The lowest neutron energy that can be reached by this technique is (with the exception of thermal activation) the cutoff energy of Van de Graaff reactions of $\sim$5 keV. Most of the measurements have been carried out with poor resolution and hence do not give much information on discrete resonances. A few exceptions have occurred in the light elements where the level spacing is sufficiently great, viz. activation measurements on F (Gabbard et al., 1959), Na (Bame and Cubitt, 1959; Le Rigoleur et al., 1966), and Al (Henkel and Barschall, 1950).

The average capture-cross-section measurements carried out by the activation technique provide an independent check on capture measurements made by other methods, and the activation technique is capable, in favorable cases, of measuring capture cross sections to an accuracy of better than 5%. The activation technique may also be the only possible method in the case of very small capture cross sections at million electron volt neutron energies. Under these circumstances the specific activity of the ground state of the compound nucleus can be observed under low background conditions away from the accelerator which permits very small cross sections to be observed; the method also clearly separates capture from inelastic scattering, which is impossible for prompt gamma-ray methods at energies of several million electron volts. Of course, only cases in which the compound nucleus has a suitable half-life can be studied. Average capture-cross-section measurements utilizing the activation technique have been reported by Bilpuch et al. (1960), Cox (1961), Harris et al. (1965), Miskel et al. (1962), Pönitz (1967), Stavisskii and Tolstikov (1960, 1961), Stavisskii and Shapar (1961), Stupegia et al. (1965, 1967), Tolstikov et al. (1963, 1964, 1967), Weston et al. (1960), Weston and Lyon (1961), and Leipunsky et al. (1958).

IV. RESULTS OF SCATTERING AND CAPTURE MEASUREMENTS

A. Individual Levels

In the introduction to this chapter, it was pointed out that measurements of scattering and capture cross sections when combined with transmission measurements enable one, in general, to determine the spins and radiation widths of neutron resonance levels. Radiation widths, but not spins, can be deduced from the total cross-section measurements of nonfissile nuclides when $\Gamma_n \ll \Gamma_\gamma$, and so, in general, radiation widths are known from this method for the first few levels in medium and heavy weight nuclei. Likewise, level spins, but not radiation widths, can be obtained from total cross-section measurements when the opposite limit is approached, i.e., $\Gamma_n \gg \Gamma_\gamma$, and the spins of many levels in light nuclei have been determined in this way.

The first object of the scattering and capture measurements, therefore, is to determine those classes of parameters which cannot be obtained from transmission measurements, namely, the spins of levels in medium and heavy weight nuclei, and the radiation widths of levels in light nuclei and at higher energies in medium and heavy nuclei. In addition, the partial cross sections can often improve on the available data, even where transmission measurements are, in principle, sufficient. For example, examination of the resonance parameters in BNL-325, Second Edition (Hughes et al., 1960; Stehn et al., 1964; Goldberg et al., 1966a,b; and Stehn et al., 1965) taken in conjunction with the original papers referred to in the text shows that a considerable body of data has already been built up. One of the most impressive aspects of a study of the literature on this work is the great improvement in the quality and in the speed of the measurements over the last few years. These measurements have all used time-of-flight spectroscopy, and the improvements referred to have been brought about partly by enormous increases in the power of the pulsed neutron sources available, and partly by the greatly improved detector systems, electronics and data handling systems; all of these developments have led to improvements in resolution and speed. Whereas the early experiments spent perhaps years in determining the parameters of a few levels, scattering and capture measurements on present day time-of-flight spectrometers using thousands of timing channels can produce good data on tens to hundreds of levels in a few days. Indeed, the main impediment to the production of even more data on resonance levels during the last few years has been the problem of automating the analysis of the observed data. This problem is rapidly being overcome, however, (see, for example, Atta and Harvey, 1962; Bhat and Chrien, 1966; Fuketa et al., 1967; Ribon et al., 1967b; Fröhner, 1966), and we can look forward to a great flow of high-quality resonance data in the near future.

Already the data obtained from the partial cross-section work has led to the determination of many radiation widths of the same nuclide, and information now exists on the distribution of radiation widths. In determining the radiation widths of 17 resonances in thorium, Asghar et al., (1966a) observed a spread of 8 (± 4) % in the widths which could not be explained by statistical or systematic errors. They have interpreted their data in terms of a 10% admixture of high-energy gamma-ray transitions which leads to fewer channels and hence a large spread in the radiation widths. Moxon (1966) determined the radiation widths of 15 levels in Co and observed a spread in the widths which could be interpreted as a χ^2 distribution with $\nu \approx 12$. Moxon points out that this distribution is consistent with the thermal capture spectrum observed by Groshev et al. (1959). By far the largest number of radiation widths in the same nucleus has been obtained from the Los Alamos team using the nuclear explosion technique (Glass et al., 1968). They have determined 58 radiation widths for ^{238}U over the neutron energy range from 20 eV to 2 keV (see Fig. 35). They have fitted their width-distribution data with a χ^2 distribution of $\nu = (44 \pm 8)$. A summary of capture areas [see Eq. (21)] for 129 resolved resonances in the kilovolt region are tabulated by Macklin and Gibbons (1965) in Table III of their review article on neutron capture at stellar temperatures.

With the improvements in neutron source intensities and sophisticated detectors, we can expect a considerable quantity of precise resonance parameters, in both zero and high-spin nuclei. We can look forward to data which will correlate resonance parameters and total and orbital angular momenta.

It was stated in Section III,A that it is generally more significant to measure capture in resonances in which the radiation width is small compared to the total width. There is, however, an exception to this, and this comes about because of the peculiar nature of capture measurements. In capture measurements on Fe up to 100 keV, Hockenbury et al. (1969) observed four times as many resonances in capture as was observed by the highest resolution transmission measurements carried out at Columbia (Rainwater et al., 1963, and also listed in BNL-325 (2nd ed.) by Goldberg et al., 1966a). Although the capture measurements were carried out with ~3 times poorer resolution, they were ~30 times more sensitive to seeing small levels than the transmission measurements. In a transmission measurement, the potential cross section of 5 to 10 b is always present, and hence all small levels must be detected against this "background" cross section; the same argument applies to a scattering measurement. Capture measurements, however, do not (to first order) detect scattered neutrons, and hence the measurement is directly sensitive to small levels. Thus capture measurements can serve to shed considerable light on level structure in nuclei. However, these small levels are in many cases caused by neutrons of higher orbital angular momenta, and extreme care must be applied in interpreting level spacings, neutron width distributions, etc., from capture measurements. Indeed, such quantities as the p-wave strength function can be obtained from these small levels, even at nuclear mass regions where such strength functions are at a minimum (Hockenbury et al., 1969; Bartolome et al., 1969; Weigmann et al., 1968).

It is interesting to note the lack of measurements of the angular distribution of resonance-energy scattered neutrons. Of course, this makes little sense below ~10 keV since the strong scattering resonances are s-wave, and hence scatter neutrons isotropically, but light and medium weight nuclei do have strong non s-wave resonances above ~10 keV. In the past, this type of measurement has been carried out with rather poor resolution with Van de Graaff accelerators (e.g., Langsdorf et al., 1957 and Block et al., 1958), and some information has been obtained on the total and orbital angular momenta of the resonances. However, with the powerful high-resolution time-of-flight spectrometers available today, it is now feasible to perform superior angular distribution measurements over a wider range of neutron energies and a first step in this direction has been reported from Harwell by Asami et al. (1969) in their determination of the l-value of the 1-keV resonance in ^{56}Fe. It is anticipated that many such measurements will be carried out in the near future.

B. Average Cross Sections

Below several kilovolts, the dominant interactions in medium and heavy weight nuclei are due to discrete s-wave resonances, and average scattering or capture measurements have little application here. In the tens to hundreds of kilovolts region the scattering cross section can exceed the capture cross section by a factor of 10 or more, and it again makes little sense to measure average scattering when it is much easier and more accurate to measure the total cross section and just subtract the capture contribution. In the case of the light elements boron and lithium, however, which have large (n,α) cross sections, this situation is reversed, and a measurement of the scattering cross section is required to obtain the absorption cross section accurately by subtraction from the total cross section. Because of the importance of these (n,α) cross sections for neutron flux measurement, considerable thought is currently being given to the problem of making these very difficult scattering measurements in the energy region where the (n,α) cross sections have a simple form, say below 100 keV.* Very recently a scattering measurement has been reported on ^{10}B below 100 keV (Asami and Moxon, 1969) which, when combined with transmission data,

---

* See, for example, Proc. Symp. Absolute Determination Neutron Flux, Oxford, 1963. EANDC-33 "U."

indicates a departure from $1/v$ behavior of the (n,α) cross section of $\sim 5\%$ at 20 keV.

Furthermore, in this higher energy range, p-wave and even d-wave contributions become important and optical model parameters can be extracted from the angular distribution of neutrons scattered over an energy region which averages over many resonances. To date, however, no measurement of this type has been carried out at a time-of-flight spectrometer although many measurements have been carried out at Van de Graaff accelerators (e.g., Langsdorf et al., 1957).

In addition to their practical application to reactor design, the measurements of average capture cross sections in the kilovolt region yield information on the strength functions and average parameters of the higher orbital angular momenta waves. In a review by Cameron et al. (1963) the authors obtain (following the capture theory of Lane and Lynn, 1957) the following formula for the average capture cross section due to resonances of spin J and partial wave l [Eq. (1) of Cameron et al. (1963)]:

$$\langle\sigma_\gamma(Jl)\rangle = \frac{(2J+1)\pi^2}{(2I+1)k^2\langle D(\lambda J)\rangle} \sum_j \frac{\langle\Gamma_n(\lambda Jlj)\rangle\langle\Gamma_\gamma(\lambda J)\rangle}{\langle\Gamma(\lambda J)\rangle} S\left(\frac{\langle\Gamma_n(\lambda Jlj)\rangle}{\langle\Gamma_\gamma(\lambda J)\rangle}\right)$$

where $D(\lambda J)$ is the average level spacing for resonances of spin J and parity λ; the sum is taken over all channel spins j; S is a correction factor because the ratio of average widths is used in the formula instead of the average of the ratio of widths; k is the neutron de Broglie wave number; and I, Γ_n, Γ_γ, and Γ have their usual meanings. As pointed out by these authors, in the capture cross section, the lower angular momenta saturate at lower neutron energies than they do in the total cross section. As a result the higher partial waves give large contributions to the capture cross section at relatively low energy (e.g., at ~ 10 keV for p-waves), and capture cross sections in the kilovolt region have been analyzed to yield information on average level spacings D, on radiative strength functions $\langle\Gamma_\gamma\rangle/D$ and neutron strength functions $\langle\Gamma_n^0\rangle/D$. In many cases excellent theoretical fits have been made to experimental capture data, but unfortunately these interpretations are subject to large uncertainties.

Theoretical fits to average capture cross sections have recently been reported from the USSR (Zakharova and Milishev, 1966), Italy (Benzi and Bortolani, 1967), and the USA (Stupegia et al., 1967). The Argonne activation capture data between 3 keV and 3 MeV have been fitted by Stupegia et al. (1967) to the theory of Moldauer (1964a,b) in which they fitted the capture cross sections of nine nuclides with seven fixed optical model parameters and with Fermi gas model level densities determined from other experiments. In this fit the authors varied the capture strength function, but for many nuclides the value of $\langle \Gamma_\gamma \rangle/D$ corresponding to a good fit differed greatly from values obtained in other experiments or from reasonable theoretical estimates. This is similar to the results of earlier fits to capture data in which low-energy average parameters had to be drastically altered to obtain fits, and thus we are still not in a position with our present knowledge to extract precise information about nuclear parameters from average capture-cross-section data.

In the past few years a considerable effort has gone into making accurate capture measurements to reduce the large spread of values reported from different laboratories. Perhaps the greatest dispute has centered on the 30 keV gold cross section, and in a recent paper, Pönitz (1967) has summarized 14 different measurements reported since 1960 for this quantity ranging from (0.880 ± 0.090) to (0.525 ± 0.100) b. Pönitz lists his own experimental determination of (0.598 ± 0.012) b, but it will require additional confirmation from other laboratories before such an accuracy will be accepted for this cross section. It is anticipated that a considerable amount of effort will continue to go into making these precise measurements.

REFERENCES

ADAMCHUK, YU. V., DANELYAN, L. S., MURADYAN, G. V., and SCHEPKIN, YU. G. (1966). *Proc. Paris Symp. Nuclear Data Reactors*, Paper CN-23/108. IAEA, Vienna.

ALBERT, R. D., and GAERTTNER, E. R. (1954). Knolls Atomic Power Lab. Rept. KAPL 1083.

ASAMI, A., MOXON, M. C., and STEIN, W. E. (1969). *Phys. Letters* **23B**, 656.

ASAMI, A., and MOXON, M. C. (1969). UKAEA Rept. AERE-R 5980.

ASGHAR, M. (1967). EANDC(UK)70'S'.

ASGHAR, M., and BROOKS, F. D. (1966). *Nucl. Inst. Methods* **39**, 68.

ASGHAR, M., CHAFFEY, C. M., MOXON, M. C., PATTENDEN, N. J., RAE, E. R., and UTTLEY, C. A. (1966a). *Nucl. Phys.* **76**, 196.

ASGHAR, M., MOXON, M. C., and CHAFFEY, C. M. (1966b). *Proc. Antwerp Conf. Nuclear Structure Studies Neutrons*. North Holland Publ., Amsterdam.

ASGHAR, M., CHAFFEY, C. M., and MOXON, M. C. (1966c). *Nucl. Phys.* **85**, 305.

ASGHAR, M., MOXON, M. C., and PATTENDEN, N. J. (1967). *Proc. Paris Symp. Nuclear Data Reactors*, Paper CN-23/31. IAEA, Vienna.

ATTA, S. E., and HARVEY, J. A. (1962). *J. Soc. Ind. Appl. Math.* **10**, 617.

BAME, S. J., and CUBITT, R. L. (1959). *Phys. Rev.* **113**, 256.

BARTOLOME, Z. M., MOYER, W. R., HOCKENBURY, R. W., TATARCZUK, J. R., and BLOCK, R. C. (1968). *Neutron Cross Sections Technol., 2nd, Washington, D.C.*, p. 795.

BARTOLOME, Z. M., HOCKENBURY, R. W., MOYER, W. R., TATARCZUK, J. R., and BLOCK, R. C. (1969). *Nucl. Sci. Eng.* **37**, 137.

BELANOVA, T. S. (1958). *Zhur. Eksperim. i Teor. Fiz.* **34**, 574 [*Soviet Phys. JETP (English Transl.)* **34**, 397 (1958)].

BELANOVA, T. S. (1960). *At. Energ. (USSR)* **8**, 549 [(1961). *Sov. J. At. Energy (English Trans.)* **8**, 462].

BELANOVA, T. S., VANKOV, A. A., MIKHAILUS, F. F., and STAVISSKII, YU. YA. (1965).
   At. Energ. (USSR) **19**, 3.
BELANOVA, T. S., VANKOV, A. A., MIKHAILUS, F. F., and STAVISSKII, YU. YA. (1967).
   Proc. Paris Symp. Nuclear Data Reactors, Paper CN-23/96. IAEA, Vienna.
BENZI, V., and BORTOLANI, M. V. (1967). Proc. Paris Symp. Nuclear Data Reactors,
   Paper CN-23/115. IAEA, Vienna.
BERGMANN, A. A., ISAKOV, A. I., MURIN, I. D., SHAPIRO, F. I., SHTRANIKH, I. V., and
   KAZARNOVSKY, M. V. (1955). Intern. Conf. Peaceful Uses Atomic Energy **4**, 135.
BERGMANN, A. A., KAPCHIGASKEV, S. P., POPOV, YU. P., and ROMANOV, S. A. (1966).
   Proc. Antwerp Conf. Nucl. Structure Studies Neutrons Paper 182. North Holland
   Publ., Amsterdam.
BETHE, H. A. (1937). Rev. Mod. Phys. **9**, 69.
BHAT, M. R. (1966). Can. J. Phys. **44**, 399.
BHAT, M. R., and CHRIEN, R. E. (1966). Proc. Antwerp Conf. Nucl. Structure Studies
   Neutrons, Paper 66. North Holland Publ., Amsterdam.
BILPUCH, E. G., WESTON, L. W., and NEWSON, H. W. (1960). Ann. Phys. **10**, 455.
BLOCK, R. C., and MOXON, M. C. (1963). Bull. Am. Phys. Soc. **8**, 513.
BLOCK, R. C., HAEBERLI, W., and NEWSON, H. W. (1958). Phys. Rev. **109**, 1620.
BLOCK, R. C., SLAUGHTER, G. G., WESTON, L. W., and VONDERLAGE, F. C. (1961). Proc.
   Saclay Symp. Neutron Time Flight Methods, EANDC, Brussels, p. 203.
BLOCK, R. C., RUSSELL, J. E., and HOCKENBURY, R. W. (1964a). Oak Ridge Rept. No.
   ORNL-**3778**, p. 53.
BLOCK, R. C., HOCKENBURY, R. W., and RUSSELL, J. E. (1964b). Wash-1048 EANDC-
   (US)-57 "U", p. 71.
BLOCK, R. C., HOCKENBURY, R. W., and RUSSELL, J. E. (1965). Oak Ridge Rept. No.
   ORNL-**3924**, p. 31.
BLOCK, R. C., HOCKENBURY, R. W., BARTOLOME, Z. M., and FULLWOOD, R. R. (1967).
   Proc. Paris Symp. Nuclear Data Reactors, Paper CN-23/126. IAEA, Vienna.
BLOKHIM, D. J., BLOKHINTSEV, D. J., BLYMKINA, YU. A., and BONDARENKO, I. I.
   (1961). At. Energ. (USSR) **10**, 5, 437.
BOGART, D., and SEMLER, T. T. (1966). Proc. Conf. Neutron Cross Section Technol.,
   Washington, D.C., 1966. CONF-660303.
BOLLINGER, L. M., COTÉ, R. E., and THOMAS, G. E. (1950). Proc. 2nd Intern. Conf. Peace-
   ful Uses Atomic Energy **15**, 127.
BOOTH, R., BALE, W. P., and MACGREGOR, M. H. (1958). Phys. Rev. **112**, 226.
BORST, L. B. (1953). Phys. Rev. **90**, 859.
BROCKHOUSE, B. N. (1953). Can. J. Phys. **31**, 432.
BROCKHOUSE, B. N., HURST, D. G., and BLOOM, M. (1951). Phys. Rev. **83**, 840.
BROOKS, F. D. (1961). Proc. Saclay Symp. Neutron Time Flight Methods. EANDC,
   Brussels.
BYERS, D. H., DIVEN, B. C., and SILBERT, M. G., (1966). Proc. Conf. Neutron Cross
   Section Technol. Washington, D.C., 1966, p. 903. CONF-660303, Book 2.
CAMERON, A. G. W., LAZAR, N. H., and SCHMITT, H. W. (1963). "Fast Neutron Physics"
   (ed. by J. P. Marion and J. L. Fowler), Part II. Wiley, New York.
CAO, M. G., TILIGNECO, E., THEOBALD, J. P., and WARTENA, J. A. (1968). Conf. Neutron
   Cross Sect. Technol. 2nd, Washington, D.C., p. 513.
CASE, K. M., DE HOFFMANN, F., and PLACZEK, G. (1953). "Introduction to the Theory
   of Neutron Diffusion," Vol. I. U.S. Govt. Printing Office, Washington, D.C.
CEULEMANS, H., and POORTMANS, F. (1961). J. Phys. Radium **22**, 707.
CEULEMANS, H., NEVE DE MEVERGNIES, M., and POORTMANS, F. (1962). Nucl. Inst.
   Methods **17**, 342.

CHAUBEY, A. K., and SEHGAL, M. L., (1965). *Nucl. Phys.*, **66**, 267.
COCEVA, C. (1963). *Nucl. Inst. Methods* **21**, 93.
COLLINS, E. R. (1954). Unpublished Harwell Committee Paper.
CORVI, F. (1962). Private communication (measurements carried out at Harwell).
COX, S. A. (1961). *Phys. Rev.* **122**, 1280.
DANELYAN, L. S., and EFIMOV, B. V. (1963). *At. Energ. (USSR)* **14**, 264.
DANELYAN, L. S., ADAMCHUK, YU. V., MOSKALEV, S. S., PEVZNER, M. I., and YASTREBOV, S. S. (1964). *At. Energ. (USSR)* **16**, 56 [(1964). *Soviet J. At. Energy (English Transl.)* **16**, 58].
DE SAUSSURE, G., WESTON, L. W., GWIN, R., RUSSELL, J. E., and HOCKENBURY, R. W. (1965). *Nucl. Sci. Eng.* **23**, 45.
DE SAUSSURE, G., WESTON, L. W., GWIN, R., INGLE, R. W., TODD, J. H., HOCKENBURY, R. W., and FULLWOOD, R. R. (1966). *Proc. Paris Symp. Nucl. Data Reactors*, Paper CN-23/48. IAEA, Vienna.
DESJARDINS, J. S., ROSEN, J. L., HAVENS, W. W., JR., and RAINWATER, L. J. (1960). *Phys. Rev.* **120**, 2214.
DIVEN, B. C., TERRELL, J., and HEMMENDINGER, A. (1960). *Phys. Rev.* **120**, 556.
DOVBENKO, A. G., KOLESOV, V. E., KONONOV, V. N., and STAVISSKII, YU. YA. (1966). *Proc. Antwerp Conf. Nucl. Structure Studies Neutrons*, Paper 199. North-Holland Publ., Amsterdam.
DRESNER, L. (1962). *Nucl. Instr. Methods* **16**, 176.
EVANS, J. E., KINSEY, B. B., WATERS, J. R., and WILLIAMS, G. H. (1958). *Nucl. Phys.* **9**, 205.
FIRK, F. W. K., SLAUGHTER, G. G., and GINTHER, R. J. (1961). *Nucl. Instr. Methods* **13**, 313.
FIRK, F. W. K., LYNN, J. E., and MOXON, M. C. (1963). *Nucl. Phys.* **9**, 205.
FOOTE, H. L. (1958). *Phys. Rev.* **109**, 1641.
FRASER, J. S., and SCHWARTZ, R. B. (1962). *Nucl. Phys.* **30**, 269.
FRIEDES, J. L. (1964). Multiple Scattering Corrections to Partial Cross Section Measurements, BNL 8028.
FRIESENHAHN, S. J., LOPEZ, W. M., FRÖHNER, F. H., CARLSON, A. D., and COSTELLO, D. G. (1968). *Neutron Cross Sections Technol.*, *2nd, Washington, D.C.*, p. 695.
FRÖHNER, F. H. (1966). TACASI-A Fortran IV Code for Least Squares Analysis of Neutron Resonance Data, General Atomics rept. GA-6906.
FRÖHNER, F. H., and HADDAD, E. (1965). *Nucl. Phys.* **71**, 129.
FRÖHNER, F. H., HADDAD, E., LOPEZ, W. M., and FRIESENHAHN, S. J. (1966). *Proc. Conf. Neutron Cross Section Technol., Washington, D.C., 1966*, P. 43, CONF-660303.
FUKETA, T., MENZEL, J. H., HOCKENBURY, R. W., and RUSSELL, J. E. (1964). Wash-1084, EANDC, (US)-57 "U", p. 93.
FUKETA, T., ASAMI, A., OHKUBO, M., NAKAJIMA, Y., KAWARASAKI, Y., and TAKEKOSHI, H. (1967). *Proc. Paris Symp. Nucl. Data Reactors*, Paper CN-23/17. IAEA, Vienna.
GABBARD, F., DAVIS, R. H., and BONNER, T. W. (1959). *Phys. Rev.* **114**, 201.
GAERTTNER, E. R., YEATER, M. L., and ALBERT, R. D. (1954). *Knolls Atomic Power Lab. KAPL* 1084.
GIBBONS, J. H., MACKLIN, R. L., MILLER, P. D., and NEILER, J. H. (1961). *Phys. Rev.* **122**, 182.
GLASS, N. W., THRESHOLD, J. K., SCHELBERG, A. D., WARREN, J. H., and TATRO, L. D. (1966). *Proc. Conf. Neutron Cross Sect. Technol. Washington, D.C. 1966*, p. 766 CONF-660303, Book 2.

GLASS, N. W., SCHELBERG, A. D., TATRO, L. D., and WARREN, J. H. (1968). *Neutron Cross Sections Technol., 2nd, Washington, D.C.*, p. 573.

GOLDBERG, M. D., MUGHABGHAB, S. F., MAGURNO, B. A., and MAY, V. M. (1966a). "Neutron Cross Sections," Vol. IIA, $Z = 21$ to 40, Suppl. No. 2. BNL-325, 2nd ed.

GOLDBERG, M. D., MUGHABGHAB, S F., PUROHIT, S. N., MAGURNO, B. A., and MAY, V. M. (1966b). "Neutron Cross Sections," Vol. IIB, $Z = 41$ to 60, Suppl. No. 2, BNL-325 2nd ed.

GROSHEV, L. V., LUTSENKO, V. N., DEMIDOV, A. M., and PELEKHOV, V. I. (1959). "Atlas of Gamma-Ray Spectra from Radiative Capture of Thermal Neutrons." Pergamon Press, Oxford.

GWIN, R., WESTON, L. W., DE SAUSSURE, G., INGLE, R. W., TODD, J. H., GILLESPIE, F. E., and HOCKENBURY, R. W. (1969). ORNL Neutron Phys. Div. Progr. Rept. for Period Ending May 31, 1968, ORNL-4280.

HADDAD, E., WALTON, R. B., FRIESENHAHN, S. J., and LOPEZ, W. M. (1964). *Nucl. Instr. Methods* **31**, 125.

HADDAD, E., FRIESENHAHN, S. J., FRÖHNER, F. H., and LOPEZ, W. M. (1965). *Phys. Rev.* **140**, B50.

HARLOW, M. V., SCHELBERG, A. D., TATRO, L. D., WARREN, J. H., and GLASS, N. W. (1968). *Neutron Cross Sections Technol., 2nd, Washington, D.C.*, p. 837.

HARRIS, K. K., GRENCH, H. A., JOHNSON, R. G., VAUGHN, F. J., FERZIGER, J. H., and SHER, R., (1965). *Nucl. Phys.* **69**, 37.

HARVEY, J. A. (1959). *Bull. Am. Phys. Soc.* **4**, 473.

HENKEL, R. L., and BARSCHALL, H. H. (1950). *Phys. Rev.* **80**, 145.

HOCKENBURY, R. W., BARTOLOME, Z. M., MOYER, W. R., TATARCZUK, J. R., and BLOCK, R. C. (1968). *Neutron Cross Sections Technol., 2nd, Washington, D.C.*, p. 729.

HOCKENBURY, R. W., and MOYER, W. R. (1969). *Nucl. Inst. Methods* **75**, 45.

HOCKENBURY, R. W., BARTOLOME, Z. M., TATARCZUK, J. R., MOYER, W. A., and BLOCK, R. C. (1969). *Phys. Rev.* **178**, 1746.

HUGHES, D. J. (1955). *J. Nucl. Energy* **1**, 237.

HUGHES, D. J., MAGURNO, B. A., and BRUSSEL, M. K. (1960). "Neutron Cross Sections," Suppl. No. 1, BNL-325.

HURST, D. G., PRESSESKY, A. J., and TUNNICLIFFE, P. R. (1950). *Rev. Sci. Instr.* **21**, 705.

ILIESCU, N., SAN, KIM HI, PIKELNER, L. B., SHARAPOV, E. I., and SIRAZHET, H. (1965). *Nucl. Phys.* **72**, 298.

ISAKOV, A. I., POPOV, YU. P., and SHAPIRO, F. L. (1961). Transl. from *Tr. Tashkentsk. Konf. po Mirnomu Ispol'z At. Energii Akad. Nauk Uz.* SSR. **1**, 64 (AEC-tr-6398 pg. 70–80) [(1960) *Soviet Phys. JETP (English Transl.)* **11**, 712].

JACKSON, H. E., and THOMAS, G. C. (1965). *Rev. Sci. Instr.* **36**, 419.

KAPCHIGASKEV, S. P. (1965). *At. Energ.* **19**, 294.

KAPCHIGASKEV, S. P., and POPOV, YU. P. (1963). *At. Energy* **15**, 120 [Trans. in (1964). *Soviet J. At. Energy* **15**, 808].

KAPCHIGASKEV, S. P., and POPOV, YU. P. (1964). *At. Energy* **16**, 256.

KASHUKEEV, N. T., POPOV, YU. P., and SHAPIRO, F. L. (1961). *J. Nucl. Energy* **14**, 76.

KING, T. J. Neutron scattering measurements in the resonance region (1969). Thesis in partial fulfillment of the requirements for the Ph.D., Rensselaer Polytechnic Institute, Troy, New York.

KING, T. J., and BLOCK, R. C. (1968). *Neutron Cross Sections Technol., 2nd, Washington, D.C.*, p. 735.

KING, T. J., FULLWOOD, R. R., MOYER, W. R., TATARCZUK, J. R., and BLOCK, R. C. (1967a). *Bull. Am. Phys. Soc.* **12**, 512.

KING, T. J., FULLWOOD, R. R., and BLOCK, R. C. (1967b). *Nucl. Instr. Methods* **52**, 321.

KOMPE, D. (1967). *Proc. Paris Symp. Nucl. Data Reactors*, Paper CN-23/10. IAEA, Vienna.

KONKS, V. A., and SHAPIRO, F. L. (1964). *Zh. Eksperim. i. Teor. Fiz.* **47**, 795 [(1965) *Soviet Phys. JETP (English Transl.)* **20**, 531].

KONKS, V. A., and FENIN, YU. I. (1966). *Proc. Antwerp Conf. Nucl. Structure Studies Neutrons*, Paper 202. North Holland Publ., Amsterdam.

KONKS, V. A., POPOV, YU. P., and SHAPIRO, F. L. (1964). *Zh. Eksperim. i Teor. Fiz.* **46**, 80 [(1964) *Soviet Phys. JETP (English Transl.)* **19**, 59].

KONONOV, V. N., STAVISSKII, YU. YA., and TOLSTIKOV, V. A. (1958). *At. Energ.* **5**, 564 [(1959/60) *J. Nucl. Energy* **11**, 46].

KONONOV, V. N., STAVISSKII, YU. YA., CHISTOZVANOV, S. R., and SHORIN, V. S. (1967). *Proc. Paris Symp. Nucl. Data Reactors*, Paper CN-23/99. IAEA, Vienna.

LAMB, W. E. (1939). *Phys. Rev.* **55**, 190.

LANE, A. M., and LYNN, J. E. (1957). *Proc. Phys. Soc. (London)* **70A**, 557.

LANGSDORF, A., LANE, A. O., and MONAHAN, J. E. (1957). *Phys. Rev.* **107**, 1077.

LEIPUNSKY, A. I., KAZACHKOVSKY, O. D., ARTYUKOV, G. Y., BARYSHNIKOV, A. I., BELANOVA, T. S., GALKOV, V. N., STAVISSKII, YU. YA., STUMBAR, E. A., and SHERMAN, L. E. (1958). *Geneva Conf., 2nd* **15**, 50.

LE RIGOLEUR, C., BLUET, J. C., and LEROY J. L. (1966). *Proc. Antwerp Conf. Nucl. Structure Studies Neutrons*. North Holland Publ., Amsterdam.

LOPEZ, W. M., FRÖHNER, F. H., FRIESENHAHN, S. J., and CARLSON, A. D. (1968). *Neutron Cross Sections Technol., 2nd, Washington, D.C.*, p. 857.

LOPEZ, W. M., HADDAD, E., FRIESENHAHN, S. J., and FRÖHNER, F. H. (1967). *Nucl. Phys.* **A93**, 340.

LYNN, J. E. (1963). Monte Carlo code written at AERE, Harwell, Private communication.

MACKLIN, R. L. (1961). Harwell Rept. AERE R-3744.

MACKLIN, R. L., and GIBBONS, J. H. (1965). *Rev. Mod. Phys.* **37**, 166.

MACKLIN, R. L., and GIBBONS, J. H. (1967). *Phys. Rev.* **159**, 1007.

MACKLIN, R. L., LAZAR, N. H., and LYON, W. S. (1957). *Phys. Rev.* **107**, 504.

MACKLIN, R. L., GIBBONS, J. H., and INADA, T. (1962). *Nature* **194**, 1272.

MACKLIN, R. L., GIBBONS, J. H., and INADA, T. (1963a). *Phys. Rev.* **129**, 2695.

MACKLIN, R. L., GIBBONS, J. H., and INADA, T. (1963b). *Nucl. Phys.* **43**, 353.

MACKLIN, R. L., GIBBONS, J. H., and INADA, T. (1963c). *Nature* **197**, 369.

MESERVEY, E. (1952). *Columbia Univ. Rept., CV* 99, 100.

MESERVEY, E. (1954). *Phys. Rev.* **96**, 1006.

MISKEL, J. A., MARSH, K. V., LINDER, M., and NAGLE, R. J. (1962). *Phys. Rev.* **128**, 2717.

MITZEL, F., and PLENDL, H. S. (1964). *Nukleonik* **6**, 371.

MOLDAUER, P. A. (1964a). *Rev. Mod. Phys.* **36**, 1079.

MOLDAUER, P. A. (1964b). *Phys. Rev.* **135**, B642.

MOORE, M. S., and REICH, C. W. (1960). *Phys. Rev.* **118**, 718.

MOORE, M. S., and SIMPSON, F. B. (1962). *Nucl. Sci. Eng.* **13**, 18.

MORTON, K. W. (1957). *J. Nucl. Energy* **5**, 320.

MOXON, M. C. (1966). *Proc. Antwerp Conf. Nucl. Structure Studies Neutrons*, Paper 88. North-Holland Publ., Amsterdam.

Moxon, M. C., and Rae, E. R. (1961). *Proc. Saclay Symp. Neutron Time-of-Flight Methods*, p. 429. EANDC, Brussels.

Moxon, M. C., and Mycock, C. (1962). Unpublished Harwell Committee Paper.

Moxon, M. C., and Chaffey, C. M. (1963). Unpublished Harwell Committee Paper.

Moxon, M. C., and Rae, E. R. (1963). *Nucl. Instr. Methods* **24**, 445.

Moxon, M. C., and Pattenden, N. J. (1967). *Proc. Paris Symp. Nucl. Data Reactors*, Paper CN-23/27. IAEA, Vienna.

Oleksa, S. (1958). *Phys. Rev.* **109**, 1645.

Pikelner, L. B., Pshitula, M. I., San, Kim Ki, Cheng, Ling Yen., and Sharapov, E. I. (1963). *Pribory i Tekhn. Eksperim.* **2**, 48, 51.

Poenitz, W. (1967). *Proc. Paris Symp. Nucl. Data Reactors*, Paper CN-23/6. IAEA, Vienna.

Poortmans, F., and Ceulemans, H. (1963). *Proc. Conf. Nucl. Phys. Reactor Neutrons*. Argonne Natl. Lab. (ANL-6797), p. 363.

Poortmans, F., and Ceulemans, H. (1966). *Proc. Antwerp Conf. Nucl. Structure Studies Neutrons*, Paper 94. North-Holland Publ., Amsterdam.

Poortmans, F., Ceulemans, H., and Neve de Mevergnies, M. (1967). *Proc. Paris Symp. Nucl. Data Reactors*. Paper CN-23/79. IAEA, Vienna.

Popov, Yu. P. (1964). *Tr. Fiz. Inst. Akad. Nauk SSSR* **24**, 111.

Popov, Yu. P., and Shapiro, F. L. (1962). *Soviet Physics JETP (English Transl.)* **15**, 683.

Rae, E. R., and Bowey, E. M. (1957). *J. Nucl. Energy. I* **4**, 179.

Rae, E. R., Collins, E. R., Kinsey, B. B., Lynn, J. E., and Wiblin, E. R. (1958). *Nucl. Phys.* **5**, 89.

Rainwater, L. J., Havens, W. W., Jr., Dunning, J. R., and Wu, C. S. (1947). *Phys. Rev.* **71**, 65.

Rainwater, L. J., Garg, J. B., and Havens, W. W., Jr. (1963). *Bull. Am. Phys. Soc.* **8**, 334.

Ribon, P., Lottin, A., Michaudon, A., and Trochon, J. (1966). *Proc. Antwerp Conf. Nucl. Structure Studies Neutrons*, Paper 165. North-Holland Publ., Amsterdam.

Ribon, P., Cauvin, B., Derrien, H., Michaudon, A., Silver, E., and Trochon, J. (1967a). *Proc. Paris Symp. Nucl. Data Reactors*, Paper CN-23/72. IAEA, Vienna.

Ribon, P., Cauvin, B., Derrien, H., Michaudon, A., and Sanche, M. (1967b). *Proc. Paris Symp. Nucl. Data Reactors*, Paper CN-23/71. IAEA, Vienna.

Robertson, J. C. (1965). *Nucl. Phys.* **71**, 417.

Romanov, S. A., and Shapiro, F. L. (1965). *Yadern. Fis.* **1**, 229 [(1965) *Soviet J. Nucl. Phys. (English Transl.)* **1**, 159].

Rosen, J. L., Desjardins, J. S., Rainwater, L. J., and Havens, W. W., Jr. (1960). *Phys. Rev.* **118**, 687.

Russell, J. E., Hockenbury, R. W., and Block, R. C. (1963). *Bull. Am. Phys. Soc.* **8**, 80.

Ryves, T. B., Robertson, J. C., Axton, E. J., Goodier, I., and Williams, A. (1966). *J. Nucl. Energy* **20**, 249.

San, Kim Hi., Pikelner, L. B., Sirazhet, H., and Sharapov, E. I. (1966). *Proc. Antwerp Conf. Nucl. Structure Studies Neutrons*, Papers 185, 188. North-Holland Publ., Amsterdam.

Sauter, G. D., and Bowman, C. D. (1965). *Phys. Rev. Letters* **15**, 761.

Sauter, G. D., and Bowman, C. D. (1967). *Nucl. Instr. Methods* **55**, 141.

SAUTER, G. D., and BOWMAN, C. D. (1968). *Neutron Cross Sections Technol., 2nd, Washington, D.C.*, p. 541.

SCHMITT, H. W. (1960). Oak Ridge Natl. Lab. Rept. ORNL-2883.

SCHMITT, H. W., and COOK, C. W. (1960). *Nucl. Phys.* **17**, 109.

SHEER, C., and MOORE, J. A. (1955). *Phys. Rev.* **98**, 565.

SIMPSON, F. B., MILLER, L. G., MOORE, M. S., HOCKENBURY, R. W., and KING, T. J. (1966). Rensselaer Polytechnic Inst. LINAC Prog. Rept. for July thru Sept. 1966, p. 31.

SINGH, P. P. (1963). *Bull. Am. Phys. Soc.* **8**, 334.

STAVISSKII, YU. YA., and TOLSTIKOV, V. A. (1960). *At. Energ.* **9**, 401 [(1962) *J. Nucl. Energy* **16**, 496].

STAVISSKII, YU. YA., and TOLSTIKOV, V. A. (1961). *At. Energ.* **10**, 508 [(1963) *J. Nucl. Energy* **17**, 579].

STAVISSKII, YU. YA., and SHAPAR, A. V. (1961). *At. Energ.* **10**, 264.

STEHN, J. R., GOLDBERG, M. D., MAGURNO, B. A., and WIENER-CHASMAN, RENATE (1964). "Neutron Cross Sections," Vol. I, $Z = 1$ to 20, Suppl. No. 2. BNL-325, 2nd ed.

STEHN, J. R., GOLDBERG, M. D., WIENER-CHASMAN, RENATE, MUGHABGHAB, S. F., MAGURNO, B. A., and MAY, V. M. (1965). "Neutron Cross Sections," Vol. III, $Z = 88$ to 98, Suppl. No. 2. BNL-325, 2nd ed.

STUPEGIA, D. C., SCHMIDT, M., and MADSON, A. A. (1965). *J. Nucl. Energy* **19**, 767.

STUPEGIA, D. C., KEEDY, C. R., SCHMIDT, M., and MADSON, A. A. (1967). *Proc. Paris Symp. Nucl. Data Reactors*, Paper CN-23/51. IAEA, Vienna.

SULLIVAN, J. G., and WERNER, G. K. (1964). Oak Ridge Natl. Lab., Private communication.

TITTMAN, J., and SHEER, C. (1951). *Phys. Rev.* **83**, 746.

TITTMAN, J., SHEER, C., RAINWATER, L. J., and HAVENS, W. W., Jr. (1950). *Phys. Rev.* **80**, 903.

TOLSTIKOV, V. A., SHERMAN, L. E., and STAVISSKII, YU. YA. (1963). *At. Energ.* **15**, 414 [(1964) *J. Nucl. Energy* **18**, 599].

TOLSTIKOV, V. A., KOLESOV, V. E., DOVBENKO, A. G., and STAVISSKII, YU. YA. (1964). *At. Energ.* **17**, 505 (Trans. in [(1965) *Soviet J. At. Energy* **17**, 1272].

TOLSTIKOV, V. A., KOROLEVA, V. P., KOLESOV, V. E., and DOVBENKO, A. G. (1967). *Proc. Paris Symp. Nuclear Data Reactors*, Paper CN-23/103. IAEA, Vienna.

UTTLEY, C. A., and JONES, R. H. (1963). Unpublished Harwell Committee Paper.

VERVIER, J. F. (1958). *Nucl. Phys.* **9**, 569.

WANG, NAI-YANG, VIZI, I., EFIMOV, V. N., KARZHAVINA, E. N., SAN, KIM HI, POPOV, A. B., SHELONTSEV, I. J., SHIRIKOVA, N. YU., and YZAVITSKII, YU. S. (1963). *J. Exp. Theor. Phys. (USSR)* **45**, 1743.

WANG, NAI-YANG, KARZHAVINA, E. N., POPOV, A. B., YAZOITSKY, YU. S., and YAO-CHI-CHUAN (1966). *Proc. Antwerp Conf. Nucl. Structure Studies Neutrons*, Paper 187. North-Holland Publ., Amsterdam.

WATERS, J. R., EVANS, J. E., KINSEY, B. B., and WILLIAMS, G. H. (1959). *Nucl. Phys.* **12**, 563.

WEIGMANN, H., and SCHMID, H. (1967). *Nucl. Phys.* **A104**, 513.

WEIGMANN, H., CARRARO, G., and BÖCKHOFF, K. H. (1967). *Nucl. Instr. Methods* **50**, 267

WEIGMANN, H., WINTER, J., and SCHMID, H. (1968). *Neutron Cross Sections Technol., 2nd, Washington, D.C.*, p. 533.

WESTON, L. W., and LYON, W. S. (1961). *Phys. Rev.* **123,** 948.
WESTON, L. W., SETH, K. K., BILPUCH, E. G., and NEWSON, H. W. (1960). *Ann. Phys.* **10,** 477.
WOOD, R. E. (1956). *Phys. Rev.* **104,** 1425.
ZAKHAROVA, S. M., and MALISHEV, A. V. (1966). *Proc. Antwerp Conf. Nucl. Structure Studies Neutrons,* Paper 201. North-Holland Publ., Amsterdam.

IV

GAMMA RAYS FROM NEUTRON CAPTURE IN RESONANCES

LOWELL M. BOLLINGER
ARGONNE NATIONAL LABORATORY
ARGONNE, ILLINOIS

I. GENERAL CHARACTERISTICS OF THE SPECTRA

The study of the gamma-ray spectra that result from capture of neutrons in resonances is the newest of the several major branches of neutron-resonance spectroscopy. It is already a mature area of study in the sense that a wide variety of experiments have been conceived and attempted. However, it is an undeveloped area of study in the sense that we are only now passing out of the exciting early stage in which qualitative results are adequate to give answers to significant questions. Also, the development of the refined apparatus and techniques that are required to observe spectra of good quality have only recently been completed, and, in particular, the revolutionary Ge-diode gamma-ray spectrometer has not been in service very long. Consequently, in this review it should be possible to give a fairly complete outline of the qualitative nature of the information that can be obtained from the gamma-ray spectra resulting from resonance capture, but the quantitative data that can be presented now might soon be outdated.*

The study of gamma rays from resonance capture owes much of its rapid development to the wealth of background knowledge on which it is possible to build. The physical insights gained through the study of the gamma rays from capture of thermal neutrons† is especially helpful. Because of the availability of intense sources of thermal neutrons, for a number of years it has been possible to measure the gamma-ray spectra with refined, but inefficient, gamma-ray spectrometers such as the magnetic pair spectrometer and the Compton spectrometer. The spectrum given in Fig. 1 is representative of the spectra that are obtained for heavy nuclides, the kinds of targets that are of principal interest for resonance capture. One sees that the spectrum is characterized by well-resolved individual lines at both the low- and the high-energy ends of the spectrum, with a broad bell-shaped maximum of unresolved lines at intermediate energies. Since these general characteristics of the experimental spectra determine the kinds of information that one tries to obtain from neutron-capture gamma rays, let us consider briefly how the spectra are formed. The process is illustrated in Fig. 2. When the incident neutron is captured, a compound nucleus‡ is formed with an excitation energy that is equal to the neutron binding energy plus the energy of the incident neutron, typically an excitation energy of about 7 MeV. For any arbitrary neutron energy, such as the thermal energy, the compound system usually is not formed in any single nuclear state since several states (which may be either above or below the binding energy) contribute to the capture cross section. The energy of excitation may be carried off either by re-emission of the incident neutron or by emission of gamma radiation. In the latter case, the radiative transition can proceed directly to the widely spaced states at low energy, resulting in the formation of the well-resolved high-energy lines seen in Fig. 1.

---

* The selection of material for inclusion in this chapter was completed in April, 1968. References to this material were brought up to date and a few footnotes were added in July, 1969. For post deadline material see "Capture Gamma-Ray Spectroscopy." Intern. Atomic Energy Agency, Vienna, 1969.

† The characteristics of the thermal-neutron-capture gamma-ray spectra have been reviewed by Kinsey and Bartholomew (1954), Kinsey (1955, 1957), Groshev *et al.* (1958a,b), and Bartholomew (1960, 1961), Motz and Bäckström (1965), and Bartholomew, Groshev *et al.* (1967).

‡ Under special circumstances, direct transitions without the formation of a compound nucleus can also occur.

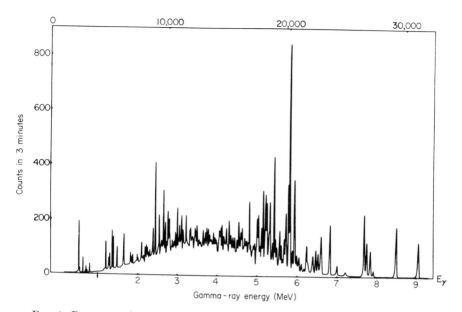

FIG. 1. Representative thermal-neutron-capture gamma-ray spectrum for a heavy nucleus, as measured with a Compton spectrometer (Groshev, 1960). The dominant reaction is $Cd^{113}(n,\gamma)Cd^{114}$.

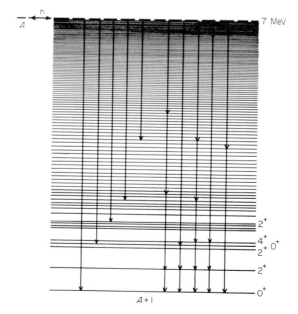

FIG. 2. Schematic representation of the radiative decay of a highly excited compound state formed by slow-neutron capture in an even-odd target nucleus.

More probably, however, the transition proceeds to one of the hundreds or thousands of closely spaced levels at intermediate energies, giving a spectrum of great complexity. Moreover, each of the intermediate states can, in turn, feed many of the states below it. Thus, for a typical heavy nucleus the usual mode of decay of the initial state is by way of a cascade in which the average number, or multiplicity, of gamma rays is about four. Since no generally applicable experimental method has been devised by which to distinguish the primary radiation by the initial state from the secondary radiation by intermediate states, the contributions from all orders of emission are superposed. To dramatize the extremely large number of ways in which the cascade can proceed, it has been said that, if all combinations of transitions were equally probable, no particular mode of decay would ever have been observed twice! Although admittedly an exaggeration of the actual degree of complexity of the capture gamma-ray spectra, this statement helps to illuminate the truth that the interpretation of the main body of the spectrum for the typical nucleus is made difficult by more than the experimental problem of resolving individual lines.

As the gamma-ray cascade nears the end of its chain, the number of states through which it can pass becomes small. Consequently, a relatively large fraction of all cascades pass through some of the low-energy states and transitions between these states are relatively intense, accounting for the strong lines in the low-energy end of the capture-gamma-ray spectrum.

The study of the gamma-ray spectra for thermal neutrons and for neutrons at resonances are similar in that the basic characteristics of the spectra are the same. As a result, in both kinds of studies one is concerned with the same three categories of data, namely, with the characteristics of the resolved high-energy gamma-ray lines, with the characteristics of the resolved low-energy lines, and with statistical information about the great mass of unresolved lines. However, the many orders of magnitude by which the intensities of the sources differ in the two kinds of measurements result in a considerable difference in emphasis. The high intensity of thermal neutrons makes it worth while for the experimenter to strive for better and better energy resolution and sensitivity in his gamma-ray spectrometer, with the result that the range of energy and the range of targets that can be studied effectively is being continually enlarged. On the other hand, the low intensity of the source in most resonance-capture experiments necessarily limits the targets that can be studied and the statistical quality of the spectra for even the most favorable targets. As a result, if his effort is to be worthwhile, the experimenter must direct his ingenuity toward making good use of the special advantage that is at his disposal, namely, the advantage of being able to vary the energy of the incident neutron.

The detailed way in which one uses the incident neutron energy as an extra variable depends, of course, on the kind of experimental information that is being sought. However, in general terms, one may list five important advantages of resonance capture over thermal capture. These are: (a) the possibility of selecting for study a wide variety of individual isotopes by observing the spectrum associated with individual resonaces in particular isotopes; (b) the possibility of selecting an initial state with a definite and *known* spin and parity; (c) the possibility of observing the spectra of a given isotope for states with *different* spins and parities; (d) the possibility of observing spectra for *many* initial states of the same spin and parity in a given nuclide; and (e) the possibility of comparing resonance and off-resonance capture, which may be different processes. Many workers in the field of neutron spectroscopy have found these distinctive features of resonance capture to be more than enough justification for the considerable experimental effort that is required to observe the gamma-ray spectra.

II. APPARATUS

Like most experiments in neutron-resonance spectroscopy, the apparatus needed for a study of the gamma rays from resonance capture fall into three major categories: a source of neutrons, a detector of neutrons or of a reaction induced by them, and an instrument in which to store the spectral information. Of these, the neutron source is of dominant importance for most experiments in neutron-resonance spectroscopy. However, in the study of resonance-capture gamma rays, the presence of the gamma-ray energy as an additional experimental variable tends to demand an equal degree of excellence in all three categories of experimental equipment, and the failure of the experimenter to satisfy this need for a balanced system usually results in experimental data of inferior quality.

A wide variety of experimental systems have been used to study resonance-capture spectra, and this multiplicity complicates our discussion. For convenience, let us separate the measurements into three categories. In one, we include all measurements in which the source of neutrons is any of the common slow-neutron spectrometers, that is, the fast chopper and the diffraction monochromator at a nuclear reactor and the pulsed accelerator that produces a broadly distributed spectrum of neutrons. To date, most of the large number of measurements of this kind have been limited to neutron energies less than 1 keV because of inadequate intensity at higher energies. Both NaI scintillators and Ge-diode detectors have been used as the gamma-ray spectrometer.

The second class of experiments contain those that make use of a pulsed electrostatic accelerator to produce a pulsed beam of neutrons in the Li(p,n) reaction. Most of the rather few measurements of this kind have made use of a large NaI scintillator to study capture in the neutron-energy range 15–300 keV.

The final class of experiments that have given significant results consists of those in which a reactor is the source of neutrons and crude neutron-energy selection is achieved by means of filters. Measurements of this kind are not useful for many purposes; but when the crude energy selection can be tolerated, the neutron-filter method has the important advantage of providing a gamma-ray spectrum that is several orders of magnitude more intense than is obtainable with more refined neutron spectrometers.

A. Measurements with Slow-Neutron Spectrometers

Let us start our discussion of apparatus by considering the class of measurements in which a slow-neutron spectrometer is the source of neutrons since a large fraction of all studies of resonance-capture spectra have been of this kind. To ensure that the general ideas involved in such a measurement are well understood, let us begin with an over-all view of a representative experimental system that might be used. Such a system is illustrated in Fig. 3. Here, the source of neutrons is a fast chopper at a reactor. Timed neutrons from the chopper are allowed to impinge on the target under investigation. In the illustration, the gamma rays that result from capture of the neutrons in the target are detected by a NaI scintillation spectrometer; today a lithium-drifted Ge-diode spectrometer would be used

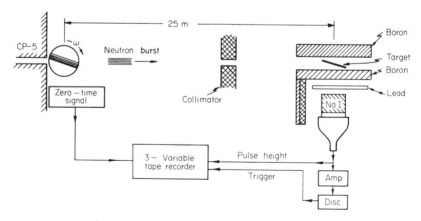

FIG. 3. Schematic representation of a time-of-flight neutron spectrometer (a fast chopper) used to measure resonance-capture spectra. (Bollinger et al., 1963.)

in most experiments. The time of flight of a captured neutron (and, hence, its energy) is obtained from the difference in time between the formation of the burst of neutrons at the chopper and the detection of gamma rays by the detector. When some arbitrary conditions on the time of flight for neutrons and on the pulse height from gamma rays are satisfied simultaneously, values of these experimental variables are written in digital form on a magnetic tape for future analysis.

The qualitative nature of the data written on the magnetic tape is illustrated in Fig. 4. Suppose we first sort out the data in such a way as to give a time-of-flight spectrum of stored counts, without placing any conditions on the value of the pulse height. The result obtained is typically a curve with resonance structure of the kind shown in the figure. One may now proceed to a more refined analysis of the data on the tape either by forming time-of-flight spectra for various conditions on the pulse height or, alternatively, by forming the pulse-height spectra associated with individual time channels or groups of time channels. One of these pulse-height spectra is shown in the insert in Fig. 4. A large part of the discussion in this chapter

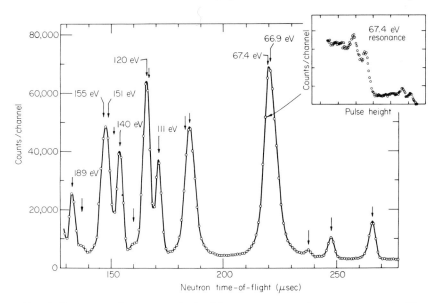

Fig. 4. Representative spectra obtained in a resonance-capture measurement with a time-of-flight system (the Argonne chopper) in which a NaI scintillator is the gamma-ray spectrometer (Bollinger *et al.*, 1963). The arrows on the time-of-flight spectrum indicate the energies of known neutron resonances. A measurement with the same neutron source and a Ge-diode gamma-ray spectrometer would have poorer neutron-energy resolution and much better gamma-ray-energy resolution.

is concerned with the interpretation of time-of-flight and pulse-height spectra of this kind.

1. Sources of Neutrons

The various sources of slow neutrons are discussed intensively in other chapters of this book and it might be thought that there is no more to be said. However, a neutron source for the study of resonance-capture gamma rays is subject to special requirements that are not emphasized elsewhere. Basically, the special circumstance to be understood is that, although the *average* counting rate of useful counts may be extremely low, at the same time the *instantaneous* rate of all counts in the detector may be so high as to cause trouble.

The nature of the experimental problem associated with high instantaneous rates may be appreciated by considering a particular example. Suppose that the gamma-ray spectrum of a resonance in a heavy nucleus is being measured by means of a low-repetition-rate accelerator such as the Nevis cyclotron, which produces 60 pps (pulses per second). Suppose further that the gamma-ray spectrometer is a Ge-diode detector and that the neutron intensity and flight path are such that the detector yields only 100 counts/day in a high-energy gamma-ray peak of interest. Then the *total* number of gamma rays detected by the Ge diode is roughly 2×10^7/day in the single resonance under consideration. Thus, roughly 3 gamma rays are detected in each accelerator pulse for the single resonance and, if the resonance is typically narrow (in time units), the 3 pulses overlap to such an extent that the gamma-ray spectrum will be meaningless. Spectral data of good quality could be obtained by reducing the neutron intensity by about two orders of magnitude, but then the number of counts in the gamma-ray peak of interest would be intolerably small. Similar considerations show that pulse pile-up is also a serious problem when the gamma-ray spectrometer is a NaI scintillator.

The tendency of a high instantaneous counting rate to result in spectra of inferior quality and the need at the same time to maximize the *average* counting rate has a significant bearing on the suitability of a neutron source for the study of resonance-capture gamma rays. In addition to the usual need for a source that gives good neutron time-of-flight resolution, two other characteristics are advantageous. First, a high-duty cycle or repetition rate is desirable because the ratio of the average rate to the instantaneous rate is directly proportional to the duty cycle. Thus, other factors being equal, one would rate the various available sources in the following order of preference: the neutron crystal monochrometer, the chopper, most pulsed accelerators, and the bomb. A second, less generally valid, rule is the desirability of using a long flight path in order that the detection system can be fully recovered from the effects of one resonance before it is required to respond to the effects of another. This requirement tends to operate to the disadvantage of pulsed sources that normally give a high intensity with a good resolution by making use of a short flight path and narrow burst width.

In the evaluation of the suitability of a neutron source, two rather different kinds of capture-gamma-ray experiments need to be understood. In one case, the information being sought is of a character such as to demand a measurement of the spectra for a large number of resonances. That is, it is necessary to achieve the best possible time-of-flight resolution for neutrons and still maintain a usable counting rate for the gamma rays of interest. The second class of experiments consists of those in which the quality of the gamma-ray spectra for one or several resonances is of more importance than the number of resonances that can be studied. For this second type of experiment, the criterion of excellence of a neutron source is the maximum value of the average counting rate that is obtainable without a significant deterioration of the pulse-height spectra because of pulse pile-up.

Since the experimental needs of different kinds of experiments tend to impose conflicting requirements on the neutron source, no one type of source is clearly dominant for capture-gamma-ray studies. The competitive situation at the present time may be summarized by saying that the crystal monochromator has important advantages in the energy range below 10 eV, the pulsed accelerator is dominant in the energy range above 100 eV, and the mechanical chopper excels in the intermediate range 10–100 eV. Representative examples of spectra obtained with a crystal monochromator have been reported by Spencer and Faler (1967), with pulsed accelerators by Jackson et al. (1966a) and Rae et al. (1967), and with choppers by Prestwich and Coté (1967b), and Beer et al. (1966). Also see Figs. 12, 28, 29, 33, and 34 later in this chapter.

2. Gamma-Ray Spectrometers

The special characteristics of the neutron-capture gamma-ray spectra place special demands on the gamma-ray spectrometer. One especially needs to bear in mind the complexity of the spectra, the high energy of some of the gamma rays of interest, and the low intensity of the source. The high energy and the complexity suggest the desirability of a refined system such as an annihilation-pair spectrometer or anticoincidence spectrometer, but the low intensity demands the relatively high sensitivity

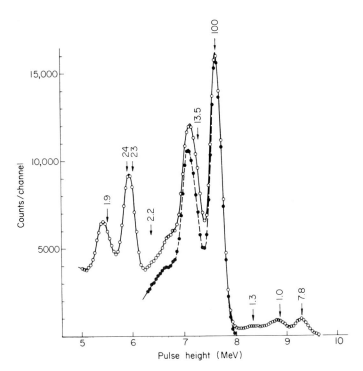

Fig. 5. Thermal-neutron-capture gamma-ray spectrum of iron, as measured with a large (15 cm thick and 20 cm diameter) NaI scintillator (Bollinger *et al.*, 1963). The dominant reaction is $Fe^{56}(n,\gamma)Fe^{57}$. The arrows on the figures show the positions of known lines in the spectrum and the numbers above the arrows are the relative intensities that were used to calculate the line shape --- for a single gamma ray.

of a single detector (or several operated in parallel) for most studies of single-gamma-ray spectra. With a few exceptions (Bostrom and Draper, 1961; Brooks and Bird, 1962), single detectors have been used.

Since the advent of the Ge-diode gamma-ray spectrometer, it has been tempting to assume that the NaI scintillator can no longer play a useful role in the study of gamma-ray spectra from resonance capture. However, as is shown in Section IV, there are still some special applications in which the high sensitivity of a large NaI crystal allows it to be more effective than the relatively small Ge diodes that are now available. For this reason and because of its historical importance, we need to consider briefly the characteristics of NaI scintillators.

An illustration of what can be achieved with a NaI detector is shown in Fig. 5, which gives a spectrum obtained with the large scintillator used with the Argonne chopper. The detector is a single NaI(Tl) scintillator, 15 cm

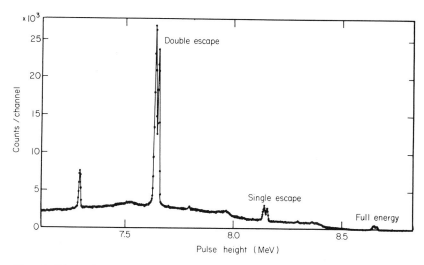

FIG. 6. Thermal-neutron-capture gamma-ray spectrum of iron, as measured with a Ge-diode spectrometer. The energy scale gives the gamma-ray energy associated with the double-escape peaks. The labeled peaks are those formed by the intense pair of gamma rays at 7634 and 7648 keV.

thick and 20 cm in diameter. It is coupled through a short light pipe to a single large photomultiplier. The line shape of the system for high-energy gamma rays, as shown in Fig. 5, is seen to have the desirable property that the full-energy peak is much more prominent than is the peak associated with the escape of one annihilation quantum and the peak associated with the escape of two annihilation quanta is almost undetectably small. Also, the width (full width at half maximum height) of only 3.5% for the full-energy peak is almost as narrow as the 3% that can be achieved with a small crystal. Finally, concerning good properties, the large surface area of the crystal enables one to make use of a very large target without a significant deterioration of line shape. The only disadvantage of the large scintillator, in comparison with a small one, is its greater sensitivity to background radiation for a fixed efficiency for detecting gamma rays from the target. In particular, large crystals characteristically exhibit a background peak at about 6.9 MeV that probably results from the capture of fast neutrons in the interior of the crystal.

In spite of its much lower efficiency, recently the lithium-drifted Ge-diode spectrometer has replaced NaI scintillators for almost all resonance-capture studies in which the gamma-ray spectra must be well resolved. The great power of the new detector may be appreciated by comparing the capture-gamma-ray spectrum of iron measured with a Ge diode (Fig. 6) with the spectrum measured with NaI. The 7-keV resolution width (at 7.6 MeV) of the Ge diode is seen to be more than thirty times smaller than that for the scintillator. However, it should not be forgotten that the large scintillator is about 10^4 times more efficient.

Several aspects of spectral measurements with Ge-diode detectors should be mentioned. One is the need for pulse-height stabilization because of the very long counting time (days or weeks) that is required to accumulate spectra of usable statistical quality. Another is the relationship between detector size and the quality of the spectra, a consideration which tends to confuse the question of what size of detector is optimum. On the one hand, a very large detector (meaning about 50 cc) appears desirable because of the urgent need for greater counting rates. On the other hand, there is some evidence (largely unpublished) that a detector of smaller size is to be preferred in some experiments because, for it, the double annihilation-quantum escape peak of interest is larger relative to the Compton background.

Until now, we have considered only the detection of single gamma rays. However, often it may be advantageous to detect coincidences between two or more gamma rays. In particular, coincidence measurements with NaI scintillation detectors have been used to increase the effective resolution of the gamma-ray spectrometer for high-energy gamma rays (Kennett et al., 1958), to improve signal to background ratio (Carpenter and Bollinger, 1960), to select certain classes of transitions (Bollinger and Coté, 1960), and to study decay schemes (Segel et al., 1964). No coincidence measurements with Ge-diode detectors have been reported yet for resonance-capture spectra.

Coincidence measurements present experimental problems that are qualitatively similar to those encountered in the detection of single gamma rays. In particular, it must be stressed that the coincidence requirement, however stringent, does not eliminate the distortion that results from pulse pile-up in a single detector. Also, the problem of pulse analysis and storage, which is already demanding when only two experimental variables are present in a measurement (a pulse height and a neutron time of flight), becomes even more severe when a third variable is added.

3. Information Storage and Sorting

Because of the low counting rates that are obtainable, the pulse-analyzer system used in an investigation of the gamma-ray spectra from resonance capture ought to be large enough and flexible enough to store simultaneously all of the information that is of interest under a given set of experimental conditions. Not long ago, this requirement was almost too severe to be satisfied, but now various forms of multiparameter analyzers offer good (but expensive) solutions to the experimental problem.

Since multiparameter analyzers are by now so widely used and understood, in our discussion here it will suffice merely to indicate those features of experiments with resonance-capture spectra that determine the basic requirements of the analyzer system. Perhaps the most distinctive feature of the measurements is that the counting rates are so low that many days of measurement are required to accumulate useful spectra. As a consequence, pulse information may be stored in some relatively inaccessible and inexpensive form, such as on magnetic tape, since data that is accumulated over a period of many days need not be in a form that permits immediate examination. In this type of recording, the values of the several variables required to describe each detected event are recorded in digital form on magnetic tape when certain elementary conditions on the variables are satisfied. Thus, the principal experimental information consists of a complete description of each individual event. At the end of a measurement, this information is sorted and displayed in such a way as to form the spectra of one variable when certain arbitrary conditions on the other variables are satisfied.

A representative example of the kind of recording system that is required for an effective study of resonance-capture gamma rays is the one developed by Rockwood and Strauss (1961) for use with the Argonne chopper It consists of two major units: a recording unit and an analyzing unit. In the original system intended for use with NaI detectors, up to three parameters characterizing any event can be recorded in digital form on magnetic tape. The two pulse heights are encoded as 8-bit numbers, and the time of flight is encoded as a 9-bit number. The 25 bits of information associated with each event are recorded on a 25-track magnetic tape that is $1\frac{1}{4}$ in. wide. Thus, there are 2^{25} possible combinations or channels available in the recorder. The data are recorded with a uniform density of about 100 events per inch of tape for a total capacity of 5×10^6 events on a 3600 ft tape. The maximum speed of recording is 175 events per second, with a pulse-pair resolution time of 300 μsec.

In the sorting unit, a flexible system of controls is available for sorting the individual events into the various spectra that are required by the experimenter. In each pass of the tape through the sorting unit, four independent spectra of one variable can be formed by placing four sets of restrictions on the values of the other two variables. The data are sorted into these four spectra and stored in four quadrants of a 1024-word core memory at the rate of about 10^6 events per minute. At the end of a pass of the tape, the four spectra may be viewed on a cathode ray tube or read out simultaneously on a typewriter, a punched paper tape, and an x-y recorder at the rate of one spectrum per five minutes. For use in experiments in which the gamma-ray spectrometer is a Ge diode, the system described above has been modified so as to permit one pulse height to be encoded as a 12-bit number (4096 channels) to satisfy the requirement for better pulse-height resolution.

Several more recently developed pulse-analyzer systems have other features that are found to be useful. Most of the commercially available tape-recording systems store the pulse information on computer-compatible magnetic tape so that the sorting operation can be carried out automatically on a large computer. This is an important advantage if the raw data require any mathematical treatment other than sorting to form spectra. On the other hand, if sorting is the only operation required, in the author's view (probably not shared by many) there is some advantage in manual sorting which allows the experimenter to critically examine each spectrum as it is formed.

Another significant advance has been incorporated in the system developed by Chrien et al. (1964) for the new Brookhaven chopper. In this system the spectral information is stored on magnetic tape also, but the logical operations of the analyzer system are controlled by a small digital computer. Thus, the analyzer has a high degree of flexibility. There is no general agreement among workers in the field as to whether or not this flexibility is an important advantage for measurements of the spectra associated with resonance capture. The use of the Brookhaven system has been described by Bhat et al. (1967).

B. Measurements with Pulsed Electrostatic Accelerators

When the neutron energy of a conventional time-of-flight spectrometer is increased beyond a few kilovolts, measurements of capture-gamma-ray spectra become very difficult because of the rapidly decreasing counting rate per unit energy interval. In principle, this decrease in rate would not need to preclude all useful measurements since adequate rates could be obtained by integrating over a wide range of energy to form an average resonance-capture spectrum that would be of considerable interest. In practice, however, average spectra have not been measured in this way, probably because fast neutrons that are scattered by the sample into the detector and its surroundings create a prohibitively large background.

The background problem that limits the energy range that can be studied by means of the conventional time-of-flight method have been effectively

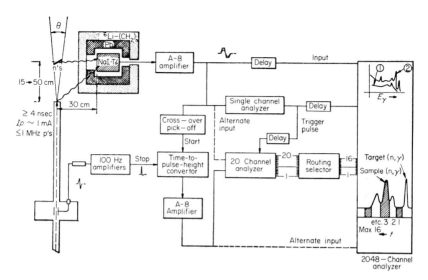

FIG. 7. Experimental arrangement used in a measurement of resonance-capture spectra with a pulsed Van de Graaff accelerator (Bird *et al.*, 1965). The inset on the lower right shows the time sequence of events detected by the NaI scintillator.

eliminated by a method that makes use of a pulsed beam from an electrostatic accelerator. The method was first used by Bergqvist and Starfelt (1961, 1962a) in Sweden and extended by various workers at Oak Ridge (Firk and Gibbons, 1961; and Bird *et al.*, 1965). A schematic representation of a typical measurement is given in Fig. 7. A pulsed beam of protons from a Van de Graaff accelerator impinges on a lithium target which yields neutrons through the ^{7}Li(p,n) reaction. The sample under investigation is placed along the direction of the incident proton beam. Thus, if the energy of the proton is just a few kilovolts above the threshold of the reaction, the kinematics of the reaction are such that the energy of the neutrons that strike the target lie within a broad band around a mean energy of about 30 keV. The gamma-ray spectrum formed by neutron capture in the sample is measured by means of a large NaI scintillation spectrometer.

Because of the finite velocity of the 30-keV neutrons, the pulsed character of the neutron source allows background counts in the detector to be effectively eliminated. This is illustrated in the lower right-hand corner of Fig. 7, where one sees that the gamma rays from the various sources arrive at the detector at different times. First come gamma rays directly from the ^{7}Li(p,n) target. Some 100 nsec later, the (n,γ) radiation of interest arrives and its spectrum is recorded. And later still, the detector begins to feel the influence of background counts from neutrons that are scattered by the sample.

Allen (1968) has recently demonstrated that similar measurements can be made with a Ge-diode detector at a pulsed Van de Graaff accelerator.

C. Measurements with Filtered Reactor Neutrons

1. Neutron Beam Experiments

Measurements in which filters are used to select a broad range of neutron energies are carried out under two rather different geometrical arrangements. In one, a collimated beam of neutrons is brought out of the reactor and passed through a filter that is placed relatively far away from the neutron-capture gamma-ray sample under investigation. Gamma rays from the sample are detected by a nearby detector—a Ge-diode detector in recent measurements. Thus the energy spectrum $C_n(E)$ of the neutrons that are captured is given approximately by

$$C_n(E) = \phi_E T_E [1 - e^{-n\sigma}](\sigma_\gamma/\sigma) \tag{1}$$

where ϕ_E is the spectrum of neutrons issuing from the reactor, T_E is the neutron transmission of the filter, n is the sample thickness, σ is the total neutron cross section of the sample, and σ_γ is the radiative-capture cross section. All of these quantities except n are functions of the neutron energy E. Usually $\phi_E \propto E^{-1}$ in the energy range of interest.

Two qualitatively different kinds of measurements with filtered beams need to be mentioned. In one, the aim is to measure the gamma-ray spectra of single resonances, and the resonance-absorption process itself is the principal mechanism for neutron-energy selection. This selection depends on the fact that, for a thin target, one or a very few low-energy resonances are responsible for most of the capture in the typical nuclide since the rate of capture at a single resonance at energy E_0 is proportional to $\sigma_0 \Gamma_\gamma E_0^{-1}$, where the peak cross section σ_0 is a rapidly decreasing function of E_0 and the total radiation width Γ_γ is approximately constant. The energy selection can be refined further by blanking out the contributions from unwanted resonances by means of resonance filters. Rather simple but useful measurements of this kind have recently been reported by Martin et al. (1966), Harvey et al. (1966), and Slaughter et al. (1966). The degree of refinement in neutron-energy resolution that might be achievable may be inferred from a paper by Roeland et al. (1958) on the use of resonance filters in a study of fission.

A filtered-beam experiment might also be useful (although none have been reported yet) for a measurement of gamma-ray spectra associated with capture in a rather broad band of energy.* In a measurement of this kind, the contribution of low-energy neutrons would be limited by selective absorption in a filter that has a $1/v$ cross section and the contribution of fast neutrons limited by the E^{-1} reactor spectrum and by the decrease of σ_0 with increasing neutron energy. It is probable that fast neutrons captured in materials other than the sample would be a serious source of background in measurements of this kind.

2. INTERNAL-TARGET GEOMETRY

The second experimental system in which neutron filters have been used involves the same basic components as the beam experiments described above, but the geometrical arrangement of the neutron source, the sample, and the detector are reversed. Here, the sample under investigation is entirely enclosed within a neutron absorber, the composite target is placed in the interior of a nuclear reactor, and the gamma rays from this target are detected by a spectrometer outside of the reactor. In the system used by Bollinger and Thomas (1967a, 1968b), the only measurements of this kind reported to date, the sample is placed in the high-flux region of a through tube that passes tangent to the core of the reactor, as shown in Fig. 8. A carefully designed collimator allows the Ge-diode gamma-ray spectrometer to view the sample but not the walls of the through tube. Neutrons in the gamma-ray beam formed by the collimator are removed by polyethylene scatterers.

With the internal-target arrangement, it is not feasible to shape the spectrum of captured neutrons as effectively as with an external beam because resonance scattering by the neutron filter is not effective in shielding the sample. Also, the filter itself must not be a prohibitively intense source of radiation for the detector. Hence, in practice the filter is limited to $1/v$ absorbers such as ^{6}Li or ^{10}B, which do no more than selectively absorb low-energy neutrons.

Because of the limited effectiveness of the $1/v$ filter in shaping the neutron spectrum, the internal-target arrangement is not very useful for the study of capture in a single resonance, except under special circumstances. On the other hand, because of its high sensitivity the internal-target method is more effective than any other for the study of certain aspects of the *average* spectra from capture in many low-energy resonances. Indeed, it appears to be the only method that is now capable of giving information of this kind with the degree of refinement that is needed to answer most physically interesting questions.

The way in which a $1/v$ neutron absorber and $1/E$ reactor spectrum combine with the energy dependence of σ_0 to form the energy spectrum of

---

* Recently Greenwood *et al.* (1969) reported on the use of a neutron beam formed by a minimum at about 2 keV in the cross section of ^{45}Sc.

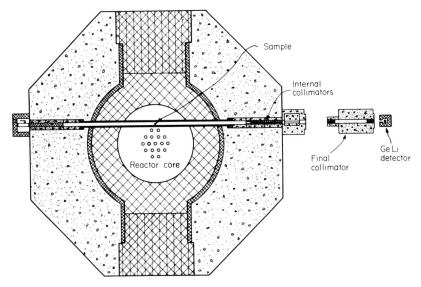

FIG. 8. An internal-target facility (Thomas *et al.*, 1967) used for the measurement of average-resonance-capture spectra (Bollinger and Thomas, 1967a, 1968b).

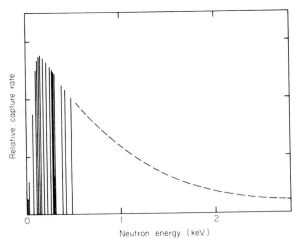

FIG. 9. Energy spectrum of neutrons captured in a thick sample enclosed in ^{10}B and immersed in a $1/E$ neutron flux. The vertical lines at low energy give the energies of known resonances in ^{195}Pt with $J = 1$.

captured neutrons is illustrated in Fig. 9. To a rough approximation the probability of capture is a function of the variable E/t^2, where t is the thickness of the absorber; that is, the basic shape of the spectrum is the same for all absorbers, but the energy at which the spectrum peaks and the width of the spectrum is proportional to t^2. An example of an average-resonance-capture gamma-ray spectrum is given in Fig. 19 in Section IV,A,4,a.

III. ANALYSIS OF SPECTRA

In a rather large fraction of measurements of the gamma rays associated with resonance capture, the experimental quantities that are being sought are either relative or absolute values of partial radiation widths. Two quite different kinds of problems are involved in the analysis of raw spectra to obtain widths. One is the problem of deducing observed intensities from the complex spectra that are measured. The second is the problem of converting these observed intensities into widths. Both kinds of problems have been discussed in some detail by Carpenter (1962).

When the gamma-ray spectra are well resolved, the observed intensity of a gamma-ray line can be obtained simply by calculating the area of the line, and most of the analyses of Ge-diode spectra that have been carried out thus far have been little more than this. However, when the spectrometer is a NaI scintillator (and even for a Ge-diode detector when the range of interest extends beyond high-energy transitions to low-energy states), a much more refined analysis is essential to obtain reliable results. The most widely used method of analysis for unresolved spectra is to carry out a least-squares fit. In such a fit, the basic physical assumption is that each "complex" spectrum is formed by a linear superposition of a set of simple spectra corresponding to individual gamma-ray lines. Then, for the most elementary complex spectra, in which the position and shape of each contributing line is known, the intensity a_{rj} of a transition from an initial state r to a final state j can be determined by making a linear least-squares fit of the experimentally measured pulse-height spectrum to a relationship of the form

$$y_{ir} \equiv (N_{ir} - B_{ir}) = \sum_j a_{rj} R_{irj} \tag{2}$$

where N_{ir} is the total number of pulses falling within the ith channel, B_{ir} is the number of background counts associated with the ith channel, and R_{irj} is the contribution to the ith channel from a transition $r \rightarrow j$ of unit intensity.

The procedure used to obtain the least-squares fit is well known (e.g., see Kendall and Stuart, 1961). First, one finds the least-squares values of the intensities a_{rj} as those values that minimize the variance V_r defined by the relationship

$$V_r = \sum_i w_{ir}\left(y_{ir} - \sum_j a_{rj}R_{irj}\right)^2 \tag{3}$$

where w_{ir} is the weight attached to the measurement at each channel i. The weights are ordinarily set equal to $(\Delta y_{ir})^{-2}$, the quantity Δy_{ir} being the standard statistical error of y_{ir}. Then, one obtains the statistical uncertainty in the intensities from the relationship

$$(\Delta a_{ij})^2 = [\chi_r^2/(n-1)](H_r)_{jj}^{-1} \tag{4}$$

where the $(H_r)_{jj}^{-1}$ are the diagonal elements of the inverse matrix defined by

$$(H_r)_{jm} = \sum_i w_{ir}R_{irj}R_{irm} \tag{5}$$

Here, n is the number of degrees of freedom in the fit, and χ_r^2 is the minimum value of V_r. However, χ_r^2 is itself a random variable subject to statistical fluctuations. Thus, when χ_r^2 is less than its expectation value, the expectation value is used in Eq. (4). As a final step, the quality of the fit is tested by comparing the experimental value of χ^2 with the probability distribution that is expected for a least-squares fit with a known number of degrees of freedom. Failure of the experimental value to be consistent with the theoretical distribution is a strong indication that the assumptions entering into the fit are not all valid or that systematic errors are present in the measurement.

Although the simple least-squares procedure outlined above is the basic one used for spectral analysis, it requires considerable amplification to make it a convenient and sensitive method for analyzing capture gamma-ray spectra. Perhaps the most serious difficulty concerns the position of the individual gamma-ray lines in the complex spectrum. As was stated above, the position of these lines must be known if we are to make use of the linear least-squares method. However, in the capture gamma-ray spectra, we often do not know the energy or even the number of the lines contributing to a particular range of the spectrum. Moreover, even when the spacing between all lines is known, the position of the lines relative to the positions of the reference lines is rarely known with sufficient accuracy. To overcome this difficulty, the line positions are found in basically the same way as are the intensities, namely, by finding the set of line positions that result in the minimum value of χ^2. However, since it is not feasible to perform the minimization analytically, an iterative technique such as variable metric minimization must be used. Known physical information about the spectra, such as the spacing between lines, is used to place constraints on the line positions. If the number of lines as well as their positions are unknown, the best that can be done is to determine the minimum number of lines that gives an acceptable value of χ^2.

A special characteristic of most resonance-capture gamma-ray spectra is that, because the neutron resonances are so close together, effectively the same set of gamma-ray energies is present in all the spectra of a given nuclide. Moreover, not only are the energies the same but, since the spectra are all recorded simultaneously in a multiparameter analyzer, the positions of the lines in the experimental data are also the same. It is clear that this physical information should be used as a constraint in the analysis. This objective can be achieved by performing the minimization for all spectra simultaneously. That is, all the intensities a_{rj} and energies E_j for the set of spectra are obtained simultaneously by minimizing the over-all variance

$$V(a_{11}, \ldots, E_1, \ldots) = \sum_r \sum_i w_{ir} \left[y_{ir} - \sum_j a_{rj} R_i(E_j) \right]^2 \qquad (6)$$

where E_j is the energy of a transition to the state j from any of the initial states r, and the subscript r on R is dropped to indicate that the line shapes are the same for all resonances. The errors associated with the a_{rj} should include a term to take account of the uncertainties in the line positions.

One of the most important sources of error in a least-squares fit of a poorly resolved spectrum is uncertainty in the correct shape of the response function R for the gamma-ray spectrometer. For convenience, it is desirable to determine the response function from some easily measured (n,γ) spectrum. To illustrate the principal problems in such a determination, let us consider the requirements for an analysis of representative NaI spectra. Because the density of low-energy states increases so rapidly with increasing excitation energy, for the typical heavy nucleus only a very limited range at the high energy end of the spectrum can be analyzed. As a result, the energy dependence of the line shape is relatively unimportant. The important problem is to determine an accurate shape for a single gamma-ray energy falling in the range of interest, often in the range 7–8 MeV. A convenient and reliable shape of this kind can be obtained from the thermal-neutron-capture gamma-ray spectrum of iron. This spectrum is complex, but a closely spaced pair of lines at 7.64 MeV is much stronger than any other line in the energy range of interest; the 14-keV separation between the pair of lines is so small that for a NaI detector they form what is effectively a single line. Since all the lines with a significant intensity have been resolved and since their relative intensities are known, the shape of the response function can be obtained by substituting the known y_i and a_k into Eq. (2).

An example of the thermal-neutron-capture gamma-ray spectrum of iron and the reference-line shape derived from it are given in Fig. 5. Because of the freedom from background and the good statistical accuracy of the data, the line shape is believed to be quite accurate. However, it has the characteristic defect that it extends over only a small range of energy.

The problem of measuring an accurate response function for the Ge-diode detector over a wide range of energy has not been undertaken yet since there has not yet been a need for the information. Until now, all that has been required is a knowledge of the width and relative heights of the full energy and the annihilation-escape peaks that are characteristic of high-energy gamma-ray spectra, and this information is easily obtained from both thermal capture and resonance capture in many materials.

Once the observed intensities a_{rj} have been extracted from the gamma-ray spectra, one is faced with the problem of converting these intensities into partial radiation widths Γ_{rj}. This problem is best understood by considering a quantitative relationship for the width, which may be written in the form

$$\Gamma_{rj} = I_{rj} \times (\Gamma_\gamma)_r \tag{7a}$$

$$I_{rj} = kA_{rj} = k(\epsilon_\gamma C_n)^{-1}a_{rj} \tag{7b}$$

Here, Γ_γ is the total radiation width and I_{rj} is the absolute intensity of the gamma-ray line expressed in units of photons per neutron captured. The factor ϵ_γ is the relative efficiency of the gamma-ray spectrometer (a function of gamma-ray energy), and C_n is the relative number of captured neutrons that form the spectrum (a function of neutron energy). The factor k is a normalizing constant.

Let us first examine the information that is required to derive relative intensities $A_{rj} = a_{rj}(\epsilon_\gamma C_n)^{-1}$ from the observed intensities a_{rj}. The relative efficiency ϵ_γ is easily obtained with an accuracy that is adequate for most experiments. For example, for a NaI scintillator it may be obtained from the published results of Monte Carlo calculations; and for a Ge-diode detector, the energy dependence of the efficiency may be obtained by calibrating the system by means of a thermal-neutron-capture gamma-ray spectrum for which the intensities have been measured reliably. There is considerable doubt about the accuracy of some of the published intensities, but those reported recently by Thomas et al. (1967) and Kane and Mariscotti (1967) provide reliable standards.

Several methods have been used to determine the relative number C_n of captured neutrons that contribute to a spectrum. In one approach, the capture rate is calculated from a knowledge of the incident neutron spectrum $\phi(E_n)$ and the parameters of the resonance under consideration. In the general case, this calculation would require a Monte Carlo calculation to take into account the effect of multiple scattering in the sample. However, in various special cases simple approximations are accurate enough. For example, when radiative capture is the dominant process the rate of capture in a resonance is proportional to $\phi(E_n)A_E\Gamma_\gamma\Gamma^{-1}$, where A_E is the area of the transmission dip and Γ is the total width of the resonance. The definition and calculation of the area A_E is discussed in detail in Chapter III.

When the experimental information being sought is limited to the relative radiation widths within a single nuclide, the problem of calculating the rate of neutron capture may be avoided by using some feature of the gamma-ray spectra as a measure of the rate of capture. Spectral characteristics that have been used for this purpose include the total intensity of gamma-ray pulses above some low-energy limit in a single detector, the intensity of pulses in a low-energy band, the total rate of coincidence counts in two gamma-ray detectors, and the intensity of strong low-energy lines. The extent to which these various intensities are reliable measures of the rate of neutron capture is discussed in Section IV,B.

When it is necessary to compare the intensities of gamma-ray transitions in several nuclides, the experimenter must find a way to obtain partial widths Γ_{rj}. As is seen in Eq. (7), this requires a knowledge of both the absolute intensities I_{rj} and the total radiation widths Γ_γ. The total radiation widths are determined by various means that are discussed in other chapters of this book. The absolute intensities may be obtained from the relative intensities A_{rj} by normalizing the data to known intensity standards. One set of standards consists of a few strong high-energy resonance-capture lines for which the absolute intensity has been determined. Lines of this kind are the ground state transition in the 33.5 eV resonance of ^{199}Hg, for which Bollinger et al. (1959) report $I_\gamma = 0.035$ photons per capture, and the 11.5 eV resonance of ^{195}Pt, for which Carpenter (1962) reports $I_\gamma = 0.048$ photons per capture. The uncertainty in these values is quite large, perhaps as much as $\pm 30\%$.

High-energy lines in thermal-capture spectra may also be used as intensity standards. Some of the published information on thermal-capture gamma-ray intensity appears to be quite unreliable, but there is no doubt that good standards could be established with relative ease now.

Certain low-energy lines appear to be especially useful (but largely unused) standards because their intensities can be known with very good accuracy. The strong transition from the first excited state to the ground state of some even–even nuclei have the additional convenience that the intensity is closely proportional to the rate of neutron capture, independent of the spin of the initial state. Evidence on this question is summarized in Section IV,B,4.

IV. EXPERIMENTAL RESULTS

In Section I of this chapter the general characteristics of neutron-capture gamma-ray spectra were described. It was shown that a typical experimental spectrum consists of three characteristic regions of energy: a high-energy region containing resolved gamma-ray lines, a region of intermediate energy containing a complex mass of unresolved lines, and a low-energy region that again contains resolved lines. Generally speaking, these three regions of a spectrum provide rather different kinds of physical information. The behavior of high-energy radiative transitions is demonstrated by the resolved high-energy lines; statistical information about transition probabilities and level densities is provided by the average characteristics of the cascades of gamma rays that produce the intermediate region of energy; and the characteristics of individual low-energy states in the compound nucleus may be determined from the resolved lines at low energy. Each of these areas of interest can, of course, be studied by measurements on the gamma rays both from capture of thermal neutrons and resonance-energy neutrons. Indeed, it is only differences in experimental methods that tend to make a distinction. Consequently, both sets of results might be treated as a single body of knowledge. However, since such a treatment would tend to be beyond the scope of this book and since several review papers have already summarized the results obtained in the study of thermal-neutron-capture gamma rays, in this chapter we will emphasize only those subjects in which resonance capture plays a significant role.

The organization of the subject matter of this chapter is complicated considerably by the rapid changes in experimental technique that have taken place during the last few years. The most important of these developments is the application (beginning in 1965) of the high-resolution lithium-drifted Ge-diode gamma-ray spectrometer to the study of resonance capture. As a consequence of the recent and sudden change from the NaI to the Ge-diode detector, one finds that most of the significant physical ideas were conceived and rather fully reported in connection with experiments in which the gamma-ray spectra were observed by means of NaI scintillation spectrometers, whereas the numerical results of these measurements are now being rapidly superseded by reports on measurements carried out with Ge-diode spectrometers.* Thus, to provide the background information with which to understand the physical significance of the new meas-

---

* A good summary of some post deadline results is given by Chrien (1969).

urements, it seems desirable to review the experiments carried out by means of both kinds of detectors, even though the actual data obtained in the NaI measurements may be obsolete.

It seems probable that almost all of the measurements that have been carried out with a *single* NaI detector will in a few years be superseded by measurements with Ge-diode detectors. However, for experiments that require two or more large detectors (such as in sum-coincidence measurements) the measurements with NaI scintillators are likely to remain unchallenged for a long time.

A. High-Energy Transitions

In this section, we will be discussing experiments that are concerned with the resolved high-energy lines in the capture-gamma-ray spectra. The intensity of these lines gives a measure of the partial widths Γ_{rj} for radiative transitions directly to some final state j from some initial state r. Because of experimental limitations, most of the experiments to date have been concerned with electric-dipole transitions, which are on the average an order of magnitude more intense than other kinds of transitions.

1. Distribution of Partial Radiation Widths

The principal motivation for studying the distribution of partial radiation widths is the same as that for studying reduced neutron widths and fission widths, namely, the hope that the shape of the distribution will give information about the nature of the emission process. Our present theoretical background for interpreting the observed distribution of widths is that provided by Porter and Thomas (1956) in their classic analysis of the reduced neutron widths. They show that the extremely broad distribution of neutron widths is a result of the complexity of the wave functions of the highly excited states formed by neutron capture. Moreover, reasonable assumptions about the nature of the complexity lead to the expectation that the distribution of the width Γ for a reaction that proceeds by way of a single exit channel is of the form $x^{-1/2}e^{-x/2}$, where $x = \Gamma/\overline{\Gamma}$. And if this form is valid for a single-channel process, the distribution of widths for a reaction that proceeds by way of ν exit channels of the same mean width must be a χ^2 distribution with ν degrees of freedom, i.e., a distribution of the form

$$\rho(x) = [\Gamma(\nu/2)]^{-1}[(\nu/2)x]^{(\nu/2)-1}e^{-\nu x/2} \tag{8}$$

where $\Gamma(\nu/2)$ is the usual Γ function. Equation (8) is written in such a form that the mean value of x is unity and the variance is $2/\nu$.

As discussed elsewhere in this volume, the experimentally measured reduced neutron widths for elastic scattering, a reaction with a single exit channel, are in excellent agreement with the Porter–Thomas distribution ($\nu = 1$). Consequently, since a radiative transition directly from a single initial state to a single final state also appears to be a reaction that proceeds by way of a single exit channel, there is good reason to expect the partial radiation widths to obey the Porter–Thomas distribution. Indeed, this expectation is so strong that there probably would be little interest in the subject had not some of the experimental data been reported as being at variance with this expectation.

Historically, the study of the distribution of widths started with relatively crude NaI-scintillator measurements of gamma-ray spectra for several neutron resonances. First, Kennett et al. (1958) and Bird et al. (1959) found that the spectra for several resonances of the same spin and parity in ^{55}Mn and ^{199}Hg, respectively, differ enough to be interpreted as evidence for a rather broad distribution of the partial radiation widths. On the other hand, Hughes et al. (1959) reported that the ground-state transition for the first five resonances of ^{183}W differ from the mean by only 20%, and this evidence was cited as an indication that the partial radiation widths are governed by a much narrower distribution than had been expected. Then Bollinger et al. (1959) showed that at least a part of the apparent uniformity resulted from a failure to observe individual transitions in the ^{183}W data; moreover, from the spectra for capture in ^{199}Hg and ^{195}Pt they observed fluctuations in the partial radiation width that were great enough to be consistent with a Porter–Thomas distribution.

The above conflicting evidence and other results of a similar nature were brought into sharp focus at the time of the Topical Conference on the Neutron Capture Reaction, Los Alamos, 1959.* Some of those present at the conference held that the experimental data indicated that the distribution in partial radiation widths was anomalously narrow, a point of view that was ably supported by Hughes (1959) in his survey paper. However, others concluded that the apparent uniformity of some of the reported widths probably resulted from experimental factors that discriminate against observation of the small widths that occur frequently in a distribution with $\nu = 1$. In particular, there appeared to be a need to avoid several kinds of systematic errors: (a) distortion of spectra by an accidental summing of pulses from two gamma rays, (b) a measurement of gamma-ray spectra with too few pulse-height channels, (c) deducing radiation widths from necessarily complex and only partially resolved gamma-ray spectra without a sufficiently quantitative and objective treatment, (d) selecting data on

---

* Abstracts of papers given at the conference appear in *Bull. Am. Phys. Soc.* **4.**

the basis of criteria that tend to discriminate against small widths, and (e) inadequate statistical techniques by which conclusions about the nature of the distributions are deduced from the small number of partial radiation widths available. It is significant that *all* of these errors tend to inhibit one from obtaining a value of ν as small as 1.

The differences in the interpretation of the early data led to a second round of measurements with NaI detectors aimed at understanding the distribution of partial radiation widths. Most of these measurements were on even–odd targets with spin $\frac{1}{2}$ and negative parity. The high-energy end of the spectra for these nuclides are especially simple because they consist of E1 transitions to widely spaced low-energy states of the compound nucleus, and the data are especially easy to interpret because the spin (0 or 1) of the initial state can usually be determined from the spectra themselves in the way outlined in Section IV,C,2. The most intensively studied target has been ^{195}Pt, for which a set of refined NaI spectra is given in Fig. 10.

In spite of the simplicity of the spectra, the second round of measurements with NaI detectors also led to irreconcilable discrepancies. This is illustrated by comparing the widths that were obtained at several laboratories for the reaction ^{195}Pt$(n,\gamma)^{196}$Pt, as is done in Table I. The several sets of data are similar in the sense that they all exhibit wide fluctuations in the widths, but the actual values of the widths obtained at different laboratories are quite different. A closer look at the data shows that, within the rather large errors, the data obtained at Argonne and Brookhaven by means of fast choppers are consistent with each other, and that they are inconsistent with the two sets of data obtained by linear accelerators. Later and presumably more refined measurements (Huynh *et al.*, 1966) at the Saclay linac with a NaI scintillator also gave results that are inconsistent with the chopper data. It is the author's opinion that the chopper data are basically sound and that the linac data are in error because, as was discussed in Section II, the gamma-ray spectra were distorted by excessively high instantaneous counting rates. This view tends to be supported by the fact that data now being obtained with Ge-diode gamma-ray spectrometers consistently are in satisfactory agreement with the second round of data obtained with NaI scintillators at choppers. Thus the author cannot agree with the view expressed by others that the NaI scintillator is incapable of yielding reliable information about the neutron capture gamma-ray spectra.

Although the measurements made with NaI scintillators are being rapidly superseded by more refined measurements with Ge-diode detectors, it seems worthwhile to outline the way in which the older data have been

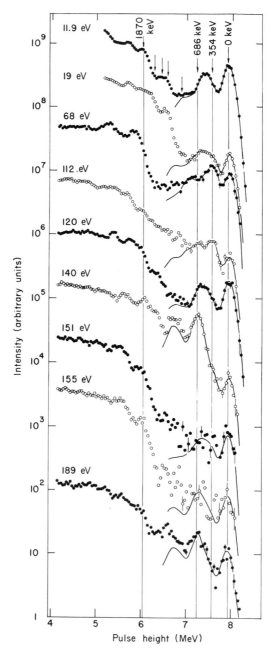

FIG. 10. Spectra from capture in resonances of ¹⁹⁵Pt, as measured with a large NaI scintillator (Bollinger *et al.*, 1963). The vertical lines on the figure give the energies of transitions to known low-energy states.

TABLE I

REPRESENTATIVE RESULTS FOR PARTIAL RADIATION WIDTHS MEASURED WITH
NaI SCINTILLATION SPECTROMETERS†

E_0 (eV)	Reference	A_0	A_1	A_2
12	ANL[a]	**1000 ± 15**	86 ± 15	25 ± 17
	BNL[b]	**1000 ± 42**	151 ± 95	<85
	Harwell[c]	**1000**	400 ± 100	<20
19	ANL	184 ± 5	91 ± 7	96 ± 7
	BNL	201 ± 42	99 ± 64	218 ± 63
	Harwell	160 ± 30	230 ± 50	250 ± 30
67.4	ANL	860 ± 25	749 ± 32	44 ± 32
	BNL	645 ± 49	754 ± 78	92 ± 78
	Harwell	450 ± 80	510 ± 100	120 ± 30
	Saclay[d]	883		
112	ANL	233 ± 12	285 ± 20	155 ± 20
	BNL	218 ± 35	299 ± 60	218 ± 63
	Saclay	0		
120	ANL	**604 ± 17**	49 ± 17	192 ± 20
	BNL	522 ± 49	105 ± 78	63 ± 63
	Saclay	**604**		
140	ANL	118 ± 10	96 ± 17	1071 ± 32
	BNL	99 ± 32	32 ± 49	1000 ± 74
	Saclay	604		
151	ANL	111 ± 12	27 ± 15	32 ± 17
	BNL	104 ± 42	99 ± 78	186 ± 95
	Saclay	327		
189	ANL	305 ± 12	44 ± 27	403 ± 34
	BNL	359 ± 56	151 ± 95	345 ± 95
	Saclay	1030		

† The columns labeled A_0, A_1, and A_2 are the widths (in arbitrary units) of transitions
to the 0^+, 2^+, and 2^+ states at 0, 356, and 689 keV, respectively. The values given in
boldface type are the values that were used to normalize the four sets of data.

[a] Bollinger et al. (1963).   [c] Brooks and Bird (1962).

[b] Chrien et al. (1962).   [d] Julien et al. (1960).

analyzed to obtain statistical information about the distribution of widths
since similar techniques will be needed to obtain the most refined information from the new data also. The primary problem in the analysis is that
the data are usually too fragmentary for large-sample statistical theory
to be valid. This problem is well illustrated by the extensive set of measurements reported by Bollinger et al. (1963), who observed a few radiative
transitions associated with each of a few resonances in ^{77}Se, ^{183}W, ^{195}Pt,
and ^{199}Hg and, in addition, analyzed some of the partial radiation widths

reported by Carpenter (1962) for resonances of ^{155}Gd, ^{173}Yb, ^{177}Hf, and ^{201}Hg. Partly to make allowance for the experimental bias that results from the difficulty in observing weak transitions and partly to determine ν in an unbiased way from the small statistical samples that were obtained, these authors used an unconventional random sampling technique in their statistical analysis. The basic idea of the method is to test various hypothetical values of ν by comparing the *physical* sample obtained in the experiment with a set of equivalent *mathematical* samples that are drawn from populations for which the value of ν is known.

First consider the case in which there is no experimental limitation of measurement, i.e., only the small size of the statistical sample causes difficulty. The mathematical samples are formed by a Monte Carlo calculation in which widths are randomly drawn from a population governed by a χ^2 distribution with a known value ν_0 of ν. Each mathematical sample is generated in such a way that it consists of several subsamples with independent mean values in exactly the same way as is the physical sample. Thus, there is no difference in the nature of the physical and mathematical samples except that the former is drawn from distributions governed by an unknown value of ν, whereas the latter is drawn from distributions for which ν is a known value ν_0. One may, therefore, test whether the physical data are consistent with the hypothesis that they are drawn from a distribution having the assumed value ν_0 by comparing the physical and mathematical samples. The comparison is made by comparing the maximum likelihood (Kendall and Stuart, 1961) value ν_p obtained from the physical

TABLE II
VALUES[a] OF ν

Target	n	ν	$\Delta\nu$	$\delta\nu$
^{199}Hg	6	1.24	1.2	0.12
^{195}Pt	24	1.56	0.51	0.14
^{183}W	10	1.12	0.8	0.3
^{77}Se	8	0.96	0.9	0.3
Mean	48	1.34	0.33	0.21
^{155}Gd	12	0.72	0.36	0.31
^{173}Yb	6	2.8	2.0	0.02
^{177}Hf	8	1.26	0.8	0.23
^{201}Hg	3	0.68	1.0	0.11
Mean	29	1.14	0.44	0.21

[a] Obtained by Bollinger *et al.* (1963). The quantity n gives the total number of widths in each statistical sample, $\Delta\nu$ is a standard statistical uncertainty dependent on the size of the statistical sample, and $\delta\nu$ is a systematic uncertainty resulting from errors in measurement.

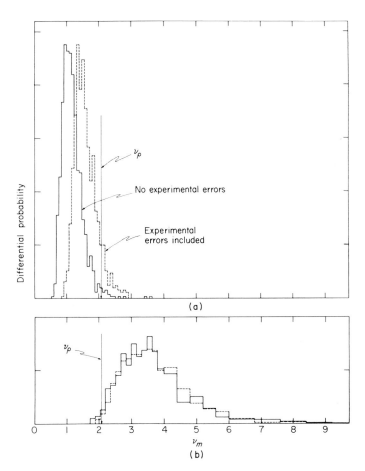

FIG. 11. Monte Carlo determination of ν for high-energy transition in ^{196}Pt. Distribution of the estimator ν_0. (a) $\nu_0 = 1$; (b) $\nu_0 = 3$.

sample with the distribution of maximum likelihood values ν_m obtained from the mathematical samples. If ν_p falls near the center of the distribution of ν_m, the physical data are consistent with the assumed value ν_0 of ν; on the other hand, if ν_p falls far out in the tails of the distributions of ν_m, then the physical data are inconsistent with ν_0.

The Monte Carlo technique of hypothesis testing is illustrated in Fig. 11 for the partial radiation widths of the ^{195}Pt resonances. The data for this target consists of three subsamples, each of which has 8 members. The mean values of the subsamples are assumed to be unknown and unrelated to each other. Then the experimental widths give $\nu_p = 1.84$. By drawing mathematical samples of the same kind from distributions with $\nu_0 = 1$,

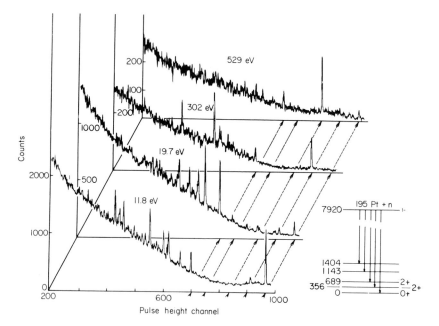

Fig. 12. Spectra from capture in resonances of ¹⁹⁵Pt, as measured with a Ge-diode spectrometer (Jackson *et al.*, 1966a). The diagonal lines on the figure indicate the energies of transitions to the low-energy states shown in the engery-level diagram.

one obtains the solid line histogram given in the upper part of Fig. 11. The value $\nu_p = 1.84$ is seen to be somewhat larger than the typical value of ν_m, but it is not entirely inconsistent with the assumption that $\nu = 1$. The lower part of Fig. 11 gives the distribution of ν_m for $\nu_0 = 3$. In this case, it is seen to be most improbable for ν_m to be as small as 1.84; hence, the experimental widths are inconsistent with the assumption that $\nu = 3$.

When the measured values of the partial radiation widths are subject to experimental error, it is no longer feasible to maintain an accurate correspondence between the physical and mathematical samples. However, one can make an approximate correction for the influence of the experimental errors. The dashed-line histograms of Fig. 11 show the results that are obtained when the experimental errors are taken into account. Note that the corrected curve is shifted quite significantly toward larger values to ν_m. This effect demonstrates quantitatively how the experimental difficulty in observing weak transitions results in a biased value of ν_p.

To obtain information about the probable range within which the true value of ν lies, the distribution of ν_m for several hypothetical values of ν_0 are compared with the physical value ν_p. From these results, confidence

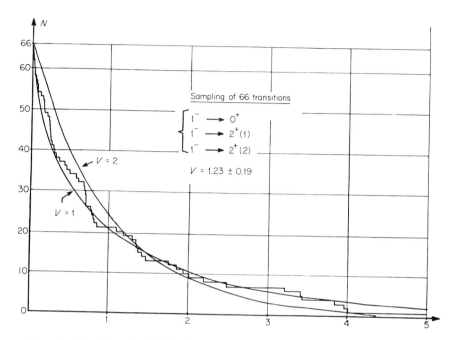

FIG. 13. Experimental distribution of partial radiation widths for the reaction $^{195}Pt(n,\gamma)^{196}Pt$ (Jackson *et al.*, 1966a). The horizontal axis gives the relative width $x = \Gamma/\bar{\Gamma}$, and the vertical axis gives the number N of transitions for which the relative width is greater than x.

limits $\nu \pm \Delta\nu$ may be determined, where $\Delta\nu$ is in the nature of a standard error for a normal error function.

The results of Bollinger *et al.* (1963) on the distribution of radiation widths are summarized in Table II. Examining these data we see that, although there are several values of ν that are considerably greater than unity, when one takes into account the uncertainties in the values, no single value can be said to be inconsistent with unity. Moreover, the mean values of ν for both of the groups of nuclides considered in Table II are small enough to make it improbable that ν could be as large as 2.

The long story of conflicting evidence about the distribution of partial radiation widths now seems to have been brought to a close, at least insofar as the reaction $^{195}Pt(n,\gamma)^{196}Pt$ is concerned, by the extensive and refined measurements by Jackson *et al.* (1966a) with a Ge-diode detector at the Saclay linac. In these measurements, the spectra for twenty-two neutron resonances with $J = 1$ were observed. Four of these spectra are shown in Fig. 12. Although transitions to a large number of low-energy states were detected, only those to the first three states were used in the analysis for

ν because they were judged to be most reliable. A plot of the distribution of the widths for these 66 transitions is shown in Fig. 13. It is immediately obvious from the plot that ν is approximately equal to the expected value $\nu = 1$. A quantitative statistical analysis yields the value $\nu = 1.23 \pm 0.19$. Thus, the high-energy transitions in ^{196}Pt are in excellent agreement with the expectations of the simple statistical model. These same ^{196}Pt data have been discussed more fully by Samour et al. (1968) in a paper that extends the Monte Carlo technique of hypothesis testing to take into account the relatively large errors of small widths in a refined way.

A second class of nuclides that has been studied to obtain information about the distribution of partial radiation widths are the even–even targets ^{238}U and ^{232}Th. These targets have the advantage that all uncertainties about the spin of the initial state are avoided, since the spin is surely $\frac{1}{2}$. On the other hand, the observed spectra are difficult to interpret because the final states are very closely spaced.

The results for ^{238}U first excited interest when Hughes et al. (1960) reported that the width associated with a 4.02-MeV transition are roughly constant for twelve resonances; when fitted by a χ^2 distribution, the data gave the result $\nu = 90$. This surprising result stimulated other measurements and the evidence concerning the distribution of widths for transitions in ^{239}U passed through several phases. In the end, the matter was effectively settled by Jackson (1964), who observed the spectra of four resonances of ^{238}U with a NaI spectrometer for which the energy resolution was good enough to allow the spectra to reveal that the apparent 4.02-MeV transition actually consists of at least three transitions. Moreover, the distribution of the derived widths were found to be entirely consistent with the Porter–Thomas distribution.

The question of exactly how many low-energy states are involved in what was once thought to be the single 4.02-MeV transition is still a subject of investigation, but there is no longer any question that the low-energy level structure is quite complex. In any case, because of the complexity, recent measurements (Chrien et al., 1967a) with Ge-diode detectors have been aimed largely at unraveling the level structure itself rather than at the study of the statistical behavior of the widths for high-energy transitions.*

Recently, Prestwich and Coté (1968) have studied the distribution of the partial radiation widths following p-wave neutron capture in ^{93}Nb. Since the distribution of widths is expected to be independent of the way in

---

* The final paper on this work (Price et al., 1968) reports the surprising result that $\nu \approx 4$ for high-energy transitions in ^{239}U. This anomaly has not yet been explained.

which the initial compound state is formed, it is gratifying that the p-wave widths are found to be in good agreement with the Porter–Thomas distribution.

2. INTERFERENCE BETWEEN RESONANCES

In the preceding section, we saw that the distribution of the experimental partial radiation widths is consistent with the interpretation that high-energy transitions proceed by way of a single exit channel. If this interpretation is correct, in the neighborhood of two closely spaced resonances the cross section for an individual high-energy transition would be expected to exhibit interference effects of the kind that are so familiar for neutron scattering. On the other hand, if it could be shown that the interference effects are not present, it would be clear evidence that the radiative process is in some way unexpectedly complex.

The first study of the expected interference effect was reported by Coté and Bollinger (1961), who with a NaI detector measured the relative intensity of the ground-state transition as a function of neutron energy in the vicinity of the 11.9- and 19.6-eV resonances of ^{195}Pt. Both of these resonances are known to have $J = 1$. The spectra recorded at the resonance energies are included in Fig. 10.

Fortunately, to study an interference effect it is not necessary to measure an actual cross section for an individual transition. A quantity that is proportional to the fraction of events in the transition of interest is equally satisfactory. Such a quantity is the ratio R_0/R_γ, where R_0 is the rate of pulses in the neighborhood of the binding energy, and R_γ is the intensity of pulses in a broad band at low energy; here it is assumed that R_γ is approximately proportional to the rate of neutron capture. The experimental dependence of R_0/R_γ on neutron energy in the neighborhood of two ^{195}Pt resonances is given in Fig. 14.

The data of Fig. 14 were interpreted in terms of a multilevel formula given by Feshbach et al. (1954). In the special case in which only two resonances are involved, the formula for the cross section σ_j of a radiative transition to a single final state j reduces to

$$\sigma_j = \frac{\Gamma_{1j}}{\Gamma_1}\left(\frac{E_1}{E}\right)^{1/2}\frac{\sigma_{01}}{1 + x_1^2} + \frac{\Gamma_{2j}}{\Gamma_2}\left(\frac{E_2}{2}\right)^{1/2}\frac{\sigma_{02}}{1 + x_2^2}$$
$$\pm 2\left[\frac{\Gamma_{1j}}{\Gamma_1}\frac{\Gamma_{2j}}{\Gamma_2}\left(\frac{E_1 E_2}{E^2}\right)^{1/2}\sigma_{01}\sigma_{02}\right]^{1/2}\frac{x_1 x_2 + 1}{(x_1^2 + 1)(x_2^2 + 1)} \qquad (9)$$

where the subscripts $r = 1$ or 2 refer to the two resonances involved, and Γ_{rj} is a partial radiation width for a transition to a final state j, Γ_r is a total

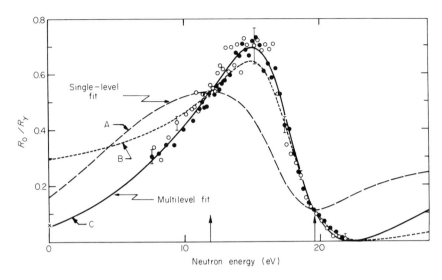

FIG. 14. Resonance-resonance interference for the ground-state transition in the reaction $^{195}\text{Pt}(n,\gamma)^{196}\text{Pt}$ (Coté and Bollinger, 1961). The vertical arrows give the energies of resonances at 11.9 and 19.7 eV. The two kinds of points are the data obtained in separate measurements.

width, E_r is a resonance energy, E is the neutron energy, σ_{0r} is a peak cross section, and $x_r = (E - E_r)(2/\Gamma_r)$. In this relationship, the first two terms are the usual single-level contributions, and the third term contains the interference effect of interest. Notice that Eq. (9) is of exactly the same form as the familiar two-level scattering formula given by Bethe (1937), except that the terms involving potential scattering are missing, and for the radiative transition the sign of the interference term is not known in advance. From Eq. (9) it can be shown that, for the interference effect to be prominent, the quantity $(\Gamma_{1j}\Gamma_{2j}\Gamma_{n1}\Gamma_{n2}) \times (E_1 E_2)^{-1}(E_2 - E_1)^{-4}$ needs to be large, where Γ_n is the neutron width.

The calculated energy dependence of R_0/R_γ under various assumptions is shown in Fig. 14. Curve A is calculated under the assumption that no interference effects are present. This calculated curve is seen to bear little resemblance to the experimental curve, especially in the neighborhood of 20 eV. Curve B is the result obtained from Eq. (9) when the ground-state transition is assumed to result entirely from contributions by the two resonances at 11.9 and 19.6 eV and the contributions from these resonances are assumed to interfere constructively in the region between the resonances. Although the calculated curve does not fit the data quantitatively, it does exhibit the correct qualitative shape. Curve C shows the excellent fit that is obtained when allowance is made for the influence of the distant resonances. Notice that the fit is improved both in the region close to thermal energy and also in the region between the 11.9- and 19.6-eV resonances. This later improvement results from the amplitude contributed by the 67.4-eV resonance.

The qualitative similarity between Curve B and the experimental data, in contrast to the dissimilarity of Curve A, is convincing evidence that most of the amplitude associated with ground-state transitions for the 11.9- and 19.6-eV resonances add constructively. The improved fit obtained by allowing the 67.4-eV resonance to contribute an amplitude strengthens this conclusion. Thus, the data are in good agreement with the hypothesis that a high-energy radiative transition proceeds by a single exit channel. Moreover, the data are inconsistent with expectation for a process that proceeds by many exit channels. However, one cannot exclude the possibility that two or three exit channels are open and that, by chance, only one channel has a significant width for the particular resonances involved in the study.

The above result for the 11.9- and 19.6-eV resonances of ^{195}Pt have been confirmed and extended by Coté and Prestwich (1966) in measurements carried out with a Ge-diode gamma-ray spectrometer. In addition to the ground state, they were able to measure the energy dependence of transitions to the states at 1404 and 1922 keV. In the energy region between the resonances the interference is constructive for the 1404-keV state and destructive for that at 1922 keV. In both cases, the energy dependence of R_j/R_γ is in good agreement with what would be expected of a single channel process.

Now that good measurements with Ge-diode detectors have become routine, it turns out that many nuclides have at least a few transitions for which interference effects can be detected—especially if the measurements are extended down to very low energy, where both neutron flux and capture cross sections are large. Examples of such nuclides are ^{238}U (Price *et al.*, 1968), ^{135}Ba (Becvar *et al.*, 1968), and ^{197}Au (Wasson *et al.*, 1969).

Interference effects of another kind are considered in Section IV,A,5.

3. Correlation of Partial Radiation Widths

Several papers have suggested that the partial radiation widths might exhibit correlation effects. On theoretical grounds, Porter and Thomas (1956) suggested that there might be a positive correlation between the widths for radiative transitions from a single high-energy state to several low-energy states of the same character. Early measurements by the Brookhaven group (Hughes *et al.*, 1959), on the other hand, seemed to indicate the existence of a negative correlation or anticorrelation between widths of this kind. The implications for the statistical theory of a negative correlation have been considered by Rosenzweig (1963).

One way of searching for the existence of correlations between widths for radiative transitions to low-energy states is to examine the distribution of the *sum* of widths. If the distribution of widths for individual transitions is a χ^2 distribution of ν degrees of freedom and if the individual transitions are uncorrelated, then the distribution of the sum of widths to n final states would be expected to be a χ^2 distribution with $n\nu$ degrees of freedom. On the other hand, a positive correlation would result in a distribution with less than $n\nu$ degrees of freedom and a negative correlation would result in a distribution with more than $n\nu$ degrees of freedom. These arguments were first applied to the partial radiation widths by Hughes (1959). He pointed out that although the Argonne group (Bollinger *et al.*, 1959) had reported that the partial widths for transitions to individual low-energy states of ^{184}W showed large fluctuations, the *sum* of widths for transitions to the ground state and first excited state was roughly constant for the first five resonances of ^{183}W. This approximate uniformity for the sum of widths has been confirmed by others. The only question is whether the result is statistically significant. This problem has been treated quantitatively by Bollinger *et al.* (1963), who used the technique of random sampling outlined in Section IV,A,1 to test the hypothesis $\nu = 2$ for the set of five sums of widths. It was found that as little fluctuation as is observed for the physical sample could occur by chance with a 5.6% probability. Thus, the uniformity of the sum of widths for the first five resonances of ^{183}W is much greater than would be expected if $\nu = 2$; however, the probability for the observed uniformity to occur by chance is too great to allow one to conclude with confidence that the partial widths are correlated.

This whole question of anticorrelation between transitions to the ground and first excited states of ^{184}W has been cleared up by measurements with a Ge-diode detector at the Saclay linac. In addition to the five resonances discussed above, Samour (1968) measured the spectra for eight resonances at higher energy. For this larger sample of widths the apparent anticorrelation simply goes away; the best estimate of ν for the sum of the two widths is now $\nu = 2.5^{+1.6}_{-1.2}$, which is entirely consistent with the value 2 that is expected if the transitions are independent.

The partial radiation widths for the resonances of ^{195}Pt may also be examined for correlation. Various conflicting conclusions were drawn from the results with measurements of NaI detectors. However, this evidence may be ignored since the recent measurements at Saclay with a Ge-diode detector are much more extensive. Jackson *et al.* (1966a) report that the distribution of the sum of widths of transition to the first three states of

^{196}Pt from twenty-two initial states is described by the value $\nu = 3.57 \pm 1.00$, a value that is in good agreement with the value 3 that is expected for uncorrelated transitions.

A test of the hypothesis of independence between various pairs of widths may also be made by examining the coefficient of correlation for the observed widths. This kind of test has the advantage that it does not depend on a knowledge of the distribution of widths for individual transitions. The results of calculations of correlation coefficients for the widths of transitions to various low-energy states in ^{78}Se, ^{184}W, and ^{196}Pt have been reported by the Argonne group (Bollinger *et al.*, 1963). They find no statistically significant evidence for a nonzero correlation; hence, the experimental data are consistent with the hypothesis of independence. However, the statistical uncertainty in the correlation coefficient is so large for the small samples studied that the test for independence is not very meaningful.

4. Average Widths for E1 Transitions

The experiments discussed up to this point have all been concerned with the partial widths of *individual* transitions. We now turn to the class of experiments for which *average* widths of transitions from many initial states to a single final state are of interest. Until recently, our principal source of information of this kind has been the capture gamma-ray spectra for thermal neutrons. With these spectra, Kinsey, Bartholomew, Groshev, and others have studied many aspects of the systematic behavior of high-energy transitions*: the dependence of average intensity on gamma-ray energy, on the level spacing at the initial state, on the multipolarity of the radiation, on the nuclear size, and on nuclear structure. These studies have been successful in giving a general outline of the behavior of high-energy radiations. However, the measurements suffer from one serious defect: The thermal-neutron-capture gamma-ray spectra are formed by significant contributions from only a few neutron resonances whereas we are interested in transition intensities averaged over many initial states. Thus, in view of the extremely broad distribution for the partial radiation widths, the thermal-neutron-capture gamma-ray spectra cannot give accurate information about the behavior of the average intensity. It is for this reason that resonance-capture measurements (in which the widths of transitions from many initial states can in principle be obtained) may be expected to play a dominant role in future studies of the average intensity.

* An extensive bibliography of papers on this subject is given by Bartholomew (1961).

The most widely used theoretical relationships for the widths of high-energy radiative transitions are those given by Blatt and Weisskopf (1952) on the basis of a single-particle model of the radiation process.* This treatment shows the dependence of the widths on gamma-ray energy E_γ, on nuclear size, and on average level spacing D at the initial state. The latter parameter plays a role because the total radiation strength is considered as being shared by all of the states in a broad range of energy. For convenience in making comparisons with the experimental data, let us write the estimates for the widths in explicit form for the four kinds of radiation (electric and magnetic dipole and quadrupole) that might be of interest in neutron-capture gamma-ray spectra. These relationships† are

$$\Gamma_{\gamma i}(\text{E1}) = 6.8 \times 10^{-2} A^{2/3} E_\gamma^3 D(E)/D_0 \tag{10a}$$

$$\Gamma_{\gamma i}(\text{E2}) = 4.9 \times 10^{-8} A^{4/3} E_\gamma^5 D(E)/D_0 \tag{10b}$$

$$\Gamma_{\gamma i}(\text{M1}) = 2.1 \times 10^{-2} E_\gamma^3 D(E)/D_0 \tag{10c}$$

$$\Gamma_{\gamma i}(\text{M2}) = 1.5 \times 10^{-8} A^{2/3} E_\gamma^5 D(E)/D_0 \tag{10d}$$

where $\Gamma_{\gamma i}$ is in eV, E_γ is the gamma-ray energy in million electron volts, A is the nucleon number, and $D(E)$ is the spacing at the initial excitation energy of levels with the same spin and parity as the radiating state. The quantity D_0 is a spacing that is characteristic of the separation between major shell-model shells. Both D and D_0 are in million electron volts. The value of D_0 originally suggested by Weisskopf was 0.5 MeV. However, recently it has been customary to use a value in the neighborhood of 15 MeV.

A second approach to the problem of electric dipole radiative transitions is one in which it is assumed that the behavior of the neutron-capture gamma-rays may be deduced from a knowledge of the photonuclear reactions at a somewhat higher energy. The basic assumption of this approach, which was introduced by Brink (1955) and developed by Axel (1962), is that the same physical processes are responsible for the E1 neutron-capture gamma-rays as the processes that form the dipole giant resonance observed in photonuclear reactions. Under this assumption, one can write down an explicit relationship between the photonuclear absorption cross section and the average width $\bar{\Gamma}_{\gamma 0}$ for the *ground-state* transition. In particular, since the central part of the giant resonance is known to be describable by a classical Lorentz line shape, the dependence of $\bar{\Gamma}_{\gamma 0}$ on the gamma-ray energy will also be expressed in terms of the Lorentz shape. Specifically, one obtains

---

* The theoretical bases of this and other models of the radiation process have been reviewed by Lynn (1967).

† The numerical constants in Eq. (10) are those given by Wilkinson (1960).

$$\frac{\bar{\Gamma}_{\gamma 0}}{D} = 1.7 \times 10^{-6} \frac{A}{g} \frac{E_\gamma{}^4 \Gamma_g}{(E_g{}^2 - E_\gamma{}^2)^2 + E_\gamma{}^2 \Gamma_g{}^2} \tag{11}$$

where E_γ and the giant-resonance energy E_g and width Γ_g are all in million electron volts and A is the atomic weight. The statistical factor g is $\frac{3}{2}$ for all dipole transitions.

The general approach of obtaining the average radiation width for the ground-state transition from the photonuclear absorption cross section is obviously correct, of course, in a region of energy where the photonuclear absorption cross section is known. The significant contribution made by Brink and Axel is the bold hypothesis that the Lorentzian shape of the photonuclear cross section extends to a much lower energy than has been experimentally verified.

To obtain numerical values of widths from Eq. (11), it is necessary to use explicit values for Γ_g and E_g and, in order to obtain a relationship that is applicable to a wide range of nuclides, Axel uses the approximate values $\Gamma_g = 5$ MeV and $E_g = 80\,A^{-1/3}$ MeV. Then, in the energy range 6–8 MeV (the range of primary interest for neutron-capture gamma rays) Eq. (11) reduces to the simple approximate relationship

$$\bar{\Gamma}_{\gamma 0}/D = KE_\gamma{}^5 A^{8/3} \tag{12}$$

where

$$K = 6.1 \times 10^{-15}$$

Here, $\bar{\Gamma}_{\gamma 0}$ and D are expressed in the same units and E_γ is in million electron volts. A comparison of Eq. (10a) and Eq. (12) shows that, although $\bar{\Gamma}_{\gamma 0}$ is directly proportional to D in both, the dependence on A and E_γ is quite different. One would hope that the difference is large enough to permit one to discriminate between the two relationships on the basis of the existing experimental data.

In addition to theories that attempt to infer the behavior of radiation widths from rather general considerations, some attempt has been made to develop theories that take into acccunt some features of nuclear structure. The most fully developed treatment of this kind is that of Lane and Lynn (1960a,b), which is considered in relationship to the experimental data in Section IV,A,5. For the time being, we will begin the discussion of the data within the framework of the more primative theories, keeping in mind that radiation widths may in principle depend on detailed characteristics of the initial and final states as well as on general parameters such as E_γ, A, and D.

a. Dependence on level spacing. The dependence of the partial radiation width on level spacing was first studied by Kinsey (1955, 1957) from E1 and M1 transitions observed in thermal-neutron-capture gamma-ray

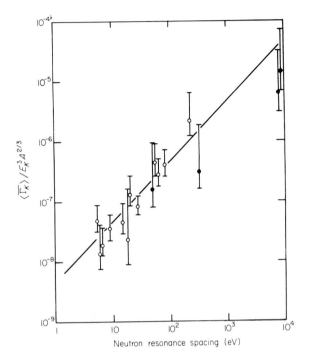

FIG 15. Dependence of the partial radiation width on level spacing (Carpenter, 1962). The open circles were obtained from resonance-capture spectra and the filled circles from selected thermal-capture spectra.

spectra. To isolate the effect of the level spacing, the dependence of intensity on A and E_γ was assumed to be given by the Weisskopf relationships. Then the corrected intensity $\Gamma_\gamma A^{-2/3} E_\gamma^{-3}$ was plotted as a function of D. In spite of the wide scatter in the data, scatter which we now understand as resulting from the broad distribution in the partial radiation widths, the corrected intensity was found to be strongly dependent on D. Within the accuracy of the data, the dependence was the expected direct proportionality. Any possible uncertainty about the correction for the variation of A and E_γ cannot influence the observed dependence on D significantly because D varies over such a great range.

Another direct test of the proportionality between $\Gamma_{\gamma j}$ and D has been made by Carpenter (1962) with the results of measurements on resonance capture. In these measurements, the gamma-ray spectra of the low-energy resonances of 12 heavy nuclides were studied with a NaI detector at the Argonne chopper. Some of the spectra used have been presented in Fig. 10.

For each spectrum, the widths for transitions to several low-energy states were determined. In all, 121 partial widths were obtained. The average values for the various targets are given in Fig. 15 by plotting $\bar{\Gamma}_{\gamma j}A^{-2/3}E_\gamma^{-3}$ versus D. Within the statistical uncertainties, the data are in good agreement with the expected proportional relationship. The agreement is equally good when $\bar{\Gamma}_{\gamma j}A^{-8/3}E_\gamma^{-5}$ is plotted against spacing.

Somewhat less direct evidence about the dependence of $\Gamma_{\gamma i}$ on D is provided by the behavior of the total radiation widths Γ_γ. It is observed experimentally that Γ_γ is approximately independent of the spin of the initial state, even though the state with higher spin has more final states accessible to it by E1 transitions. Brink (1955) has pointed out that this independence of spin gives convincing evidence about the dependence of $\Gamma_{\gamma i}$ on the level spacing D. The argument is most easily made by writing down an expression for the total radiation width. If the partial radiation width is assumed to be of the form $\Gamma_{\gamma j} = \Phi(E_\gamma)\Psi(D_a)$, where $\Phi(E_\gamma)$ is a function of the gamma-ray energy, and $\Psi(D_a)$ is a function of the level spacing D_a at the initial state a, then the total radiation width becomes

$$\Gamma_\gamma = \sum_j \bar{\Gamma}_{\gamma j} = \Psi(D_a) \sum_J \int_0^Q \rho(E,J)\Phi(Q - E)\,dE \tag{13}$$

where Q is the excitation energy of the initial state and $\rho(E,J)$ is the density of states with angular momentum J at an excitation energy E. The sum Σ_J is over all J values accessible by E1 transitions from the initial state.

In writing Eq. (13) in the form of an integral, we have in effect adopted a statistical model in which no single transition is important. Let us now carry this concept further by assuming that the density $\rho(E,J)$ is separable into a product of a function ρ_E of E and a function ρ_J of J. Then the integral is independent of J. Consequently, the total radiation width is independent of D_a only if

$$\Psi(D_a) \propto \left(\sum_J \rho_J \right)^{-1} \tag{14}$$

Let us now further assume $\rho_J \propto (2J + 1)$, a relation at least approximately valid when J is small (Lynn, 1967). Then Eq. (14) implies $\Gamma_{\gamma i} \propto D_a$.

The most sensitive case with which to test the proportionality between $\Gamma_{\gamma j}$ and D is a target with spin $\frac{1}{2}$, since for it the ratio of level spacings for the two possible initial states is 3. The total radiation widths are found to be almost the same for resonances with $J = 0$ and $J = 1$ in ^{107}Ag, ^{109}Ag and other targets with spin $\frac{1}{2}$, which implies that $\Gamma_{\gamma j} \propto D_a$, as has been emphasized by Kinsey (1957).

The above discussion of the total radiation widths is somewhat unsatis-

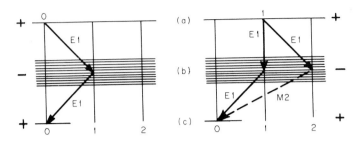

FIG. 16. Decay scheme for 2-step cascades from initial states with $J^\pi = 0^+$ and 1^+ to a 0^+ final state.

factory because it requires too much knowledge about the characteristics of the states accessible to E1 transitions. This objection can be avoided by using a kind of information that is similar to the radiation width but much less widely appreciated, namely, the relative intensity of the 2-step cascades that proceed in *exactly two steps* from the initial state to some final state of the compound nucleus. The use of this kind of information for spin assignments of resonances was introduced by Bollinger and Coté (1960) and exploited by Bollinger et al. (1964); but the alternative use of the data to investigate the radiation process itself has not yet been explored.

As the simplest case, let us consider the 2-step cascades that start from the compound states formed by s-wave capture in a target with spin half and positive parity, i.e., from initial states with $J^\pi = 0^+$ or 1^+. The 2-step cascades from these states to the ground state are illustrated in Fig. 16 where it is assumed that cascades involving transitions other than electric dipole radiation may be neglected. In this case, the total intensity I_J of the 2-step cascades must satisfy the relationship

$$I_J = K \sum_b (\bar{\Gamma}_{ab}/\Gamma_a)(\bar{\Gamma}_{bc}/\Gamma_b) = [K\Psi(D_a)/\Gamma_a] \sum_b \Phi(Q - E_b)(\Gamma_{bc}/\Gamma_b) \quad (15)$$

where K is the number of neutrons captured, Γ_a and Γ_b are the total radiation widths for states a and b, and Γ_{ab} and Γ_{bc} are the partial radiation widths for the transitions $a \rightarrow b$ and $b \rightarrow c$. From Fig. 16 it is immediately clear that the two initial states being considered have the same set of intermediate states. Thus, all of the factors that determine the value of the sum in Eq. (15) are the same for the two initial states except for any factor that depends on the character of the initial state itself. In particular, if we again assume $\bar{\Gamma}_{\gamma i} = \Phi(E_\gamma)\Psi(D_a)$, then the expression for I_J takes on the second form given in Eq. (15). Now the sum is independent of the initial state. Consequently, the ratio of intensities for 2-step cascades from the two initial states is a direct measure of the dependence of $\Psi(D_a)$ (hence of Γ_{ab}) on D_a. Notice that no assumptions have been made about the properties of the nuclear states except the implicit one that the density of states is large.

The data required to provide at least a qualitative test of the ideas presented above have been obtained by Bollinger et al. (1964). They show that the intensity of 2-step cascades to the ground states can easily be measured by recording the spectrum of the sum of pulse heights for coincident counts in two NaI scintillators. In a spectrum of this kind, the intensity of sum pulses with heights in the neighborhood of the binding energy is a direct measure of the intensity of 2-step cascades if the first excited state of the compound nucleus is not too close to the ground state. Examples of sum-coincidence spectra are given later in this chapter (Figs. 26 and 27).

The experimental intensities of 2-step cascades for the resonances of ^{113}Cd, a target with spin half and positive parity, are given in Table III.

TABLE III

RELATIVE INTENSITIES OF 2-STEP CASCADES TO THE 0^+ GROUND STATE
IN THE REACTION ^{113}Cd (n,γ) ^{114}Cd[a]

E_0 (eV)	Relative intensity	Assigned J value
18.5	101 ± 9	1
64	91 ± 9	1
85	98 ± 8	1
108	110 ± 8	1
143	226 ± 34	0
159	79 ± 15	1
193	245 ± 22	0

[a] Bollinger et al. (1964).

The resonances are seen to fall into two classes for which the relative intensities differ by a factor of about three. Bollinger et al. (1964) interpret this difference as resulting from a dependence of intensity on the level density at the initial state under the assumption $\rho_E \propto (2J + 1)$ and on this basis they use the relative intensities to make spin assignments. Unfortunately, there is no independent information with which to check these assignments.

We may now treat the same experimental data for ^{113}Cd from a somewhat different point of view. The spin assignment of the resonances does not depend on the functional dependence of Γ_{ab} on J_a, but only on the assumption that the dependence is moderately strong and monotonic in a known direction. It appears to be meaningful, then, first to use the *qualitative* characteristics of the data to make spin assignments and then to use the same data to make a quantitative determination of the way in which Γ_{ab} depends on D_a. However, since D_a, K, and Γ_a were not carefully measured, it does not seem worthwhile to make a careful assessment of the implications of the existing data. Nevertheless, since the relative intensity reported in Table III is closely proportional to I_J/K and since Γ_a would be expected to be almost independent of J, the data of Table III show that the partial radiation widths are approximately proportional to $(2J_a + 1)^{-1}$. Thus, if $D_a = (2J_a + 1)^{-1}$, the sum-coincidence data imply that the partial radiation widths are approximately proportional to D_a. It is now technically feasible to make sum-coincidence measurement with enough accuracy to allow a sensitive test of the dependence of $\Gamma_{\gamma i}$ on D_a to be made.

  b. Dependence on gamma-ray energy and nuclear size. Much of the experimental data that are available for an examination of the dependence of partial radiation widths on gamma-ray energy and on nuclear size are the same widths as were used in the previous section to examine the dependence on level spacing. However, since the range of variation of E_γ and A are relatively small, the statistical uncertainties in the experimental values of $\Gamma_{\gamma i}$ are now a serious limitation.

  It has long been recognized that the Weisskopf estimates do not give a very satisfactory description of the energy dependence of the experimental partial radiation widths when the energy is varied over a wide range. For example, Kinsey and Bartholomew (1954) pointed out that a comparison of the widths for E1 transitions at about 8 MeV and at very low energies suggests an energy dependence that is much steeper than the E_γ^3 relationship. Similarly, Cameron (1959) found that the average value of $\Gamma_{\gamma i} E_\gamma^{-3}$ required to explain the absolute value of the total radiation widths is smaller by a factor of about 25 than would be inferred from the widths of the high-energy gamma rays. Both of these observations suggest that Eq. (11) gives a more valid description of the energy dependence of E1 radiation widths than does Eq. (10a). However, neither kind of information can be accepted as entirely convincing. The first kind makes use of widths for low-energy states that are probably influenced by details of nuclear structure; the second kind is too indirect. On the other hand, if only high-energy transitions are considered, Carpenter (1962) finds that the statistical uncertainties are so great that the widths obtained from the neutron-capture data are consistent with both the E_γ^3 dependence given in Eq. (10a) and the E_γ^5 dependence given in Eq. (11).

  The influence of nuclear size on the widths for all primary transitions observed in the thermal-neutron-capture gamma-ray spectra has been

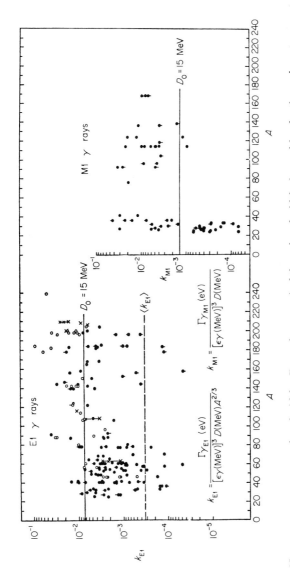

Fig. 17. Neutron capture γ-ray reduced widths. Dependence on A of the reduced widths k_{E1} and k_{M1} for thermal-neutron-capture gamma rays (Bartholonew, 1961). The dashed line labeled $\langle k_{E1} \rangle$ gives the average value obtained by Cameron (1959) from an analysis of total radiation widths.

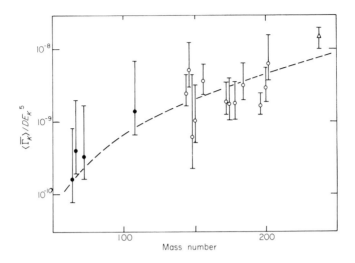

Fig. 18. Dependence on A of average values of partial radiation widths for resonance-capture gamma rays (Carpenter, 1962). The filled circles were obtained from selected thermal-neutron-capture spectra.

studied by Bartholomew (1961) by plotting values of the reduced width as a function of A, where by definition the reduced widths k_{E1} and k_{M1} for E1 and M1 transitions, respectively, are

$$k_{E1} = \Gamma_{\gamma i} \, (\text{E1}) \, D^{-1} E_\gamma^{-3} A^{-2/3} \tag{16a}$$

$$k_{M1} = \Gamma_{\gamma i} \, (\text{M1}) \, D^{-1} E_\gamma^{-3} \tag{16b}$$

It is customary to have $\Gamma_{\gamma i}$ in electron volts and both D and E_γ in million electron volts. As is shown in Fig. 17, some 180 values of k_{E1} were available to Bartholomew, and they form a broad band of points for which the order of magnitude of the mean value is independent of A, as expected from the single-particle model. However, closer examination seems to reveal a tendency for the mean value to increase somewhat with increasing A, in agreement with what would be expected under the giant-resonance description of the radiation widths, as summarized by Eq. (12).

Carpenter (1962) has used his average widths from resonance capture in an attempt to discriminate between Eq. (10a) and (12) on the basis of their different dependence on A. The comparison with Eq. (12) is given in Fig. 18, where the quantity $\Gamma_{\gamma i} D^{-1} E_\gamma^{-3}$ is seen to be consistent with the $A^{8/3}$ dependence on A. Another figure given by Carpenter (1962) shows that the data are in somewhat poorer agreement with the $A^{2/3}$ dependence of Eq. (10a). However, the statistical uncertainties in the average widths are too great to permit one to make a definite choice between Eq. (10a) and (12).

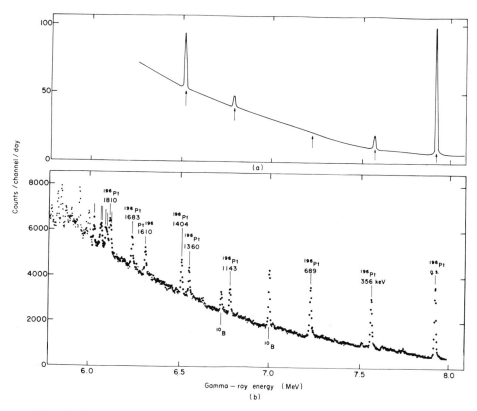

Fig. 19. Average-resonance-capture spectrum for the reaction ^{195}Pt$(n,\gamma)^{196}$Pt in comparison with the spectrum of a single resonance (Bollinger and Thomas, 1967a). The energies associated with the peaks are the energies of the final states for the high-energy transitions [(a) 12-eV resonance ^{195}Pt; (b) average spectrum normal platinum].

The above inconclusive results for the dependence of partial radiation widths on E_γ and A suggest that a major change in experimental technique is needed for an effective study of the average values of partial widths. At least a partial answer to the experimental problem has recently been reported by Bollinger and Thomas (1967a), who make use of a filtered-neutron technique to select neutron capture in a broad but controlled range of energy, as outlined in Section II,C,2. In these measurements, a Ge-diode is used to observe directly the spectrum formed by capture in many resonances, and from these average spectra the average values of partial radiation widths are deduced.

An example of an average gamma-ray spectrum obtained by the filtered-neutron method is given in Fig. 19. This spectrum for the reaction $^{195}Pt(n,\gamma)^{196}Pt$ is seen to be of excellent quality: It has good statistical accuracy and the background from the ^{10}B absorber and other sources does not mask the ^{196}Pt lines of interest. The main interest of the spectrum, however, is the impressive degree of uniformity of the high-energy transitions in ^{196}Pt, in marked contrast to the wild fluctuations that are characteristic of the high-energy portion of the spectrum for a single resonance.

If the widths of the high-energy transitions in ^{196}Pt are assumed to depend only on the gamma-ray energy E_γ, the average-capture spectra may be used to determine the energy dependence even though E_γ varies over only a rather small range. For comparison with the theories outlined earlier, let it be assumed that the radiation width is proportional to E^α. Then for the high-energy transitions in ^{196}Pt, $\alpha = 4.9 \pm 0.5$. This energy dependence is in excellent agreement with the E_γ^5 dependence inferred from the shape of the giant resonance, and it is clearly inconsistent with the E_γ^3 dependence of Eq. (10a). The relationship between the experimental data and the energy dependence inferred from the shape of the giant resonance is given in Fig. 20.

Although the widths of the ^{196}Pt transitions are in good agreement with the E_γ^5 law, it needs to be mentioned that the data and Eq. (12) are not concerned with identical physical phenomena. The data consist of the widths of transitions from a fixed set of initial states to a variable final state whereas the theoretical relationship is concerned with transitions to the ground state from a variable initial state. Recent (p,γ) measurements for light targets suggest a way of explaining why this difference may be unimportant. For a number of light targets it has been found (e.g., see Allas *et al.*, 1964) that the giant resonances built on excited states are similar to the resonance built on the ground state, except that the resonance curve is shifted toward higher energies by an amount that is equal to the energy of the excited state. If this behavior is assumed to exist for heavy nuclides also, then it is obvious that the radiation width depends only on the gamma-ray energy, independent of whether it is the energy of the initial state or the final state that is varied.

At this time, it is too early to know whether or not the E_γ^5 law is valid for all nuclides. However, this seems unlikely, in view of the results obtained by Axel *et al.* (1963) in measurements of the intensity of gamma rays resonantly scattered from zirconium, tin, lead, ^{206}Pb, and ^{209}Bi in the energy range 5–12 MeV. For lead and bismuth the intensity was found to exhibit broad resonances of such a character that, if the level density is a monotonic function of excitation energy, the radiation width could not be described accurately by a simple function of A and E_γ such as Eq. (11).

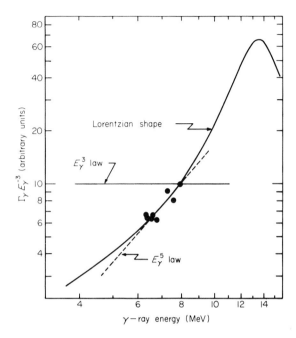

FIG. 20. Dependence of $\Gamma_\gamma E_\gamma^{-3}$ on E_γ for the reaction ^{195}Pt(n,γ)^{196}Pt, as determined from an average-capture spectrum (Bollinger and Thomas, 1967a). The vertical scale of the Lorentzian giant resonance has been adjusted to fit the data.

On the other hand, since the characteristics of lead and bismuth are in many respects unusual, the existing gamma-ray scattering results may not give a very reliable indication of the degree to which the energy dependence of Γ_γ for the typical heavy nucleus may be expected to deviate from a simple dependence on E_γ and A.

Up to this point, we have been concerned only with the manner in which the partial radiation widths *vary*. However, the *absolute values* are also of interest. It has long been accepted that the Weisskopf relationships Eq. (10) do not give a good estimate of the absolute values of partial radiation widths unless the parameter D_0 is arbitrarily adjusted. The best fit of Eq. (10a) to the data of Carpenter (1962) requires $D_0 = 16.5$ MeV. Perhaps a more satisfactory way of summarizing the experimental data within the framework of the single-particle model is to give the average value of the reduced width k_{E1}. Carpenter (1962) obtains the result $k_{E1} = 0.0039$ from measurements on resonance capture in heavy even–odd targets. This value is in good agreement with the value 0.0032 obtained for the same quantity by Bartholomew (1961) from the widths of some 180 E1 transitions observed in the thermal-neutron-capture gamma-ray spectra.

Turning now to Eq. (11), we observe that all of its parameters are determined by experimental data that are independent of the radiation widths. If the ideas that lead to Eq. (11) have a degree of validity, one would expect that the absolute values of widths predicted by it would be at least roughly correct. This expectation is satisfactorily fulfilled, in that the experimental widths measured by Carpenter (1962) are on the average only about a factor of 2 smaller than those given by Eq. (12). The experimental widths imply that the proportionality constant* in Eq. (12) is 3.2×10^{-15} instead of the predicted value 6.1×10^{-15}

In view of the reasonable nature of the arguments that lead to Eqs. (11) and (12) and the satisfactory agreement between these relationships and the experimental data, it seems probable that Carpenter's data as plotted in Fig. 18 provide the most reliable *general* relationship we have for the average width of high-energy (7–8 MeV) E1 transitions in heavy nuclides. For convenient reference, we summarize the experimental data with the equation

$$\bar{\Gamma}_{\gamma j} = 3.2 \times 10^{-15} D E_\gamma^5 A^{8/3} = 1.1 \times 10^{-5} D (E_\gamma/7)^5 (A/100)^{8/3} \quad (17)$$

where $\bar{\Gamma}$ and D are in the same units and E_γ is in million electron volts. As may be seen from Fig. 18, for most even–even compound nuclides this relationship is satisfied within about a factor of 2; however, for some it might be in error by as much as a factor of 3.

Many of the concepts outlined above in connection with the giant-resonance description of radiation widths have recently been synthesized and clarified by Rosenzweig (1968). He examined quantitatively both the energy dependence and the absolute values of the partial radiation widths in ^{196}Pt in relation to the measured photoabsorption cross sections for ^{196}Pt. The two kinds of data were found to be in satisfactory agreement. Moreover, the suggestion that each excited state has built on it a giant resonance with a universal shape and magnitude was made reasonable by means of arguments based on dipole sum rules.

5. Influence of Nuclear Structure

Up to this point we have examined the dependence of radiation widths on several variables under the assumption that the details of the structure of the states involved may be ignored. Let us now inquire whether there is

---

* The proportionality constant given by Carpenter (1962) is ten times larger than it should be.

any experimental evidence to the contrary. The evidence of this kind is still rather fragmentary and the interpretation of what information there is is hampered by the extreme fluctuations that are characteristic of the Porter–Thomas distribution. This problem of random fluctuations is especially severe for the thermal-capture spectra, which are usually dominated by contributions from a very few resonances. Even so, a number of effects that are hard to reconcile with a purely statistical model of the radiation process have been observed in the thermal spectra, and this has led to the development of theories of capture processes in which both the initial and final states are rather simple (Bockelman, 1959; Lane and Lynn, 1960a,b). The theories have been reviewed by Clement (1966) and Lynn (1967).

To help focus attention on the kinds of experimental information that might reveal the influence of nuclear structure, let us briefly review the predictions of the theory of Lane and Lynn. Starting from nuclear-dispersion theory, they develop a formalism for radiative capture in the resonance region of neutron energy. This treatment takes into account not only the usual compound-nucleus formation that leads to an initial radiative state of great complexity, but it also includes processes in which dipole radiation is emitted while the incident s-wave neutron in the field of the target nucleus undergoes a single-particle transition to a p-wave orbit in the final state. The amplitude for the emission of any primary gamma ray is found to be made up of three parts: an "internal" compound-nucleus resonance part, an "external" or "channel" resonance part, and a "hard-sphere" part. The internal resonance contribution exhibits the familiar statistical behavior that is insensitive to nuclear structure, whereas both channel resonance and hard-sphere capture depend on the single-particle character of the final state. The channel resonance part exhibits the same kind of resonance behavior as the internal resonance part, which makes it hard to separate the two in any straightforward way; however, in principal the presence of channel resonance capture can be recognized either by the anomalously large intensity of a transition or by a dependence of its radiation width on the reduced neutron width. Hard-sphere capture, which has a nonresonance behavior, also results in anomalously strong transitions and, in addition, it may be recognized by a characteristic neutron-energy dependence for the gamma-ray intensity; interference between the amplitude for resonance and hard-sphere capture produces asymmetries of the same kind as are observed in the neutron-scattering cross section.

The three experimental characteristics that are of interest to the Lane and Lynn theory (the gamma-ray intensity, the correlation with neutron width, and the neutron-energy dependence) serve as a convenient framework within which to discuss all of the experimental evidence for nuclear-structure effects in the neutron-capture gamma radiation, whether or not there is reason to believe that the mechanism responsible for the observed effect is understood.

a. Anomalously strong transitions. The intensity of radiative transitions is the only kind of information about single-particle effects that can be obtained from a study of thermal capture. Since Bartholomew (1961) has given a good summary of most of the data of this kind, we need mention only the evidence that has a close relationship to resonance-capture experiments. Perhaps the most widely discussed indication of a dependence of thermal-neutron capture on nuclear-structure is what appears to be the anomalously large strengths of certain high-energy transitions in the spectra for nuclides in the range $24 < A < 70$. A qualitative examination of these spectra seems to show that the strong transitions are associated with final states that have a large component of p-wave neutron orbital. Since this behavior suggests that the prominent lines are formed by direct capture, it is somewhat surprising that the absolute values of the reduced widths for these transitions are not unusually large in comparison with those of heavier nuclei (Bartholomew, 1961).

A more quantitative way to study the nonstatistical behavior of the capture gamma rays is to compare the strengths of (n,γ) transitions with the strengths of the corresponding (d,p) transitions, for which the intensity depends on the single-particle character of the final state in a well-understood way. The strengths of radiative transitions in the thermal-capture spectra of at least 11 light nuclides ranging from ^{25}Mg to ^{67}Zn exhibit a strong correlation with the strengths of (d,p) transitions in which one unit of orbital angular momentum is transferred ($l_n = 1$), as discussed by Groshev *et al.* (1958a), Bockelman (1959), and Groshev and Demidov (1966); most of the data are summarized in Fig. 21. This striking behavior is strong support for the reality of single-particle effects. However, the data do not tell us whether hard-sphere capture or channel resonance capture is the dominant process.

For a smaller number of light nuclides (^{54}Fe, ^{56}Fe, ^{62}Ni, ^{64}Zn targets) the thermal-capture spectra contain intense primary transitions to final states that are not markedly single particle in character; consequently, there is a negative correlation between the strengths of (n,γ) and (d,p) transitions. This behavior has been attributed by Groshev and Demidov (1966) and Ikegami and Emery (1964) to a "doorway-state" reaction, in which the incident neutron excites a pair of nucleons in the target nucleus before the radiative transition occurs. It should be noted, however, that the observed anticorrelation in a few nuclides probably is not inconsistent with the

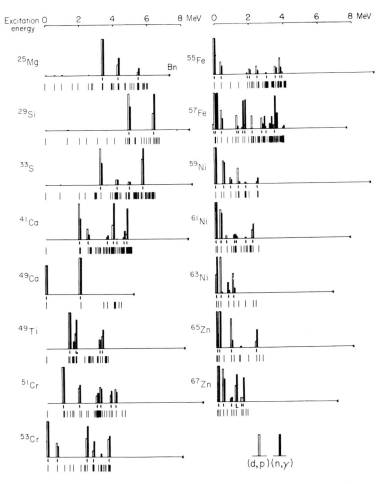

FIG. 21. Correlation between the strengths of (n,γ) transitions and (d,p) transitions with $l_n = 1$ (Groshev and Demidov, 1966). The energy scale at the top of the figure gives the energy of the final state.

statistical behavior of a compound-nucleus reaction, for which the nature of the final state is relatively unimportant.

Measurements of the gamma-ray spectra from resonance capture could provide a valuable extension of our knowledge of single-particle effects for the light nuclides but, because of the difficulty of the measurements, relatively little has been learned yet in this way. Several studies have been undertaken with a pulsed Van de Graaff accelerator as the source of neutrons and a large NaI scintillator as the gamma-ray spectrometer. Firk and

Gibbon (1961) and Bird *et al.* (1962, 1963) measured the spectra of many light targets for capture of neutrons in a broad energy band in the neighborhood of 30 keV. The most interesting of these spectra is that for capture in ^{63}Cu. For it the resonance-capture spectrum, which is an average spectrum formed by contributions from many resonances, does not have the anomalously strong high-energy transitions that are so prominent in the thermal-capture spectrum. This result has recently been refined by Allen (1968), who used a Ge-diode spectrometer at a pulsed Van de Graaff to obtain a resolved average spectrum formed by neutron capture in the broad energy range 10–60 keV. The intensities of the γ-ray lines are found to vary as E_γ^3, in striking contrast to the thermal-capture spectrum, for which transitions to 3 low-energy states carry 40% of the intensity (Groshev *et al.*, 1958b). A similar, though less marked, difference between the spectra for thermal capture and capture in a single resonance (590 eV) of ^{63}Cu was observed by Wasson and Draper (1965) in measurements with the Yale linac. Bergqvist and Starfelt (1963) have discussed the change in the copper spectrum and the absolute intensities of the high-energy lines. These results indicate that channel resonance capture is relatively inimportant for ^{63}Cu and that the strong lines in the thermal spectrum come from hardsphere capture. In contrast, the spectrum formed by capture in a single *s*-wave resonance (42 keV) of ^{207}Pb contains the same strong transitions to the *p*-shell ground state that dominates the thermal-capture spectrum (Biggerstaff *et al.*, 1967). This result suggests that channel resonance capture is important for ^{207}Pb, but this inference needs to be checked by measurements at other resonances. The spectra from capture of very energetic neutrons (7.4 MeV) in Pb, Bi, Ni, and Fe also exhibit dominant high-energy transitions that seem to imply the operation of some special mechanism (Bergqvist *et al.*, 1966), but a discussion of these results is beyond the scope of this book.

For those nuclides for which the spacing between resonances is small enough to permit a chopper or a linac to be used in resonance-capture measurements, somewhat more complete spectral information can be obtained, since Ge-diode spectra can be obtained for individual resonances. Prestwich and Coté (1967a) have used a chopper to study resonance capture in ^{55}Mn and ^{59}Co in an effort to extend the information on the correlation between (n,γ) and (d,p) transitions. Three resonances of ^{55}Mn were studied. The spectra for the two $J = 2$ resonances, and also the thermal-capture spectrum, show a significant positive correlation between the strengths for (n,γ) and (d,p) transitions; but the spectrum for the $J = 3$ resonance does not exhibit this correlation. The authors point out that this difference is not unexpected, since the same final states are not necessarily accessible by E1 transitions from the two kinds of initial states; consequently, to obtain full correlation the (d,p) spectrum for an even-odd nucleus should be compared to an (n,γ) spectrum that is formed by equal contributions from the two initial spin states that can be formed by s-wave capture. For ^{59}Co it was possible to study only the single resonance at 132 eV. In its spectrum the (n,γ) transitions to states fed by (d,p) transitions with $l_n = 3$ are only 0.12 ± 0.08 times as strong as (n,γ) transitions to states with $l_n = 1$. This dependence on l_n is what is expected from single-particle effects, since only the $l_n = 1$ transitions are associated with final states that have a strong component of p-wave. Contrary to expectation, however, the strengths of the (d,p) transitions with $l_n = 1$ are not correlated with the strengths of the corresponding (n,γ) transitions; it is not clear whether the absence of the correlation results from a failure to include in the (n,γ) spectrum contributions from both possible initial spin states or whether it results from random fluctuations in the widths associated with internal resonance capture.

Turning now to heavier substances, a widely discussed indication of a nuclear-structure effect is the seemingly anomalous strength of high-energy transitions in the spectra for some nuclides in the range $130 < A < 209$. Most of these spectra are too complex to be fully resolved and the experimental evidence for a nonstatistical behavior of the radiation widths is provided by the over-all shapes of the spectra. As has been recognized for a long time, some nuclides such as Ta exhibit the bell-shaped spectrum that is qualitatively expected for a simple statistical model, but others such as Au have a prominent high-energy "bump" that seems to be inconsistent with the model (Groshev *et al.*, 1958). Examples of these spectral shapes are given in Fig. 22. Starfelt (1964) has shown that the anomalous bump is definitely inconsistent with a statistical model in which the radiation width is a monotonic function of energy, and Groshev *et al.* (1958a) pointed out that the energy of the bump changes with the nuclear species in the way that would be expected of direct transitions to the $3p$ shell-model states.

The suggestion that the bump in the spectra of heavy nuclides is formed by transitions to p-wave final states seems not to be consistent with more recent experimental findings. By means of measurements with a NaI gamma-ray spectrometer at a pulsed Van de Graaff accelerator, Bergqvist and Starfelt (1962a,b) showed that the gross shapes of the spectra from capture in Ag, Sn, I, Cs, W, and Au are effectively independent of the energy of the incident neutron over the range 0–300 keV. In more recent measurements of the same kind by Bergqvist *et al.* (1963), the incident neutron energy was extended to 3.2 MeV, and the only influence on each gamma-ray

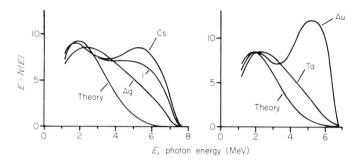

FIG. 22. Gross shapes of capture-gamma-ray spectra (Starfelt, 1964). The curves marked "theory" were obtained assuming that $\Gamma_{\gamma j} \propto E_{\gamma}^3$. A better fit to the Ag and Ta spectra is obtained under the assumption $\Gamma_{\gamma j} \propto E_{\gamma}^5$.

spectrum was the addition of a long, weak tail in the energy range greater than the neutron binding energy. Similarly, in a study of the closely related reaction ^{197}Au (d,pγ) Bartholomew *et al.* (1967a) found that the energy of a bump at about 5 MeV in the gamma-ray spectrum is independent of the initial excitation energy over a wide range of energy. These observations contradict the idea that the bump is formed by transitions to fixed final states because for such transitions the energy of the bump would increase linearly with the energy of the initial state. An alternative interpretation of the spectra in terms of an M1 giant resonance is discussed in Section IV,A,6.

Rae *et al.* (1967) have recently reported evidence for nuclear-structure effects in the resolved gamma-ray spectra from resonance capture in ^{199}Hg, one of the nuclides often said to exhibit an anomalous bump. They used a Ge-diode spectrometer at the RPI linac to measure the spectra for two 1$^-$ resonances. For these two resonances, they found that the average reduced width $\Gamma_{\gamma j}/E_{\gamma}^3$ of E1 transitions to the four states (with $J^\pi = 0^+$ or 2^+) below 1300 keV in ^{200}Hg are six times smaller than the average reduced width for transitions to states of higher energy. This result tends to strengthen the idea that some nuclear-structure effect is responsible for the unresolved bump for the heavy nuclides. However, the author of this chapter is not convinced that the reported difference in the intensities of the transitions in ^{200}Hg could not be explained by the combined effects of unresolved lines (causing an underestimation of the number of transitions to the higher states) and a reasonable degree of statistical fluctuation.

Unexpected regularities in the easily resolved portion of the gamma-ray spectra from thermal-neutron capture in several deformed even-even heavy nuclides have recently excited interest. The first evidence of this kind came from studies of thermal capture in ^{176}Hf and ^{178}Hf by Namenson *et al.*

(1966) and capture in ^{182}W, ^{184}W and ^{186}W by Martin *et al.* (1966). The evidence has been reviewed by Namenson and Bolotin (1967). In these data, high-energy transitions to the $\frac{1}{2}^-$ and $\frac{3}{2}^-$ states of the $\frac{1}{2}^-[510]$ Nilsson band and to the $\frac{3}{2}^-$ state of the $\frac{3}{2}^-[512]$ band are observed. The arresting feature of the data is that for all five nuclides the high-energy transitions to the $\frac{1}{2}^-[510]$ band are very much stronger than the transitions to the $\frac{3}{2}^-[512]$ band, as is shown in Table IV; also, except in ^{187}W, the $\frac{1}{2}^-$ state of

TABLE IV

RELATIVE INTENSITIES OF HIGH-ENERGY RADIATIVE TRANSITIONS IN THE
ISOTOPES OF TUNGSTEN

Emitting nucleus	Final state			Relative intensity	
	Energy (keV)	Nilsson band	J^π	Thermal[a] capture	Resonance[b] capture
^{183}W	0	$\frac{1}{2}^-[510]$	$\frac{1}{2}-$	100	100 ± 25
	47	$\frac{1}{2}^-[510]$	$\frac{3}{2}-$	34	99 ± 25
	209	$\frac{3}{2}^-[512]$	$\frac{3}{2}-$	<1	63 ± 25
^{185}W	0	$\frac{3}{2}^-[512]$	$\frac{3}{2}-$	10	90 ± 25
	23	$\frac{1}{2}^-[510]$	$\frac{1}{2}-$	100	100 ± 25
	93	$\frac{1}{2}^-[510]$	$\frac{3}{2}-$	16	77 ± 25
^{187}W	0	$\frac{3}{2}^-[512]$	$\frac{3}{2}-$	5	75 ± 35
	147	$\frac{1}{2}^-[510]$	$\frac{1}{2}-$	100	100 ± 30
	205	$\frac{1}{2}^-[510]$	$\frac{3}{2}-$	139	138 ± 30

[a] Martin *et al.* (1966).

[b] Bollinger and Thomas (1967b).

the $\frac{1}{2}^-[510]$ Nilsson band is fed more strongly than the $\frac{3}{2}^-$ state of the same band. A somewhat similar pattern was observed by Prestwich and Coté (1967c) for the reaction ^{166}Er$(n,\gamma)^{167}$Er. It is not surprising that this unusual degree of regularity was for a time considered to be convincing evidence for a specific nuclear-structure effect. Direct capture does not explain the data, since the (n,γ) intensities are not correlated with the (d,p) transition strengths. Namenson and Bolotin (1967) suggested that the observed intensity pattern might result from the Λ selection rule that, in the collective model, would govern transitions from an initial state in which the incident $s_{\frac{1}{2}}$ neutron is strongly coupled to the 0^+ deformed even–even target.

Several studies of resonance-capture spectra have been carried out in an effort to check the intensity pattern observed in thermal capture. In all of these studies a Ge-diode was the gamma-ray spectrometer. The first such study was reported by Spencer and Faler (1967), whose source of neutrons was a crystal monochromator at the MTR reactor. They obtained spectra of good quality for the 4.1-eV resonance in ^{182}W and the 18-eV resonance in ^{186}W. The spectra are similar to those for thermal capture, an understandable result since these resonances have a dominant influence on the thermal cross sections of ^{182}W and ^{186}W. The measurements on the tungsten isotopes have been extended to a few resonances at higher energy by Prestwich and Coté (1967b) and Beer et al. (1966), using fast choppers at reactors and by Rae et al. (1967), using an electron linac. These results throw considerable doubt on the significance of the regularity in the thermal spectra since not all of the resonance spectra exhibit the intensity pattern found in the thermal spectra.

Even more convincing evidence that nuclear-structure does not have a significant influence on the high-energy transitions of the tungsten isotopes is provided by the results of Bollinger and Thomas (1967b), who used the filtered-neutron method outlined in Section II,C,2. The average spectra formed by capture in many resonances were measured for samples of ^{182}W, ^{184}W, and ^{186}W. In all of these spectra, the transitions to the $\frac{1}{2}^-$[510] and $\frac{3}{2}^-$[512] Nilsson bands are of equal intensity within the experimental uncertainty of about 30%, as is shown in Table IV. Thus, it seems almost certain that the neutron-capture spectra of the tungsten isotopes are not influenced to a significant degree by nuclear-structure and that the regularities observed in the thermal spectra result from an uninteresting statistical fluctuation.

Before closing this discussion of heavy nuclei, it seems worthwhile to inquire to what extent high-energy transitions may be *independent* of the character of relatively simple low-energy final states, that is, to what degree of refinement is the statistical model valid for at least some transitions of this kind? The most refined experimental answer to this question comes from the filtered-neutron studies by Bollinger and Thomas (1967a) of the average spectra from capture in many resonances of ^{195}Pt. These same data were used in the preceding section in the discussion of the energy dependence of radiation widths. When the widths for transitions to the 8 states of lowest energy in ^{196}Pt are assumed to vary as $E_\gamma{}^\alpha$, the rms deviation of the measured widths from the least-squares fit is only 9%, and this deviation is consistent with what would be expected on the basis of the number of resonances that contribute to the average spectra. Thus, the structure of the final state appears to have almost no influence on the average width of the high energy transitions in ^{196}Pt.

b. Correlation with the neutron width. Until recently the effort to find the predicted correlation (Lane and Lynn, 1960a) between the width $\Gamma_{\gamma j}$ of radiative transitions and the reduced neutron width $\Gamma_n{}^0$ was quite unsuccessful. However, several experiments gave results that are worth mentioning. For the resonances of ^{77}Se, Bollinger et al. (1963) obtained an interestingly strong correlation between $\Gamma_n{}^0$ and the sum of the widths for radiative transitions to the ground and first excited states of ^{78}Se; however, the observed correlation (+0.84) cannot be considered statistically significant, since spectra were measured for only four resonances.

Another interesting set of data is that of Prestwich and Coté (1967b) for capture in four resonances of ^{182}W. They find no significant correlation between $\Gamma_n{}^0$ and the radiation widths for the three high-energy transitions studied, but there is almost complete correlation between $\Gamma_n{}^0$ and the ratio $\Gamma_\gamma[512]/\Gamma_\gamma[510]$ for transitions to the $\frac{1}{2}^-$ states of the [512] and [510] Nilsson bands. Here again, the statistical sample is too small to establish the reality of the correlation, expecially since the ratio under consideration would be expected to be influenced by random fluctuations to a greater extent than the radiation width $\Gamma_\gamma[512]$ itself.

If an unresolved region in the capture-gamma spectrum of a heavy nucleus is formed by transitions for which the radiation amplitudes have components that are proportional to the reduced neutron-width amplitude, then the gross intensity of transitions in the region of interest should be correlated with the reduced neutron width. This expectation has the important experimental implication that, since individual lines do not need to be resolved, a NaI scintillation spectrometer may be used in measurements aimed at detecting the correlation. In an experiment of this kind, Wasson and Draper (1963) measured the capture spectra of twelve resolved resonances of ^{197}Au. From these they determined the average intensity $\bar{I}_\gamma$ of what are now known (Prestwich and Coté, 1967d) to be about 12 unresolved gamma-ray lines in the energy range $6.0 < E < 6.5$ MeV. No correlation between $\bar{I}_\gamma$ and $\Gamma_n{}^0$ was observed. Similarly, the ^{195}Pt spectra given in Fig. 10 were reexamined by the author in a search for a correlation between $\Gamma_n{}^0$ and the average gamma-ray intensity in the range 5–6 MeV, the region of the spectrum that is considered to be anomalously strong. No correlation was found for the five spectra that are usable. Thus, the correlation between $\Gamma_n{}^0$ and the intensities of most transitions in many heavy elements is weak, if present at all.

This peaceful and uninteresting situation with regard to correlations was recently disrupted by Beer et al. (1968) when they reported the detection of a statistically significant positive correlation between the reduced neutron width and partial radiation widths in the reaction ^{169}Tm (n,γ) ^{170}Tm. They studied fifteen high-energy transitions in the spectra for seven neutron resonances observed with a Ge-diode detector at the Brookhaven chopper; three of these spectra are shown later in Fig. 29. The average coefficient of correlation between the reduced neutron width $\Gamma_n{}^0$ and the radiation widths $\Gamma_{\gamma j}$ was found to be $R = +0.27$. To determine the statistical significance of this value, the data were interpreted in terms of the theory of Lane and Lynn (1960a) by assuming that the radiation amplitude is made up of a channel resonance-capture term and a compound-nucleus term, giving

$$\Gamma_{\gamma r j}{}^{1/2} = A_j(\Gamma_{nr}^0)^{1/2} + \Gamma_{crj}{}^{1/2} \tag{18}$$

where A_j is a parameter independent of the initial state r, and $\Gamma_{crj}^{1/2}$ is a random compound-nucleus amplitude with a mean value of zero. Defining ρ_j as the correlation between $\Gamma_{\gamma r j}^{1/2}$ and $(\Gamma_{nr}^0)^{1/2}$, a Monte Carlo technique similar to that described in Section IV,A,1 was then used to show that the value $R = +0.27$ implies that $0.2 < \rho^2 < 0.8$ with 80% confidence and that there is only a 0.1% probability for there to be no correlation. This positive correlation was cited as evidence for the detection of channel-resonance capture. Although the statistical analysis presented by Beer *et al.* (1968) is quite convincing, it is somewhat disturbing that their nonzero correlation is caused entirely by the influence of one resonance—the one with the largest radiation width.*

Since the amplitude for channel resonance capture depends on the single-particle character of the final state, when channel resonance capture is significant the radiation widths are expected to be correlated with the intensities of (d,p) transitions as well as with the reduced neutron widths. The expected relationship is contained in the parameter A_j in Eq. (18), which should be proportional to the $l_n = 1$ component of the (d,p) cross section. Thus, under the channel-resonance-capture interpretation of the ^{169}Tm (n,γ) ^{170}Tm data, there should be a positive correlation between the strengths for (n,γ) and (d,p) transitions to the final states for which a nonzero value of $\rho_j{}^2 = A_j{}^2\langle\Gamma_n{}^0\rangle/\langle\Gamma_{\gamma j}\rangle$ was measured. Bollinger and Thomas (1968a) have attempted to test this idea by using a neutron-filter technique to obtain accurate values of the average widths of radiative transitions from many initial states to individual final states in ^{170}Tm. No correlation between the (n,γ) and (d,p) intensities was observed, even though the final states of the transitions compared were the same as those involved in the

* Measurements have now been extended to additional resonances by Jackson [quoted by Bollinger (1968)] and Chrien *et al.* [see the short contribution following the paper by Bollinger (1968)]. These measurements strongly reinforce the validity of a positive correlation between the reduced neutron width and partial radiation widths in ^{169}Tm$(n,\gamma)^{170}$Tm, since they both show that a second resonance (153 eV) with a very large reduced neutron width has exceptionally large partial radiation widths.

transitions that exhibit a correlation with the neutron widths. This negative result suggests strongly that the correlation between radiation and neutron widths in ^{169}Tm$(n,\gamma)^{170}$Tm does not result from channel resonance capture. On the other hand, this inference may be premature, in view of doubts raised by the complexity of the nuclear reaction and the incompleteness of the (d,p) data involved. In any case, the general question of the extent to which channel resonance capture is significant for most nuclides is still an unsolved experimental problem.

c. Interference effects. When the possibility of hard-sphere capture is taken into account (Lane and Lynn, 1960a) the cross section $\sigma_j(E)$ for a primary radiative transition to a final state j varies with the neutron energy E according to a relationship of the form

$$\sigma_j(E) \propto \frac{1}{\sqrt{E}} \left| a_j + \sum_r \frac{(\Gamma_{rn}^0 \Gamma_{rj})^{1/2}}{E_r - E + \frac{1}{2}i\Gamma_r} \right|^2 \tag{19}$$

where a_j is the amplitude for hard-sphere direct capture and the terms in the summation are amplitudes for resonance capture; the quantities Γ_r, Γ_{rn}^0, Γ_{rj}, and E_r are, respectively, the total width, the reduced neutron width, the width for a radiative transition to a state j, and the resonance energy for an initial state r. It is assumed that the spacing between the levels is much greater than the widths. The hard-sphere and resonance amplitudes add constructively and destructively to form a capture cross section σ_j with an asymmetric interference pattern of the kind that is familiar from neutron scattering.

In principle the observation of the expected interference effect in the neutron-energy dependence of the cross section for individual radiative transitions should provide an unambiguous measure of the amplitude for the hard-sphere part of direct capture. However, in practice such measurements are difficult and usually the data are not very sensitive to a small cross section for direct capture, since the interference effects associated with hard-sphere capture are obscured by the ordinary resonance–resonance interference discussed in Section IV,A,2. The uncertainty caused by resonance–resonance interference is especially hard to remove because nuclear levels below the binding energy, about which we have little information, can have a large influence on the cross section above the binding energy.

Several unsuccessful attempts to observe hard-sphere capture in ^{197}Au(n,γ) have been reported. Wasson and Draper (1963) used a NaI annihilation-pair gamma-ray spectrometer at the Yale linac to measure the cross section for the ground-state transition over the neutron-energy range that contains the prominent resonances at 4.9 and 60 eV. As in other studies of interference effects, the experimentally measured variable was the ratio R_0/R_γ, the fraction of all detected gamma rays that go to the ground state. This ratio was found to exhibit interference effects, but they could be accounted for in terms of ordinary resonance-resonance interference. In another study of ^{197}Au, Dorchoman et al. (1964) were also unsuccessful in detecting hard-sphere capture in an experiment carried out at the Dubna pulsed reactor. Since it appeared possible that the looked-for asymmetry caused by direct capture was obscured in the measurements with NaI detectors because individual transitions were not well resolved, Wetzel et al. (1967) used a Ge-diode detector at the Yale linac to compare the thermal-neutron-capture spectrum with the spectrum from capture in the 4.9-eV resonance. No significant difference was detected. And finally, Wasson et al. (1968) have used a Ge-diode detector at the Brookhaven chopper in a study of the neutron-energy dependence of twenty-four transitions in ^{197}Au$(n,\gamma)^{198}$Au over the neutron-energy range 0–60 eV. Interference effects were observed for many of the transitions, but all of these effects could be explained in terms of resonance-resonance interference involving the well-understood resonances at 4.9 and 60 eV and a postulated (but reasonable) resonance at -20 eV.

In a similar measurement, Wasson et al. (1966) have studied the neutron-energy dependence of nine high-energy transitions in the reaction ^{59}Co$(n,\gamma)^{60}$Co over the neutron-energy range up to and including the prominent 4^- resonance at 132 eV. They report that they observe convincing evidence for an interference effect produced by hard-sphere capture. The ground-state transition was found to have a clear-cut interference pattern that is consistent with what would be expected if the hard-sphere-capture cross section at a neutron energy of 1 eV is $(9.2 \pm 2.4) \times 10^{-3}$ barns. This value is in good agreement with that calculated from a relationship given by Lane and Lynn (1960a). Now the key question is, can the observed interference pattern be explained in any other way? This question cannot be answered in a completely straightforward manner because we do not know enough about nearby levels (especially those below the binding energy) to show directly that resonance–resonance interference is negligibly small. However, Wasson et al. (1969) argue that it is highly unlikely that a nearby level is responsible for the observed interference since seven of the nine transitions studied do not exhibit any interference pattern, that is, since a single-level description of the cross section is accurate for the seven transitions. The present author does not find this argument compelling because it does not seem to make adequate allowance for possible fluctuations in the partial radiation widths of ignored resonances and because the

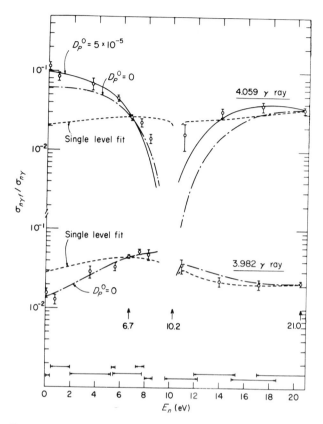

FIG. 23. Interference effects for the 3.982- and 4.059-MeV gamma rays from $^{238}U(n,\gamma)^{239}U$ (Price *et al.*, 1968). The quantity $D_p{}^0$ is the amplitude for direct hard-sphere capture. The arrows at 6.7, 10.2, and 21.0 eV give the energies of resonances, of which the one at 10.2 eV is negligibly weak.

single-level behavior itself is hard to understand. In the latter connection, one would expect the cross section of some of the transitions to deviate from the single-level shape because of contributions from 3^- initial states (which cannot contribute to transitions to the 5^+ ground state); further-more, if direct capture is present for the ground-state transition, it would also be expected to influence some of the transitions to higher states since the ground state is not the only one that has a significant single-particle component.

Another recent experiment that has been interpreted as revealing hard-sphere capture is one performed by Price *et al.* (1968). They measured the neutron-energy dependence of the cross section for the 4059- and

3982-keV gamma rays for the ^{238}U(n,γ) reaction over the energy range 0–21 eV, a range that contains prominent resonances at 6.7 and 21 eV. The cross sections for both transitions show a sharp departure from the shape described by a sum of single-level Breit–Wigner terms. The data for the 3982-keV gamma-ray can be explained in terms of the interference between nearby resonances. But the energy dependence of the cross section for the 4059-keV transition cannot be explained in terms of known resonances unless hard-sphere capture is assumed to be present. As is shown in Fig. 23, the best fit is obtained for a hard-sphere-capture cross section of 0.0016 b at 1 eV, a value that is an order of magnitude smaller than is calculated from the theory of Lane and Lynn. This direct-capture interpretation is reinforced by the fact that the state fed by the 4059-keV transition is known to have a strong p-wave single-particle component whereas the state fed by the 3982-keV transition has little if any single-particle character. However, from a strictly experimental point of view, the lack of information about the influence of states below the binding energy again would seem to be a source of uncertainty.

6. M1 AND E2 TRANSITIONS

As an introduction to the expected magnitudes of the widths of M1 and quadrupole transitions, let us write the Blatt–Weisskopf relationships Eq. (10) in the form of ratios

$$\Gamma_{\gamma j}^{(E1)}/\Gamma_{\gamma j}^{(M1)} = 70(A/100)^{2/3} \tag{20a}$$

$$\Gamma_{\gamma j}^{(E1)}/\Gamma_{\gamma j}^{(E2)} = 1320(100/A)^{2/3}(7/E_\gamma)^2 \tag{20b}$$

$$\Gamma_{\gamma j}^{(E1)}/\Gamma_{\gamma j}^{(M2)} = 92000(7/E_\gamma)^2 \tag{20c}$$

where the gamma-ray energy E_γ is in million electron volts.

With a few exceptions, only E1 transitions are strong enough to have been detected in the high-energy portion of the resonance-capture spectra that have been measured in the past. However, this limitation is beginning to be removed. The primary reason for the increased experimental sensitivity comes from the introduction of the Ge-diode gamma-ray spectrometer. For it the summing of several cascade gamma-ray pulses to form a false peak, a serious problem in an observation of weak transitions with a large NaI detector, is very improbable and at the same time the good energy resolution of the Ge-diode detector allows weak transitions to be detected above a continuous background.

The rather fragmentary experimental information about high-energy M1 transitions that has been obtained from resonance capture is summarized*

---

* A more recent summary is given by Bollinger (1968).

TABLE V

SUMMARY OF INFORMATION ON M1 TRANSITIONS[†]

Emitting nucleus	$\bar{E}_\gamma$ (MeV)	$\bar{k}_{M1} \times 10^3$	$r(E/M)$	Error factor	Ref.
Single-particle estimate		1.3	70		
Thermal, all A	~7	4			a
Thermal, $110 < A < 175$	~7	7	12	2	a
^{28}Al	7.2	5	0.24*	3	b
^{92}Zr	7.7	8	4*	2	c
^{106}Pd	7.7	11*	4.5	1.5	d
^{117}Sn, ^{119}Sn	6.6	150	0.3*	2	e, f
^{136}Ba	8.0	49	1.9*	2	g
^{168}Er	7.0	20*	6.5	1.2	h
^{207}Pb	5.7	14	11*	2.5	i
^{239}U	4.7	30	5	2	j

† The values marked with an asterisk were calculated from the experimentally measured quantity in the way described in the text. The "error factor" is the factor by which
the *measured quantity* is uncertain. The single-particle estimate is obtained from Eq. (10)
with $D_0 = 16.5$ MeV.

   [a] Bartholomew (1961).   [f] Slaughter *et al.* (1966).
   [b] Bergqvist *et al.* (1967).   [g] Becvar *et al.* (1968).
   [c] Jackson (1964).   [h] Bollinger and Thomas (1968b).
   [d] Thomas and Bollinger (1969).   [i] Biggerstaff *et al.* (1967).
   [e] Harvey *et al.* (1963).   [j] Bergqvist (1965).

in Table V. For comparison with the results from thermal capture and with
the Blatt–Weisskopf relationship, the experimental data are first expressed
in terms of the average value of the reduced width k_{M1} for transitions, a
quantity that was defined in Eq. (16). For some of the data, the experi-
mentally measured quantity is the ratio $\bar{\Gamma}_{\gamma j}(E1)/\bar{\Gamma}_{\gamma j}(M1) \equiv r(E/M)$. If
so, the value of $\bar{k}_{M1}$ given in the table is calculated from $r(E/M)$ under the
assumption that E1 widths are given by Eq. (17), the relationship that
summarizes the results of Carpenter (1962). Similarly, when the ratio
$r(E/M)$ is not measured, it is calculated from $\bar{k}_{M1}$ by making use of Eq. (17).
The errors are rough estimates of the factors by which the *measured*
quantities are uncertain. These estimates take account of all sources of error,
including purely experimental factors and Porter–Thomas fluctuations.

The data of Table V were obtained in a variety of ways. In his survey
of all of the widths derived from thermal-neutron-capture spectra,
Bartholomew (1961) obtained the results $k_{E1} \simeq 0.003$ and $k_{M1} \simeq 0.004$
(see Fig. 17). From these values we obtain $r(E/M) \simeq 16(A/100)^{2/3}$. The
factor of 4 difference between this and the single-particle estimate is perhaps
related to the suggestion by Wilkinson (1960) that the single-particle

estimate of the width for E1 transitions should be divided by 4 to make allowance for an "effective charge" effect.

Since the above experimental value of $\bar{k}_{M1}$ was determined largely from measurements on light targets, let us determine a value for a range of nuclides that is more typical of those studied in slow-neutron resonance spectroscopy. Figure 17 gives widths for 17 M1 transitions in nuclides in the mass range $110 < A < 175$. These data yield the value $\bar{k}_{M1} \simeq 0.007$. For roughly the same range of targets and many more transitions, the data of the figure give the value $\bar{k}_{E1} \simeq 0.003$. In view of the uncertainties caused by Porter–Thomas fluctuations, these values of $\bar{k}_{M1}$ and $\bar{k}_{E1}$ do not differ significantly from the values obtained from all nuclides.

Resonance-capture measurements have provided relatively accurate values of $\bar{k}_{M1}$ for a few nuclides. In measurements with a large NaI gamma-ray spectrometer at a pulsed Van de Graaff accelerator, Bergqvist et al. (1967) studied the gamma-ray spectra from neutron capture in resolved resonances of a number of nuclides in the $2s$–$1d$ shell. Although a preliminary report (Bergqvist et al., 1965) of these measurements cited evidence for a very large difference in $r(E/M)$ for even–odd and odd–odd nuclides, the final report (Bergqvist et al., 1967) appears to give definite information about M1 widths only for high-energy transitions in ^{28}Al. In the table the value of k_{M1} for ^{28}Al has a rather large uncertainty attached to it because the number of final states involved in the observed γ-ray lines is not well known.

Jackson (1963) used a chopper and a NaI gamma-ray spectrometer to measure the intensities of M1 transitions from the reaction ^{91}Zr$(n,\gamma)^{92}$Zr. The most reliable data consist of widths for transitions from four initial states with $J^{\pi} = 2^{+}$ or 3^{+} to the 2^{+} first excited state.

Thomas and Bollinger (1968) studied capture in ^{105}Pd by selecting the energy range of captured neutrons by means of filters and by measuring the gamma-ray spectrum with a Ge-diode spectrometer. The neutron filter was chosen so as to emphasize capture in the low-energy resonances, in this way limiting the contribution from p-wave capture. In the average resonance-capture spectrum that was measured, the pertinent information consists of relative intensities of M1 transitions (from many initial states) to 1^{+} or 4^{+} keV and E1 transitions to a 3^{-} state at 2090 keV. From these data one obtains the value $r(E/M) = 2.5$, a very small value. Similar measurements on the reaction ^{167}Er$(n,\gamma)^{168}$Er yield a larger value of 6.5. Further information on ^{167}Er(n,γ) is given in Section IV,D,3.

Measurements by Harvey et al. (1963) with a large NaI scintillator at the ORNL chopper showed that M1 transitions in at least some of the tin isotopes are unusually strong, although there was uncertainty about the multipolarity of some of the transitions studied. Recent measurements (Harvey *et al.*, 1966; Slaughter *et al.*, 1966) in which the neutron energy was selected by means of a neutron filter and the gamma-ray spectrometer was a Ge-diode detector have clarified the situation. In an effort to select data in an objective way, the transitions used to calculate the value of $\bar{k}_{M1}$ given in Table V are restricted to positively identified M1 transitions in the even–odd isotopes of tin. These data consist of transitions to the $s_{\frac{1}{2}}$ and $d_{\frac{3}{2}}$ low-energy states from the positive-parity states excited by the 111-eV resonance in $^{116}Sn(n,\gamma)$ and the 360-eV resonance in $^{118}Sn(n,\gamma)$, four transitions in all. The extremely large value $\bar{k}_{M1} = 0.15$ obtained in this way seems to be reliable except for the unlikely possibility that the s-wave resonance in ^{118}Sn at 360 eV, which contributes the main part of $\bar{k}_{M1}$, overlaps a p-wave resonance. Recent measurements with the Brookhaven chopper (Bhat *et al.*, 1968) give approximately the same result.

An anomolously strong M1 ground-state transition has also been observed (Urbanec, 1965) in the 24-eV resonance of the reaction $^{135}Ba(n,\gamma)^{136}Ba$. More recent measurements at the Dubna pulsed reactor (Becvar *et al.*, 1968) showed that this one transition is atypically strong for ^{136}Ba. Nevertheless, the average value of k_{M1} deduced from the spectra of six resonances (thirteen transitions in all) is also very large.

Strong M1 transitions from resonance capture in ^{206}Pb have been observed by Biggerstaff *et al.* (1967) in measurements with a pulsed Van de Graaff. The value of k_{M1} given in the table is calculated from the widths for the three positively identified M1 transitions, namely, transitions to the first three excited states of ^{207}Pb in the spectrum from capture in the 25-keV resonance.

Finally, for the heavy nucleus ^{239}U Bergqvist (1965) has obtained values of the ratio $r(E/M)$ and of the average reduced widths by comparing high-energy M1 transitions observed in the spectra for individual low-energy s-wave resonances (Jackson, 1964) with transitions of the same energy in the spectrum for predominantly p-wave capture in a broad band of energy in the neighborhood of 33 keV. The latter spectrum was measured with a NaI detector at a pulsed Van de Graaff. The concept of an M1 giant resonance was introduced as a possible explanation of the relatively large value of $\bar{k}_{M1}$.

The conclusions that may be drawn from the results of Table V are to some extent a matter of taste. First consider $\bar{k}_{M1}$. If all the values listed in the table are accepted as reliable within the indicated errors, then clearly $\bar{k}_{M1}$ is not independent of the nuclear species, as the single-particle model predicts it should be. On the other hand, since this conclusion rests rather heavily on the strong transitions associated with only one resonance of ^{118}Sn, it appears that the conclusion is not quite certain yet.

Another conclusion drawn from the table is that it is by no means safe to assume, as has been customary in the past, that transitions strong enough to be detected in resonance capture are E1 transitions. Judging from the small values of $r(E/M)$ in the table, it is clear that Porter–Thomas fluctuations cause some M1 transitions in some nuclides to be stronger than many E1 transitions. On the other hand, perhaps the data of Table V give a misleading impression of the typical magnitude of $r(E/M)$; the fact that almost all of the values of $\bar{k}_{M1}$ obtained from resonance capture are greater than the average from thermal capture suggests that the resonance-capture values are atypically large, perhaps as a result of bias in the selection of materials for investigation.

A possible cause for the apparent dependence of the strength of M1 transitions on nuclear structure could be the presence of an M1 giant resonance at relatively low energy. This possibility has been considered at length by Starfelt (1964) in an attempt to explain the anomalous bump in the neutron-capture gamma-ray spectra of some heavy nuclides. The hypothesis that the bump results from direct capture was considered and rejected in Section IV,A,5a. In his analysis of the spectra Starfelt tried a variety of expressions for the level density and the partial radiation width. It proved possible to fit the spectrum for Ta and for Ag with reasonable assumptions about the energy dependence of the level density and the widths for E1 transitions, but the bump at high energy in the spectrum of Cs and especially Au could not be reproduced, as is shown in Fig. 22. However, the Cs and Au spectra can be fit satisfactorily by assuming the existence of a giant resonance at 5.5 MeV with a width of 1.5 MeV and a strength that is only 1% as great as that of the well-known E1 giant resonance at higher energy.

The difficulty in explaining the over-all shapes of the Cs and Au spectra appears to be a reliable indication that some special nuclear effect is operative and the suggested giant resonance at 5.5 MeV is an appealing explanation. In their various publications Bergqvist and Starfelt (following a suggestion by Mottelson) point out that the energy and nuclei for which the anomalous bump is observed are consistent with what would be expected for an M1 giant resonance associated with the spin-orbit splitting of the g, h, and i nuclear shells. However, Bartholomew et al. (1967b) have shown that what is often called the anomalous bump for the spectrum from thermal-neutron capture in ^{199}Hg is formed by E1 radiation, not M1.

The experimental observation of high-energy quadrupole transitions is only now beginning to be technically feasible and to date only one such transition has been reported for a resonance-capture experiment.* In a measurement that makes use of a neutron filter to select neutron capture in ^{105}Pd within a broad band of energy, Thomas and Bollinger (1969) found that the average reduced width of E2 transitions to the 0$^+$ final state of ^{106}Pd is forty-five times weaker than the reduced width for E1 transitions to the 3-state at 2.09 MeV. Thus for ^{106}Pd, at least, the ratio of E1 to E2 strengths is very much smaller than is expected from Eq. (19c).

B. Statistical Characteristics of Gamma-Ray Cascades

The starting point for a discussion of the capture-gamma-ray cascade in a heavy nucleus is a purely statistical model in which a very large number of transitions are assumed to be present, with all individual transitions weak in comparison with the sum. Under this extreme statistical description, any given characteristic of the cascade can be deduced if one has a full knowledge of the way in which level densities and transition probabilities depend on the spins, parities, and energies of the states involved. Conversely, experimental information about the cascade can lead to an understanding of transition probabilities and nuclear-level densities. A considerable effort has gone into the interpretation of thermal-neutron-capture gamma-ray spectra with this objective in mind. In particular, extensive calculations have been carried out to explain the general shape of the main body of the spectrum and to interpret the absolute value and the dependence on A of the total radiation width Γ_γ. The results of some of these investigations were used in Section IV,A. For a further discussion the reader is referred to the original literature (e.g., Groshev et al., 1960; and Strutinski et al., 1960) and to the various reviews of thermal-neutron-capture gamma rays (e.g., Bartholomew, 1961). In the present chapter, we will restrict our attention to those kinds of information for which measurements at several resonances are essential. In particular, we will examine the relationship between the capture gamma-ray spectra and the degree to which various average intensities vary from resonance to resonance.

1. TOTAL RADIATION WIDTHS

It has long been recognized that the total radiation width varies only slightly from resonance to resonance for most nuclides that have been studied. Indeed, in most cases the experimental errors are large enough to mask any real variation that may be present. This high degree of uniformity is in itself an indication of the essential validity of the statistical description

---

* Since this was written, E2 transitions have been detected in many nuclides, but the results have not yet been reported in written papers.

of the capture gamma-ray spectrum since the absence of random fluctuations shows that many transitions are present.

Although the most striking characteristic of the total radiation width is its uniformity from resonance to resonance, there are a number of nuclides for which variations have been reported. Some examples are given in Table VI. The first apparently real variations in Γ_γ were obtained from

TABLE VI

EXAMPLES OF NUCLIDES FOR WHICH THE TOTAL RADIATION WIDTH Γ_γ AND THE MULTIPLICITY ν DEPEND ON THE RESONANCE SPIN J[†]

Target	E_0 (eV)	J	Γ_γ $(10^{-3}$ eV)	ν
^{115}In	1.5	5	72 ± 2^a	4.4 ± 0.2^e
	3.9	4	81 ± 4^a	5.6 ± 0.4^e
	9.1	5		4.2 ± 0.4^e
^{177}Hf	1.1	3	67 ± 2^b	
	2.4	4	60 ± 1^b	
^{199}Hg	33	1	$295 \pm 10^{c,d}$	
	130	0	200 ± 30^d	
^{201}Hg	43	2	$290 \pm 15^{c,d}$	3.44 ± 0.16^c
	71	1	$460 \pm 30^{c,d}$	3.00^c
	210	1	$430 \pm 50^{c,d}$	
	313	2	320 ± 65^d	

[†] The absolute value of ν for the 71-eV resonance of ^{201}Hg was normalized to $\nu = 3$. The J values were obtained from Reference f.

[a] Landon and Sailor (1955).   [d] Huynh (1965).
[b] Igo and Landon (1956).   [e] Draper and Springer (1960).
[c] Carpenter and Bollinger (1960).   [f] Goldberg et al. (1966).

transmission measurements with the Brookhaven crystal monochromator. For ^{115}In, Landon and Sailor (1955) reported $\Gamma_\gamma = 0.072 \pm 0.002$ for $E_0 = 1.45$ eV, and $\Gamma_\gamma = 0.081 \pm 0.004$ for $E_0 = 3.85$ eV. Similarly, for ^{177}Hf Igo and Landon (1956) obtained $\Gamma_\gamma = 0.067 \pm 0.002$ eV for $E_0 = 1.10$ eV, and $\Gamma_\gamma = 0.060 \pm 0.001$ eV for $E_0 = 2.39$ eV. Although the spins of these resonances were not known at the time, it was suggested that the observed differences indicate that Γ_γ somehow depends on the spin of the initial compound state. Since later measurements showed that the spins of each pair of resonances are in fact different (Table VI), we are encouraged to speculate how the spin J_r of the resonance state might influence Γ_γ.

A J_r dependence could come about in two ways, both of which result from a failure of level density $\rho(E,J)$ to be proportional to $(2J + 1)$. As is shown by the arguments connected with Eq. (14), this proportionality is a necessary condition for Γ_γ to be entirely independent of J_r. The $(2J + 1)$ condition on $\rho(E,J)$ surely is not satisfied by the low-energy states of the compound nucleus and as a result the sum of partial widths for high-energy transitions to these states could be greater for one initial spin state than for the other. An effect of this kind is present for capture in ^{177}Hf, since below 1.5 MeV there are 9 states accessible by E1 transitions from a 3^- initial state and only 3 states accessible from a 4^- initial state. From the measurements of Carpenter (1962), we know that each of these high-energy transitions contributes on the average about 0.5% of the total radiation width. Thus, assuming that $\rho(E,J) \propto 2J + 1$ for $E > 1.5$ MeV and assuming $\Gamma_{\gamma j} = \text{const}/\rho_r$, on the average the total radiation width for a resonance with $J_r = 3$ would be expected to be about 3.3% greater than the width for a resonance with $J_r = 4$; a much larger difference than this might well result from the same kind of effect, if the $2J + 1$ law continues to be invalid above 1.5 MeV.

Although detailed information about low-energy states is required for a quantitative evaluation of the dependence of Γ_γ on J_r, simple qualitative arguments give considerable insight into how the J_r dependence would be expected to be influenced by the general characteristics of the emitting nucleus. First, the J_r dependence (associated with transitions to low-energy states) would be expected to be weak for odd–odd nuclides, since for them there is no energy gap and statistical equilibrium in the level density is probably established rapidly. Second, for even–even compound states with negative parity, arguments like those given above for ^{177}Hf lead to the following expectation: for all values of the target spin I the radiation width for the odd value of $J_r = I \pm \frac{1}{2}$ is greater than the width for the even value of J_r. Third, when the initial even–even state has positive parity the influence of M1 transitions to low-energy positive-parity states in competition with E1 transitions to negative-parity states at a higher energy tends to diminish the J dependence of Γ_γ. And finally, other factors being constant, the relative magnitude of the J dependence would be expected to be a monotonic increasing function of $\bar{I}_{\gamma 0}$, where $\bar{I}_{\gamma 0}$ is the average value of the absolute intensity [defined at Eq. (7)] for an E1 transition to the ground state.

A second and more general explanation of how Γ_γ can depend on J_r may be obtained as an implication of a level density (Bloch, 1954) that is proportional to $(2J + 1) \exp[- (2J + 1)^2/8\sigma_m{}^2]$, where σ_m is a parameter that is independent of energy. For this form of the dependence of density on J, calculations based on Eq. (13) show that Γ_γ increases with increasing J, if $\Gamma_{\gamma j} = \text{const}/\rho_r$. For example, if we use the reasonable value $\sigma_m = 3$ (Huizenga and Vandenbosch, 1960), then $\Gamma_\gamma(I + \frac{1}{2})/\Gamma_\gamma(I - \frac{1}{2}) \approx 1.04$ for

$I = \frac{9}{2}$ (the ratio is almost independent of I). Thus, the difference in the widths of the two spin states for s-wave capture is smaller than the typical experimental error for Γ_γ, but the change is in the wrong direction to help explain the observed difference in the widths for $^{115}\text{In}(n,\gamma)$.

In addition to systematic effects of the kind we have been considering above, for some nuclides one would expect that there would be random fluctuations in the total radiation width caused by random fluctuations in the widths of a few individual transitions. Moreover, one would expect there to be a strong correlation between the magnitude of Γ_γ and qualitative aspects of the capture gamma-ray spectra. The most clear-cut example of this kind is ^{201}Hg, a target that was thoroughly studied by Carpenter and Bollinger (1960). The total radiation width was found to fluctuate by about a factor of two. For the resonances with large widths the spectra are dominated by a few strong transitions to final states at low energy. On the other hand, these high-energy transitions are relatively weak when the total radiation width is small. A quantitative treatment of the data showed that the observed variation in Γ_γ can be accounted for in terms of fluctuations in the intensity of the individual transitions.

The correlation between the values of Γ_γ and the known characteristics of the capture gamma-ray spectra for the resonances of ^{201}Hg suggests that the spectra are a generally reliable indication of the degree of fluctuations in Γ_γ for all targets. As with the J_r dependence, the relative magnitude of the fluctuation in Γ_γ would be expected to be a monotonically increasing function of the average value $\bar{I}_{\gamma 0}$ of the absolute intensity of high-energy primary E1 transitions. On this basis, one expects the total radiation widths for most of the light nuclides, when they are measured, to exhibit fluctuations at least as great as those observed for ^{201}Hg, that is, fluctuations of a factor of 2. On the other hand, the large variations in Γ_γ that have been reported in the past for nuclides such as ^{195}Pt, ^{197}Au, and ^{235}U are most surprising, since for them the individual high-energy transitions are rather weak. Some of these reported variations have been shown by recent measurements to have been in error.

2. Multiplicity

Relatively little effort has gone into the study of the gamma-ray multiplicity for resonance capture, perhaps because little of fundamental interest is learned from such studies. The most extensive set of results has been given by Draper and Springer (1960), who have measured the multiplicity of a total of twenty-three resonances in fifteen nuclides. Other results have been reported by Carpenter and Bollinger (1960). For the resonances of ^{167}Er, ^{176}Lu, ^{177}Hf, and ^{199}Hg the multiplicity at the few resonances studied is found to be constant within the experimental error of 5 or 10%. On the other hand, what appear to be real differences from resonance to resonance are reported for ^{115}In, ^{149}Sm, and ^{201}Hg. The results for ^{115}In and ^{201}Hg are given in Table VI. It is interesting to observe that for both targets the change in multiplicity is associated with a variation in the total radiation width, but the sign of the correlation is different in the two cases. The negative correlation for the ^{201}Hg data is easily understood. The large values of Γ_γ result from strong transitions to low-lying states of ^{202}Hg, and the multiplicity associated with these high-energy transitions is small—in the neighborhood of two; thus, the average multiplicity for all transition is diminished. The positive correlation between multiplicity and total radiation width for the resonances of ^{115}In is not so simply understood. It could result from the high values of the spin of the first and second excited states in ^{116}In, both of which have long lifetimes, or it could result from the way in which the level density of ^{116}In depends on excitation energy and spin.

Recently Coceva *et al.* (1968) have systematically investigated the dependence of multiplicity ν (or rather a quantity closely related to the multiplicity) on the spin of the initial state for several even–even compound nuclides with relatively high spin. The ratio $\bar{\nu}/\nu(\nu - 1)$ was found to increase by about 10% when J increases by one unit. For even–even compound nuclides, this result can easily be understood qualitatively in terms of a model in which the initial part of the cascade is governed by simple statistical relationships and the final part is nonstatistical because of the band structure of low-energy states. The multiplicity of the statistical part is independent of the spin of the initial state, but the multiplicity of the nonstatistical part depends on the average amount by which the spin at the start of the cascade differs from the spin of the ground state. When the initial spin J is not too small, this model implies that $\Delta\nu/\Delta J \approx \frac{1}{2}$, assuming that the nonstatistical part of the cascade is dominated by quadrupole transitions. When the initial J value is small (0 or 1), the value of $\Delta\nu/\Delta J$ would be expected to be much smaller.

Additional information about the spin dependence of the multiplicity is given in Section IV,C,2 in a discussion of the use of the multiplicity to determine the spin.

3. AVERAGE GAMMA-RAY INTENSITY

A quantity that is closely related to gamma-ray multiplicity is the average intensity (per neutron capture) of gamma rays with an energy greater than some lower limit. The principal interest in this quantity is the practical one of using it as a measure of the relative rate of neutron capture. More specifically, one often finds it expedient to use the rate of detector

TABLE VII

RELATIVE INTENSITY (PER NEUTRON CAPTURE) OF NaI PULSES GREATER
THAN 2 MeV FOR RESONANCE-NEUTRON-CAPTURE GAMMA RAYS
FROM $^{195}Pt(n,\gamma)^{196}Pt^a$

Resonance energy (eV)	Relative intensity	J
12	1014 ± 14	1
19	1017 ± 17	1
66 + 67	1001 ± 16	1 and 0
112	957 ± 23	1
120	999 ± 13	1
140	974 ± 19	1
151 + 155	1005 ± 13	1 and 0
189	1026 ± 32	1

[a] Moxon and Rae (1962). The $J = 1$ and $J = 0$ resonances in the unresolved pairs contribute roughly equal intensities.

pulses greater than some fixed bias as an approximate measure of the relative rate of neutron capture. The justification for this procedure is usually based on the experimental facts that the gamma-ray multiplicity, the total radiation width, and the gross features of the gamma-ray spectra are approximately constant from resonance to resonance for most targets.

The ratio of the gamma-ray counting rate to the rate of neutron capture has been carefully determined for one case, the resonances of platinum. The results are given in Table VII. There it is seen that the gamma-ray counting rate is accurately proportional to the rate of neutron capture in the resonances of ^{195}Pt. This result does not depend sensitively on the bias level, although both systematic and random differences in the spectra have a larger effect as the bias level increases. It should be born in mind that the proportionality constant linking γ-ray counting rate to neutron-capture rate must depend on the spin of the initial state for some nuclides, since for some the multiplicity depends on the spin.

4. INTENSITY OF LOW-ENERGY LINES

The intensity of low-energy lines formed by transitions between low-energy states was one of the first characteristics of the resonance-capture gamma-ray spectra to be studied. Most of the results reported may be understood in terms of a fully statistical picture of the gamma-ray cascade, for which the intensity of the low-energy lines depends only on the spin and parity of the initial compound state. For many nuclides the observed intensity is the same from resonance to resonance within an experimental accuracy of 5 or 10%; examples of constancy from resonance to resonance have been reported by Fenstermacher et al. (1957, 1959) for capture in

^{149}Sm and ^{167}Er and by Jackson and Bollinger (1961) for capture in ^{189}Os and ^{195}Pt. The most extensive of these data is that for the target ^{189}Os, for which the intensity of the ground-state transition from the 2+ first excited state (155 keV) was measured for eleven resonances. The rms deviation of the intensity about its mean was found to be only 3.5%. This high degree of constancy is exactly what is expected on the basis of a statistical model of the cascade, since for ^{189}Os(n,γ)^{190}Os individual transitions are weak and the probability of populating a low-energy 2+ state is expected to be almost the same for the two initial spin states ($J_r = 0$ or 1), almost independent of the multiplicity ν of the cascade.

When the intensity of a low-energy line is known to be constant from resonance to resonance, this intensity may serve as a convenient and sensitive measure of the rate of capture in a given target. The strong line resulting from decay of the first excited state of even–even product nuclides is particularly useful for this purpose.

In measurements with NaI scintillators only two nuclides were found to have low-energy lines for which the intensity differs from resonance to resonance. These cases are capture in ^{115}In (Draper et al., 1958) and ^{177}Hf (Fenstermacher et al., 1959). In both cases the relative intensities of certain lines vary by roughly a factor of 2. The data from capture in ^{177}Hf can be explained, because here both the spins of the initial states and the level scheme of the low-energy states of ^{178}Hf are known. The greatest variation in intensity is observed for the decay of the 6+ state at about 660 keV. The intensity of the 330 keV line formed by this decay is 2.8 ± 0.8 times greater when the initial state has $J = 4$ than it is when the initial state has $J = 3$. This difference in intensity is qualitatively reasonable since, when the multiplicity is about 4, the number of paths by which a 6+ state may be populated is greater when the initial state has $J = 4$ than when it has $J = 3$. The J dependence of the intensity of the 330-keV gamma ray has been used by Huizenga and Vandenbosch (1960) to determine a value of the level density parameter σ_m.

The above explanation of the spin dependence in the population of the 6+ state of ^{178}Hf obviously implies that all nuclides have some low-energy lines whose intensities depend on the spin of the initial state, and recent measurements with Ge-diode detectors are beginning to find these lines. Since this subject is mainly of interest in connection with the determination of the spin, the recent data are summarized in Section IV,C,2.

C. Determination of Resonance Parameters

Up to this point our discussion of experimental results has been concerned with subjects that are not very closely related to what is usually thought of as neutron spectroscopy; that is, we have not been much interested in the parameters of the initial compound states formed by neutron capture. However, now that we have some understanding of the characteristics of the capture gamma-ray spectra themselves, let us see how the spectra may be used to determine the properties of the initial state. Historically, the order of events was just the reverse of this. The first measurements [at Yale, Fenstermacher *et al.* (1957)] were largely directed toward using the low-energy capture gamma-ray spectra to determine isotopic and spin assignments of resonances. It was only later that the fundamental nature of the gamma radiation itself began to be investigated.

1. ISOTOPIC ASSIGNMENTS

Any characteristic feature of the gamma-ray spectra associated with a given target nuclide may be used to determine the nucleus responsible for a resonance. Nuclear properties that might be useful are: (a) characteristic low-energy gamma rays, (b) characteristic high-energy gamma rays, (c) the neutron binding energy, (d) the gamma-ray multiplicity, (e) the general shape of the spectrum, and (f) the lifetime of low-energy nuclear states. All of these properties can be useful in particular cases but, as we shall see, only the first three of them are generally useful.

The first efforts at isotopic identification were those by the Yale group (Fenstermacher *et al.*, 1957; Draper *et al.*, 1958). Their measurements showed that many targets have capture gamma-ray spectra that contain strong, easily observed low-energy lines. Moreover, the intensity of these lines, although not always constant, do not fluctuate greatly from resonance to resonance. Consequently, the characteristic low-energy lines are often valuable for isotopic identification and by now many experimenters have made use of them.

The early measurements were made with a single NaI scintillator. Later, Carpenter and Bollinger (1960) demonstrated that the serious background present in most measurements with a single detector could be effectively eliminated by recording the spectra associated with coincident pulses in two scintillators. Jackson and Bollinger (1961) systematically explored this technique and came to the conclusion that large scintillators are better than small scintillators for measuring the low-energy spectra required for isotopic identification, even if the characteristic gamma rays of interest are at very low energies. Examples of low-energy spectra for several isotopes of cadmium are given in Fig. 24. Notice that the gamma-ray lines corresponding to the transition from the first excited state to the ground state of the even–even compound nuclides are particularly prominent—a feature that is observed in all the nuclides of this kind that have been studied. Also,

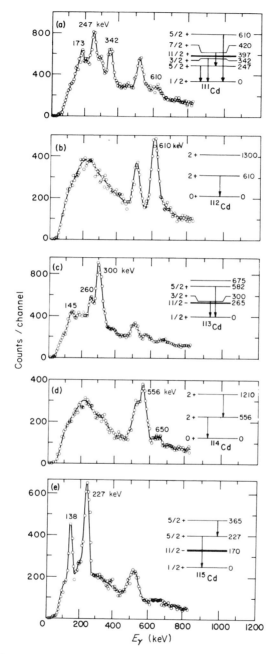

FIG. 24. NaI measurements of low-energy resonance-capture spectra for isotopic identification (Jackson and Bollinger, 1961).

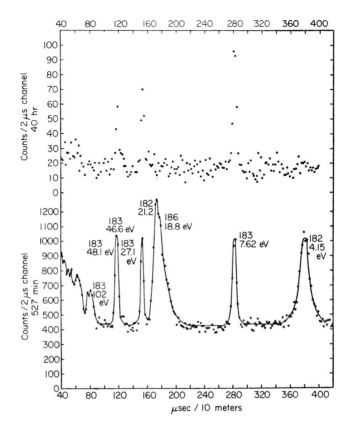

Fig. 25. Time-of-flight spectra for high and low biases on a NaI detector (Fox *et al.*, 1958). The high-bias curve (upper) isolates resonances of ¹⁸³W.

notice that all of the spectra are distinctive enough to permit one to identify the isotopes contributing to a neutron time-of-flight peak formed by unresolved resonances in two or three isotopes.

The Ge-diode detector may also be expected to provide low-energy spectra with which to make isotopic identifications with high sensitivity. However, no systematic investigation of this technique has been reported yet.

A second nuclear property that has been widely used for isotopic identification is the neutron binding energy. Early measurements that make use of this property are those by Fox *et al.* (1958) on tungsten. The resonances of the even–odd isotope ¹⁸³W were isolated merely by recording the time-of-flight spectrum of pulses greater than the binding energy of the even–even isotope. A demonstration of the efficacy of this simple measurement is given by the data of Fig. 25.

In the early determinations of isotopic assignment on the basis of binding energy a single gamma ray was detected. This technique works well for isolating the resonances of an isotope that has a binding energy that is much greater than that of any of the others. However, detection of a single gamma ray is not very useful for most isotopes because of the dependence of the spectra on the spin and parity of the initial state and because of the random fluctuations in partial radiation widths. To overcome this difficulty, Jackson and Bollinger (1961) recorded the spectra of the *sum* of coincident pulses in two scintillators. These spectra are largely free of random fluctuations. Thus, the usefulness of the spectra for isotopic identification depends only on the degree to which the upper energy limits of the spectra are different for the different isotopes in the target. For many targets the upper limit can be determined in advance from the neutron binding energy and from the known spins and parities of the low-energy states in relationship to those of the initial state. Indeed, the upper limit is often just the neutron binding energy. Figure 26 gives examples of sum-coincidence spectra for the isotopes of cadmium, a case for which the binding energies differ enough to permit the spectra to be quite useful for isotopic identification.

The spectra given in Fig. 26 were obtained with 4- by 4-in. scintillators. For the sum-coincidence method of measurement to be fully useful, it is important that both scintillators be somewhat larger. In the measurements made with the Argonne fast chopper it has become standard practice for both scintillators to be 8 in. in diameter and 6 in. thick. Examples of spectra obtained with these large NaI crystals are presented in Fig. 27. The improved quality of the spectra obtained with the large detectors is obvious.

Characteristic high-energy gamma-ray lines were rarely useful for isotopic identification when the detector had to be a NaI scintillator, because the poor resolution of the detector and Porter–Thomas fluctuations of widths made it difficult to observe and identify the lines reliably. With the Ge-diode detector the limitation imposed by the fluctuation is much less severe, since many characteristic lines may be observed in most spectra. Thus almost any Ge-diode spectrum that has reasonable statistical accuracy can be identified from the high-energy γ-rays. However, it should be borne in mind that adequate statistical accuracy is obtained in measurements with Ge-diode detectors only by sacrificing neutron-energy resolution, except at very low neutron energy.

One interesting conclusion that seems to emerge from this review is that the NaI scintillator is still useful for measurements in which the primary aim is the isotopic identification of a large number of resonances. This usefulness is most clear-cut for the even–odd targets, since their strong low-energy lines and high binding energies cause the poor resolution of the NaI detector to be less of a handicap.

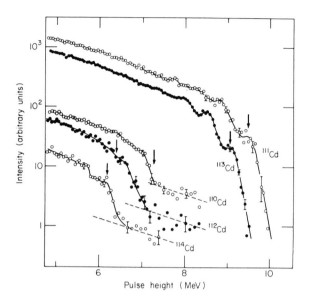

FIG. 26. Sum-coincidence spectra for isotopic identification (Jackson and Bollinger, 1961).

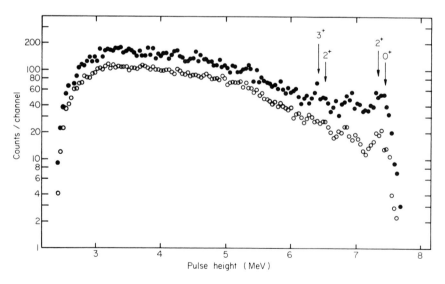

FIG. 27. Sum-coincidence spectra used for spin assignment of ^{183}W resonances (Bollinger *et al.*, 1964). The vertical lines show the energies of transitions to known low-energy states of ^{184}W [○, $J = 1$ resonance (27 eV); ●, $J = 0$ resonance (48 eV)].

The other nuclear properties suggested above as possibly useful for isotopic identification are not applicable to most targets. The multiplicity and the general shape of the spectra at most enable one to separate resonances into broad classes of isotopic types, a separation that can be made more easily on the basis of the binding energy. The final property, the lifetime of the low-energy states, would appear to be a powerful method in those cases to which it is applicable; however, there are not very many nuclides for which there are states that are known to have the necessary characteristics.

To conclude this section on isotopic identification, let us briefly compare the gamma-ray methods with the older method of transmission measurements on separated isotopes. The gamma-ray methods have several advantages: (a) a sample of a normal element may be used, (b) in many cases the isotopic assignment may be made more rapidly since only one run is required to gain information on all isotopes, (c) the gamma-ray method is sometimes more sensitive since the samples of "separated isotopes" used in transmission measurements are often small and relatively impure isotopically. The principal disadvantage of the gamma-ray method (aside from the large array of equipment required to implement it) is that the method is not very sensitive for some nuclides. Thus, one sees that there is no generally valid choice of one method over the other. Both the gamma-ray and transmission methods are useful and should be looked upon as complementing each other.

2. Spin Assignment

The first efforts to use the neutron-capture gamma-ray spectra for spin assignments were measurements on low-energy spectra at the Yale linac. As we have already seen in Section IV,B, most of the lines in these low-energy spectra are quite insensitive to the characteristics of the initial compound state because of the complexity of the gamma-ray cascade. Nevertheless, substantial variations in the intensities of certain low-energy lines from capture in ^{115}In (Draper et al., 1958) and ^{177}Hf (Fenstermacher et al., 1959) were measured and the intensities are correlated with resonance spins that have since been determined in other ways. The idea of using low-energy gamma rays for spin assignment then lay dormant until the development of the Ge-diode γ-ray spectrometer.

As was explained in Section IV,B,3, in the reaction ^{177}Hf$(n,\gamma)^{178}$Hf the low-energy line whose intensity varies from resonance to resonance comes from the decay of the 6^+ state at 660 keV in ^{178}Hf, and the dependence of this line on resonance spin has been explained by Huizenga and Vandenbosch (1960) in terms of a simple statistical model of the gamma-ray

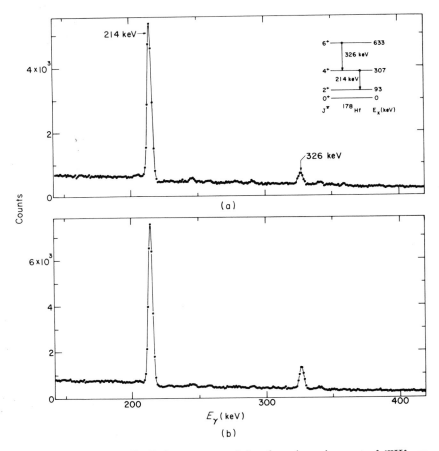

FIG. 28. Low-energy Ge-diode spectra used for the spin assignment of ^{177}Hf resonances (Wetzel and Thomas, 1970). Relative to the $4^+ \rightarrow 2^+$ transition, the $6^+ \rightarrow 4^+$
transition is 1.9 $\pm$ 0.2 times stronger for resonances with $J = 4$ than it is for resonances
with $J = 3$. (a) ^{177}Ht($n_{res,\gamma}$)^{178}Hf, 1.098 eV, $J = 3$; (b) ^{177}Hf($n_{res,\gamma}$)^{178}Hf, 2.380 eV, $J = 4$.

cascade. This model implies that there are some low-energy lines in all
nuclides whose intensities depend significantly on the resonance spin. Thus,
as has been emphasized by Poenitz (1966), it is clear that the low-energy
spectra could in principle be generally useful for spin determination. The
only question is whether the lines of interest are strong enough to be measured easily with a Ge-diode detector; the spectra are too complex to permit
a NaI spectrometer to be competitive for most nuclides.

Becvar et al. (1968) have recently used the Dubna pulsed reactor to
show that there is no difficulty in obtaining Ge-diode spectra of acceptable

quality for the reaction ^{135}Ba(n,γ)^{136}Ba; the intensities of the low-energy lines were observed to depend on resonance spin in a way that can be explained qualitatively by a simple statistical model of the cascade. Similarly, the low-energy spectra following neutron capture in many resonances of ^{95}Mo, ^{105}Pd, ^{135}Ba, ^{167}Er, ^{177}Hf, ^{179}Hf, ^{183}W, and ^{189}Os have been measured by Wetzel and Thomas (1970) at the Argonne chopper. All of these spectra contain lines whose intensities exhibit the expected dependence on the spin of the resonance. Typically, the intensities of the detectable lines that are most sensitive to the spin vary by a factor or 2. An example of a variation of this magnitude is shown in Fig. 28, where low-energy spectra for the reaction ^{177}Hf(n,γ)^{178}Hf are given.

The first important advance in the use of spectral measurements for spin assignments was the recognition by Landon and Rae (1957) of the value of high-energy transitions. In particular, they emphasized the significance of the ground-state transition from the 0^- and 1^- states formed by capture in a target with spin $\frac{1}{2}$ and negative parity. For such a nucleus, the ground-state transition from an initial state with $J = 1$ can be an E1 transition, whereas a ground-state transition cannot occur at all from a state with $J = 0$. Hence the observation of a ground-state transition with any nonzero intensity is conclusive evidence for the assignment $J = 1$ to the initial state. As a result of our knowledge of the breadth of the distribution of partial radiation widths, we now recognize that a failure to observe a ground-state transition is not equally informative; the transition might be present but undetectably weak.

The idea of using a ground-state transition for spin assignments was applied by Landon and Rae (1957) to the 34-eV resonance of ^{199}Hg. Although these first data were quite crude, the ground-state transition was detected. This first positive assignment of $J = 1$ soon led to many other measurements of the same kind.

Similar ideas have been used in making spin assignments for other classes of transitions on the basis of gamma-ray intensities. For example, the 1^- states formed by capture in an even–odd target with $I^\pi = \frac{1}{2}^-$ can be identified by observing transitions to a 2^+ final state, on the assumption that the M2 transition from a 0^- initial state is undetectably weak. Similarly, the 1^- states formed by neutron capture in a target with $I^\pi = \frac{3}{2}^-$ can be identified by observing transitions to a 0^+ final state and the 3^- states formed by capture in a target with $I^\pi = \frac{7}{2}^-$ by observing transitions to a 2^+ final state.

Measurements on the resonances of ^{195}Pt by means of a NaI detector provide a good demonstration of both the value and the weakness of using single gamma rays for spin assignments. The spectra for the ^{195}Pt resonances were given in Fig. 10 and the partial radiation widths in Table I. These data show that there are many resonances for which the ground-state transition is so strong that the spin assignment can be made rapidly and with certainty. For example, at the 12-eV resonance a nonzero intensity for the ground-state transition can be observed in less than 30 sec in measurements with the Argonne chopper. On the other hand, for some other resonances, runs of a week or more are required to detect any of the high-energy transitions. Moreover, weak transitions of this kind are not expected to be rare since, for a Porter–Thomas distribution, the fraction of widths smaller than Γ is approximately equal to $0.8(\Gamma/\bar{\Gamma})^{1/2}$ when Γ is small.

The Ge-diode detector makes the high-energy γ-ray spectra much more useful for spin assignments. One important improvement is that transitions to *many* low-energy states can be observed, with the result that Porter–Thomas fluctuations are a less serious source of uncertainty. Also, the ability to observe transitions to closely-spaced low-energy states permits the high-energy spectra to be useful for spin assignments in odd–odd as well as even–even compound nuclides. Both of these advantages are illustrated in Fig. 29 by the spectra obtained by Lone et al. (1968) in a study of the reaction ^{169}Tm$(n,\gamma)^{170}$Tm with a Ge-diode gamma-ray spectrometer at the Brookhaven chopper. The target ^{169}Tm is a material for which $I^\pi = \frac{1}{2}^+$ and in ^{170}Tm there are in the energy range 0–600 keV seven states for which the spins and parities are (starting with the ground state) 1^-, 2^-, 0^-, 2^-, 2^-, 1^-, 1^-. Thus, the detection of a transition to any one of the four low-energy states with $J^\pi = 0^-$ or 2^- is definite evidence that the initial state has spin 1 (assuming s-wave capture). And the observation of a transition to a 1^- low-energy state accompanied by an *absence* of detectable transitions to *all* of the 0^- and 2^- states gives some evidence that the initial state has spin 0. As is shown by Fig. 29, on this basis the spin of most of the low-energy resonances of ^{169}Tm can be assigned with confidence, even though the statistical accuracy of one of the spectra (29 eV) is poor.

We have seen that the observation of high-energy gamma rays is a useful tool for spin assignments. However, it has two weaknesses: (a) It is applicable only to nuclides for which the spins and parities favor high-energy E1 transitions; and (b) Even when the method is applicable, it does not give an unambiguous assignment when no intensity is detected for spin-determining transitions. The 2-step cascade method of spin assignment complements the information gained from the single high-energy gamma rays in both respects. The general idea of this method, which was introduced by Bollinger and Coté (1960), is that the relative intensity of the cascades that proceed by *exactly two* transitions from the initial state

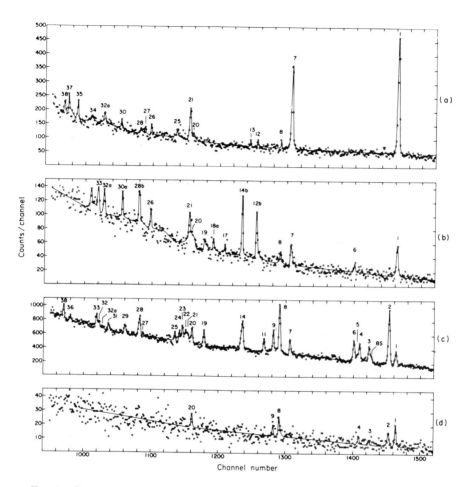

Fig. 29. Capture gamma-ray spectra for resonances of ^{169}Tm$(n,\gamma)^{170}$Tm, as measured with a Ge-diode spectrometer (Lone *et al.*, 1968). Running time was 4854 min, resolution width 12 keV, and sample 192 gm of Tm$_2$O$_3$. (a) 14-eV resonance with $J = 0$; (b) 17 eV, $J = 1$; (c) 3.9 eV, $J = 1$; (d) 29 eV, $J = 1$. The gamma-ray line labeled 1 is at 6594 keV and 7 is at 6003 keV. The complete absence of lines 2, 3, 4, and 5 in the spectrum for the 17-eV resonance is the basis for the assignment $J = 0$, and the presence of these lines in the spectra for the 3.9- and 29-eV resonances requires the assignments $J = 1$.

to some final state at low energy often depends on the spin and parity of the initial state. This idea has already been considered in Fig. 16 for the 0^+ and 1^+ initial states formed by capture in a target with spin $\frac{1}{2}$ and positive parity. In this case, the ratio of intensities was seen to depend entirely on the level spacing at the initial state. The ratio is 3 for the case considered

TABLE VIII

CALCULATED VALUES OF THE RELATIVE INTENSITY OF 2-STEP CASCADES FROM AN
INITIAL STATE WITH SPIN J_r AND POSITIVE PARITY TO
THE INDICATED FINAL STATES[a]

J_r	J^π of final state			
	0^+	2^+	4^+	6^+
0	100	100	0	0
1	33	67	0	0
2	20	60	20	0
3	0	29	29	0
4	0	11	33	11
5	0	0	18	18

[a] Bollinger et al. (1964). Only E1 transitions in a fully statistical model are considered.

because the level spacing is proportional to $(2J + 1)^{-1}$. The experimental data given in Table III for the target ^{113}Cd were seen to be consistent with this expectation.

The ideas outlined above for a target with $J^\pi = \frac{1}{2}^+$ may easily be extended to the other even–odd targets with positive parity. The expected dependence of the 2-step intensity on the spin of the initial state is given in Table VIII. The J dependence of the intensity of transitions to the 0^+ or to the 2^+ final state are seen to be strong enough to allow the 2-step-cascade method to be useful for all stable even–odd targets with positive parity. Note that these targets are a class for which single high-energy transitions to low-energy states are usually too weak to be detected easily.

When we turn to the class of targets for which the single gamma rays are most useful for spin assignments—the even–odd targets with negative parity—it is not entirely obvious that the 2-step cascade can be employed. The principal worry is that the intensity might be too small, because the 2-step cascade must now include an M1 transition. In practice it turns out, for the few cases that have been studied, that the intensity of the E1 → M1 cascade is not much different from the intensity of cascades formed by two E1 transitions. A possible explanation of this similarity is that the total radiation width Γ_b associated with an intermediate state of positive parity is much smaller than the width for states of negative parity, with the result that the ratio Γ_{ab}/Γ_b in Eq. (15) is roughly the same for E1 and M1 transitions originating at the intermediate state b. In this way, a cascade of E1 → M1 gamma rays from an initial state with negative parity could be roughly as intense as a case of E1 → E1 gamma rays from an initial state with positive parity.

In principal, the 2-step cascade method of spin assignment should also be applicable to odd–even targets. However, each case would have to be discussed individually since the low-energy level structure of the odd–odd product nuclei does not have the regularity needed to allow us to state general conclusions.

A property of central importance to the usefulness of the 2-step cascade is that its intensity is largely free of random fluctuations because many intermediate states contribute. In the general case, it is not easy to specify quantitatively the magnitude of the fluctuations that are expected since the dependence of the cascade intensity on the energy of the intermediate state is not well understood. However, if the intermediate state is required to lie within a relatively narrow band in the neighborhood of half of the binding energy, then the energy dependence may be ignored and the relative scatter in the cascade intensity is $\Delta I/I = (4/n)^{1/2}$, where n is the number of intermediate states involved and the widths of the individual transitions are assumed to satisfy the Porter–Thomas distribution. For the typical heavy nucleus the value of n corresponding to a 1-MeV band at half the binding energy is expected to be quite large—several hundred. Hence $\Delta I/I \ll 1$.

The 2-step cascade method of spin assignment has been systematically and successfully applied to a variety of targets by Bollinger et al. (1964). The general ideas outlined above were verified, but two experimental problems were found to be bothersome. First, there is the problem of achieving a counting rate that is high enough to make the method attractive. This problem was solved by using two very large scintillators (8 in. diameter × 6 in. thick). The second and more fundamental problem is that, although random fluctuations in the intensity of 2-step cascades are unimportant for most targets, there are a few targets for which the fluctuations cannot be ignored. It was shown that a qualitative feeling for the relative importance of such fluctuations can be obtained (Bollinger and Coté, 1961) by observing the spectra in one scintillator when the sum of pulse heights is required to be approximately equal to the neutron binding energy. For a well-behaved target like ^{113}Cd this distribution is roughly flat, with no significant structure. For the target ^{201}Hg, on the other hand, most of the counts in the spectrum are concentrated in a few lines and the intensities of these lines fluctuate greatly from resonance to resonance. One sees immediately, then, that the sum-coincidence data might not give a very reliable indication of the spin of the resonances of ^{201}Hg. At the same time, one also sees that the individual cascades causing the greatest fluctuations are those for which the intermediate state is fairly near the ground state. Thus, the magnitude of the random fluctuations can be reduced by requiring the intermediate state to lie within a band of energy in the neighborhood of half the binding energy, a condition that can be imposed by placing appropriate requirements on the pulse heights for the individual scintillators. Experimentally, the fluctuations remaining after limiting the allowed energy of the intermediate state are found to be small for ^{77}Se, ^{95}Mo, ^{105}Pd, ^{111}Cd, ^{113}Cd, and ^{183}W; significant fluctuations remain for ^{199}Hg and ^{201}Hg, however.

The most reliable way to make spin assignments by means of the 2-step method is to measure the spectra of many resonances and to observe that the relative intensities of 2-step cascades to some low-energy state are separated into two distinct groups. However, in principle the spin assignment of a single resonance can be obtained from its 2-step cascade spectrum alone, since the *shape* of the spectrum depends on the spin. This may easily be seen for an even–odd target nucleus with $I^\pi = \frac{1}{2}^+$, for which the ratio of intensities of 2-step cascades to the first excited state (2+) and ground state (0+) is 1.0 if the initial state has $J = 0$, and 2.0 if the initial state has $J = 1$ (see Table VIII). This difference is large enough to permit the spin of a single resonance to be determined with ease if the ground and first excited states are separated by more than say 200 keV and if one knows the effective line shape of the detection system for 2-step cascades.

Let us conclude this discussion of 2-step cascades by considering some representative experimental data, that for resonances of ^{183}W, a target for which the spins of the resonances ($J = 0$ or 1) have been determined in other ways. Several of the sum-coincidence spectra for this target are presented in Fig. 27. A minor complication for these data is that the peak at the high-energy end of these spectra results from 2-step transitions to both the ground state (0+) and the first excited state (2+), which are separated by only 111 keV. Elementary considerations similar to those used in Fig. 16 lead one to expect that this double peak in a spectrum for a 0$^-$ initial state should be twice as intense as the peak for a 1$^-$ initial state. The experimental results are given in Table IX. When all 2-step events are accepted, the expected dependence on spin is observed although there is a bothersome degree of fluctuation in intensity for states of the same spin. The results given in column 3 show that most of this fluctuation can be eliminated by requiring the intermediate state to lie in the range (2–5.5 MeV).

Another kind of 2-step cascade measurement has been suggested by Ignat'ev *et al.* (1963) as being useful for spin determinations. This is an angular-correlation measurement. The expected difference in the correlation is strong enough to allow a rather easy measurement to distinguish between initial states with spin 0 and 1 in even–even compound nuclides.

TABLE IX

SPIN ASSIGNMENT OF RESONANCES OF ^{183}W FROM 2-STEP CASCADE DATA[a]

Resonance energy (eV)	Relative intensity		Assigned J
	A	B	
7.6	263 ± 5	100 ± 3	1
27	197 ± 4	105 ± 3	1
41	163 ± 8	97 ± 6	1
46	204 ± 5	111 ± 4	1
48	290 ± 9	210 ± 8	0
65	209 ± 13	95 ± 9	1

[a] Bollinger et al. (1964). The intensities in column A are those for all 2-step cascades to the 0^+ and 2^+ states at 0 and 111 keV in ^{184}W. For column B the intermediate state is restricted to the range 2.0–5.5 MeV.

This kind of measurement appears to be more difficult to carry out and less sensitive than the 2-step method that depends on intensity alone, but it does have the advantage that it is independent of Porter–Thomas fluctuations.

When resonances are close enough together, the resonance–resonance interference discussed in Section IV,A,2 can be useful for the determination of spins. If the cross section for an individual gamma-ray line exhibits the interference pattern in the neighborhood of two resonances, the resonances surely have the same spin. The absence of interference effects between the 14 and 17 eV resonances of ^{169}Tm have been used by Lone et al. (1968) to establish the inequality of the spins of the two states, in agreement with the assignments inferred from γ-ray intensities (Fig. 29).

In addition to well-understood processes of the kind described above, various characteristics of the spectra might exhibit a positive correlation with the spin of a resonance. If so, once the effect has been demonstrated empirically for resonances with known spins, the correlation may be used to determine the spins of other resonances in the same target. This approach has been used by Corge et al. (1961) in a study of ^{109}Ag; the ratio of the intensity of pulses in the range (6.2–6.8 MeV) to the intensity of low-energy pulses is found to be much greater for resonances with $J = 1$ than for resonances with $J = 0$.

Possibly the most important development in the use of capture-gamma spectra for spin assignments is the recent discovery by Coceva et al. (1968) that the γ-ray multiplicity depends significantly and systematically on the spin of the initial compound state. In their investigation, NaI scintillators were used to measure the quantity $r = I_s(h)/I_c$, where I_c is the intensity of

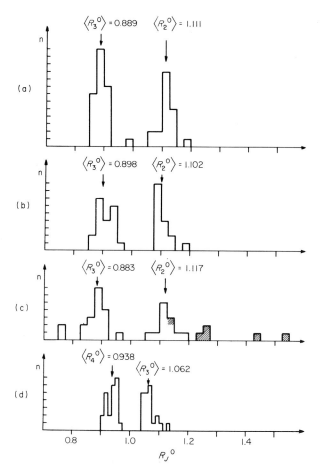

FIG. 30. Histograms of measured values of the ratio $r = I_s(h)/I_c$ for neutron resonances in the even–odd isotopes of Mo, Ru, Pd, and Hf (Coceva *et al.*, 1968). The quantity $R_J{}^0$ is a normalized value of r for a resonance with spin J. (a) ^{105}Pd(n,γ); (b) ^{99}Ru(n,γ), ^{101}Ru(n,γ); (c) ^{95}Mo(n,γ); ^{97}Mo(n,γ); (d) ^{177}Hf(n,γ).

coincident counts in two detectors and $I_s(h)$ is the intensity of counts with pulse height greater than h in a single detector. When $h = 0$, $r = \bar{\nu}/\nu(\nu - 1)$.

Coceva *et al.* (1968) have studied the behavior of r for the even–odd isotopes of Mo, Ru, Pd, and Hf, and Asghar *et al.* (1968) has extended the measurements to the fissionable nuclide ^{235}U. The ratio of r for the two possible spin states is found to depend somewhat on the bias level h for the single detector. Typically, $r(I - \frac{1}{2})/r(I + \frac{1}{2}) = 1.1$ when h is small and it increases to 1.2 for $h \approx 2$ MeV. Although the values of r for the two spin states do not differ much, they do differ enough to allow the measured values to separate cleanly into two groups, depending on the spin of the initial state. This separation is shown in Fig. 30. In view of the ease and accuracy with which the quantity r can be measured, for many nuclides the method developed by Coceva et al. (1968) appears to be the most effective means yet developed for the determination of the spin of low-energy resonances. The only reservation about the method is whether it will turn out to be as useful for most nuclides as it is for the few even–odd targets that have been studied to date. On the basis of the arguments given in Section IV,B,2 and the insensitivity of the average γ-ray intensity to the spin of the ^{195}Pt resonances (Section IV,B,3), it seems unlikely that the new method will be very useful for even–odd targets with spin $\frac{1}{2}$.

Let us conclude this section on the determination of spin by attempting to assess the relative merits of the various methods that have been discussed. This is done in tabular form in Table X. The columns labeled "percent

TABLE X

COMPARISON OF METHODS FOR THE DETERMINATION
OF SPINS OF NEUTRON RESONANCES[a]

| Method | Detector | Percent nuclides | | Percent reliability | Relative sensitivity |
		e–e	o–o		
Low-energy singles	Ge(Li)	90	50	100	10
	NaI	10	0	100	100
High-energy singles	Ge(Li)	50	50	95	1
	NaI	20	0	90	100
Sum coincidence	NaI	80	10	100	10
Angular correlation	NaI	20	0	100	1
Multiplicity	NaI	80	50	?90	100

[a] The symbols e–e and o–o refer to even–even and odd–odd product nuclides.

nuclides" give rough estimates of the percentage of nuclides for which each method is useful for even–even and odd–odd compound nuclides. All of the methods appear to be quite reliable, but the method that depends on the intensity of single high-energy γ rays suffers more than the others from uncertainties caused by Porter–Thomas fluctuations, and we do not yet know enough about the method that depends on multiplicity to be sure how reliable it is for most nuclides. The last column gives rough estimates of the relative sensitivity of each method for spin determinations, where the "sensitivity" is defined as being inversely proportional to the time required to determine the spin with a high degree of certainty by means of measurements with a given neutron source and neutron time-of-flight resolution.

3. Parity Assignment

Because of the centrifugal barrier for neutrons with orbital angular momenta greater than zero, most of the resonances that have been studied at low energy may safely be assumed to have been excited by s-wave neutrons. However, the states that could be excited by p-wave neutrons are surely present and, as more and more sensitive measurements are made, narrow resonances that might be p-wave resonances are beginning to be detected. These resonances are often so narrow that their parity cannot be identified by observations on the scattered neutrons; the cross section for resonance scattering is undetectably small. Thus, the parity of the initial state can be positively identified only by observations on the gamma rays emitted.

To date most of the identifications of p-wave resonances by means of gamma-ray spectra depend on the intensity of high-energy transitions. Jackson (1963) has given a full account of the identification of a total of twenty-three resonances of ^{91}Zr, ^{93}Nb, and ^{95}Mo. The most clear-cut assignments are those for some of the resonances of ^{95}Mo. For this target the initial compound state has $J^{\pi} = 1^{-}, 2^{-}, 3^{-}$, or 4^{-} for p-wave excitation and 2^{+} or 3^{+} for s-wave excitation. The observation of a ground-state transition is almost conclusive evidence for $J^{\pi} = 1^{-}$, since we expect (see Section IV,A,6) that E2 or M2 transitions from other kinds of states would be undetectably weak in comparison with an E1 transition. The parity (and spin) assignments of two resonances in ^{95}Mo were made on the basis of the observation of a ground-state transition.

Unfortunately, the observation of the ground-state transition is effective in identifying only a small fraction of the p-wave resonances that are expected to be present in a target like ^{95}Mo. However, additional information may be obtained from the intensity of the transition to the $2+$ state. If the transition is very strong, one may be confident that it is an E1 transition from a state with $J = 1^{-}, 2^{-}$, or 3^{-}; two p-wave resonances of ^{95}Mo were identified on this basis. However, if the transition is weak, it could be either an M1 or an E1 transition; hence, the parity cannot be determined.

As we have just seen, the fluctuations of individual partial radiation widths are too large to allow the individual transitions to be very effective in identifying the parity of the initial state when the competing radiations are E1 and M1. However, if one can determine the *sum* of intensities of transitions to several low-energy states of positive parity, this sum may

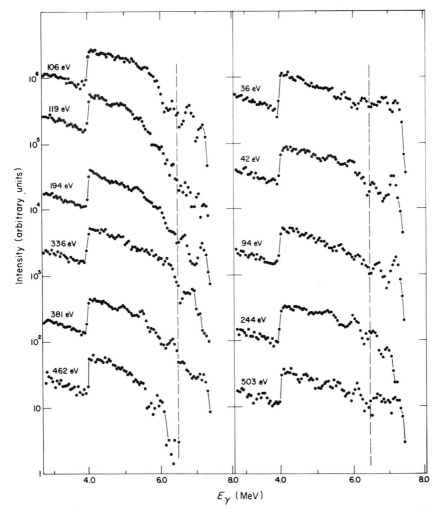

FIG. 31. Capture gamma-ray spectra for resonances of ^{93}Nb $+$ n as measured with a NaI scintillator (Jackson, 1964). The spectra on the left are associated with *s*-wave resonances and those on the right with *p*-wave. The abrupt drop in intensity at about 4 MeV is caused by a rejection of three-fourths of all pulses below that level, a procedure adopted to conserve analyzer memory (magnetic tape) for the recording of the larger pulses.

be sufficiently free of random fluctuations to allow one to distinguish between E1 transitions from states of negative parity and M1 transitions from states of positive parity. Jackson convincingly demonstrates that this idea is valid for the resonances of ^{91}Zr. Both the available experimental data and theoretical calculations indicate that the first ten low-energy states of ^{92}Zr have positive parity. Thus, one would expect the high-energy portion of the spectrum to be much more intense for p-wave than for s-wave resonances. Experimentally, the sum of intensities of transitions to the first 10 low-energy states of ^{92}Zr are observed to fall into two distinct classes for which the intensities differ by a factor of about 12. The parities of 9 resonances were assigned on the basis of these sums of intensity.

The third target studied by Jackson is ^{93}Nb, a case for which the assignment of parity is made on the basis of the entirely empirical observation that the average intensity of high-energy gamma rays falls into two distinct classes. It is found that a large intensity is always associated with a small reduced neutron width. Thus the resonances with a strong intensity may reasonably be assigned to be p-wave resonances. Five p-wave resonances were assigned on this basis, as is shown in Fig. 31.

In addition to various ways of assigning parity on the basis of gamma-ray intensity, some p-wave resonances can be identified from the angular distribution of primary gamma rays with respect to the incident neutron. The first measurement of this kind was carried out at the Harwell linac by McNeill et al. (1965) who studied the 62-eV resonance in ^{124}Sn. This is known to be a p-wave resonance from the symmetrical shape observed in total cross-section measurements.

For an even–even target nucleus, primary dipole transitions from a $J = \frac{1}{2}$ initial state are expected to be isotropic, independent of the parity, so that an angular distribution measurement cannot distinguish between s-wave and p-wave capture. On the other hand, the gamma rays from a $J = \frac{3}{2}$ state formed by p-wave capture (it cannot be formed by s-wave capture) will usually exhibit a considerable asymmetry for pure dipole transitions, as is illustrated in Fig. 32.

In the Harwell study, the gamma-ray spectra were measured with a NaI detector. The high-energy part of the observed spectrum is dominated by a line formed by unresolved contributions from transitions to $\frac{3}{2}^{+}$ and $\frac{1}{2}^{+}$ states are 30 and 210 keV, respectively. The angular distribution of this composite line is effectively isotropic, suggesting that $J = \frac{1}{2}$. However, as may be seen from Fig. 32, there is some possibility that the observed intensity is isotropic because the two components in the unresolved line have asymmetries that tend to cancel. Hence, the interpretation of the experimental result was somewhat uncertain.

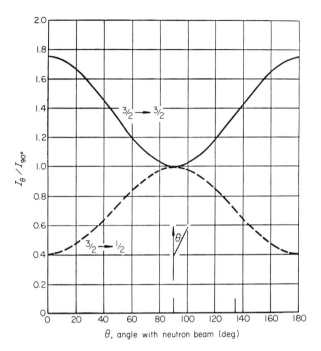

FIG. 32. Calculated angular distributions for radiative transitions to states with $J = \frac{3}{2}$ and $\frac{1}{2}$ from a $J = \frac{3}{2}$ state formed by p-wave capture (Slaughter *et al.*, 1966).

More recently, Slaughter *et al.* (1966) have made similar angular-distribution measurements in which neutron-resonance capture is selected by means of a filtered-beam technique. Because of the relatively high capture rates obtainable in this way, it was possible to obtain spectra of good quality with a Ge-diode spectrometer for several resonances in even–even isotopes of tin. The 45.8-eV resonance of ^{118}Sn was shown to be a p-wave interaction with $J = \frac{3}{2}$ and the 62-eV resonance of ^{124}Sn was established with certainty as having $J = \frac{1}{2}$. These results have recently been confirmed by Bhat *et al.* (1967) in the first angular-distribution measurements with a Ge-diode detector at a time-of-flight neutron spectrometer.

The other spectral properties that were discussed in connection with spin assignments might also be useful for the determination of parity; these properties include the intensity of 2-step cascades, the intensity of low-energy gamma rays, and the γ-ray multiplicity. There is already a strong indication that for some nuclides, at least, the quantity $I_s(h)/I_c \rightarrow \langle \nu \rangle / \langle \nu(\nu - 1) \rangle$ measured by Coceva *et al.* (1968) depends on the parity as well as the spin of the resonance. The evidence for this is given in Fig. 30, where the data points for the p-wave resonances of ^{95}Mo and ^{97}Mo are seen to be displaced from the points for the s-wave resonances. On the basis of the available information, it is not clear whether this displacement results mainly from a difference in the pulse-height spectrum of single gamma rays or from a difference in the multiplicity.

D. Study of Bound States

Since a consideration of the physics of bound nuclear states is beyond the scope of this book, our discussion here will be restricted to the technological subject of *how* resonance capture is used to study low-energy bound states. Such studies are closely related to similar studies with thermal-neutron-capture gamma rays. The principal difference is that the capturing nuclide can be known with certainty in resonance capture whereas several nuclides usually contribute to thermal capture. Other advantages of resonance capture are that the spin and parity of the initial state can be known, the spin and parity can be varied, random fluctuations in the intensity of the high-energy gamma rays help to improve the resolving power of the gamma-ray spectrometer, and the background resulting from radioactive decay of the ground state of the product nucleus can be made negligibly small.

In spite of many potential advantages in comparison with thermal capture, before the advent of the Ge-diode spectrometer resonance capture was a very weak tool for the study of bound states; the resolution of the NaI scintillator was simply not adequate. As a consequence, resonance capture was almost unused to study bound states until recently.

As in the case of thermal-neutron-capture spectra, in resonance capture two clearly distinguishable kinds of gamma rays are useful in a study of bound states. These are the high-energy gamma rays, which result from primary transitions to the states of interest, and low-energy gamma rays, which result from secondary transitions between the states of interest.

1. Low-Energy Gamma Rays

For most of the targets that can be studied by resonance capture, the low-energy spectra are so complex that measurements with a single NaI scintillator are not likely to be very useful; coincidence measurements are required. Probably because of the low counting rates that are obtainable, the only reported coincidence measurement aimed primarily at a study of low-energy states is that by Segel et al. (1964) for capture in the resonances of ^{199}Hg. They established the presence and probable character of two new states in ^{200}Hg.

Although the Ge-diode detector is undoubtedly better than NaI for an observation of low-energy spectra, even it is relatively ineffective. In addition to the problem of resolving the complex spectra, data of good quality are difficult to obtain because the tails of the high-energy gamma-ray lines seriously overlap the low-energy region of interest. As a result, there have been few if any published reports aimed primarily at the study of bound states by means of measurements on low-energy gamma rays following resonance capture.

2. High-Energy Gamma Rays

The high-energy gamma rays in neutron-capture spectra are a particularly useful source of information about low-energy nuclear states because they give the energy of the states directly, with very little interpretation. Under the assumption that the high-energy lines result from primary transitions, the energies of the states are simply $B_n - E_\gamma^*$, where B_n is the neutron binding energy and E_γ^* is a gamma-ray energy that has been corrected for the recoil energy lost to the emitting nucleus; numerically, $E_\gamma^* = E_\gamma + 0.537 \times 10^{-6}E_\gamma^2/A$, where E_γ^* and the gamma-ray energy E_γ are in kiloelectron volts and A is the atomic weight of the emitting nucleus. The assumed primary character of the high-energy lines is open to question, perhaps, for the light nuclides and for other nuclides with large level spacings. However, for a nuclide that has a small enough level spacing to allow resonance capture to be studied with a slow-neutron spectrometer, the high density of states near the binding energy rules out the formation of secondary high-energy gamma-ray lines of detectable intensity, according to the statistical description of the de-excitation process. All of the experimental evidence supports this expectation.

Since the advent of the Ge-diode spectrometer, high-energy gamma rays in resonance-capture spectra have begun to yield the expected useful information about low-energy states. In contrast, the resolution of the NaI scintillator is so poor that few, if any, states were discovered by observing high-energy gamma rays, although NaI measurements did give some worthwhile information about the character of states that were known to exist. Hence, unless a NaI detector is explicitly mentioned, in the remainder of this discussion it will be assumed that the gamma-ray spectrometer is a Ge-diode.

An important limitation on the usefulness of the gamma-ray spectra formed by thermal-neutron capture is the confusion caused by isotopic or chemical impurities in most samples; because of the great variation in capture cross sections and the wide fluctuations in the Porter–Thomas distribution, a very small impurity can cause relatively strong gamma-ray lines. This difficulty is effectively eliminated in resonance capture, since

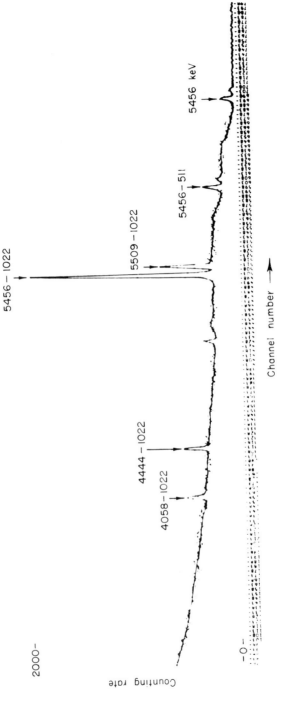

FIG. 33. Gamma-ray spectrum from capture in the 96-eV resonance of ^{199}Pt + n as measured with a Ge diode spectrometer (Jackson *et al.*, 1966b) (5.34 keV/channel).

at a typical low-energy neutron resonance the capture cross section of the isotope of interest is many orders of magnitude greater than that of any other isotope in the sample. An example of this advantage of resonance capture is given by the results of Jackson *et al.* (1966b) for capture in the 96-eV resonance of ^{198}Pt. This spectrum (Fig. 33) is seen to be entirely free of lines from capture in other isotopes such as ^{195}Pt, which dominates the thermal-capture spectrum.

It is now widely recognized that the intensities of high-energy gamma-rays from thermal-neutron capture are not a reliable indication of the spin and parity changes involved in a transition, because of Porter–Thomas fluctuations. If the spectra of several resonances are measured, however, the data become much more useful since (for most nuclides) the average reduced width of transitions from many initial states indicates the nature of the radiation, in practice, whether it is E1 or M1 since radiation of higher multipolarity is rarely strong enough to be detected. And if the nature of the radiation is known, the parity of the final state can be determined with certainty and limits can be placed on the spin. For example, if a gamma-ray line is strong enough to indicate that it is an E1 transition, the final-state parity is $\pi_j = -\pi_r$ and the spin is $J_j = J_r - 1$, J_r, or $J_r + 1$. Moreover, a measurement of spectra for resonances having each of the two spin values formed by s-wave capture gives additional information about J_j. A limitation on the effectiveness of this procedure is that a rather large number of spectra are needed to yield average widths that are accurate enough to distinguish between E1 and M1 transition with certainty. For example, suppose that E1 transitions are ten times stronger than M1 transitions on the average and that spectra are measured for five resonances of the same spin. Even under these favorable circumstances the average intensities of roughly 15% of the E1 transitions are too weak to be clearly distinguishable from M1 transitions.

If one does have spectra for enough resonances to identify E1 transitions reliably and especially if the resonances have known spin and parity, then the data can add greatly to our knowledge of the spins and parities to the low-energy states. An excellent example of this is provided by the data of Jackson *et al.* (1966a) for capture in ^{195}Pt. The spectra for twenty-two resonances with $J^\pi = 1^-$ and seven resonances with $J^\pi = 0^-$ were measured. The strong high-energy transitions may confidently be assumed to be electric dipole and hence they immediately establish the parity of the final states as positive. Moreover, transitions from initial states with $J = 0$ identify final states with $J = 1$ and transitions from initial states with $J = 1$ allow one to conclude that the spin of the final state is either 0, 1, or 2. Even though this kind of information is less complete than one would hope for, it is often more complete than has been obtained in other ways.

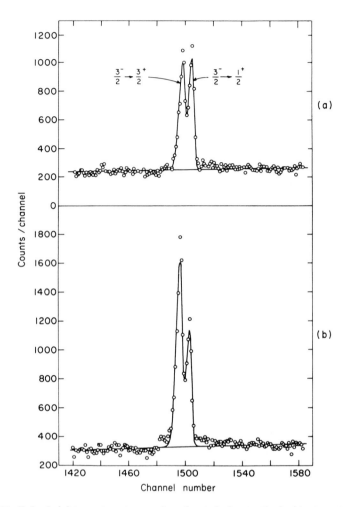

F<small>IG</small>. 34. Pulse-height spectra at several angles (relative to the incident neutron) following neutron capture in the p-wave 46-eV resonance of ^{118}Sn (Bhat *et al.*, 1968). The peaks are formed by transitions to states at 0 and 23 keV. (a) $\theta = 90°$; (b) $\theta = 135°$.

For the few nuclides that have strong p-wave resonances at low energy, the angular distribution of primary capture gamma rays may be used to determine the spins of final states. This technique has been pioneered by the Brookhaven chopper group with measurements on the even–even targets ^{118}Sn (Bhat *et al.*, 1968) and ^{98}Mo (Mughabghab *et al.*, 1968). When the p-wave resonance has $J = \frac{1}{2}$, the angular distribution is isotropic and hence gives no information about the final states. However, when the resonance has $J = \frac{3}{2}$, the distribution exhibits the asymmetry shown in

TABLE XI

ANGULAR DISTRIBUTION OF DIPOLE γ RAYS ORIGINATING IN A $\frac{3}{2}$ RESONANCE[a]

$\frac{3}{2} \to \frac{1}{2}$	$\frac{3}{2} \to \frac{3}{2}$	$\frac{3}{2} \to \frac{5}{2}$
$I(\theta) = (1/16\pi)(2 + 3\sin^2\theta)$	$I(\theta) = (1/20\pi)(7 - 3\sin^2\theta)$	$I(\theta) = (3/80\pi)(6 + \sin^2\theta)$
$I(90°)/I(135°) = 10/7$	$I(90°)/I(135°) = 8/11$	$I(90°)/I(135°) = 14/13$

[a] Bhat *et al.* (1968).

Fig. 32 or given by the relationships in Table XI. In the measurements on ^{118}Sn and ^{98}Mo, the distributions were determined from the spectra recorded at two angles: 90° and 135° relative to the direction of the incident neutrons. An example of such spectra is given in Fig. 34, where the influence of the asymmetric angular distributions is seen to be quite pronounced. Data of this kind have been used to identify final states with $J = \frac{1}{2}, \frac{3}{2},$ or $\frac{5}{2}$.

3. AVERAGE RESONANCE-CAPTURE SPECTRA

As we have seen above, Porter–Thomas fluctuations in the gamma-ray spectra measured at individual resonances limit their usefulness for the determination of spins and parities of low-energy states. To a large extent this difficulty can be overcome by measuring *directly* the spectrum formed by capture of neutrons in many resonances. In the only measurements of this kind reported to date,* Bollinger and Thomas (1967a,b, 1968b) make use of a neutron-filter technique at an internal-target facility at a nuclear reactor, in the way outlined in Section II,C,2. The average resonance-capture technique has no advantage over a measurement of spectra at many individual resonances except for the important one that it is in fact possible to measure directly an average spectrum of good quality, whereas it is not yet technically feasible to obtain an equivalent average by measurements at individual resonances.

The information that can be obtained from the intensities of lines in average resonance-capture spectra may be understood in terms of a simple statistical model for both the capture and radiation processes. For simplicity, assume that the rate of capture in a single resonance is proportional to $\sigma_0{}^m\Gamma_\gamma$, where σ_0 is the peak cross section and $\frac{1}{2} < m < 1$. This familiar relationship is accurate when the sample is thin and is at least qualitatively meaningful for a thick sample when the average recoil-energy loss of a scattered neutron is much larger than both Γ and Δ, where Δ is the Doppler width. Under this assumption about the rate of capture and the further assumption that dipole transitions dominate, the intensity $S_j(J)$ of primary transitions from initial states with spin J to a single final state j is

* See also recent work of Greenwood *et al.* (1969).

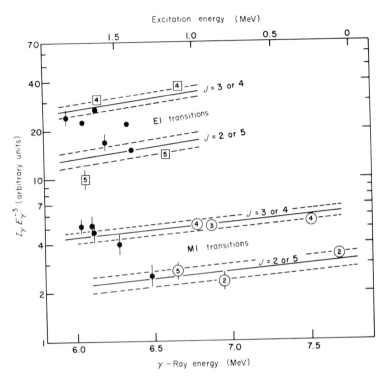

F$_{\text{IG}}$. 35. Plot of $I_\gamma E^{-3}$ versus E_γ for transitions observed in the average resonance-capture spectrum of the ^{167}Er(n,γ)^{168}Er reaction (Bollinger and Thomas, 1968b). The numbers within the data points give the known spins of the final states. ($\bigcirc$, positive π; $\square$, negative π; $\bullet$, unknown π.)

$$S_j(J) \; \propto \; \langle \Gamma_{\gamma j}/D \rangle_J \langle n\sigma_0 \rangle^m \; \propto \; \langle \Gamma_{\gamma j}/D \rangle_J \langle ng(\Gamma_n/\Gamma) \rangle^m \qquad (21)$$

where g is the usual statistical factor $\frac{1}{2}(2J + 1)(2I + 1)^{-1}$ and I is the spin of the target nucleus.

For Eq. (21) to be useful, one must now make the physical assumption that the gamma-ray strength function $\langle \Gamma_{\gamma j}/D \rangle$ is a smoothly varying function of gamma-ray energy, independent of the spin of the initial and final states (see Section IV,A,4 for supporting evidence). Then it is obvious from Eq. (21) that the intensity $S_j(J)$ is insensitive to the spin J of the initial state for most nuclides. Thus, the intensity of all transitions to a final state is just proportional to $Q_J \langle \Gamma_{\gamma j}/D \rangle$, where Q_J is the number of paths by which the final state can be fed and $\langle \Gamma_{\gamma j}/D \rangle$ is different for E1 and M1 radiation. For example, in the reaction ^{167}Er(n,γ)^{168}Er, in which 3+ and 4+ states are formed by s-wave capture, final states $J = 3$

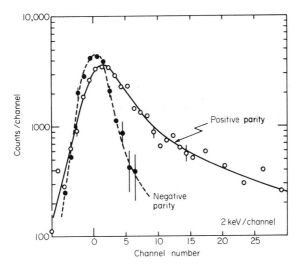

FIG. 36. Dependence of gamma-ray line shape on the parity of the final state for primary transitions in an average-resonance-capture spectrum of the reaction ^{105}Pd $(n,\gamma)^{106}$Pd (Bollinger and Thomas, 1968b). The initial states have positive parity for s-wave capture.

and 4 have $Q = 2$ and states with $J = 2$ or 5 have $Q = 1$. Also, it is expected that $\bar{\Gamma}_{E1} \gg \bar{\Gamma}_{M1}$. Thus, in a plot of gamma-ray intensity I_γ versus E_γ, the points are expected to lie along four curves corresponding to the four possible combinations of Q and $\langle \Gamma_{\gamma j}/D \rangle$.

The ideas outlined above have been experimentally verified by Bollinger and Thomas (1968b) in an investigation of low-energy states in ^{168}Er, a nucleus about which a great deal was known from previous work. Some of the results obtained are given in Fig. 35, where each data point associated with a final state of known spin and parity falls in the expected way near one of the four solid lines. These lines were drawn in such a way as to satisfy the expectation that the intensity of a curve for states with $J = 3$ or 4 is twice as great as the intensity of the curve for states of the same parity with $J = 2$ or 5. Since everything about the intensities of the transitions to final states of known spin and parity behave in the way that is expected, the intensities of transitions to previously unknown states may be used with confidence to obtain information about their spins and parities.

Similar resonance-capture measurements have been made for about thirty-five nuclides ranging from ^{74}Ge to ^{239}U (Bollinger and Thomas, 1969). The data for most of the nuclides seem to be consistent with the model outlined above, although for some of them (those with resonance spacings greater than 100 eV) Porter–Thomas fluctuations prevent the data from providing a sensitive test of the model. On the other hand, the data for the odd–odd nuclide ^{170}Tm are inconsistent with the model in the sense that the γ-ray lines exhibit fluctuations in intensity that are much larger than the expected Porter–Thomas fluctuations. This discrepancy is not yet understood. At this early stage the data may be summarized by the statement that the average γ-ray intensity gives the parity of a final state with certainty for most nuclides and sets a limit on the spin for many.

Although the intensities of high-energy transitions are the main source of spectroscopic information in the average resonance-capture spectra, the γ-ray *line shape* can also be used to determine the parities of low-energy states under certain circumstances. An example of this use is given in Fig. 36, where the lines associated with negative- and positive-parity final states in the reaction ^{105}Pd$(n,\gamma)^{106}$Pd are seen to be strikingly different. This phenomenon results from the influence of p-wave capture, whose dependence on neutron energy differs from that of s-wave capture. The parity dependence of the line shape is large enough to be detected easily only when the p-wave strength function is relatively large, i.e., mainly for nuclides in the mass range $85 < A < 130$.

Two other characteristics of the average resonance-capture spectra are useful for nuclear spectroscopy. First, the spectra are relatively insensitive to impurities in the sample since (as is implied by Eq. 21) all atoms contribute comparable intensities of gamma rays. Second, the data are uniformly sensitive in the sense that, when a measurement is sensitive enough to reveal the presence of *any* state with a given spin and parity in a well-behaved nucleus like ^{168}Er, then *all* states of that kind in a given range of energy can be detected. This uniform sensitivity is unusual and useful, since in most nuclear reactions the transition probability depends on the structure of the nuclear states involved.

REFERENCES

ALLAS, R. G., HANNA, S. S., MEYER-SCHÜTZMEISTER, L., and SEGEL, R. E. (1964) *Nucl. Phys.* **58**, 122.

ALLEN, B. J. (1968). *Nucl. Phys.* **A111**, 1.

ASGHAR, M., MICHAUDON, A., and PAYA, D. (1968). *Phys. Lettes* **26B**, 664.

AXEL, P. (1962). *Phys. Rev.* **126**, 671.

AXEL, P., MIN, K., STEIN, N., and SUTTON, D. C. (1963). *Phys. Rev. Letters* **10**, 299.

BARTHOLOMEW, G. A. (1960). *In* "Nuclear Spectroscopy, Part A" (F. Ajzenberg-Selove, ed.), p. 304. Academic Press, New York.

BARTHOLOMEW, G. A. (1961). *Ann. Rev. Nucl. Sci.* **11**, 259.

BARTHOLOMEW, G. A., EARLE, E. D., FERGUSON, A. J., and BERGQVIST, I. (1967a). *Phys. Letters* **24**, 47.

BARTHOLOMEW, G. A., GUNYE, M. R., and EARLE, E. D. (1967b). *Can. J. Phys.* **45**, 2063.

BARTHOLOMEW, G. A., GROSHEV, L. V., *et al.* (1967c). *Nucl. Data Sect. A*, **3**, 367.

BECVAR, F., VRZAL, J., LIPTAK, J., and URBANEC, J. (1968). Joint Inst. for Nucl. Res., preprint P3-3696.

BEER, M., BHAT, H., CHRIEN, R. E., LONE, M. A., and WASSON, O. A. (1966). Slow-Neutron Capture Gamma-Ray Spectroscopy, Argonne Natl. Lab. Rept. ANL-7282, p. 459.

BEER, M., LONE, M. A., CHRIEN, R. E., WASSON, O. A., BHAT, M. R., and MUETHER, H. R. (1968). *Phys. Rev. Letters* **20**, 240.

BERGQVIST, I. (1965). *Nucl. Phys.* **74**, 15.

BERGQVIST, I., and STARFELT, N. (1961). *Nucl. Phys.* **22**, 513.

BERGQVIST, I., and STARFELT, N. (1962a). *Nucl. Phys.* **39**, 353.

BERGQVIST, I., and STARFELT, N. (1962b). *Nucl. Phys.* **39**, 529.

BERGQVIST, I., LUNDBERG, B., and STARFELT, N. (1963). *Proc. Intern. Conf. Nucl. Phys. Reactor Neutrons*, Argonne Natl. Lab. Rept. ANL-6797, p. 220.

BERGQVIST, I., and STARFELT, N. (1963), *Arkiv Fysik* **23**, 435.

BERGQVIST, I., BIGGERSTAFF, J. A., GIBBONS, J. H., and GOOD, W. M. (1965). *Phys. Letters* **18**, 323.

BERGQVIST, I., LUNDBERG, B., NILSSON, L., and STARFELT, N. (1966). *Phys. Letters* **19**, 670.

BERGQVIST, I., BIGGERSTAFF, J. A., GIBBONS, J. H., and GOOD, W. M. (1967). *Phys. Rev.* **158**, 1049.

BETHE, H. A. (1937). *Rev. Mod. Phys.* **9**, 69.

BHAT, M. R., BORRILL, B. R., CHRIEN, R. E., RANKOWITZ, S., SOUCEK, B., and WASSON, O. A. (1967). *Nucl. Instr. Methods* **53**, 108.

BHAT, M. R., CHRIEN, R. E., WASSON, O. A., BEER, M., and LONE, M. A. (1968). *Phys. Rev.* **166**, 1111.

BIGGERSTAFF, J. A., BIRD, J. R., GIBBONS, J. H., and GOOD, W. M. (1967). *Phys. Rev.* **154**, 1136.

BIRD, J. R., MOXON, M. G., and FIRK, F. W. K. (1959). *Nucl. Phys.* **13**, 525.

BIRD, J. R., GIBBONS, J. H., and GOOD, W. M. (1962). *Phys. Letters* **1**, 262.

BIRD, J. R., GIBBONS, J. H., and GOOD, W. M. (1963). *In* "Proceedings of the Conference on Direct Interactions and Nuclear Reaction Mechanisms," p. 305. Gordon and Breach, New York.

BIRD, J. R., BIGGERSTAFF, J. A., GIBBONS, J. H., and GOOD, W. M. (1965). *Phys. Rev.* **138**, B20.

BLATT, J. M., and WEISSKOPF, V. F. (1952). "Theoretical Nuclear Physics." Wiley, New York.

BLOCH, C. (1954). *Phys. Rev.* **93**, 1094.

BOCKELMAN, C. K. (1959). *Nucl. Phys.* **13**, 205.

BOLLINGER, L. M. (1968). *In* "Nuclear Structure," p. 317. Intern. Atomic Energy Agency, Vienna.

BOLLINGER, L. M., and COTÉ, R. E. (1960). Argonne Natl. Lab. Rept. ANL-6146, p. 1.

BOLLINGER, L. M., and COTÉ, R. E. (1961). *In* "Neutron Time-of-Flight Methods" (J. Spaepen, ed.), p. 199. Euratom, Brussels.

BOLLINGER, L. M., and THOMAS, G. E. (1967a). *Phys. Rev. Letters* **18**, 1143.

BOLLINGER, L. M., and THOMAS, G. E. (1967b). *Bull. Am. Phys. Soc.* **12**, 105.
BOLLINGER, L. M., and THOMAS, G. E. (1968a). *Bull. Am. Phys. Soc.* **13**, 722.
BOLLINGER, L. M., and THOMAS, G. E. (1968b). *Phys. Rev. Lette·s* **21**, 233.
BOLLINGER, L. M., and THOMAS, G. E. (1969). (To be published.)
BOLLINGER, L. M., COTÉ, R. E., and KENNETT, T. J. (1959). *Phys. Rev. Letters* **3**, 376.
BOLLINGER, L. M., COTÉ, R. E., CARPENTER, R. T., and MARION, J. P. (1963). *Phys. Rev.* **132**, 1640.
BOLLINGER, L. M., COTÉ, R. E., and JACKSON, H. E. (1964). *In* "Comptes Rendus du Congrès International de Physique Nucleàire, Paris, 1964" (Editions du Centre National de la Recherche Scientifique, Paris) Vol. II, p. 673.
BOSTROM, C. O., and DRAPER, J. E. (1961). *Rev. Sci. Instr.* **32**, 1024.
BRINK, D. M. (1955). Doctoral thesis, Oxford Univ., quoted by Kinsey, B. B., "Handbuch der Physik," Vol. XL, p. 316.
BROOKS, F. D., and BIRD, J. R. (1962). *In* "Neutron Physics," p. 109. Academic Press, New York.
CAMERON, A. G. W. (1959). *Can. J. Phys.* **37**, 322.
CARPENTER, R. T. (1962). Argonne Natl. Lab. Rept. ANL-6589.
CARPENTER, R. T., and BOLLINGER, L. M. (1960). *Nucl. Phys.* **21**, 66.
CHRIEN, R. E. (1969). *In* "Neutron Capture Gamma-Ray Spectroscopy," p. 627. Intern. Atomic Energy Agency, Vienna.
CHRIEN, R. E., BOLOTIN, H. H., and PALEVSKY, H. (1962). *Phys. Rev.* **127**, 1680.
CHRIEN, R. E., RANKOWITZ, S., and SPINRAD, R. J. (1964). *Rev. Sci. Instr.* **35**, 1150.
CHRIEN, R. E., BHAT, M. R., WASSON, O. A., and PRICE, D. L. (1967a). *Bull. Am. Phys. Soc.* **12**, 105.
CLEMENT, C. F. (1966). Slow-Neutron Capture Gamma-Ray Spectroscopy, Argonne Natl. Lab. Rept. ANL-7282, p. 417.
COCEVA, C., CORVI, F., GIACOBBE, P., and CARRARO, G. (1968). *Nucl. Phys.* **A117**, 586.
CORGE, C., HUYNH, V. D., JULIEN, J., MORGENSTERN, J., and NETTER, F. (1961). *In* "Neutron Time-of-Flight Methods" (J. Spaepeu, ed.), p. 167. Euratom, Brussels.
COTÉ, R. E., and BOLLINGER, L. M. (1961). *Phys. Rev. Letters* **6**, 695.
COTÉ, R. E., and PRESTWICH, W. V. (1966). Slow-Neutron Capture Gamma-Ray Spectroscopy, Argonne Natl. Lab. Rept. ANL-7282 p. 501.
DORCHOMAN, D., KARDON, B., KISH, D., and SAMOSVAT, G. (1964). *J. Exp. Theoret. Phys. (U.S.S.R.)* **46**, 1578.
DRAPER, J. E., and SPRINGER, T. E. (1960). *Nucl. Phys.* **16**, 27.
DRAPER, J., FENSTERMACHER, C., and SCHULTZ, H. L. (1958). *Phys. Rev.* **111**, 906.
FENSTERMACHER, C. A., BENNETT, R. G., WALTERS, A. E., BOCKELMAN, C. K., and SCHULTZ, H. L. (1957). *Phys. Rev.* **107**, 1650.
FENSTERMACHER, C., DRAPER, J., and BOCKELMAN, C. (1959). *Nucl. Phys.* **10**, 386.
FESHBACH, H., PORTER, C. E., and WEISSKOPF, V. F. (1954). *Phys. Rev.* **96**, 448.
FIRK, F. W. K., and GIBBONS, J. H. (1961). Neutron Time-of-Flight Methods, p. 213. European Atomic Energy Community, Brussels.
FOX, J. D., ZIMMERMAN, R. L., HUGHES, D. J., PALEVSKY, H., BRUSSEL, M. K., and CHRIEN, R. E. (1958). *Phys. Rev.* **110**, 1472.
GOLDBERG, M. D., MUGHABGHAB, S. F., PUROHIT, S. N., MAGURNO, B. A., and MAY, V. M. (1966). "Neutron Cross Sections," Vols. IIB and IIC. Brookhaven Natl. Lab. Rept. BNL 325, 2nd ed., Supp. No. 2.
GREENWOOD, R. C., HARLAN, R. A., HELMER, R. G., and REICH, C. W. (1969). *In* "Neutron Capture Gamma-Ray Spectroscopy," p. 607. Intern. Atomic Energy Agency, Vienna.
GROSHEV, L. V. (1960). *In* "Proceedings of the International Conference on Nuclear Structure" (D. A. Bromley and E. W. Vogt, eds.), p. 568. Univ. of Toronto Press, Toronto.

GROSHEV, L. V., and DEMIDOV, A. M. (1966). I. V. Kurchatov Inst. for Atomic Energy preprint IAE-1037.

GROSHEV, L. V., DEMIDOV, A. M., LUTSENKO, V. N., and PELEKHOV, V. I. (1958a). *Proc. U. N. Intern. Conf. Peaceful Uses At. Energy, 2nd, Geneva, 1958.* **15,** 138.

GROSHEV, L. V., DEMIDOV, A. M., LUTSENKO, V. N., and PELEKHOV, V. I. (1958b). "Atlas of γ-Ray Spectra from Radiative Capture of Thermal Neutrons" (English edition). Pergamon Press, London.

GROSHEV, L. V., DEMIDOV, A. M., and PELEKOV, V. I. (1960). *Nucl. Phys.* **16,** 645.

HARVEY, J. A., SLAUGHTER, G. G., BIRD, J. R., and CHAPMAN, G. T. (1963). Proceedings of the International Conference on Nuclear Physics with Reactor Neutrons, Argonne Natl. Lab. Rept. ANL-6797, p. 230.

HARVEY, J. A., MARTIN, M. J., and SLAUGHTER, G. G. (1966). Slow-Neutron Capture Gamma-Ray Spectroscopy, Argonne Natl. Lab. Rept. ANL-7282, p. 507.

HUGHES D. J. (1959). Brookhaven Natl. Lab. Rept. BNL-4464.

HUGHES, D. J., BRUSSEL, M. K., FOX, J. D., and ZIMMERMAN, R. L. (1959). *Phys. Rev. Letters* **2,** 505.

HUGHES, D. J., PALEVSKY, H., BOLOTIN, H. H., and CHRIEN, R. (1960). *In* "Proceedings of the International Conference on Nuclear Structure, Kingston" (D. A. Bromley and E. W. Vogt, eds.), p. 771. Univ. of Toronto Press, Toronto; North-Holland Publ., Amsterdam.

HUIZENGA, J. R., and VANDENBOSCH, R. (1960). *Phys. Rev.* **120,** 1305.

HUYNH, V. D. (1965). Thesis, Rept. CEA-R2810.

HUYNH, V. D., DE BARROS, S., JULIEN, J., LE POITTEVIN, G., MORGENSTERN, J., NETTER, F., and SAMOUR, C. (1966). *In* "Nuclear Structure Study with Neutrons" (M. Nève de Mévergnies, P. Van Assche, and J. Vervier, eds.), p. 529. North-Holland Publ., Amsterdam.

IGNAT'EV, K. G., KIRPICHNIKOV, I. V., and SUKHORUCHKIN, S. I. (1963). *J. Exptl. Theoret. Phys.* **45,** 875.

IGO, G., and LANDON, H. H. (1956). *Phys. Rev.* **101,** 727.

IKEGAMI, H., and EMERGY, G. T. (1964). *Phys. Rev. Letters* **13,** 26.

JACKSON, H. E. (1963). *Phys. Rev.* **131,** 2153.

JACKSON, H. E. (1964). *Phys. Rev.* **134,** B931.

JACKSON, H. E., and BOLLINGER, L. M. (1961). *Phys. Rev.* **124,** 1142.

JACKSON, H. E., JULIEN, J., SAMOUR, C., BLOCH, A., LOPATA, C., MORGENSTERN, J., MANN, H., and THOMAS, G. E. (1966a). *Phys. Rev. Letters* **17,** 656.

JACKSON, H. E., JULIEN, J., SAMOUR, C., CHEVILLON, P. L., MORGENSTERN, J., and NETTER, F. (1966b). *In* "Lithium-Drifted Germanium Detectors," p. 154. Intern. At. Energy Agency, Vienna.

JULIEN, J., CORGE, C., HUYNH, V. D., NETTER, F., and SIMIC, J. (1960). *J. Phys. Radium* **21,** 423.

KANE, W. R., and MARISCOTTI, M. A. (1967). *Nucl. Instr. Methods* **56,** 189.

KENDALL, M. G., and STUART, A. (1961). "Advanced Theory of Statistics," Vol. 2. Griffin, London.

KENNETT, T. J., BOLLINGER, L. M., and CARPENTER, R. T. (1958). *Phys. Rev. Letters* **1,** 76.

KINSEY, B. B. (1955). *In* "Beta- and Gamma-Ray Spectroscopy" (K. Siegbahn, ed.), p. 795. North-Holland Publ., Amsterdam.

KINSEY, B. B. (1957). "Handbuch der Physik," Vol. XL. Springer-Verlag, Berlin.

KINSEY, B. B., and BARTHOLOMEW, G. A. (1954). *Phys. Rev.* **93,** 1260.

LANDON, H. H., and SAILOR, V. L. (1955). *Phys. Rev.* **98,** 1267.

LANDON, H. H., and RAE, E. R. (1957). *Phys. Rev.* **107,** 1333.

LANE, A. M., and LYNN, J. E. (1960a). *Nucl. Phys.* **17,** 563.

LANE, A. M., and LYNN, J. E. (1960b). *Nucl. Phys.* **17**, 586.

LONE, M. A., CHRIEN, R. E., WASSON, O. A., BEER, M., BHAT, M. R., and MUETHER, M. R. (1968). *Phys. Rev.* **174**, 1512.

LYNN, J. E. (1967). "Theory of Neutron Resonance Reactions." (Clarendon), Press, Oxford, England.

MARTIN, M. J., HARVEY, J. A., and SLAUGHTER, G. G. (1966). *Bull. Am. Phys. Soc.* **11**, 336.

MARTIN, M. J., HARVEY, J. A., and SLAUGHTER, G. G. (1966). Slow-Neutron Capture Gamma-Ray Spectroscopy, Argonne Natl. Lab. Rept. ANL-7282, p. 14.

McNEILL, K. G., McCONNELL, D. B., and FIRK, F. W. K. (1965). *Can. J. Phys.* **43**, 2156.

MOTZ, H. T., and BÄCKSTROM, G. (1965). *In* "Alpha-, Beta-, and Gamma-Ray Spectroscopy" (K. Siegbahn, ed.), p. 769. North-Holland Publ., Amsterdam.

MOXON, M. C., and RAE, E. R. (1962). Private communication.

MUGHABGHAB, S. F., BHAT, M. R., WASSON, O. A., GARBER, D. I., and CHRIEN, R. E. (1968). *Bull. Am. Phys. Soc.* **13**, 721.

NAMENSON, A. I., and BOLOTIN, H. H. (1967). *Phys. Rev.* **158**, 1206.

NAMENSON, A. I., JACKSON, H. E., and SMITHER, R. K. (1966). *Phys. Rev.* **146**, 844.

POENITZ, W. P. (1966). *Z. Physik* **197**, 262.

PORTER, C. E., and THOMAS, R. G. (1956). *Phys. Rev.* **104**, 483.

PRESTWICH, W. V., and COTÉ, R. E. (1967a). *Phys. Rev.* **155**, 1223.

PRESTWICH, W. V., and COTÉ, R. E. (1967b). *Phys. Rev.* **160**, 1038.

PRESTWICH, W. V., and COTÉ, R. E. (1967c). *Phys. Rev.* **162**, 1112.

PRESTWICH, W. V., and COTÉ, R. E. (1967d). Private communication.

PRESTWICH, W. V., and COTÉ, R. E. (1968). *Phys. Rev.* **174**, 1421.

PRICE, D. L., CHRIEN, R. E., WASSON, O. A., BHAT, M. R., BEER, M., LONE, M. A., and GRAVES, R. (1968). *Nucl. Phys.* **A121**, 630.

RAE, E. R., MOYER, W. R., FULLWOOD, R. R., and ANDREWS, J. L. (1967). *Phys. Rev.* **155**, 1301.

ROCKWOOD, C. C., and STRAUSS, M. G. (1961). *Rev. Sci. Instr.* **32**, 1211.

ROELAND, L. W., BOLLINGER, L. M., and THOMAS, G. E. (1958). *Proc. Intern. U. N. Conf. Peaceful Uses At. Energy, 2nd, Geneva, 1968* **15**, 440.

ROSENZWEIG, N. (1963). *Phys. Letters* **6**, 123.

ROSENZWEIG, N. (1968). *Nucl. Phys.* **A118**, 650.

SAMOUR, C. R. M. (1968). Thesis, Paris.

SAMOUR, C., JACKSON, H. E., JULIEN, J., BLOCH, A., LOPATA, C., and MORGENSTERN, J. (1968). *Nucl. Phys.* **A121**, 65.

SEGEL, R. E., SMITHER, R. K., and CARPENTER, R. T. (1964). *Phys. Rev.* **133**, B583.

SLAUGHTER, G. G., MARTIN, M. J., and HARVEY, J. A. (1966). Slow-Neutron Capture Gamma-Ray Spectroscopy, Argonne Natl. Lab. Rept. ANL-7282, p. 290.

SPENCER, R. R., and FALER, K. T. (1967). *Phys. Rev.* **155**, 1368.

STARFELT, N. (1964). *Nucl. Phys.* **53**, 397.

STRUTINSKI, V. M., GROSHEV, L. V., and AKIMOVA, M. K. (1960). *Nucl. Phys.* **16**, 657.

THOMAS, G. E., and BOLLINGER, L. M. (1969). *Bull. Am. Phys. Soc.* **14**, 515.

THOMAS, G. E., BLATCHLEY, D. E., and BOLLINGER, L. M. (1967). *Nucl. Instr. Methods* **56**, 325.

URBANEC, J. (1965). *Zh. Esperim. i Theo. Fiz.* **49**, 80.

WASSON, O. A., and DRAPER, J. E. (1963). *Phys. Letters* **6**, 350.

WASSON, O. A., and DRAPER, J. E. (1965). *Phys. Rev.* **137**, B1175.

WASSON, O. A., BHAT, M. R., CHRIEN, R. E., LONE, M. A., and BEER, M. (1966). *Phys. Rev. Letters* **17,** 1220.

WASSON, O. A., CHRIEN, R. E., BHAT, M. R., LONE, M. A., and BEER, M. (1969). *Phys. Rev.* **173,** 1170.

WETZEL, K. J. and THOMAS, G. E. (1970). *Phys. Rev.* In press.

WETZEL, K. J., BOCKELMAN, C. K., and WASSON, O. A. (1967). *Nucl. Phys.* **92,** 696

WILKINSON, D. H. (1960). *In* "Nuclear Spectroscopy," Part B (F. Ajzenberg-Selove, ed.), p. 852. Academic Press, New York and London.

V

MEASUREMENTS ON FISSILE NUCLIDES

M. S. MOORE*

IDAHO NUCLEAR CORPORATION
IDAHO FALLS, IDAHO

* Present Address: Los Alamos Scientific Laboratory, Los Alamos, New Mexico.

I. INTRODUCTION

A. Development of Fission Theory

1. THE LIQUID-DROP MODEL

Almost immediately after the discovery of fission, by Hahn and Strassmann (1939) and by Meitner and Frisch (1939), there was an intensive experimental effort made which revealed many of the properties of fission as we know them today. The stage was set for the first of three early monumental papers which treated the mechanism of nuclear fission in analogy with the fission of a liquid drop. Bohr and Wheeler (1939) derived expressions for the change in potential energy of a uniformly charged incompressible fluid drop in which the attractive nuclear forces, saturated in the interior of the nucleus, have the effect at the surface of a surface tension which balances the repulsive coulomb forces. Deformations of the liquid drop were treated by a spherical harmonic expansion about a spherical equilibrium state, as

$$r(\theta) = R[1 + \alpha_0 + \alpha_2 P_2 (\cos \theta) + \alpha_3 P_3 (\cos \theta) + \cdot \cdot \cdot]. \tag{1}$$

In calculating the surface energy plus the coulomb energy of the distorted drop, Bohr and Wheeler showed that the coefficient of $\alpha_2{}^2$, on which the stability of the drop depends, varies linearly as Z^2/A, where Z is the charge on the drop and A is the mass. It is thus natural to define as a fissionability parameter, the ratio

$$x = (Z^2/A)/(Z^2/A)_{\text{crit}}, \tag{2}$$

where $(Z^2/A)_{\text{crit}}$ is that value for which the drop is unstable for any infinitesimal deformation. Bohr and Wheeler concluded that heavy nuclei, which are fissionable with about 6 MeV of excitation (the binding energy of one added neutron), correspond to fissionability parameters of about 0.74. In this range, the excitation sets up oscillations which can eventually lead to fission of the drop into two or more fragments.

Bohr and Wheeler also considered the statistical mechanics of the fission process in order to arrive at an estimate of the fission width. They concluded that

$$\Gamma_f \approx (D/2\pi)N, \tag{3}$$

where Γ_f is the fission width, D is the level spacing in the compound nucleus, and N is the "number of levels in the transition state available with the given excitation." In presently accepted terminology, the number N is to be identified with the number of fission channels available.

The concepts first formulated by Bohr and Wheeler have proved very useful. Of particular interest are detailed calculations on the mechanism of fission of a charged liquid drop, first done by Frankel and Metropolis (1947). Perhaps the most extensive of these calculations have been made by Swiatecki and co-workers. (See, for example, Swiatecki, 1956a,b, 1958; Cohen and Swiatecki, 1962; Nix and Swiatecki, 1965; Nix, 1966.) As illustrated in Fig. 1, the potential energy of the drop can be conveniently expressed as a function of the normal coordinates of motion, the deformation parameters α_n in the spherical harmonic expansion. Only the α_2 and α_4 coordinates are shown in Fig. 1, but in general the contour maps would be multidimensional. In Fig. 1, the spheroidal equilibrium shape of the drop is represented by the potential well near the origin. The valley represents the potential energy hollow which leads to division of the drop into two fragments. The saddle point is shown for such division; classically, the excitation energy of the drop must exceed the saddle point before fission can occur. The division illustrated in Fig. 1 would lead to symmetric fission of the liquid drop since it involves even values of α_n. Asymmetric fission occurs for saddle points along odd α_n coordinates.

More recent studies have been reviewed by Swiatecki (1965). Some of these amount to mathematical experiments carried out with digital computers. They have not only clarified present understanding of the static deformation energy of the charged liquid drop but also have given considerable insight into the dynamics and kinematics of liquid-drop fission. In view of the successes of these studies in describing fission of nuclei, it is too easy to forget that the liquid-drop model of nuclear fission is no more than a model, and it cannot be expected to describe those aspects of nuclear fission in which, say, shell structure plays an important part. The limita-

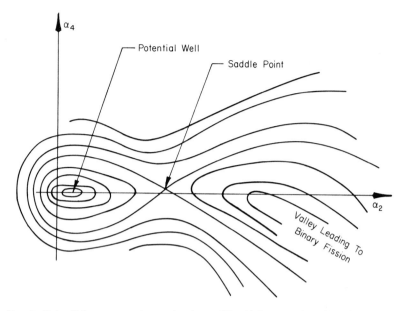

FIG. 1. Potential energy contours of a charged liquid drop as a function of deformation parameters α_2 and α_4, after Swiatecki (1958). The equilibrium configuration is represented by the potential well near the origin. If the excitation energy is in excess of the saddle point, fission of the drop into two (symmetric) fragments is permitted.

tions of the liquid-drop model have not really been explored. In this connection, it would be desirable to make further experimental studies of the fission of charged liquid drops, such as those of Ryce and Wyman (1964) (see also Ryce, 1966). Such studies might serve to bridge the gap between the theory of fission of a charged liquid drop and the observed characteristics of fission of heavy nuclei.

2. THE COLLECTIVE MODEL

A second paper which contributed significantly to modern ideas about fission was that of Hill and Wheeler (1953), who attempted to utilize the collective model (Bohr, 1952; Bohr and Mottelson, 1953) to obtain a more nearly complete description of nuclear fission than that based on the liquid-drop model. Hill and Wheeler pointed out that a quantum description of the nuclear deformation potential should include many sheets similar to the surface shown in Fig. 1. Whenever the sheets cross, (which occurs when the deformations are symmetric), crossover or slippage can occur from one potential sheet to another. The deformation to the saddle-point configuration that involves an exchange of energy between nuclear excitation and collective vibration by "radiationless transition" or slippage from one sheet to another.

Hill and Wheeler also introduced a quantitative estimate of the energy dependence of the fission cross section at threshold (in the region of excitation near which classical crossing of the saddle point is possible). The fission barrier consists of coulomb repulsion modified by nuclear forces. Hill and Wheeler assumed that at energies near the top the barrier shape resembles an inverted harmonic oscillator potential. The probability of crossing such a barrier is given by

$$P = \left[1 + \exp \frac{2\pi(E_b - E)}{\hbar\omega} \right]^{-1}, \tag{4}$$

where E and E_b are the nuclear excitation energy and the barrier height, respectively, and $\hbar\omega$ is a constant determined by the curvature of the barrier. Hill and Wheeler estimated this constant to be about 0.8 MeV. This expression can be combined with Eq. (3), to give an estimate of the variation of the fission "strength function" Γ_f/D with excitation energy. This can in turn be related to the fission cross section. The average fission width for a single channel is usually assumed to vary as

$$\langle\Gamma_f\rangle = \frac{D}{2\pi} \left\{ 1 + \exp\left[\frac{2\pi(E_b - E)}{\hbar\omega} \right] \right\}^{-1}. \tag{5}$$

3. The Channel Theory of Fission

The third and perhaps most far-reaching of the early ideas on fission was introduced by Aage Bohr at the time of the first Geneva Conference (Bohr, 1956). The concept is a simple one, yet quantitative enough to be useful in a number of ways. It states that for excitations near threshold, the nucleus in passing over the saddle point is "cold" or not highly excited. Most of the excitation energy has become bound up in potential energy of deformation, and the quantum states representing the motion are widely separated, being restricted to low-energy vibrations and rotations similar to those which the compound system would exhibit near its ground state. Bohr suggested (i) that the nuclear angular momentum at the saddle point should be concentrated on a collective rotation, leading to a characteristic angular distribution of the fragments, and (ii) that the differences in the fission threshold for states of different spin and parity in the compound nucleus could lead to peculiar selection rules in the fission process.

Bohr investigated the systematics of the "quantum states at the saddle point" (transition states or fission channels) by considering them in analogy to low-lying level structure of stable deformed nuclei. Bohr assumed that the nuclear shape at the saddle point is axially symmetric so that the quantum number K, representing the component of nuclear angular momentum on the nuclear symmetry axis, could be used to characterize the channel. Bohr considered the fragment angular distribution from photofission and from fast neutron induced fission, and he showed that the observed anisotropies are in both cases correctly described by the model.

Of particular interest to low-energy neutron spectroscopy are Bohr's comments concerning slow neutron induced fission:

> Slow neutron capture by an even–odd nucleus of spin I_0 leads to a compound nucleus of even–even type with spin $I = I_0 \pm \frac{1}{2}$, and with same parity π as the target nucleus.
>
> As follows from the above discussion of the spectrum at saddle point, only one of these spin–parity combinations is contained in the rotational band associated with the lowest nucleonic configuration ($K = 0$), viz. the state with $(-1)^I = \pi$.
>
> Therefore, the fission threshold is expected to differ appreciably, by as much as a million electron volts, for the two types of compound levels formed. This may imply a rather different ratio of fission to capture for the two level systems, especially if the neutron binding energy is smaller than the larger of the threshold values.
>
> The circumstance that slow neutron fission is expected to proceed primarily through a definite quantum state at saddle point also implies large fluctuations in the fission width among the levels of same I. The fluctuations should be similar to those observed for the neutron width, which likewise depends on the presence in the compound nucleus wave function of a definite component (representing elastic neutron scattering).
>
> The parity of the target nucleus may further affect the mass distribution of the fission fragments. Thus for the levels with $(-1)^I = \pi$, the nuclear wave function is symmetric or antisymmetric in the asymmetry coordinate, for $\pi = +1$ and -1, respectively. In the latter case, symmetric fission is expected to be inhibited, as in the case of the photofission (cf. above).
>
> If slow neutron fission of aligned nuclei could be studied, large angular anistropies of the fragments would be expected. For the states with $(-1)^I = \pi$, which pass through the saddle point with $K = 0$, the distribution is directly related to the distribution of M-values of the compound states. For the other set of levels with $(-1)^I = -\pi$, the distribution in addition depends on the K-value of the fissioning nucleus.
>
> Some even–even target nuclei have also been observed to undergo slow neutron fission. For such targets the compound states all have $I = \frac{1}{2}$ and even parity, and the fission threshold depends on the energy of the lowest $K = \frac{1}{2}$ configuration at saddle point. If the intrinsic parity of this configuration is odd, symmetric fission should again be inhibited.

Since Bohr first proposed the channel theory of fission, there have been a number of reviews of nuclear-fission theory and experiment. One of the most useful and comprehensive is that of Hyde (1964) and also see, Hyde (1960a,b). Reviews, primarily of the developments of fission theory, have been made by Halpern (1959), Huizenga and Vandenbosch (1962), Wheeler (1963), and Wilets (1964). At the Salzburg Symposium on the Physics and Chemistry of Fission (held by the International Atomic Energy Agency in 1965), many new and interesting experiments were reported. The contributions to this conference were discussed as part of the complete body of information on nuclear fission in two excellent reviews by Fraser and Milton (1966) and by Gindler and Huizenga (1968).

The scope of the present review is somewhat more restricted. It includes only those aspects of neutron-induced fission which can be expected to depend on the energy of the incident neutron. An attempt has been made to summarize the experimental measurements and theoretical developments up to the present time in order to show to what extent the channel theory of Bohr can be used in describing the fission of nuclei by low energy neutrons. The specific questions to be considered are the following: (i) Does the average fission width, or the capture to fission ratio, depend on the spin of the compound nucleus resonances? (ii) Does the fragment mass distribution depend on spin or parity of the resonances? (iii) Are there other phenomena which appear to depend on resonance spin, parity, or other fission-channel quantum numbers such as K?

B. Fission as a Multistage Process

Of basic importance to the theoretical treatment of fission is the assumption that the fission process takes place in sequential steps or stages. The stages in fission, while they are not strictly independent, are convenient to use as a series of real or artificial boundaries separating the various aspects of the fission process in both time and space.

For low-energy neutron-induced fission, the first of these stages is almost always the formation of the compound nucleus where the excitation energy afforded by the binding energy of the original incident neutron is shared among all the nucleons. It is with this stage that certain observed resonance properties such as neutron widths, spins, and resonance spacings are connected.

The second stage is the crossing of the saddle point (see Fig. 1), by way of one or a few simple modes of motion, which might be designated as fission channels. The time required for the fissioning nucleus to deform to the saddle point can be estimated from the observed sizes of the fission widths of the resonances as about 10^{-14} to 10^{-12} sec. This time is probably long compared to the time of descent down the other side of the barrier. The average fission width is assumed to be related to the fission channel structure at the saddle point, as are the fluctuations in the widths and the existence of interference among the fission resonances. The angular distributions of the fission fragments can also be related to the channel properties at the saddle point.

The next stage is the transition from the saddle point to scission. Scission is defined as that point beyond which the nuclear interaction between the fragments vanishes. It follows that when the scission point has been reached, the fission-fragment mass and charge distributions have been established. The time required for the transition from the saddle point to scission remains an open question. The qualitative success of the statistical model of Fong (1956) in explaining fission mass distributions shows that over at least part of the way, the same sort of slippage occurs as in the approach to the saddle point from the compound nucleus configuration.

After scission the nascent fission fragments separate under the influence of long range coulomb forces. The actual process of scission may involve a violent snapping action, permitting the release of some of the prompt neutrons. Sometimes a light charged nucleus, most probably an α-particle, is released between the two heavy fragments at scission, in what is known as a ternary-fission event. It often is assumed that at scission the nascent fragments have a negligible amount of kinetic energy so that all the experimentally observed fragment kinetic energy is a result of coulomb repulsion. As the fragments separate, they reorient themselves from the presumably highly deformed configuration at the scission point, and emit most or all of the prompt neutrons and prompt fission-gamma and conversion-electron radiation. The time required for prompt neutron and gamma emission can be estimated as about 10^{-14} sec since the prompt gammas, thought to be mostly fast E2 transitions, compete to some extent with prompt neutron emission.

Finally, the neutron-rich fragments undergo sequential beta decays (occasionally releasing delayed neutrons), accompanied by delayed gamma and conversion-electron emission, and by characteristic X rays. The time scale of this process is much longer than any other, ranging from a lower limit of 10^{-3} sec upward.

One notes that the correlation suggested by Bohr (1956), that of resonance spin, fission width, and mass distribution, covers almost the entire sequence of stages in fission. Thus low-energy neutron spectroscopy of the properties of the neutron resonances might be an especially fruitful approach to understanding the fission process. The organization of the material to be presented here follows, as closely as is practicable, the stages in the fission process. In Section II, a review is given of the experimental methods which have been found useful for measurements on fissile nuclides. In Section III, properties of resonances are considered in detail, including the determination of cross sections and cross-section ratios and resonance parameters, including spins. The treatment of angular distributions of fission fragments and of other experiments which give information on the channel spectrum at the saddle point is given in Section IV. Experiments which appear to be related to the properties of the scission point are described in Section V. These include studies of fission-fragment mass and kinetic-energy distributions, of prompt neutron-emission and ternary-fission probabilities, and of prompt gamma and electron emission. In Section VI, the question of a spin dependence of the mass distribution and of the fission-width distribution is considered in the light of the experimental evidence presented.

II. EXPERIMENTAL METHODS USED IN LOW-ENERGY NEUTRON SPECTROSCOPY OF FISSILE NUCLIDES

A. Objectives of Experimental Measurements

For nonfissile nuclides, low-energy neutron spectroscopy includes measurements of total, scattering, and radiative-capture cross sections, as well as studies of individual gamma transitions as functions of neutron energy. The existence of fission as an important process leads to a wealth of further possible experiments. Fission into two or more fragments is accompanied by prompt and delayed fission neutrons, gamma and conversion-electron radiation, as well as by postfission beta decay. At higher energies where neutrons of angular momentum greater than zero begin to contribute, or if aligned targets of fissile material are used, angular distributions of the fragments are of interest. All of these and their relations to one another can be studied in detail as a function of incident neutron energy.

Even though there is a large number of possible experiments which are appropriate to neutron spectroscopy of fissile nuclides, the techniques for carrying out such experiments are not very different from those which have become standard for low-energy neutron spectroscopy. There is one exception: the detection of fission fragments. Measurements of total and scattering cross sections of the fissile nuclides are carried out by the techniques used for nonfissile nuclides, with little or no modification. The measurement of radiative capture is more difficult since one must be able to differentiate between capture and fission radiation generally by relying on a coincidence between the gamma ray and the fission fragment. In any of the other complex experiments which can be devised to give information on fission, detection of the fragment or of the prompt neutrons emitted by the fragment is invariably required.

Perhaps the simplest measurement is that of the fission cross section as a function of neutron energy. This measurement, that illustrates many of the methods of more complex experiments, involves the determination of the number of fission events which occur in a sample of known thickness in a known neutron flux. The customary way of determining the neutron flux is by observing the rate of a reaction which is assumed to be known, such as the charged particles from the $^{10}B(n,\alpha)$, $^{6}Li(n,\alpha)$, or $^{3}He(n,p)$ reactions. The cross sections for these standard reactions are reasonably well known for neutron energies near 0.025 eV. In the energy region of a few tens or hundreds of electron volts, the dependence of the cross section on neutron energy is very well given by a $1/v$-functional form, where v is the neutron velocity. It should be pointed out, however, that at higher energies uncertainties in the standard cross sections and thus in the neutron flux can become a limiting factor in the accuracy of fission cross-section measurements.

Two methods have been used to determine the number of fission events: fission-fragment counting and fission-neutron counting. The gaseous ionization chamber is still a popular device for fission-fragment detection. The gaseous scintillation chamber is a natural extension of the ionization chamber, in which the light emitted in optical transitions is detected by means of a photomultiplier. Both these detectors were reviewed by Lamphere in 1960. Since that time, the solid state detector has been developed, and has contributed significantly to the field. The spark chamber has been used for detection of fission fragments from highly radioactive samples, since its response to ionizing radiations other than fission fragments can be made effectively zero. The same is true of fission-track detectors, thin sheets of mica or other material which can be etched after radiation damage by heavy charged particles. Fission neutron detection has been frequently done by means of large volume organic scintillators, detecting either recoil protons scattered by the fast neutrons, or neutron capture-gamma radiation from a high-cross-section additive to the scintillant.

Analysis of a measurement of the energy dependence of the total and partial cross sections for a given nuclide gives information on a number of properties of the resolved resonances in the compound nucleus: the neutron, fission, and radiative-capture widths, the resonance spacings, and, in principle, the resonance spins. Fission-fragment detection with high resolution of the fragment energy and velocity will give mass and kinetic-energy distributions. Angular distributions of the fragments, emitted either from fission of aligned nuclei or from fission induced by other than s-wave neutrons, give information on saddle-point states. Coincidence measurements of fission fragments with prompt neutrons, gammas, and conversion electrons permits one to distinguish between capture and fission events and to study the appropriate emitted radiations.

In addition to experimental measurements designed to reveal as much as possible about a given fissioning nuclide, the systematics of fission are of interest. One should like to be able to make such measurements on as great a variety of samples as possible. In this connection, the techniques of preparing samples of rare heavy isotopes are important. Until recently, there were only three fissile nuclides that were common enough that samples for neutron spectroscopy were readily available. These nuclides are: ^{233}U, ^{235}U, and ^{239}Pu. Less common nuclides for which low-energy neutron spectroscopy has been reported are: ^{229}Th, ^{232}U, ^{238}Pu, ^{241}Pu, and ^{242m}Am, which are available in only limited quantities. In the near future, it is expected that the list may be extended to include many other transplutonic nuclides. Nevertheless, one might well ask, "Why is this list so short?" The answer to this question is found in the restrictions imposed by the techniques of neutron spectroscopy. With the exception of beams from nuclear explosions, neutron beam intensities are so low that reasonably large amounts of material (ranging from a few milligrams to several grams) are required for use as samples. Many of the heavy nuclei which are fissile have half-lives which are too short to permit the preparation of adequate samples.

B. Fission-Fragment Detection

1. THE GASEOUS IONIZATION CHAMBER

The gaseous ionization chamber as a general utility instrument in nuclear physics has been discussed exhaustively (see, for example, Rossi and Staub, 1949; Wilkinson, 1950; Staub, 1953; Fulbright, 1958; Curran and Wilson, 1965). For details of the theory and operation of such chambers the interested reader is referred to the existing literature. The following discussion will be limited to problems peculiar to neutron spectroscopy of fissile nuclides, i.e., to detection of fragments of fission usually induced by neutron beams of relatively low intensity. A review of a similar nature was made by Lamphere (1960).

A successful fission-fragment counter for use with such neutron beams must have a low or negligible efficiency for the counting of products of natural radioactivity of fissile nuclides. In particular, even for ^{233}U and ^{239}Pu, the number of α-particle pulses from decay is many orders of magnitude higher than the number of fission-fragment pulses produced by neutron-spectrometer beams. Although the alpha pulses are much smaller than fission-fragment pulses, gaseous ionization chambers are so slow (having a pulse-rise time of perhaps 0.25 μsec) that alpha-pulse pileup can create a serious background problem. The amount of fissile material which can be used in a given chamber is thus limited to $\sim$100 mg for ^{233}U and $\sim$30 mg for ^{239}Pu. Advantage can be taken of the relative ionization rates of fission fragments and α-particles in order to reduce the background effects of the latter. Figure 2 shows the specific ionization of a typical light and heavy fission fragment (Lassen, 1949) and of an α-particle of representative energy, as a function of range in argon, as calculated from the work of Holloway and Livingston (1938), corrected to argon according to Whaling (1958). The shape of the curves for fission fragments is markedly different from that for the α-particle. These curves suggest that a parallel plate ionization chamber, designed to detect fission fragments in a high alpha field, should have plates which are separated by no more than $2.5/p$ cm, where p is the absolute pressure in atmospheres. If discrimination against background is the only consideration, a smaller separation may be preferred.

The differences in the theoretical treatment of the slowing down of α-particles and fission fragments have been highlighted in reviews by Fano (1963) and Northcliffe (1963). Over most of its range, the α-particle is stripped of electrons, and the energy loss per unit of path length is described by the Bethe stopping relation (Bethe, 1950) (see also Livingston and Bethe, 1937; Bethe and Ashkin, 1953). Atomic electrons are much more tightly bound to fission fragments, and the stopping is expected to follow the theoretical treatment of Lindhard et al. (1963a,b) and Lindhard and Winther (1964).

Lindhard, following the approach outlined by Bohr (1948) (see also Bohr and Lindhard, 1954), separates the stopping cross section for heavy ions in matter into two components: a nuclear-stopping cross section which varies comparatively slowly with particle energy and for high energies and large deflections has the form of Rutherford scattering, and an electronic-stopping cross section which describes the energy loss by ionization and excitation of atoms in the stopping material due to the passage of the heavy ions. Over most of the range, the stopping is dominated by the electronic cross section. Lindhard assumes that the effective charge of the fragment is determined by its velocity: the fragment keeps only those electrons whose orbital velocity is greater than the velocity of the fragment. Lindhard then uses the Thomas–Fermi model of the atom to evaluate the effective charge,

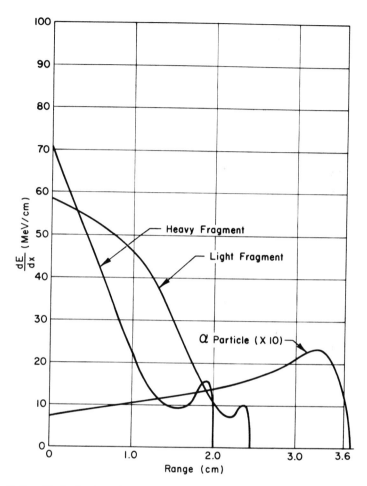

Fig. 2. Specific ionization of typical fission fragments, after Lassen (1949), and of a 5 MeV α-particle as a function of range, in argon at STP. The α-particle curve, which lies a factor of 10 lower than shown, has been converted from measurements by Holloway and Livingston (1938) of α-particles in air, according to ratios of relative stopping powers of argon and air given by Whaling (1958).

finding it to be proportional to the fragment velocity. (See also Lassen, 1951; Bell, 1953.)

In his theoretical treatment, Lindhard introduces the dimensionless energy and range parameters ϵ and ρ, defined as

$$\epsilon = \frac{EaM_2}{Z_1 Z_2 e^2 (M_1 + M_2)} \tag{6}$$

and

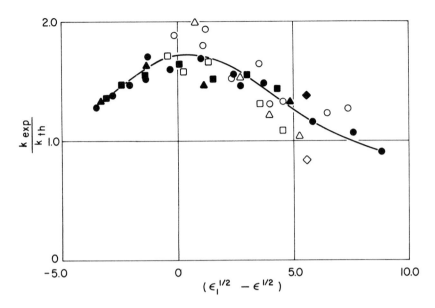

FIG. 3. The ratio of the measured proportionality factor $k_{\exp} = (d\epsilon/d\rho)_e/\epsilon^{1/2}$ [defined in Eq. (9)] to the theoretical proportionality constant k_{th} as defined in Eq. (10). The deviation of the experimental and theoretical proportionality factors is attributed to electronic shell effects in the fragment. The critical velocity $\epsilon_1^{1/2}$ corresponds to the velocity of the M_1- or N_1-shell electron in the median light or heavy fragment, respectively (Moore and Miller, 1967). Argon: light ●, heavy ○; krypton: light ▲, heavy △; xenon: light ■, heavy □; nickel: light ◆, heavy ◇.

$$\rho = RNM_2(4\pi a^2)\,\frac{M_1}{(M_1 + M_2)^2},\tag{7}$$

where E is the particle energy, and R is the range, N is the number of atoms of stopping material per unit volume, Z_1 and Z_2 are the nuclear charges of the incident particle and of the atoms of the stopping material, respectively, and M_1 and M_2 are the respective atomic masses. The parameter a is a Thomas–Fermi penetration length, equal to $0.8853a_0(Z_1^{2/3} + Z_2^{2/3})^{-1/2}$, where $a_0 = \hbar^2/me^2$, the Bohr radius in hydrogen. In this notation, the rate of energy loss of the fragment is given by

$$d\epsilon/d\rho = (d\epsilon/d\rho)_e + (d\epsilon/d\rho)_n,\tag{8}$$

where the subscripts e and n refer to electronic and nuclear stopping, respectively.

If the effective charge on the fragment is proportional to the velocity, the electronic-stopping cross section is given by

$$(d\epsilon/d\rho)_e = k\epsilon^{1/2}, \tag{9}$$

where the proportionality constant k can be estimated as

$$k \approx \frac{0.0793 Z_1^{(1/6)} (Z_1 \cdot Z_2)^{(1/2)} (M_1 + M_2)^{3/2}}{(Z_1^{(2/3)} + Z_2^{(2/3)})^{(3/4)} M_1^{(3/2)} M_2^{(1/2)}}. \tag{10}$$

The ratio of the energy loss due to nuclear collisions to that due to electronic collisions is estimated to be inversely proportional to the square of the effective charge on the fragment (Bethe and Ashkin, 1953). The nuclear stopping is thus inversely proportional to $E^{1/2}$, and becomes an important effect only near the end of the range where E is small. Integration of Eq. (8) with the appropriate normalization gives the following expression:

$$\rho(\epsilon) = (2/k)[\epsilon^{1/2} - \Delta(k,\epsilon)], \tag{11}$$

where $\Delta(k,\epsilon)$ represents a range correction due to nuclear scattering approximately given by

$$\Delta(k,\epsilon) = (\pi/10k)^{1/2} \tan^{-1}[(10k\epsilon/\pi)^{1/2}]. \tag{12}$$

Experimental measurements have been made which indicate that the Lindhard treatment may not be completely satisfactory for the slowing of fission fragments (Noshkin, 1963; Aras et al., 1965; Mulás and Axtmann, 1966). Figure 3 shows a comparison of an experimentally determined proportionality constant of range and energy, compared to the theoretical value calculated from Eq. (10), found by Moore and Miller (1967). The discrepancy between theory and experiment is attributed to electronic shell structure of the fragment, which is not treated by the Thomas–Fermi model of the atom. (See also, Bergström and Domeij, 1966.) Electronic shell effects have also been observed by Moak et al. (1967) in the charge distributions of slowed heavy ions of ^{79}Br and ^{127}I in the energy region appropriate to fission fragments.

Satisfactory results in calculating range-energy relations for fission fragments in heavy gaseous absorbers have been obtained by using the Lindhard formalism as outlined above, correcting the proportionality constant k given in Eq. (10) as shown in Fig. 3.

In a gaseous ionization chamber used for detecting electrons, the total ionization has been found to be proportional to the electron energy (Curran et al., 1949). For heavy charged particles, this is no longer true; the total ionization I is related to the particle energy E as

$$E = wI + \Delta \tag{13}$$

where w is the energy loss per ion pair, which has been measured by Jesse and Sadauskis (1955) to be the same for electrons and α-particles in argon.

The quantity Δ, sometimes called the ionization defect, was measured by Cranshaw and Harvey (1948) to be approximately 85 keV for α-particles in argon. For fission fragments in argon, Schmitt and Leachman (1956) showed that the quantity Δ ranges from 5 to 10 MeV. The ionization defect is attributed to energy loss by the fragment in nonionizing collisions at low fragment velocities near the end of the range. These are in the region where the nuclear stopping is dominant, and where ionization of the medium by electronic stopping has virtually ceased. For fission fragments, the proportionality constant w in Eq. (13) was found by Schmitt and Leachman (1956) to be a few percent larger for heavy than for light fragments. This can possibly be attributed to differences in the relative electronic and nuclear stopping of energetic recoiling atoms of the stopping medium.

The efficiency of a gaseous ionization chamber approaches 100% since the deposit of fissile material is inside the chamber and the fragments are emitted back to back. Because of the high initial specific ionization of the fragments, the foils of fissile material must be extremely thin ($\lesssim 100$ $\mu g/cm^2$), especially if pulse height information or absolute fission counting is desired. On the other hand, resolved neutron beam intensities are usually rather low so that in order to attain adequate counting statistics in a reasonable length of time, a larger amount of material is needed. A possible compromise is a multiple-foil chamber, the pulses from each independent section being mixed after background discrimination. If only relative fission cross sections are needed, or if the efficiency of the chamber is to be determined independently, then it may be possible to use fairly thick fission foils in the interest of improving the counting rate. Foil thicknesses up to 1 mg/cm² have been used (Havens et al., 1959). The counter efficiency in this case can be estimated to be about 85%. For ^{233}U, where the bias point must be set considerably higher because of alpha pileup, Weston et al. (1967) used a chamber with foil thickness of 0.5 mg/cm², which had an efficiency of 61%. Even this value was only achieved by also requiring fission-fragment gamma-ray coincidences.

With a fission-fragment detector which has high energy resolution, one can study fission-fragment mass distributions by analyzing the fragment kinetic energies. For binary fission induced by low-energy neutrons, the masses of the fragments are very simply related to the kinetic energies (before fission-neutron emission) by the equation

$$M_1E_1 = M_2E_2, \tag{14}$$

where M_1 and M_2 are the masses of the two fragments before prompt neutron emission, and E_1 and E_2 are their respective kinetic energies. The sum $M_1 + M_2$ is approximately equal to a constant (the mass of the original

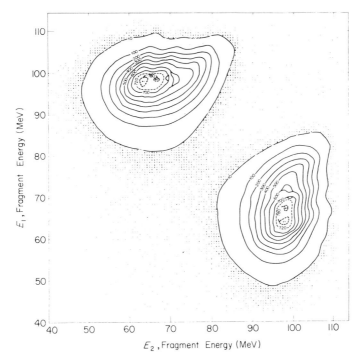

FIG. 4. Typical spectrum of correlated fission-fragment pulse heights for thermal-neutron-induced fission of ^{235}U, as obtained by Williams *et al.* (1964). The analyzer matrix contains 128 × 128 channels; the energy scales are approximate.

fissioning compound nucleus), but the sum $E_1 + E_2$ is not. Most of the potential energy of the nascent fragment configuration at the scission point is eventually transformed into fragment kinetic energy, which is found to vary between ~130 and 200 MeV for the slow neutron fission of ^{235}U. In order to determine mass distributions from pulse-height analysis of the fragment kinetic energies, it is necessary to record both E_1 and E_2 as a single event in a two dimensional pulse-height analyzer. A contour plot of such data, obtained by Williams *et al.* (1964) with solid state detectors, is shown in Fig. 4. Prompt fission-neutron emission, which is known to occur mostly after the fragments are well separated and have attained their final kinetic energies (Terrell, 1959), tends both to broaden the kinetic-energy spectra and to cause a disproportionate shifting of the distribution. Corrections for this effect require a knowledge of the number of prompt neutrons emitted by each fragment and, ideally, of the energies of these neutrons. (Britt *et al.*, 1963a.)

A conceptually simpler approach is the measurement of the fission-fragment velocities by time-of-flight methods (Milton and Fraser, 1958, 1961, 1962; Stein and Whetstone, 1958; Fraser et al., 1963; Whetstone, 1963, 1964). Here, if one can assume that the prompt fission neutrons are evaporated isotropically from the moving fragment, there is essentially no shift in the average velocity of the fragment before and after neutron emission (Stein, 1957). There is about 1% broadening in the fragment-velocity distribution. A more elaborate experiment is the simultaneous measurement of combinations of energies and velocities of the coincident fragments (Britt and Whetstone, 1964; Schmitt et al., 1965c; Andritsopoulos, 1967).

The gaseous ionization chamber is almost never the optimum detector for such complex measurements, however. The parallel plate chamber shows poor pulse-height-resolution characteristics because the pulse height depends upon the position in the chamber at which the ions are formed. The introduction of a third electrode, often called the Frisch grid (Frisch, 1949), is a method of attaining much higher resolution. The pulse is obtained as the electrons travel from the grid to the collection plate. Thus the grid serves as the virtual source of all the electrons, and the pulse height is proportional to total ionization. Gridded ionization chambers are capable of attaining resolutions (full-width at half-maximum) of 1% for α-particles (Bunemann et al., 1949). For high-resolution fission-fragment detection, a gridded ionization chamber was used by Roeland et al. (1958) in studies of fission-product mass distributions by pulse-height analysis. A typical pulse-height distribution for ^{235}U fission fragments in a Frisch grid chamber is shown in Fig. 5a (Roeland, 1958). In recent years, the gridded ionization chamber has been almost completely supplanted by the more convenient solid state detector.

In high-resolution fission-fragment studies, the thickness of the target of fissile material is severely restricted because of the high initial specific ionization of the fragments (Fig. 2). Furthermore, since both fragments from a given event are to be detected with high resolution, the thickness of the backing material must also be kept to a minimum. Backings can be obtained which are ~ 5 $\mu g/cm^2$ in thickness; on such backings, thicknesses of ~ 10 $\mu g/cm^2$ of fissile material appear to give reasonably good results.

2. The Gaseous Scintillation Counter

Nonionizing processes such as optical transitions in the gas of an ionization chamber account for about 20–40% of the energy loss of an ionizing particle (Northrop et al., 1958; see also Fano, 1946). For noble gases the light emitted is mostly in the ultraviolet. With an ultraviolet-sensitive

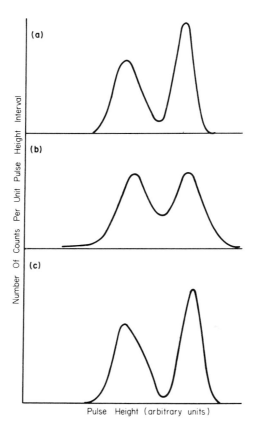

Fig. 5. Comparison of typical resolution properties of various detectors for fission fragments from (^{235}U + n). The criterion for resolution in fission fragment counting is most often taken to be the ratio of the light-fragment peak to valley. Curves in (a) Frisch grid chamber, (b) gas scintillation chamber, and (c) solid state detector were obtained by Roeland (1958), Huddleston (1958), and Deruytter *et al.* (1962), respectively. The unpublished data from which (c) was drawn were supplied by Dr. Deruytter.

photomultiplier, or with the use of a wavelength shifter (a fast fluorescent material coating the chamber) matched to the sensitive region of a conventional photomultiplier, the scintillations can readily be counted (Eggler and Huddleston, 1956; Northrop and Nobles, 1956; see also reviews by Huddleston, 1960; Lamphere, 1960). The heavier noble gases give larger pulse heights, and are usually preferred for that reason. Nobles (1956) found that the relative pulse heights observed for α-particles in argon, krypton, and xenon vary in the ratios 1:2.8:10, and that a gain of about a factor of 15 could be obtained with *p*-quaterphenyl as a wavelength shifter. Other common wavelength shifters are sodium salicylate, tetraphenyl-butadiene, and diphenyl-stilbene. It is found that the ratio of light obtained from one gas to another is almost a constant for the various phosphors, and that the ratio of yield from one phosphor to another is approximately the same for the various gases (Northrop *et al.*, 1958). Organic materials as wavelength shifters have certain disadvantages. When used in a static system they appear to have a poisoning effect, causing a noticeable decrease in pulse height over a few hours, and some wavelength shifters show a pronounced dependence of the conversion efficiency on temperature (Nobles, 1956).

Probably the most systematic studies of the fluorescence of rare gases excited by nuclear charged particles have been carried out by Koch (1960) and by Bennett (1962). Much of this work has been included in a review by Birks (1964). Koch's investigations of the ultraviolet-emission spectra of spectroscopically pure rare gases as commonly used in gas-scintillation fission detectors indicated that the predominant light in all cases is due to a nitrogen contaminant, even when it is present in only very minute concentrations (≤ 10 ppm). Koch found that xenon and (to a much lesser extent) krypton show an appreciable yield which can be attributed to the normal noble gas spectra, but spectroscopically pure argon and helium show only nitrogen spectra. It is interesting to note that xenon also showed an intense ultraviolet continuum, which Koch attributed to the formation and deexcitation of excited xenon diatomic molecules in three-body collisions. While argon showed band structure of the N_2 neutral molecule, helium showed structure of the N_2^+ singly ionized molecule. These results and the observed relative light yields of the various gases (Nobles, 1956; Northrop *et al.*, 1958) can be correlated with the destruction of metastable states in the noble gases. Koch found that the highest light yield in the case of helium was obtained with a concentration of 50 ppm of nitrogen (and ~ 1 ppm Hg vapor which was also present in the system). Higher nitrogen concentrations, increasing the probability of nitrogen–nitrogen collisions, tend to shift the spectrum into the infrared and to permit nonradiative transitions. This effectively reduces the visible and ultraviolet light yield.

Koch also investigated the time constants of the scintillation pulses, finding a pronounced pressure dependence of that fraction of the light yield which arises from destruction of the metastable states. For xenon at atmospheric pressure, the average pulse width is estimated to approach 2×10^{-9} sec. Even for argon, with a nitrogen concentration of 2×10^{-3}, at atmospheric pressure the metastable state destruction time is reduced to less than 2×10^{-8} sec. Birks (1964) concluded that nitrogen contamination also gives a slower component and effectively reduces the scintillation efficiency. For this reason, he recommends a fast wavelength shifter or an ultraviolet-sensitive phototube rather than nitrogen additive.

Reasonably large pulses can be obtained from gas scintillators. Koch determined that the conversion yield of samples of "pure" helium and "pure" xenon (with ∼10 ppm N_2) are 15 and 10% of NaI(Tl), respectively, for 5 MeV α-particles. Northrop *et al.* (1958) found that the conversion yield of xenon with a diphenylstilbene wavelength shifter is equal to that of NaI(Tl). This amounts to about 3% of the total energy deposited for 5 MeV α-particles (Boicourt and Brolley, 1954). Koch also found that the application of a transverse electric field produced large gains in the light output, by enabling electrons from ionization to acquire enough energy to excite the gas. This phenomenon is very similar to amplification in a proportional counter, except that considerably lower fields are required. Koch observed an amplification factor of 60 at a field of 600 V/cm. The pulse-rise time is about 5 μsec at 300 V/cm, considerably longer than for the primary pulses.

As a fission-fragment detector, the gas-scintillation counter has the following advantages: (i) The pulse-rise time, estimated as 10^{-9} sec, and the pulse duration of ∼2 × 10^{-8} sec enable one to avoid the alpha pileup problem even for samples of reasonably high specific alpha activity. The probability of pileup varies as the square of the pulse width, so the gas scintillator, compared to a conventional ionization chamber with a pulse width of 2 × 10^{-6} sec, can be used successfully with samples having a factor of 10^4 higher background rates. Higher nitrogen concentrations, although giving smaller pulse heights, may give even faster pulses. Bollinger *et al.* (1963) successfully used a specially designed helium gas-scintillation counter with 2.0% nitrogen in a measurement of the fission cross section of ^{229}Th, with a sample having an activity of 2 × 10^9 alpha disintegrations/sec. (ii) The gas-scintillation chamber is a high-gain system, and discrimination against background can be done at the anode of the photomultiplier, e.g., by means of tunnel diodes which are ideally suited to the problem. (iii) Saturation effects, which are observed in solid or liquid scintillators for highly ionizing particles such as fission fragments (see Horrocks, 1963), do not seem to be present in the low-density medium of a gas scintillator. Nobles (1956) has investigated the relative pulse height for protons, deuterons, and α-particles of 2–5 MeV and found that the pulse height is dependent only on the particle energy and not on the type of particle. Koch (1960), repeating the work of Boicourt and Brolley (1954), concluded from studies of relative pulse heights for α-particles and fission fragments that the pulse height is still dependent only on particle energy. Since there is no observable pulse-height defect in gas-scintillation chambers compared to ionization chambers [see Eq. (13)], it can be concluded that atomic excitation is the dominant mechanism of energy loss for nuclear stopping in gases, for both the primary and secondary scattered heavy particles near the end of the range. (iv) The efficiency for detecting fission fragments from an internal foil again approaches 100%, as in the conventional ionization chamber. (v) The gas-scintillation chamber can be operated in a high gamma field since it is relatively insensitive to gamma radiation. Koch (1960) concluded that discrimination of gamma-induced background from fission fragments can readily be done even in radiation fields as high as 2000 R/hr.

The resolution properties of gas-scintillation chambers are reasonably good compared to other scintillation techniques although they are not as good as gridded ionization chambers. With gas scintillators, resolutions up to 4% have been reported for α-particles (Sayres and Wu, 1957). The resolution is primarily governed by the light-collection efficiency and the phototube characteristics. The pulse-height distribution for fission fragments can be reasonably good, as shown in Fig. 5b (Huddleston, 1958) although the resolution is not adequate for precision fission-fragment kinetic-energy studies.

3. The Solid-State Counter

The development of the semiconductor detector (solid-state analog to the gas-ionization chamber) has significantly changed the field of experimental nuclear physics. Reviews of the theory and practice of fabricating and using solid-state detectors have been made by Dearnaley and Whitehead (1961), Miller et al. (1962), Dearnaley and Northrop (1963), Gibson et al. (1965), and Goulding (1966). Historically, the first solid-state counters used successfully for fission-fragment detection were the germanium diodes described by Walter et al. (1960). At the present time, however, commercially available varieties of thin, silicon detectors are generally used for fission-fragment counting. The silicon-junction detector consists of a p-type silicon crystal into the surface of which is diffused phosphorus or some other n-type material to form a p–n junction very near the surface. The surface-barrier detector consists of n-type silicon, the surface of which is chemically etched, oxidized, and coated with a gold or similar metallic film, to produce a p-type layer. Both procedures produce a semiconductor diode. A reverse biasing voltage produces a depletion layer at the junction, and the passage of charged particles through this layer creates charge carriers which are swept apart by the electric field just as in a conventional gaseous ion chamber. The field strength in the depletion layer of a solid-state counter decreases linearly with distance from the junction, and in this respect the solid-state counter is more nearly analogous to a cylindrical gaseous ionization chamber, where the field is not constant but drops off linearly with distance from the center wire.

Solid-state detectors represent an improvement over gaseous ionization detectors in two important respects: they are much faster, and they show better resolution properties. Since the medium is a solid, the particle ranges are short and the collecting field intensities can be made high ($\sim 10^4$ V/cm). The mobility of both electrons and holes in silicon is high enough that charge collection times range between 10^{-8} and 10^{-9} sec. Furthermore, it only requires ~ 3.5 eV to produce an electron-hole pair in silicon (Miller *et al.*, 1962) so that much more charge is generated in stopping a given particle in such a solid-state device than in a gaseous ion chamber. Thus the resolution characteristics are expected to be much improved. (See Fano, 1947, 1963.)

The best solid-state detectors have remarkable properties, which make them well suited for fission-fragment detection. Resolutions as high as 0.3% have been reported for 5 MeV α-particles (Blankenship and Borkowski, 1960). Pulse-rise times are comparable to gas-scintillation counters, $\sim 3 \times 10^{-9}$ sec for high fields and low-resistivity barriers (Dearnaley and White-head, 1961). For fission fragments the rise times are found to be considerably larger ($\sim 4 \times 10^{-8}$ sec), perhaps due to the great density of electron-hole pairs formed, and a consequent local space charge which tends to neutralize the collecting field (Melkonian, 1961; Dearnaley, 1961). The fission-fragment pulse-height distribution is remarkably good, as shown in Fig. 5c (Deruytter *et al.*, 1962). The best α-particle spectra would lead one to expect the resolution of the fission spectrum to be better than it is. Deshpande (1967) attributes the broadening in resolution to generation-recombination noise which would be particularly high for fission fragments with long pulse-rise times.

In the surface-barrier counter, there is a dead layer up to 100 Å thick, consisting of a gold film and a few atoms of etched or oxidized silicon. Although in principle the dead layer can be reduced to virtually zero, in practice, the losses in energy due to surface absorption are never negligible. Konecny and Hetwer (1965) found that the surface losses are somewhat larger for fission fragments than extrapolations from α-particle studies would indicate.

The recent surge of interest in fission is due in large measure to the resolution properties of these detectors, which are well suited to studies of kinetic-energy distributions of the fragments. The geometrical efficiency of the counters for fission fragments can be made almost as high as 100% if the fissionable material is deposited directly on the surface of the counter. This procedure, however, can be accompanied by unwanted side effects, such as loss of resolution due to lack of collimation. A pulse-height defect, similar to that discussed for gaseous ionization chambers, is also noted for fission fragments detected with solid-state counters. (Schmitt *et al.*, 1961, 1965a; Britt and Wegner, 1963a.) Surface-barrier detectors have also been noted to exhibit a mass-dependent pulse-height multiplication phenomenon (Britt and Wegner 1963b; Walter, 1964; Axtmann and Kedem, 1965) when operated at high field strengths. Such effects might be expected, in analogy to the behavior observed in ionization chambers or gas-scintillation counters operated at high fields. Quantitatively, the mechanism of pulse-height multiplication in solid-state detectors is not completely understood. The effect appears to be related to the surface treatment and thickness of the front electrode (Walter, 1964). It is observed only for heavy particles such as fission fragments; α-particle response at the same field strengths is found to be linear. (Britt and Wegner, 1963b.)

One disadvantage of these counters in fission detection is that they are relatively susceptible to radiation damage. Klingensmith (1961) found that surface-barrier detectors exposed to fast neutrons showed the formation of satellite peaks (much poorer in resolution and at lower pulse heights than the original peaks) and an increase in noise level and reverse current. These effects first became noticeable after an integrated dose of perhaps 3×10^{11} n/cm^2. Babcock (1961) observed similar fast neutron damage to diffused-junction counters. Partial, though by no means complete, recovery was noted by annealing. Dearnaley and Whitehead (1961) carried out experiments to determine radiation damage induced in surface-barrier counters by alpha bombardment. They observed much the same effects as did Klingensmith (1961), the resolution deterioration becoming apparent after $\sim 2 \times 10^9$ α-particles/cm^2. Britt and Benson (1964) carried out studies of radiation damage by fission fragments on both p–n junction and surface-barrier detectors. The radiation damage, as measured by the rate of change of the pulse-height defect, was found to increase linearly as the integrated flux, and to be much more severe for surface-barrier than for diffused-junction detectors. There appears to be no threshold for damage by fission fragments; the resolution of the detectors begins to deteriorate almost immediately, after $\sim 10^6$ fragments/cm^2 for surface-barrier detectors (Miller, 1963). Under the usual operating conditions, diffused-junction counters seem to be more resistant to radiation damage. Up to 10^7 fragments/cm^2 produce only slight changes in resolution and pulse height (Miller, 1965).

The mechanism of the slowing of fission fragments in solid-state detectors is expected to be reasonably well described by the formalism of Lindhard *et al.* (1963a,b). The separation made by Lindhard into two terms, representing nuclear and electronic-stopping cross sections, is physically rather significant for fission fragments in solid-state detectors: it is the nuclear-stopping cross section which is responsible for the pulse-height defect, high rate of radiation damage, and loss of energy resolution of solid-state detectors for fission fragments. This was discussed theoretically by Lindhard and Thomsen (1962), and it has since been demonstrated in several ways. Axtmann and Kedem (1965) made experimental studies of the pulse-height defect for slowed fission fragments, showing that the defect persists essentially unchanged over a change of fragment energies from 25–100 MeV. Haines and Whitehead (1965, 1966) have shown that the magnitude of the defect in silicon can be calculated adequately by assuming it to be due to nuclear stopping near the end of the range, for both primary and secondary scattered particles. Finally, studies of channeled heavy ions in silicon (Moak *et al.*, 1966) provide conclusive evidence that electronic and nuclear stopping are physically distinguishable. The phenomenon of channeling has been discussed by Lindhard (1965), see also Bergström and Domeij (1966). In order for an ion to channel through a crystal, the crystal planes must be aligned parallel to the direction of the ion beam (within about a degree), such that the ions can travel through the crystal along a reasonably wide path which does not intersect any of the lattice points. Nuclear scattering occurs only at very long range (essentially at atomic distances) and serves only to keep the ion in the channel. For channeled ions in silicon detectors, where electronic stopping is the only mechanism possible (nuclear stopping being effectively prohibited), the pulse-height defect disappears, and the resolution improves dramatically (Moak *et al.*, 1966). Since the creation of lattice vacancies and interstitials is minimized for such channeled ions, it is reasonable to expect that virtually no radiation damage occurs (see Lindhard and Thomsen, 1962).

Solid-state counters are relatively small—the largest with high resolution have areas of only a few square centimeters. Thus, unless an array of matched counters is used (see, for example, Mehta, 1963), the amount of fissile material constituting the sample is much more restricted than for ionization or scintillation chambers.

Since in general the detectors show a mass dependent pulse-height defect, they must be calibrated before they are used as fission spectrometers. It is also important that they be checked periodically while they are being used so that corrections can be made for possible radiation damage. Schmitt and Pleasonton (1966) have made systematic studies of the response of solid-state counters to fission fragments, and they have listed a set of parameters from which the performance of a given counter may be judged. Schmitt et al. (1965a) have also given a practical formula which permits one to calibrate his detector against a ^{235}U thermal neutron fission spectrum or a ^{252}Cf spontaneous fission source. Schmitt recommends the equation:

$$E = (a + a'm)x + b + b'm, \tag{15}$$

where E is the fragment kinetic energy, x is the pulse height, m is the fragment mass, and a, a', b, b' are calibration constants of the detector, defined in Table I.

TABLE I

CONSTANTS IN THE ENERGY CALIBRATION EQUATION[a]

$$E = (a + a'm)x + b + b'm$$

^{252}Cf	^{235}U
$a = \dfrac{24.0203}{P_L - P_H}$	$a = \dfrac{30.9734}{P_L - P_H}$
$a' = \dfrac{0.03574}{P_L - P_H}$	$a' = \dfrac{0.04596}{P_L - P_H}$
$b = 89.6083 - aP_L$	$b = 87.8626 - aP_L$
$b' = 0.1370 - a'P_L$	$b' = 0.1345 - a'P_L$

[a] The constants are expressed in terms of the observed pulse heights P_L and P_H that correspond to the midpoints between the $\frac{3}{4}$-maximum points in the light (L) and heavy (H) groups, respectively. Constants are given for both ^{252}Cf spontaneous fission and ^{235}U thermal-neutron-induced fission. These constants were deduced by Schmitt et al. (1965a) from a systematic study of the response of silicon solid-state detectors to fission fragments and to accelerated heavy ions of bromine and iodine. They are expected to be valid at fragment energies above about 30 MeV.

The problems involved in the use of solid-state detectors as fission-fragment spectrometers have so far been discussed in the context of conventional experiments with neutrons. In such an experiment, the intensity of the neutron beam is relatively low, and the pulse heights of individual fission events are recorded and analyzed. There is another class of fission experiments in which solid-state detectors have been used successfully: those in which a nuclear detonation serves as the neutron source (Hemmendinger, 1965, 1967; Diven, 1965, 1966; see also Hemmendinger et al., 1967; Seeger and Bergen, 1967). Here, the neutron beam intensity is so high that individual events are generally not resolved. The important properties of the detector are the following: (i) linear response to the total energy deposited in the detector, up to very high instantaneous currents, (ii) fast pulse rise and fall time with no long decay constants such as are observed with some scintillants, and (iii) good resolution and resistance to radiation damage, such that a strict proportionality is maintained between the average number of events per unit time and the detector current throughout the experiment. Silbert and Moat (1964) have carried out elaborate studies of solid-state detectors with both fission fragments and positive ion beams from a Van de Graaff accelerator. It was found that diffused junction detectors are suitable for use in experiments involving nuclear explosions. The detector output current must be limited to $2 B/\rho$ A where B is the bias voltage and ρ is the resistivity (Hemmendinger et al., 1965). For the detectors used in these experiments, the output current limit is 2.5 A. For a representative detector, Silbert (1968) observed about 3% loss of pulse height with 10^8 fragments/cm^2.

4. The Spark Chamber

Bowman and Hill (1963) have studied the properties of a corona-type spark chamber (Chang and Rosenblum, 1945; see also Meyer, 1963) as a fission-fragment detector for use with high alpha-activity samples. In the Rosenblum detector, the chamber is in a continuous state of corona discharge, and the probability for sparking is found to depend to a high degree on the ionization density of the incident particle. Bowman and Hill, by varying the voltage and pressure in the chamber, demonstrated that the counter could be operated in such a manner that it responded to fission fragments with a reasonably high efficiency ($\sim 30\%$) while the response to α-particles was a factor of 3×10^7 lower. Under these operating conditions the spark chamber was found to be nearly completely insensitive to beta and gamma radiation.

The primary advantages of the spark chamber as a fission-fragment detector have been demonstrated by measurements of the fission cross sections of ^{238}Pu and ^{232}U, where the quantity of fissile material in the chamber amounted to tens or hundreds of milligrams (Stubbins et al., 1967; Auchampaugh et al., 1968). Since the chamber has almost negligible response to α-particle background, the alpha rates from such samples, approaching 10^{12}/sec, are not particularly troublesome. Bowman and Hill (1963) reported that the output pulse characteristics of the spark chamber are reasonably good, with rise times of 5 nsec and decay times of 0.3 μsec. The pulses are large enough that only passive circuitry is required; the detector has no limitation on size; and it operates on ordinary air as the filling gas.

The chamber gives no pulse-height information, of course. While its success-ful use to discriminate against alpha radiation shows that the ionization density is important in inducing sparks, it is probable that the 30% effi-ciency is largely a geometrical effect, bearing little relation to the mass or kinetic energy distribution of the fragments.

5. Fission-Fragment Track Counting

A solid state analog to the spark chamber also exists. It has been found that heavy charged particles such as fission fragments produce submicro-scopic tracks or "spikes" in insulating solids such as mica (Silk and Barnes, 1959). Price and Walker (1962a,b,c,d) found that fission-fragment tracks in mica can be etched by hydrofluoric acid, producing channels which are visible in an optical microscope, and that only heavy particles ($A \gtrsim 30$) are recorded. Fleischer et al. (1965) have reviewed this technique. It was used by Price et al. (1963) to study ternary fission where three heavy frag-ments are emitted. It is an ideal detector for the study of fission induced by high-energy charged particles (Maurette and Stephan, 1965; Konshin et al., 1965) since it does not respond to such particles. It was also used by Flerov et al. (1963, 1964, 1965) in studies of short half-life spontaneous fission. This technique is similar to that involving nuclear emulsions as detectors (see, for example, Manley, 1962; deCarvalho et al., 1964, 1965) with the additional advantage as a fission-fragment detector of a complete insensitivity to other radiations. The major disadvantage to neutron spec-troscopy is that the record is permanent, requiring that a complete and often tedious experiment must be performed at each neutron energy of interest.

C. Fission Neutron and Gamma Detection

Experimental measurements always have associated with them system-atic errors and resolution effects which make it difficult to compare different measurements by different experimenters. One way of avoiding the problem is to require that the experimentalist perform a complete experiment, one in which all the quantities of interest are measured simultaneously, with the same systematic errors and the same resolution. Even though such an experi-ment would be highly desirable, to suggest that it be done is clearly out of the question. Perhaps a reasonable request is that several quantities be measured, permitting ratios of interest to be formed. One such experiment is the simultaneous measurement, as a function of neutron energy, of the fission reaction rate (with a thin sample), the radiative capture rate, and the flux of neutrons in the beam. Another experiment of interest is the measurement of the number of prompt neutrons emitted from a thick sample, the transmission of the sample, and the flux of neutrons in the beam. These two measurements have been combined into a single experiment using the same equipment, and carried out at the same resolution, by Brooks at Harwell (Brooks, 1961, 1966). From this experiment, one determines the energy dependence of the fission, capture, and total cross sections, and of η, the number of neutrons emitted per neutron absorbed.

Any discussion of the cross section of the fissile nuclides should include η, and $\bar{\nu}$, the average number of neutrons emitted per fission. These quantities are related by the expression

$$\eta = \bar{\nu}/(1 + \alpha), \tag{16}$$

where $\alpha = \sigma_{n\gamma}/\sigma_{nf}$ is the ratio of the cross section for radiative capture to the cross section for fission. Several years ago, it was established that $\bar{\nu}$ varies at most by perhaps a few percent in the resonance region (Auclair et al., 1955, 1956; Bollinger et al., 1956), so that, to a good approximation, the fission reaction rate can be determined by counting fission neutrons. The samples can be made considerably thicker, and the statistical accuracy of the data will be correspondingly higher. Direct measurements of η or α are very important to multilevel analysis of cross-section data. If no asymmetries due to interference were present in fission, the capture-to-fission ratio should be reasonably symmetric about the resonance energy. However, if the fission cross section is enhanced on one side of the resonance and reduced on the other side by interference effects, then a slope would be observed in the measured ratio and in the quantity η. The observation of such a slope in measurements of η for ^{239}Pu (Bollinger, 1957; Bollinger et al., 1958) was one of the earliest demonstrations that interference exists in fission.

Fission-neutron detection has been accomplished by two different techniques. Saplakoglu (1958) and Brooks (1961) used organic scintillation spectrometers that detect recoil protons from fast neutrons scattered in the scintillator. Since the pulse shape in certain organic scintillators is different for electrons (produced by gamma background) and for heavier charged particles (recoil protons), Brooks (1959) developed a pulse-shape discrimination circuit for distinguishing the two types of events. (See also Owen, 1958; Brooks et al., 1960; Rethmeyer et al., 1961; Roush et al., 1964.) The second method of counting fast neutrons involves the use of the large volume scintillator tank, first developed by Reines et al. (1954). It has been used in measurements of η and $\bar{\nu}$ (Diven et al., 1956, 1958; Diven and Hopkins, 1961; Hopkins and Diven, 1962, 1963). For fission-neutron detection, the scintillator tank is generally loaded with cadmium or some other isotope with a high absorption cross section for slow neutrons. If a slow neutron is captured in the Cd absorber, the event will be accompanied by the characteristic capture radiation of Cd. The tank can also be used to detect radiative capture of neutrons by a sample inside the tank. In this case, the tank can be used unloaded (Diven, 1958; Diven et al., 1960). It can also be loaded with boron or lithium which will capture neutrons without releasing any appreciable energy in the form of gamma rays (Gibbons et al., 1961; Block et al., 1961). The tank designed by Block was used by de Saussure et al. (1965a, 1967) and Weston et al. (1967) in measuring α for ^{235}U and ^{233}U below 100 eV.

D. Preparation of Samples of Fissile Isotopes

The three most common fissile nuclides are: ^{233}U, ^{235}U, and ^{239}Pu. Kilogram quantities of all three materials are available for purposes of neutron spectroscopy. The fissile nuclide ^{235}U has perhaps been studied more intensively than any other because of its early importance as the only naturally occurring fissile nuclide. Samples of ^{235}U are prepared by isotopic separation from natural uranium, generally by electromagnetic enrichment or by gaseous diffusion methods. Production of ^{239}Pu or ^{233}U involves reactor irradiation of ^{238}U or ^{232}Th to form ^{239}U or ^{233}Th which subsequently beta-decays to ^{239}Pu or ^{233}U, which is then extracted chemically from the thorium feed material.

The heavier plutonium and transplutonium isotopes are produced by the reactor irradiation of ^{239}Pu. The most common fissile isotope among these is 13-yr ^{241}Pu. Before it is useful as a sample for neutron spectroscopy, the ^{241}Pu must be separated by electromagnetic enrichment from the other plutonium isotopes. Perhaps as much as 50 g of ^{241}Pu are presently available in enrichments exceeding 90%.

The above nuclides are all even–odd and, as might be expected, they show similar fission properties. The observed resonance behavior is possibly due to the very simple channel spectrum of the even–even compound nuclei formed. For this reason, it is of interest to examine the fission properties of a fissile target nucleus which is not even–odd. The targets ^{232}U and ^{238}Pu seem to be two of the few such cases which fission with high probability. Because of the position of these nuclides on the beta stability curve, the binding energy of an additional neutron is higher for these than for other even–even targets, high enough to permit the compound nuclei to decay by fission. A reasonably large quantity ($\sim$1 g) of ^{232}U has been prepared by the reactor irradiation of ^{231}Pa followed by chemical separation of the ^{232}U which is formed. Similar quantities of ^{238}Pu have been obtained by reactor irradiation of ^{237}Np, followed by electromagnetic enrichment of the ^{238}Pu produced. Smaller amounts of ^{238}Pu have been produced by chemically separating plutonium from purified reactor-produced curium in which the 163 day ^{242}Cm isotope has been allowed to decay.

Two other fissile isotopes have been obtained in limited quantities. About 4 mg of ^{229}Th, a daughter product of ^{233}U alpha decay, have been prepared by chemical separation of Th from old ^{233}U samples. Milligram quantities of the odd–odd isotope ^{242m}Am have been produced by reactor irradiation of ^{241}Am and subsequent enrichment of the long-lived ^{242m}Am isomer in a special electromagnetic separator (Love *et al.*, 1965).

The preparation of samples for fission measurements involves not only the producing of the material but also the fabrication and assaying of the sample itself. Both very thin foils and rather thick foils or plates are of interest, depending upon whether the experiment requires the observing of fission fragments or not. The techniques of target preparation of both thick and thin foils has been discussed by Kobisk (1965). Techniques of fabricating thin films of fissile material are not very different from techniques of making samples for alpha and beta spectroscopy, as reviewed by Parker and Slätis (1965).

The thinnest samples require backing foils to support the deposit of fissile material. The backing foils must also be thin if the experiment requires that both fission fragments be detected in coincidence. The thinnest self-supporting backing foils are those made of carbon or of a synthetic hydrocarbon such as nylon, collodian, or VYNS. The minimum thickness which is self-supporting ranges from about 5 to 10 μg/cm^2 (see Yaffe, 1962; Parker and Slätis, 1965; Viola and O'Connell, 1965). The fissile material must then be deposited on the thin backing foil by some technique which will not rupture the backing. For uranium or plutonium, vacuum evaporation of one of the more volatile salts (the fluoride or chloride) can be used. Another technique is that of electrospraying a solution of the material to be studied (Carswell and Milsted, 1957; Bruninx and Rudstam, 1961; Lauer and Verdingh, 1963; Verdingh and Lauer, 1964; Bertolini *et al.*, 1965).

Often, coincidences between fragments are not required, and then the backing foils can be much thicker. In this case, uniform deposits of fissile material can be prepared by any of several techniques. Perhaps the oldest and most popular of these is vacuum evaporation. Since generally only a small quantity of material is available, preparation of a target by vacuum evaporation may include some means of limiting the solid angle over which the evaporated material is distributed. The use of electron-bombardment heating has greatly increased the effectiveness of vacuum evaporation techniques for preparing metallic and oxide targets of the heavy elements (see Menti et al., 1964; Dar et al., 1964; Blackburn and Haller, 1965).

One quantitative method of preparing uniform samples of fissile material involves a modified technique of electrodeposition, called "molecular plating" (Parker and Falk, 1962; Parker et al., 1964; Getoff and Bildstein, 1965; Getoff et al., 1967). This technique involves the deposition of heavy element nitrates dissolved in an appropriate organic solvent. Plating efficiencies approaching 100% have been obtained for Th, U, and Pu deposits (Parker et al., 1964; Getoff and Bildstein, 1965).

Thicker metallic foils of fissile materials are generally prepared by rolling (Kobisk, 1965). The uniformity of thick samples is just as important as the uniformity of thin samples, but for somewhat different reasons. Nonuniformity of the thin films introduces broadening by inhomogeneous degradation of the fission-fragment energies. Nonuniformity of samples which are not thin to the neutron beam at resonance peaks can introduce uncertainties in the measured cross sections (Simpson et al., 1962).

The assaying of samples involves a measurement of several variables: total amount of material, thickness, uniformity, isotopic composition, and chemical purity. Van Audenhove et al. (1963) have discussed the problem of fabricating and assaying boron samples for flux standards in neutron spectroscopy. Measurements of fission cross sections to accuracies of a fraction of 1% will require equally high quality sample preparation. For example, if the sample is to be made by vacuum evaporation, there should be no molten material in contact with the crucible material (see Van Audenhove, 1965), and ideally, the deposit should be weighed in the vacuum apparatus to avoid contamination and inaccuracies from handling, oxidation, absorption, etc. Assaying of uniformity can be done outside the vacuum. For thin samples of heavy elements which undergo alpha decay, scanning of the deposit with a high-resolution α-particle detector is a useful technique. Thicknesses and uniformities of thin backing films can be determined by measuring the transmission of α-particles through the film (Farouk et al., 1965). Fleischer (1966) has suggested the use of heavy-particle track detectors which appears to be ideal for checking foil uniformity. In this way, a permanent record of the uniformity can be obtained and examined optically at any later time, even if the foil itself is no longer in existence.

The preparation of large samples of the rarer fissile isotopes, ranging from several hundred milligrams to several grams, often involves the use of remote handling facilities (see Berreth, 1965). Heating from radioactive decay of short-lived samples such as ^{238}Pu or ^{242}Cm can amount to several watts of power to be dissipated. The chemistry of samples of this type is made much more difficult by the radiolytic decomposition of solutions containing high concentrations of short-lived heavy isotopes.

In the future, it can be expected that limited quantities of curium and californium isotopes will be produced. The techniques of measuring small quantities of highly active samples must also evolve since the handling of gram quantities of the heavier curium or californium isotopes presents an even more formidable problem, as a result of the prompt neutrons from spontaneous fission. In this connection, it appears that the use of nuclear explosions as neutron sources has several unique advantages. Only a small amount of sample material is necessary to make a high quality measurement of the fission or capture cross section, and the samples can be highly radioactive without jeopardizing the success of the measurement (Diven, 1966). In addition, with nuclear explosions, there is a possibility of making superheavy nuclides with long enough half-lives and in sufficient quantity to make such neutron spectroscopy feasible (Bell, 1966).

III. CROSS-SECTION MEASUREMENTS ON THE COMMON FISSILE ISOTOPES IN THE RESONANCE REGION

A. Low-Energy Neutron-Resonance Analysis

1. LIMITATIONS OF THE SINGLE-LEVEL FORMULA

In slow-neutron spectroscopy, probably more effort has been expended on measurements of the cross sections of the fissile nuclides than on studies of any other group. It is also very probable that the resonance properties of the fissile nuclides are not as well understood as those of any other group. From this, one may correctly infer that the effects introduced by fission are not simple in nature. And it is for just this reason that they are physically interesting enough to warrant such intensive study.

In the presence of fission the usual techniques of resonance analysis (discussed for example by Harvey, 1966a,b; Fröhner et al., 1966; or Haddad et al., 1966) are no longer adequate. The total neutron cross section of any of the common fissile isotopes shows marked asymmetries in the resonance shapes, similar to those in the elastic scattering and strongly suggestive of additional interference effects. These asymmetries are found to arise in the fission component of the cross section. The radiative-capture component shows only very slight asymmetries, which are again attributable to the fission interference. It is usually adequate to treat the radiative-capture component of the resonances by the single-level Breit–Wigner formula (Breit and Wigner, 1936).

The difference in treatment of fission and capture resonances, or of resonances of fissile and nonfissile nuclei, may be easily summarized: cross sections of nonfissile nuclei usually can be described by the single level Breit–Wigner formula; cross sections of fissile nuclei usually cannot. With the single-level Breit–Wigner formula as applied to resonances in nonfissile nuclei, only three parameters are required to describe any given resonance level.[*] These parameters are Γ_n, the neutron width; Γ_γ, the radiative-capture width; and E_0, an effective characteristic energy of the resonance level. It is well known that interference occurs in the scattering cross section. However, if the neutron widths are small and the levels are widely spaced, such that the level-level interference in the scattering can be neglected, the interference can be treated according to the single-level Breit–Wigner formula as interference of the resonance scattering with potential scattering. The absorption component of the total cross section (in this case all radiative capture) can be assumed to show no detectable interference effects.

2. Statistical Distributions of Resonance Widths

Porter and Thomas (1956) have suggested that distributions of partial widths for various nuclear processes should fall into the class of distributions known as chi-squared distributions:

$$P(x,\nu) = \frac{\nu/2}{\Gamma(\nu/2)} \left(\frac{\nu x}{2}\right)^{\nu/2-1} \exp\left(-\frac{\nu x}{2}\right), \qquad (17)$$

where $x = \Gamma_\alpha/\langle\Gamma_\alpha\rangle$ is the ratio of an individual partial width for channel α to the average width. The function $\Gamma(\nu/2)$ in the denominator of Eq. (17) is the gamma function, and ν is the number of degrees of freedom. Porter and Thomas investigated the distributions of neutron and radiative-capture widths and found that the neutron-width distribution for levels of the same spin in the compound nucleus follows a chi-squared distribution for one degree of freedom. For fissile nuclides, the neutron-width distributions are unchanged; such a distribution is shown for the neutron widths of the compound system (^{241}Pu + n) in Fig. 6 (Simpson et al., 1966). The radiative capture widths are reasonably constant from level to level of the same spin. following a chi-squared distribution for a large number of degrees of freedom. The number of degrees of freedom of a chi-squared distribution of

---

[*] In this simplified discussion, based on the R-matrix definition of parameters (see Lane and Thomas, 1958), it is assumed that (i) the spin of the resonance level is known, i.e., the statistical weight factor g is not a free parameter, and (ii) the l-value of the incident neutron is known, so that penetrability factors can be evaluated explicitly. It is also assumed that the magnitude of the potential scattering is not a free parameter.

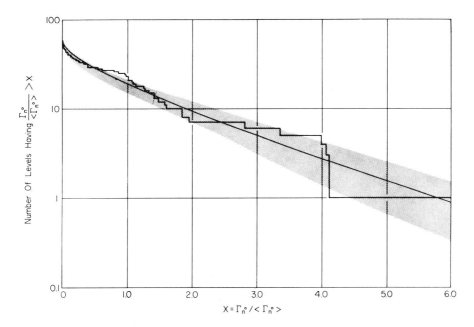

FIG. 6. Distribution of neutron widths from the analysis of resonance structure of ^{241}Pu observed in the "Petrel" nuclear explosion experiment (Simpson *et al.*, 1966). The shaded section shows the expected standard deviation for a set of sixty resonances whose neutron width distribution is described by a chi-squared distribution for one degree of freedom.

widths is to be interpreted as the number of channels or modes of deexcitation available to a given process. The phenomenon of s-wave neutron scattering by resonances of the same spin is a single-channel process. Radiative capture, on the other hand, is a many-channel process, presumably with each mode of radiative deexcitation of the compound nucleus corresponding to a separate channel. Coté and Bollinger (1961) were among the first to demonstrate both fluctuations in partial radiative capture widths and interference among resonances, by examining only a single mode of radiative deexcitation. These interference effects disappear when many channels are summed together (i.e., when total radiative capture is measured). The radiative capture width Γ_f is in reality a sum of many partial capture widths, and the interference effects tend to cancel out in the summation. For this reason the resonance structure of nonfissile nuclei is particularly simple and can be described by the single-level Breit–Wigner formula.

For fissile nuclei at least one additional level parameter is required, the

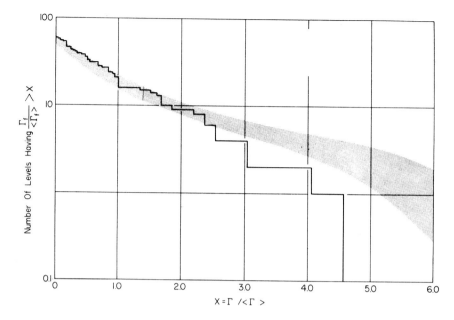

FIG. 7. Distribution of fission widths from the analysis (Simpson *et al.*, 1966) of resonance structure of ^{241}Pu observed in the "Petrel" nuclear explosion experiment. The shaded section shows the expected standard deviation of fission widths of a set of sixty resonances of two different spin-state populations which may have channel properties appropriate for ^{241}Pu (see Moore and Simpson, 1966).

fission width Γ_f. If two or more fission channels are open, then the fission width is a sum of two or more partial widths, each of which is an additional parameter to be used in describing the data. Each of the partial-width distributions is expected to follow a chi-squared distribution for one degree of freedom, such as that shown for the neutron widths in Fig. 6. However, if certain of the fission channels are only partially open, then there may be several distributions with different average partial widths. Such a composite distribution is shown in Fig. 7 for the fission widths of (^{241}Pu + n) (Simpson *et al.*, 1966). In such cases, plotting the distributions as shown in Fig. 7 may not be the most useful approach. The number of effective channels can instead be estimated from the observed fluctuations by using an analytical expression suggested by Wilets (1962):

$$\nu_{\text{eff}} = \frac{(\Sigma \nu_\alpha)^2}{(\Sigma \nu_\alpha{}^2)} = \frac{2\langle \Gamma_f \rangle^2}{\langle \Gamma_f{}^2 \rangle - \langle \Gamma_f \rangle^2}, \tag{18}$$

where $\langle \Gamma_f \rangle$ is the average fission width and ν_α can be interpreted as the degree of openness of channel α. This expression gives the same value for

ν_{eff} as the chi-squared distribution only if the ν_α are equal to zero or one. There is a further complication in the estimating of the number of fission channels for the common fissile nuclides. Here the targets are even–odd, and one expects two populations of resonances, one for each of the two possible spin states. Furthermore, the two populations may very well have different average fission widths. It is thus almost essential to be able to separate the resonances of different spins before applying detailed statistical tests.

One can conclude, either from a calculation of the number of effective degrees of freedom (Eq. (18)] or by plotting a distribution of fission widths as shown in Fig. 7 (Simpson et al., 1966) that the number of fission channels is not very large. (The distribution shown in Fig. 7 appears to be almost linear, which is characteristic of $\nu_{\text{eff}} = 2$ with $\nu_1 = \nu_2$.) This implies that the asymmetries which are observed in the fission resonances are due to interference among the resonances in these few channels.

The spacings between resonances seem to be adequately represented by the Wigner distribution:

$$P(x) = (\pi/2)x \exp(-(\pi/4)x^2) \tag{19}$$

where $x = S/D$ is the ratio of the spacing S of any given pair of levels to D, the average spacing, and $P(x)$ is the probability that the relative spacing lies between x and $x + dx$. Rosenzweig (1958) (see also Rosenzweig and Porter, 1960) has shown that the Wigner distribution is a reasonable approximation to the distribution of eigenvalues of a matrix whose elements are random numbers.

3. MULTILEVEL ANALYSIS

In order to describe the asymmetries of the fission resonances, a multi-level formula must be used which correctly takes into account the inter-ference effects. The multilevel formulas which are often used (Vogt, 1958; Reich and Moore, 1958; Adler and Adler, 1962) can be derived as special-izations of the general dispersion theory of Wigner and Eisenbud (1947). These formulas are described in some detail in the Appendix. There are difficulties involved in using multilevel formulas to analyze a given set of cross-section data. The parameters (partial widths) for the various levels are treated as free parameters, the values of which are to be extracted from the data. However, since there is level-level interference in fission, these parameters can no longer be varied independently as with the single-level formula. Thus the problem is more complex than would be indicated by the addition of one or two extra parameters for each level.

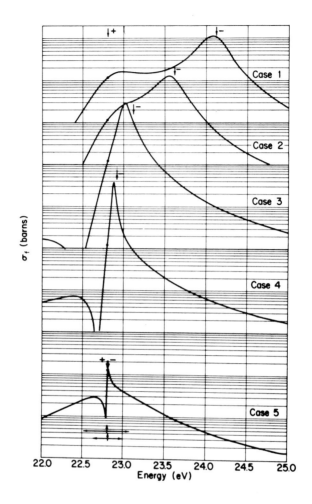

FIG. 8. The fission cross section of two interfering resonances as a function of the spacing between them. The resonance parameters, except for the spacing, are those chosen by Lynn (1964) to demonstrate this effect. The R-matrix fission widths of these levels are shown below case 5. The wider of the two levels is held fixed at 22.8 eV, and the narrower one is moved relative to it. The resonance energies are denoted by arrows; the dot on each curve at 22.8 eV is independent of the position of the other resonance, and has the value of 12 barns. The signs shown above each resonance give the relative value of the product $(\Gamma_n \Gamma_f)^{1/2}$. When the two resonances coincide in energy, the quasiresonance disappears, and the cross section shows a single resonance with a width equal to the sum of the "partial widths." The appearance of quasiresonances has been discussed in detail by Lynn (1964, 1965a, 1967) and by Garrison (1966).

Lynn (1964) has pointed out a further complication in the interpretation of fission resonances. This effect is demonstrated in Fig. 8, for two interfering resonances. If the spacing between two such resonances is significantly smaller than the widths, then the appearance of two interfering resonances changes dramatically. Instead of two resonances whose widths are related to the conventional nuclear parameters, one observes a single narrow spike, whose width is related both to the difference in the usual widths and to the spacing. This narrow spike or quasiresonance is superimposed on a broad, almost featureless background resonance, whose width is related to the sum of the two conventional widths. Lynn (1964, 1965a, 1967, 1968) has studied this effect in considerable detail, and suggests that in some cases it may lead to erroneous results in the interpretation of fission resonance data. If only the single narrow spike is observed, a conventional analysis to determine the average fission width will give much too low a result. However, in most of the cases of interest here, the effect occurs only rarely, a result of the level-repulsion effect in the Wigner distribution of resonance spacings [Eq. (19)]. On the basis of level systematics, Michaudon (1965) has estimated that for levels in ^{235}U, perhaps 5% may lie close enough together to show some of the anomalous interference discussed by Lynn. For ^{241}Pu, Moore and Simpson (1966) calculated that the fraction of resonances showing this effect is less than 5%. For ^{239}Pu, the number is almost negligible. Only for ^{233}U, where the resonance widths are considerably larger, is it expected that an appreciable number of the observed peaks may show interference effects due to closely overlapping resonances of the same spin.*

The multilevel approach of Adler and Adler (1962) is ideally suited to fitting the Lynn doublets or quasiresonances wherever they exist. The Adler formula is a specialization, suitable for fission analysis, of the generalized Kapur–Peierls (1938) formula, and is formally very similar to the Humblet–Rosenfeld (1961) S-matrix representation which Lynn has used in his calculations. The parameters defined in the Adler formula (or in the Humblet–Rosenfeld formula) correspond to the observed widths, regardless of resonance overlap or the appearance of Lynn quasiresonances for close spacings. It thus appears that the Adler formula may be most suitable for analysis of fission resonances since the complication pointed out by Lynn is handled in the most natural way. It is certainly true that the Adler

---

* In his 1964 paper, Lynn estimated that 35% of the ^{233}U low-energy resonances are not observed, noting that Michaudon (1963) estimated that 20% of the resonances in ^{235}U are also not observed. However, most of this is the result of Doppler and resolution broadening and overlapping of resonances of different spins. In these cases, quasiresonances do not appear, and the level parameters are not misleading to the averages.

formula is much to be preferred if the ultimate aim of resonance analysis is to obtain a set of parameters which will reproduce the measured cross sections in a simple analytical way. One hopes that, in general the aims are not this limited. In particular, in the present review, it is the intention to explore the connection between fission widths, fission product mass distributions, and resonance spins. If such an investigation is to retain its validity, it is essential that the fission width be simply defined in terms of the reduced widths which are characteristics of the compound nucleus. It is generally conceded that in the case of slow neutron fission, the definitions of Wigner and Eisenbud retain this simplicity, while the Kapur–Peierls and Humblet–Rosenfeld definitions do not.

4. The Concept of a Fission Channel

The definition of a fission channel in the analysis of fission cross sections is a phenomenological one. There have been observed interferences among the fission resonances and large variations in the fission widths. Analysis of such resonances by the R-matrix approach can be done with the use of a multilevel few-channel formula.

This view of fission as a few-channel process is at variance with a more conventional definition in which one might treat the entire fission process as a black box. Here one observes that a compound nucleus resonance is formed by bombarding a fissile target nucleus with neutrons of the appropriate energy. Next, one notices that there are hundreds of different possible reaction products emitted, the fission fragments. Each of the fragments, with the multiplicity of observed final spins and parities, might be expected to constitute a possible fission channel. Such a definition is not a reasonable one for fission by low-energy neutrons since the resonance interference and the fission width distributions show that only a few channels are effective. The only consistent view is to accept the ideas of the Bohr model (Bohr, 1956) that the spectrum of quantum states energetically available at the saddle point of the fission barrier determines the number of fission channels. The final mass and quantum state distribution is determined at a different stage after the saddle point is crossed but before the fissioning nucleus separates into fragments at the scission point.

To avoid confusion, let us define a *fission channel* according to the needs of low-energy neutron-resonance analysis, as one of a few modes by which fission can occur—the *alternatives* of Wigner and Eisenbud (1947). These are the channels which are reflected in the fission-width distribution. It will only be assumed that these are to be identified with the channels determined by the spectrum of quantum states available to the compound nucleus at the saddle point, according to the model proposed by Bohr (1956). But those modes of motion which lead to the very large number of possible final states eventually assumed by the fragments will not be designated as fission channels.

B. Cross Sections in the Thermal-Energy Region

Most of the high precision measurements on the cross sections of the fissile nuclides have been done in the region commonly known as thermal-neutron energies (neutron energy about 0.0253 eV, or neutron velocities of 2200 m/sec). As a consequence, the thermal cross section has generally been accepted as a normalization point. Frequently, measurements of cross sections of the fissile isotopes are made relative to one another [such as the ratio of $\sigma_f(^{235}\text{U})/\sigma_f(^{239}\text{Pu})$] which tends to cancel out systematic errors. In consequence, the determination of a "best" set of standard cross sections from all the available measurements requires a careful evaluation of the interrelation of the various experimental quantities. One of the most recent and probably the most thorough of these evaluations is that of Westcott et al. (1965). (See also Evans and Fluharty, 1960; Leonard, 1962, 1967; and Sher and Felberbaum, 1962, 1965.)

The energy region below 1 eV is also of interest to neutron spectroscopy, for it is here that the data are most sensitive to the effects of negative-energy resonances, whose characteristic energy lies below the neutron-binding energy. It was at one time noted that the thermal (0.0253 eV) cross section for the common fissile nuclides is so large as to seem anomalous. Only a small portion of the thermal cross section can be accounted for by the observed positive energy resonances, and in order to account for the observed thermal value, particularly strong negative-energy resonances were postulated. These resonances were required to have neutron widths much larger than any of the observed positive-energy resonances. Vogt (1960), as a result of a careful multilevel analysis of the low-lying level structure of the three common fissile nuclides, concluded that the parameters of the assumed negative-energy levels were not outside the limits of the expected values.

Representative measurements of the low-energy total cross section of ^{233}U are shown in Fig. 9a, of the fission cross section Fig. 9b, and of the radiative-capture cross section in Fig. 9c. Direct measurements of η for ^{233}U are shown in Fig. 9d. In order to remove the $1/v$ dependence, the cross-section data in Fig. 9 are all plotted as $\sigma(E)E^{1/2}$ versus E, where E is the neutron energy. Solid lines in Fig. 9 are the results of the multilevel analysis carried out by Moore and Reich (1960). This analysis assumed the exist-

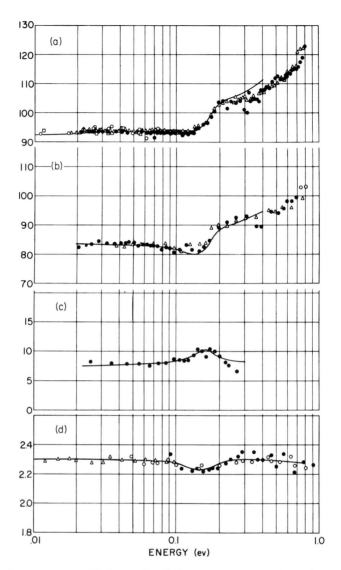

Fig. 9. Low-energy, total fission, and radiative-capture cross section and eta measurements for (^{233}U + n) as a function of neutron energy. The cross-section data are multiplied by the square root of the neutron energy in order to remove the $1/v$ dependence. The solid curves represent the results of multilevel analysis of these data (Moore and Reich, 1960). (a) $\sigma_t E^{1/2}(^{233}$U); (b) $\sigma_f E^{1/2}(^{233}$U); (c) $\sigma_c E^{1/2}(^{233}$U); (d) $\eta(^{233}$U). Data designated as MTR–FC (●) in (a), (b), and (c) are those of Moore et al. (1960). In (a), ORNL–FC (△) are Block et al. (1960) and COL–CS (○□) are Safford et al. (1960a). In (d), Harwell–CS (○) are Sanders et al. (1957) and BNL–SC (△) are Palevsky et al. (1956). In (b), MTR–CS (△) are Miller and Moore (1957) and in (d) MTR–CS (●) are Smith and Magleby (1958).

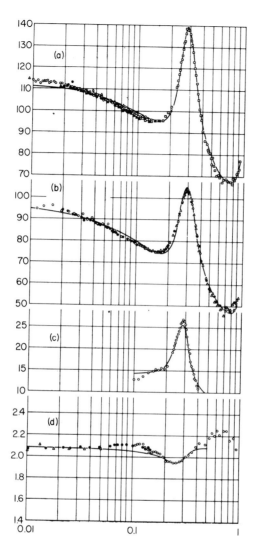

Fig. 10. Low-energy total fission, and radiative-capture cross sections and eta measurements for (^{235}U + n) as a function of neutron energy. The cross-section data are multiplied by the square root of the neutron energy in order to remove the $1/v$ dependence. The solid curves represent the results of multilevel analysis of only the data above 0.1 eV (Vogt, 1958). Data designated as BNL–CS [(a) □, (b) △, (c) ○] are those of Shore and Sailor (1958); MTR–FC [(a) ●] are Simpson *et al.* (1960); ORNL–FC [(a) ▲] are Block *et al.* (1960); ANL–FC [(a) ○] are Saplakoglu (1961); COL–CS [(a) △] are Safford *et al.* (1959); HAN–CS [(b) ●] are Leonard (1955); BRI–SC [(b) ○] are Deruytter (1961); MTR–CS [(d) ○] are Smith and Magleby (1957); Harwell–CS [(d) △] are Skarsgard and Kenward (1958); and BNL–SC [(d) ●] are Palevsky *et al.* (1956). (a) $\sigma_t E^{1/2}(^{235}U)$; (b) $\sigma_f E^{1/2}(^{235}U)$; (c) $\sigma_c E^{1/2}(^{235}U)$; (d) $\eta(E)^{235}U$.

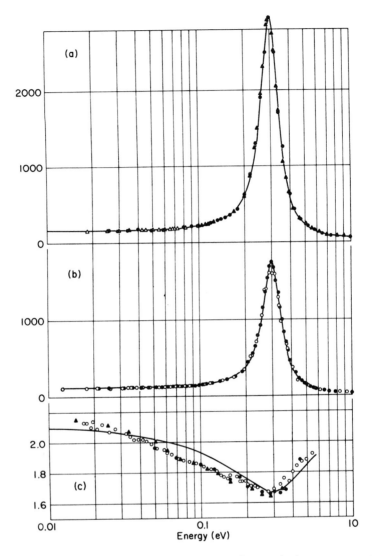

Fig. 11. Low-energy total and fission cross sections and eta measurements for (^{239}Pu + n) as a function of neutron energy. The cross-section data are multiplied by the square root of the neutron energy in order to remove the $1/v$ dependence. The solid curves represent the results of multilevel analysis of only the cross-section data (Vogt, 1960). Data designated as COL–CS ($\triangle$) are those of Safford and Havens (1961); Han–CS ($\bullet$) are Leonard (1956) in (a) and (b), and Leonard *et al.* (1956) in (c); ANL–FC ($\blacktriangle$) are Bollinger *et al.* (1958); Harwell–CS ($\bigcirc$) are Richmond and Price (1956) in (b) and Skarsgard and Kenward (1958) in (c). (a) $\sigma_t E^{1/2}(^{239}\text{Pu})$; (b) $\sigma_f E^{1/2}(^{239}\text{Pu})$.

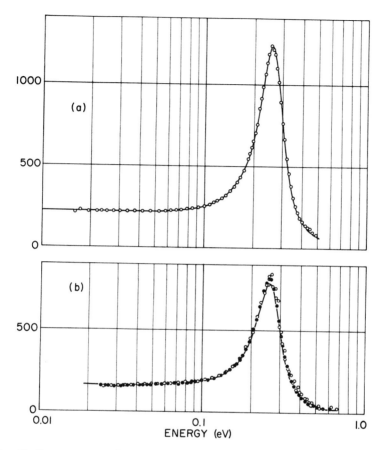

FIG. 12. Low-energy total and fission cross sections for (^{241}Pu + n) as a function of neutron energy. The data are multiplied by the square root of the neutron energy in order to remove the $1/v$ dependence. The solid curves represent the results of multilevel analysis of only the total cross-section data (Simpson and Moore, 1961). Data designated as MTR–FC ($\bigcirc$) are those of Simpson and Schuman (1961) in (a) and of Watanabe and Simpson (1964) in (b); Han–CS ($\bullet$) are data of Leonard and Friesenhahn (1959). (a) $\sigma_t E^{1/2}(^{241}\text{Pu})$; (b) $\sigma_f E^{1/2}(^{241}\text{Pu})$.

ence of a large $1/v$ component, to which was attributed about 90% of the total and fission cross sections in this region. This component would appear as a constant on a $\sigma(E)E^{1/2}$ plot. Vogt's analysis of ^{233}U differed from this in that he explicitly assumed several broad, largely unresolved resonances which contributed an effective background-fission cross section. Neither analysis required any negative energy resonances; since the scattering is nearly constant (Oleksa, 1958), no large negative-energy resonance is expected. Fluharty *et al.* (1966) and Smith and Fast (1966) reported an analysis based on the Adler approach (see Adler and Adler, 1962). Here, the parameters of the small resonance near 0.2 eV show a negative contribution to the fission cross section. These parameters are consistent with the *R*-matrix parameters obtained by Moore and Reich (1960). The cross section which has been removed from the fission component appears primarily as radiative capture, shown in Fig. 9c.

Some representative measurements of the total and fission cross sections of ^{235}U are shown in Figs. 10a and 10b (p. 389). The radiative-capture cross section over the first resonance was determined by Shore and Sailor (1958), as shown in Fig. 10c. Some direct relative measurements of the variation of η for ^{235}U are shown in Fig. 10d. The solid lines in Fig. 10 are the results of the multilevel analysis of Vogt (1960), who fitted only the data of Shore and Sailor (above 0.1 eV). In Vogt's analysis a moderately large negative energy resonance was postulated at -0.95 eV, which interferes with both the 0.28 eV resonance shown and the 1.1 eV resonance. Schermer *et al.* (1968) have since determined, by polarization experiments, that the spins of the 0.28 and the 1.1 eV resonances differ, and that the single negative-energy resonance assumed by Vogt must be treated as two resonances, one for each spin. According to Deruytter (1961), two such negative-energy resonances are necessary to describe the lowest energy data. The low-energy scattering cross section of ^{235}U (Foote, 1958) also shows the effects of one or more large negative-energy resonances.

The total and fission cross sections of ^{239}Pu in the thermal-energy region are shown in Fig. 11a and 11b. The change in scale (by a factor of 40) should be noted for the cross-section data shown in Fig. 11 compared to those in Figs. 9 and 10. The solid curves in Fig. 11 represent again the results of a multilevel analysis of Vogt (1960), in which a moderately large negative-energy resonance was assumed at -1.2 eV. Vogt found that interference is not required between this resonance and the observed resonance at 0.292 eV.

Precision cross-section data on ^{241}Pu in the thermal-energy region have also been obtained as shown in Fig. 12 although relative eta measurements are lacking. The solid curves in Fig. 12 represent the results of the multilevel analysis made by Simpson and Moore (1961). Interference is required between the resonance at 0.264 eV and a small negative-energy resonance which was postulated at -0.16 eV.

C. The Low-Energy Resonance Region and Multilevel Analysis

Multilevel analyses have been limited to the neutron-energy region below 100 eV. This is primarily because both Doppler broadening (introduced by thermal motion of the target atoms) and resolution broadening tend to obscure the details of the fission-resonance structure (asymmetries and interference dips) above 100 eV. For most of the fissioning compound nuclei, the level spacing in the resonance region is about 1 eV. Multilevel analysis is primarily shape analysis. Particular attention is given to the shape in the wings of the resonances and the amount of cross section between resonances. Resolution and Doppler broadening corrections are important even at energies as low as a few electron volts. The Doppler width is customarily defined as

$$\Delta = (4mEkT/M)^{1/2} \qquad (20)$$

where Δ is the half-width of the function at $1/e$ of the maximum, m and M are the masses of the neutron and target nucleus, respectively, E is the neutron energy, k is Boltzmann's constant, and T is the effective temperature of the target material, Debye corrected (Lamb, 1939). For neutron energies above a few electron volts, it is usual to assume that Doppler broadening corrections can be made as outlined by Solbrig (1961a,b) by treating the atoms of the sample as a gas and neglecting the detailed effects of crystalline binding. For the fissile nuclides in the form of metallic samples at room temperature the Doppler width is ~0.2 at 100 eV neutron energy, somewhat larger than the average resonance width. At neutron energies higher than 1 keV, the Doppler width becomes comparable to the level spacings, and the effects of individual levels can no longer be discerned. For most neutron spectrometers, the width of the instrumental resolution function is very nearly a constant in neutron time-of-flight, and so increases as the $\frac{3}{2}$ power of the neutron energy, much faster than the Doppler width. In principle, it is possible to choose such a long flight path that the only source of resonance broadening is the Doppler motion of the sample atoms although, in practice, there is always some energy at which the resolution of the measurement is dominated by the instrumental resolution. For most of the measurements on the common fissile nuclides, this crossover point occurs between 10 and 100 eV. For even the highest resolution spectrometers (associated with electron linear accelerators, charged particle accelerators, and nuclear explosion sources), the resolution function becomes comparable to the Doppler function between 100 eV and 1 keV.

1. ^{233}URANIUM

The resonances in ^{233}U are more closely spaced and have larger fission widths than for any of the other fissile isotopes. Even in regions where the experimental resolution is good, perhaps half, or more than half, the resonances are obscured to some extent by overlapping. Most of the over-

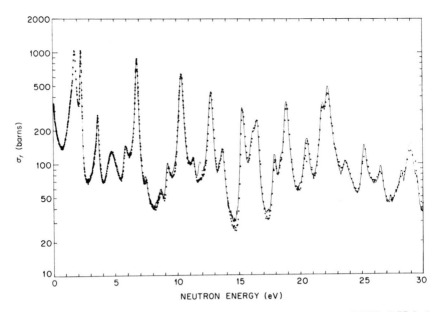

FIG. 13. The total cross section of ^{233}U below 30 eV. Data shown as ORNL–RPI (—) are those of Weston *et al.* (1967); they are compared to those (●) of Pattenden and Harvey (1963). (This figure was supplied by Dr. Weston prior to publication.)

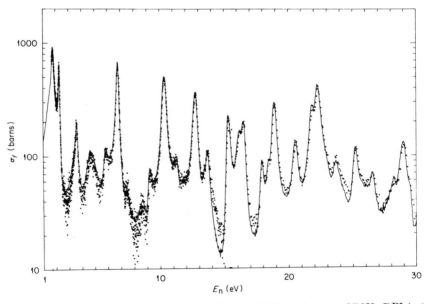

FIG. 14. The fission cross section of ^{233}U below 30 eV. Data shown as ORNL–RPI (—) are those of Weston *et al.* (1967); they are compared to those (●) of Nifenecker *et al.* (1963). (This figure was supplied by Dr. Weston prior to publication.)

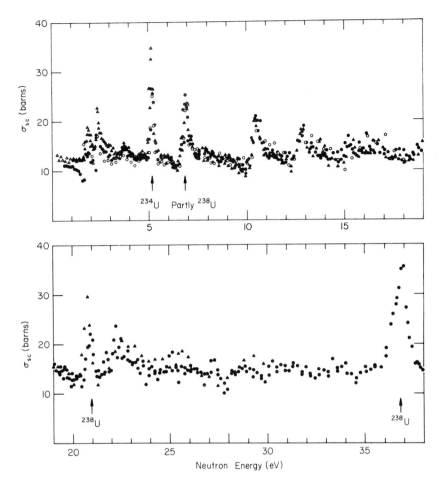

FIG. 15. The elastic scattering cross section of ^{233}U below 36 eV. Data shown are those of Simpson *et al.* (1968) (●), Sauter and Bowman (1968) (▲), Moore and Simpson (1962) (○), and Oleksa (1958) (△). (The data of Sauter and Bowman were supplied by Dr. Bowman prior to publication.)

lapping doublets which occur are expected to be characteristic of opposite spins, which in principle permits an analysis with R-matrix parameters. In practice, obscuring of the resonances leads to large uncertainties in the analysis, and several of the most prominent peaks are likely to be the result of overlapping resonances of the same spin (Lynn, 1964).

Plotted in Figs. 13–15 are the total, fission, and scattering cross sections of ^{233}U below ∼30 eV. The variation of eta for ^{233}U is shown in Fig. 16.

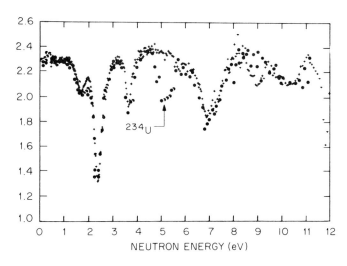

FIG. 16. The number of neutrons emitted per neutron absorbed (eta) for ^{233}U below 12 eV. Data shown as present results ($+$) were calculated from the alpha measurements of Weston *et al.* (1967); data indicated as J. R. Smith *et al.* (●), are the eta measurements reported by Smith and Fast (1966) and include data obtained by Smith and Magleby (1958). (This figure was supplied by Dr. Weston prior to publication.)

Analyses of data comparable to those shown in Figs. 13 and 14 have been reported by Vogt (1960), by Moore and Reich (1960), by Pattenden and Harvey (1960), by Adler *et al.* (1964), by Nifenecker and Perrin (1965), and, above 20 eV, by Bergen and Silbert (1968). Multilevel (*R*-matrix) resonance parameters for the low-energy ^{233}U resonances are shown in Table II. These parameters include the latest of those reported in the second supplement to the Neutron Cross-Section Compilation, BNL-325 (Stehn *et al.*, 1965). Added are multilevel parameters of Bergen and Silbert (1968), and, for comparison, single-level parameters of Nifenecker and Perrin (1965). In general, the parameters obtained by analyses of the data on ^{233}U, as shown in Table II, do not agree very well, which emphasizes the lack of uniqueness in the multilevel approach. However, there are similarities which should be pointed out. Two kinds of resonances seem to be present: narrow resonances which can be treated as though they interfere with one another in one or two fission channels, and somewhat broader resonances which do not seem to interfere with the others. It has been conjectured that two groups of resonances of this kind may be characteristic of different spin states in the compound nucleus. Unfortunately, the scattering data do not permit the spins of the resonances to be determined.

TABLE II
Resonance Parameters[a]

$I^\pi = \frac{5}{2}+$							$^{233}_{92}\text{U}$ $1.62 \times 10^5 y$
E_0 (eV)	Γ (mV)	Γ_γ (mV)	Γ_f (mV)	θ (deg)	$2_g\Gamma_n{}^0$ (mV)	Group	Reference
$\begin{bmatrix} -0.3 \\ 0.10 \end{bmatrix}$		(45)	960	0	0.044	B	Pattenden 60
		56	994	0	0.0586	B	Vogt 60
$\begin{bmatrix} 0.188 \\ 0.150 \end{bmatrix}$		(40)	63	0	0.00046	A	Pattenden 60
		30	60	10	0.00002	A	Moore 60
$\begin{bmatrix} 1.61 \\ 1.56 \\ 1.45 \end{bmatrix}$		(45)	600	150	0.142	B	Pattenden 60
		54	420	240	0.121	A, B	Moore 60
		54	716	100	0.152	B	Vogt 60
$\begin{bmatrix} 1.823 \\ 1.785 \\ 1.755 \\ 1.76 \end{bmatrix}$	309	(45)	264		0.354		Nifenecker 65
		(45)	210	150	0.225	A	Pattenden 60
		36	186	150	0.189	A	Moore 60
		49	231	150	0.235	A	Vogt 60
2.191	160		115		0.048		Nifenecker 65
$\begin{bmatrix} 2.321 \\ 2.307 \\ 2.305 \\ 2.30 \end{bmatrix}$	70.9	(45)	66.4		0.141		Nifenecker 65
		(45)	50	295	0.115	A	Pattenden 60
		34.6	48	308	0.100	A	Moore 60
		47	49	270	0.116	A	Vogt 60
3.418	312		267		0.025		Nifenecker 65
$\begin{bmatrix} 3.678 \\ 3.64 \\ 3.65 \\ 3.64 \end{bmatrix}$	182	(45)	137		0.058		Nifenecker 65
		(45)	155	20	0.074	A	Pattenden 60
		53	149	−19	0.067	A	Moore 60
		48	212	+1	0.084	A	Vogt 60
$\begin{bmatrix} 4.819 \\ 4.79 \\ 4.825 \\ 4.80 \end{bmatrix}$	857	(45)	809		0.135		Nifenecker 65
		(45)	950	240	0.187	B	Pattenden 60
		80	850	−20	0.123	A, B	Moore 60
		60	740	225	0.164	B	Vogt 60
$\begin{bmatrix} 5.995 \\ 5.85 \\ 5.82 \end{bmatrix}$	421	(45)	376		0.064		Nifenecker 65
		(45)	195	230	0.041	A	Pattenden 60
		80	316	180	0.055	A	Moore 60
6.723	750		705		0.153		Nifenecker 65
$\begin{bmatrix} 6.912 \\ 6.85 \\ 6.82 \\ 6.77 \end{bmatrix}$	146	(45)	101		0.293		Nifenecker 65
		(45)	165	80	0.41	A	Pattenden 60
		55	146	0	0.350	A	Moore 60
		50	160	0	0.203	A	Vogt 60

[a] Resonance parameters for (^{233}U + n) from single-level or multilevel (R-matrix) analyses. They are presented here only for purposes of comparison. The resonance phases which may be appropriate in a multilevel calculation have been included as an angle following the notation of Vogt (1958, 1960). Recommended values for the parameters are not given; any single set of parameters would be expected to give a better representation of the cross section than would an average of individual parameters. Complete multilevel analyses of the ^{233}U resonance cross sections have also been reported by Adler et al. (1964) and by de Saussure et al. (1967). These parameters have not been included in this table because of differences in the definition of the parameters. A shortened notation is used for referencing; the more complete notation is given in the text.

TABLE II (Continued)

$I^\pi = \frac{5}{2}+$							$^{233}_{92}\mathrm{U}$ $1.62 \times 10^5 y$
E_0 (eV)	Γ (mV)	Γ_γ (mV)	Γ_f (mV)	θ (deg)	$2_g\Gamma_n^0$ (mV)	Group	Reference
⌈7.66	322	(45)	277		0.018		Nifenecker 65
⎢7.57		(45)	90	200	0.015	A	Pattenden 60
⌊7.6		48	125	180	0.008	A	Moore 60
⌈8.71	399	(45)	354		0.025		Nifenecker 65
⎢8.78		(45)	700	15	0.067	B	Pattenden 60
⌊8.7		40	300	180	0.012	A	Moore 60
⌈9.48	559	(45)	514		0.066		Nifenecker 65
⎢9.30		(45)	195	165	0.047	A	Pattenden 60
⌊9.2		50	180	180	0.022	A	Moore 60
⌈10.54	347	(45)	302		0.500		Nifenecker 65
⎢10.41		(45)	235	5	0.51	A	Pattenden 60
⌊10.47		85	270	0	0.480	A	Moore 60
⌈11.41	594	(45)	549		0.092		Nifenecker 65
⌊11.48		(45)	175	145	0.035	A	Pattenden 60
12.29	1102		1057		0.080		Nifenecker 65
⌈12.98	318	(45)	273		0.370		Nifenecker 65
⌊12.88		(45)	250	200	0.40	A	Pattenden 60
⌈13.94	429	(45)	384		0.122		Nifenecker 65
⌊13.89		(45)	300	350	0.125	A	Pattenden 60
⌈ ⎰15.56	64		19		0.163		Nifenecker 65
⎢ ⎱15.69	230	(45)	185		0.126		Nifenecker 65
⌊15.46		(45)	190	315	0.265	A	Pattenden 60
⌈16.48	672	(45)	627		0.320		Nifenecker 65
⌊16.32		(45)	450		0.270		Pattenden 60
⌈16.81	182	(45)	137		0.120		Nifenecker 65
⌊16.67		(45)	140		0.126		Pattenden 60
⌈18.27	158	(45)	113		0.0508		Nifenecker 65
⌊18.10		(45)	150		0.076		Pattenden 60
⌈18.72	155	(45)	110		0.036		Nifenecker 65
⌊18.60		(45)	115		0.044		Pattenden 60
⌈19.26	350	(45)	305		0.390		Nifenecker 65
⌊19.09		(45)	240		0.43		Pattenden 60
⌈20.535		40	320	0	0.32	B	Bergen 68
⎢20.92	481	(45)	436		0.191		Nifenecker 65
⌊20.76		(45)	470		0.265		Pattenden 60
22.00	482		437		0.066		Nifenecker 65
⌈21.885		40	190	0	0.49	A	Bergen 68
⎢22.23	205	(45)	160		0.168		Nifenecker 65
⌊22.00		(45)	205		0.250		Pattenden 60
⌈22.33		40	355	180	1.34	B	Bergen 68
⎢22.69	445	(45)	400		0.635		Nifenecker 65
⌊22.50		(45)	390		0.68		Pattenden 60
⌈22.94		65	950	0	0.61	C	Bergen 68
⌊23.17	1910		1865		0.376		Nifenecker 65

TABLE II *(Continued)*

E_0 (eV)	Γ (mV)	Γ_γ (mV)	Γ_f (mV)	θ (deg)	$2_g\Gamma_n^0$ (mV)	Group	$^{233}_{92}U$ $1.62 \times 10^5 y$ Reference
23.61		40	540	180	0.185	B	Bergen 68
24.15	332	(45)	287		0.074		Nifenecker 65
23.90		(45)	900		0.280		Pattenden 60
24.63	574	(45)	529		0.059		Nifenecker 65
25.245		65	220	0	0.25	C	Bergen 68
25.68	410	(45)	365		0.177		Nifenecker 65
25.48		(45)	290		0.230		Pattenden 60
25.84		40	40	180	0.013	A	Bergen 68
26.30		40	550	0	0.115	B	Bergen 68
26.31	812		767		0.095		Nifenecker 65
26.63		40	330	180	0.19	A	Bergen 68
27.00	542		497		0.112		Nifenecker 65
27.28		40	250	0	0.0074	B	Bergen 68
27.69		65	725	180	0.14	C	Bergen 68
28.14	503		458		0.003		Nifenecker 65
28.35	40		200	0	0.09	A	Bergen 68
28.64	1390		1345		0.186		Nifenecker 65
29.11		40	420	0	0.5	B	Bergen 68
29.54	580		535		0.286		Nifenecker 65
29.55		40	250	0	0.105	A	Bergen 68

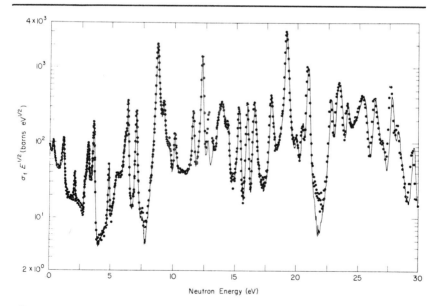

Fig. 17. The fission cross section of ^{235}U below 30 eV. Data shown as ORNL–RPI (—) are those of de Saussure *et al.* (1967); remaining data (●) are those of Bowman *et al.* (1966). (This figure was supplied by Dr. de Saussure prior to publication.)

2. ²³⁵URANIUM

Shown in Figs. 17–19 are the low-energy partial cross sections for ²³⁵U. Analyses of ²³⁵U cross-section data have been reported by Shore and Sailor (1958), by Vogt (1958, 1960), by Michaudon (1964) (see also Michaudon *et al.*, 1965a), by Ignatiev *et al.* (1964) (see also Kirpichnikov *et al.*, 1964),

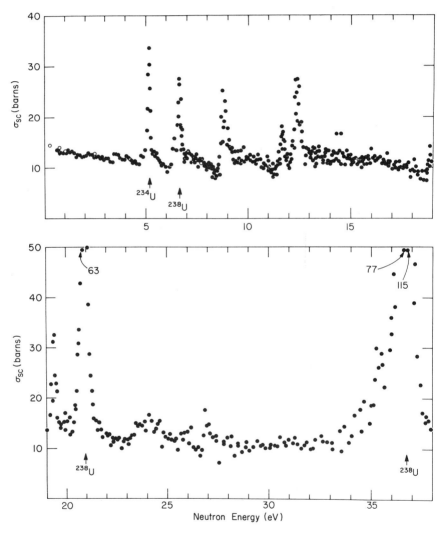

FIG. 18. The elastic scattering cross section of ²³⁵U below 40 eV. The data (●) are those of Simpson *et al.* (1968); and (○) are those of Foote (1958).

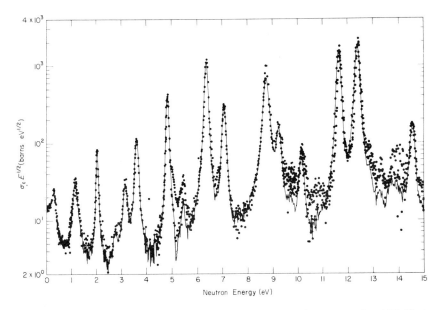

FIG. 19. The capture cross section of ^{235}U below 15 eV (de Saussure *et al.*, 1966). Data shown as ORNL–RPI (—) are those of de Saussure *et al.* (1966); HAR (●) are the eta measurements of Brooks (1966) from which capture cross sections were calculated. (This figure was supplied by Dr. de Saussure prior to publication.)

by Adler and Adler (1966a), by Wang *et al.* (1965), by Drawbaugh and Gibson (1966, 1967), and by Lubitz (1967). Two of these analyses (Michaudon, 1964; Wang *et al.*, 1965) suggested that both the fission and radiative-capture widths of the resonances appear to fall into two groups, which were interpreted as characteristic of different spin states of the compound system. Scattering data obtained by Poortmans *et al.* (1966, 1967) have indicated spin values for the 8.8 and 12.4 eV resonances in ^{235}U. These assignments, which are the same as those made from the scattering data of Simpson *et al.* (1968) in Fig. 18, as well as tentative spin assignments made from polarization measurements by Schermer *et al.*, 1968, are included in the summary of low-energy resonance parameters in Table III. For the most part, the resonance parameters given in Table III are those compiled by Stehn *et al.* (1965) in BNL-325. Added are some results of an analysis by Wang *et al.* (1965). Parameters from least-squares single-level analyses by Drawbaugh and Gibson (1967) and by Lubitz (1967) are not included in Table III nor are the results of a multilevel analysis by Adler and Adler (1966a), because of differences in the approach and in the interpretation of the parameters.

TABLE III

$I = \frac{7}{2}-$				
E_0 (eV)	Γ (mV)	$2g\Gamma_n$ (mV)	Γ_γ (mV)	Γ_f (mV)
⌈ <0				
| −2.0			(40)	187
| −1.45	259		(33)	223
⌊ −0.95			27.6	169.4
−0.02	97		(34)	63
⌈ 0.28				
| 0.30 ± 0.01			36 ± 3	99 ± 8
| 0.273			29	99
⌊ 0.282	114.7		32.2	82.5
⌈ 1.14				
| 1.14 ± 0.01			43 ± 5	129 ± 13
| 1.140			44	124.6
⌊ 1.138	148		42	106
⌈ 2.0				
| 2.026 ± 0.004	54 ± 7	0.0087 ± 0.0009	40	14 ± 2
| 2.03	48 ± 3	0.0078 ± 0.0003	38 ± 4	10 ± 1
| 2.04 ± 0.01			38 ± 4	10 ± 2
| 2.035			35	12
⌊ 2.036	41.4		34.6	6.8
⌈ 2.84 ± 0.02	173 ± 53	0.0033 ± 0.0007	40	133 ± 40
| 2.83		0.012 ± 0.002		
⌊ 2.84 ± 0.04			(40)	160
⌈ 3.136 ± 0.006	123 ± 15	0.027 ± 0.003	31 ± 4	92 ± 12
| 3.14	167 ± 15	0.0280 ± 0.0015	51 ± 9	116 ± 15
| 3.14 ± 0.02			44 ± 5	79 ± 10
⌊ 3.16			31.1	155
⌈ 3.584 ± 0.006	81 ± 7	0.047 ± 0.004	31 ± 4	50 ± 5
| 3.61	93 ± 8	0.048 ± 0.002	46 ± 5	47 ± 5
| 3.60 ± 0.02			40 ± 4	43 ± 4
⌊ 3.599	81.4		37	45
⌈ 4.81 ± 0.01	34 ± 20	0.066 ± 0.005	29.5 ± 19	4.5 ± 3
| 4.84	41 ± 3	0.060 ± 0.003	37 ± 4	3.8 ± 0.5
| 4.84 ± 0.02			(40)	4
⌊ 4.847	27.8		25.5	2.3
⌈ 5.45	70 ± 20	0.023 ± 0.003	46 ± 11	24 ± 7
| 5.45		~0.009		
⌊ 5.45 ± 0.02			(40)	23
5.82	103 ± 49	0.012 ± 0.004	40	63 ± 30

[a] Resonance parameters for (^{235}U + n), as obtained by single-level or multilevel (R-matrix) analysis. The multilevel parameters of Adler and Adler (1966b) are not presented, because of differences in definition. These parameters are given here primarily for comparison, and the resonance phases which may be appropriate to a multilevel calculation have been included as angles, following the notation of Vogt (1958, 1960).

$2g\Gamma_n^0$ (mV)	θ (Deg)	Spin, etc.	$^{235}_{92}\mathrm{U}$ $7.13 \times 10^8\ y$ Reference
		Mostly $J = 4(3)$	Schermer 68
			Ignatiev 64
3.056	0		Shore 58
1.488	0		Vogt 60
0.00072			Shore 58
		$J = 3(4)$	Schermer 68
			Ignatiev 64
0.00563	278		Vogt 60
0.00516	270		Shore 58
		$J = 4(3)$	Schermer 68
	114		Ignatiev 64
0.01613	104		Vogt 60
0.0143	90		Shore 58
		Probably $J = 3(4)$	Schermer 68
0.0061 ± 0.0006			Wang 65
0.0055 ± 0.0002			Michaudon 64
			Ignatiev 64
0.00537			Vogt 60
0.00530			Shore 58
0.0020 ± 0.0004			Wang 65
0.0071 ± 0.0006			Michaudon 64
	-70	$\eta/\nu = 0.80 \pm 0.15$	Ignatiev 64
0.015 ± 0.002			Wang 65
0.0158 ± 0.0008			Michaudon 64
			Ignatiev 64
0.01823	225		Vogt 60
0.025 ± 0.002			Wang 65
0.025 ± 0.001			Michaudon 64
	120		Ignatiev 64
0.0243	90		Shore 58
0.030 ± 0.002			Wang 65
0.0270 ± 0.0013			Michaudon 64
		$\eta/\nu = 0.095 \pm 0.010$	Ignatiev 64
0.025			Shore 58
0.0099 ± 0.0013			Wang 65
~ 0.0038			Michaudon 64
	0	$\eta/\nu = 0.36 \pm 0.04$	Ignatiev 64
0.005 ± 0.002			Wang 65

It is expected that any given set of parameters will give a better representation of the measured cross section than will an average of parameters for individual resonances; hence, no attempt has been made to obtain a recommended set of parameters. A shortened notation has been used for the referencing in this table; the more complete notation is given in the text. Spin assignments by Schermer et al. (1968) are uncertain in that they depend on the sign of the magnetic moment of ^{235}U.

TABLE III

E_0 (eV)	Γ (mV)	$2g\Gamma_n$ (mV)	Γ_γ (mV)	Γ_f (mV)

$I = \frac{7}{2} -$

E_0 (eV)	Γ (mV)	$2g\Gamma_n$ (mV)	Γ_γ (mV)	Γ_f (mV)
$\begin{bmatrix} 6.20 \pm 0.01 \\ 6.19 \\ 6.20 \pm 0.08 \end{bmatrix}$	106 ± 31	0.031 ± 0.006 0.035 ± 0.004	38 ± 17 (40)	68 ± 23 270 ± 70
$\begin{bmatrix} 6.40 \pm 0.01 \\ 6.38 \\ 6.39 \pm 0.03 \\ 6.40 \end{bmatrix}$	63 ± 15 45 ± 3 42	0.30 ± 0.02 0.260 ± 0.015	52 ± 13 33 ± 3 (40) (33)	11 ± 3 12 ± 2 11 9
$\begin{bmatrix} 7.095 \pm 0.015 \\ 7.07 \\ 7.07 \pm 0.3 \end{bmatrix}$	52 ± 9 64 ± 5	0.115 ± 0.006 0.126 ± 0.006	31 ± 5 36 ± 5 (40)	21 ± 4 28 ± 4 25
$\begin{bmatrix} 8.78 \\ 8.77 \pm 0.02 \\ 8.78 \\ 8.80 \pm 0.04 \\ 8.795 \end{bmatrix}$	113 ± 17 133 ± 13 93	1.15 ± 0.05 1.18 ± 0.05	40 ± 7 50 ± 10 (40) (33)	73 ± 12 82 ± 10 49 60
$\begin{bmatrix} 9.30 \pm 0.03 \\ 9.28 \\ 9.26 \pm 0.05 \end{bmatrix}$	109 ± 43 160 ± 40	0.147 ± 0.014 0.20 ± 0.04	37 ± 15 65 ± 35 (40)	72 ± 30 95 ± 40 25
$\begin{bmatrix} 9.73 \pm 0.06 \\ 9.74 \end{bmatrix}$	127 ± 74	0.047 ± 0.015 0.026 ± 0.005	40	87 ± 51
$\begin{bmatrix} 10.20 \pm 0.03 \\ 10.18 \\ 10.16 \pm 0.05 \end{bmatrix}$	88 ± 23 95 ± 10	0.066 ± 0.006 0.064 ± 0.006	41 ± 14 37 ± 19 (40)	47 ± 13 58 ± 14 7
$\begin{bmatrix} 10.65 \pm 0.06 \\ 10.80 \end{bmatrix}$	120	~ 0.025 0.016 ± 0.008	40	80
11.05	60	~ 0.026	40	20
$\begin{bmatrix} 11.66 \pm 0.04 \\ 11.66 \\ 11.68 \pm 0.04 \end{bmatrix}$	68 ± 13 40 ± 5	0.62 ± 0.03 0.59 ± 0.04	59 ± 11 36 ± 6 (40)	9 ± 2 3.5 ± 0.6 6
$\begin{bmatrix} 12.4 \\ 12.39 \pm 0.04 \\ 12.39 \\ 12.40 \pm 0.04 \end{bmatrix}$	65 ± 9 69 ± 6	1.29 ± 0.05 1.29 ± 0.06	42 ± 6 44 ± 6 (40)	28 ± 4 24 ± 3 15
$\begin{bmatrix} 12.82 \pm 0.04 \\ 12.85 \\ 13.00 \pm 0.08 \end{bmatrix}$	100 ± 40 83 ± 13	0.046 ± 0.006 0.040 ± 0.015	39 ± 18 23 ± 12 (40)	61 ± 24 60 ± 15 13
$\begin{bmatrix} 13.28 \pm 0.05 \\ 13.28 \\ 13.40 \pm 0.05 \end{bmatrix}$	90 ± 30	0.057 ± 0.009 0.055 ± 0.015	45 ± 17 (40)	51 ± 16 4
$\begin{bmatrix} 13.67 \pm 0.10 \\ 13.71 \end{bmatrix}$	135 ± 48	0.055 ± 0.023 0.040 ± 0.015	40	95 ± 34
$\begin{bmatrix} 13.98 \pm 0.05 \\ 13.98 \\ 14.1 \pm 0.1 \end{bmatrix}$	214 ± 102	0.40 ± 0.09 0.34 ± 0.07	40 (40)	174 ± 83 74

$2g\Gamma_n{}^0$ (mV)	θ (Deg)	Spin, etc.	$^{235}_{92}\text{U}$ $7.13 \times 10^8 \, y$ Reference
0.013 ± 0.002			Wang 65
0.0140 ± 0.0015			Michaudon 64
	180		Ignatiev 64
0.118 ± 0.008			Wang 65
0.103 ± 0.006			Michaudon 64
	180	$\eta/\nu = 0.21 \pm 0.02$	Ignatiev 64
0.100	90		Shore 58
0.043 ± 0.002			Wang 65
0.047 ± 0.002			Michaudon 64
		$\eta/\nu = 0.38 \pm 0.03$	Ignatiev 64
		$J = 3$	Poortmans 66
0.39 ± 0.02			Wang 65
0.40 ± 0.02			Michaudon 64
		$\eta/\nu = 0.55 \pm 0.03$	Ignatiev 64
0.257	90		Shore 58
0.048 ± 0.005			Wang 65
0.066 ± 0.012			Michaudon 64
		$\eta/\nu = 0.38 \pm 0.10$	Ignatiev 64
0.015 ± 0.005			Wang 65
0.0083 ± 0.0015			Michaudon 64
0.021 ± 0.002			Wang 65
0.020 ± 0.002			Michaudon 64
		$\eta/\nu = 0.15 \pm 0.05$	Ignatiev 64
~ 0.007			Wang 65
0.0050 ± 0.0025			Michaudon 64
~ 0.008			Wang 65
0.182 ± 0.009			Wang 65
0.173 ± 0.012			Michaudon 64
		$\eta/\nu = 0.13 \pm 0.04$	Ignatiev 64
		$J = 4$	Poortmans 66
0.367 ± 0.014			Wang 65
0.368 ± 0.020			Michaudon 64
		$\eta/\nu = 0.27 \pm 0.04$	Ignatiev 64
0.013 ± 0.002			Wang 65
0.0111 ± 0.0012			Michaudon 64
		$\eta/\nu = 0.25 \pm 0.12$	Ignatiev 64
0.016 ± 0.002			Wang 65
0.015 ± 0.004			Michaudon 64
		$\eta/\nu = 0.1$	Ignatiev 64
0.015 ± 0.006			Wang 65
0.011 ± 0.004			Michaudon 64
0.11 ± 0.03			Wang 65
0.090 ± 0.020			Michaudon 64
		$\eta/\nu = 0.65 \pm 0.10$	Ignatiev 64

TABLE III

E_0 (eV)	Γ (mV)	$2g\Gamma_n$ (mV)	Γ_γ (mV)	Γ_f (mV)

$I = \frac{7}{2}-$

E_0 (eV)	Γ (mV)	$2g\Gamma_n$ (mV)	Γ_γ (mV)	Γ_f (mV)
⎡ 14.50 ± 0.06	73 ± 12	0.18 ± 0.04	40	33 ± 5
⎢ 14.54	52 ± 8		29 ± 9	23 ± 7
⎣ 14.5 ± 0.1			(40)	40
⎡ 15.42 ± 0.05	93 ± 27	0.24 ± 0.01	49 ± 14	44 ± 13
⎢ 15.40	98 ± 15	0.25 ± 0.02	49 ± 13	49 ± 10
⎣ 15.5 ± 0.1			(40)	33
⎡ 16.08 ± 0.05	41 ± 10	0.35 ± 0.02	31 ± 7	10 ± 3
⎢ 16.08	56 ± 7	0.37 ± 0.02	37 ± 6	19 ± 4
⎣ 16.2 ± 0.1			(40)	17
⎡ 16.66 ± 0.06	74 ± 19	0.30 ± 0.01	30 ± 8	44 ± 12
⎢ 16.66	138 ± 15	0.28 ± 0.02	52 ± 15	86 ± 15
⎣ 16.7 ± 0.1			(40)	37
16.90 ± 0.10	53	∼0.05	40	13
⎡ 18.05 ± 0.06	108 ± 26	0.36 ± 0.03	36 ± 10	72 ± 18
⎢ 18.05	160 ± 20	0.36 ± 0.03	70 ± 22	90 ± 20
⎣ 18.2 ± 0.1			(40)	37
⎡ 18.6 ± 0.1	107	∼0.07	40	67
⎣ 18.98		0.084 ± 0.020		
⎡ 19.30 ± 0.05	108 ± 10	3.2 ± 0.13	48 ± 6	60 ± 7
⎢ 19.29	105 ± 10	3.10 ± 0.15	50 ± 8	52 ± 7
⎣ 19.4 ± 0.1			(40)	24
⎡ 20.10 ± 0.08	97 ± 41	0.09 ± 0.03	40	57 ± 24
⎣ 20.15		0.13 ± 0.02		
⎡ 20.62 ± 0.06	62 ± 10	0.25 ± 0.04	40	22 ± 4
⎣ 20.62	92 ± 10	0.19 ± 0.02	59 ± 12	33 ± 6
⎡ 21.13 ± 0.05	60 ± 15	1.22 ± 0.12	37 ± 11	23 ± 6
⎣ 21.06	70 ± 6	1.58 ± 0.10	47 ± 7	21 ± 3
21.8 ± 0.1	170	∼0.02	40	130
22.13		0.006 ± 0.003		
22.4 ± 0.1	170	∼0.02	40	130
⎡ 22.99 ± 0.06	77 ± 17	0.46 ± 0.04	42 ± 11	35 ± 7
⎣ 22.93	92 ± 10	0.45 ± 0.03	50 ± 9	42 ± 6
⎡ 23.43 ± 0.15	47 ± 0.14	0.55 ± 0.14	40	7 ± 2
⎣ 23.40	37 ± 4	0.69 ± 0.06	29 ± 4	8.0 ± 1.5
⎡ 23.68 ± 0.07	140 ± 30	0.89 ± 0.16	54 ± 15	86 ± 31
⎣ 23.61		0.95 ± 0.10		
⎡ 24.25 ± 0.07	70 ± 23	0.33 ± 0.11	40	30 ± 10
⎣ 24.26		0.345 ± 0.040		
⎡ 24.41 ± 0.15	80 ± 24	0.14 ± 0.04	40	40 ± 12
⎣ 24.56		∼0.12		
⎡ 25.16 ± 0.16	113 ± 31	0.22 ± 0.06	40	73 ± 20
⎣ 25.19		∼0.16		
⎡ 25.56 ± 0.10	62 ± 17	0.61 ± 0.19	40	22 ± 6
⎣ 25.58		∼0.66		

406

$2g\Gamma_n^0$ (mV)	θ (Deg)	Spin, etc.	$^{235}_{92}$U 7.13 $\times$ 10^8 y Reference
0.047 ± 0.009			Wang 65
0.0330 ± 0.0025			Michaudon 64
		$\eta/\nu = 0.50 \pm 0.20$	Ignatiev 64
0.061 ± 0.003			Wang 65
0.064 ± 0.005			Michaudon 64
		$\eta/\nu = 0.45 \pm 0.04$	Ignatiev 64
0.087 ± 0.006			Wang 65
0.092 ± 0.005			Michaudon 64
		$\eta/\nu = 0.30 \pm 0.03$	Ignatiev 64
0.074 ± 0.003			Wang 65
0.0685 ± 0.0050			Michaudon 64
		$\eta/\nu = 0.48 \pm 0.05$	Ignatiev 64
~ 0.013			Wang 65
0.085 ± 0.006			Wang 65
0.085 ± 0.007			Michaudon 64
		$\eta/\nu = 0.48 \pm 0.05$	Ignatiev 64
~ 0.02			Wang 65
0.019 ± 0.005			Michaudon 64
0.73 ± 0.03			Wang 65
0.705 ± 0.035			Michaudon 64
		$\eta/\nu = 0.38 \pm 0.04$	Ignatiev 64
0.020 ± 0.006		$\sigma_0\Gamma_f = 3.5$	Wang 65
0.029 ± 0.005		$\sigma_0\Gamma_f = 5.0 \pm 1.5$	Michaudon 64
0.055 ± 0.008		$\sigma_0\Gamma_f = 5.8$	Wang 65
0.042 ± 0.004		$\sigma_0\Gamma_f = 4.35 \pm 0.30$	Michaudon 64
0.26 ± 0.03		$\sigma_0\Gamma_f = 33.0 \pm 1.0$	Wang 65
0.344 ± 0.020		$\sigma_0\Gamma_f = 29.0 \pm 1.5$	Michaudon 64
~ 0.0051			Wang 65
0.0013 ± 0.0006			Michaudon 64
~ 0.0055			Wang 65
0.096 ± 0.008		$\sigma_0\Gamma_f = 13.0 \pm 0.3$	Wang 65
0.094 ± 0.006		$\sigma_0\Gamma_f = 11.7 \pm 0.6$	Michaudon 64
0.11 ± 0.03		$\sigma_0\Gamma_f = 4.5$	Wang 65
0.142 ± 0.012		$\sigma_0\Gamma_f = 8 \pm 1$	Michaudon 64
0.18 ± 0.03		$\sigma_0\Gamma_f = 30.3$	Wang 65
0.195 ± 0.020		$\sigma_0\Gamma_f = 22 \pm 4$	Michaudon 64
0.067 ± 0.022		$\sigma_0\Gamma_f = 7.5$	Wang 65
0.070 ± 0.008		$\sigma_0\Gamma_f = 8.0 \pm 1.5$	Michaudon 64
0.028 ± 0.009		$\sigma_0\Gamma_f = 3.9$	Wang 65
~ 0.0242			Michaudon 64
0.044 ± 0.012		$\sigma_0\Gamma_f = 7.4$	Wang 65
~ 0.0316			Michaudon 64
0.12 ± 0.04		$\sigma_0\Gamma_f = 11.0$	Wang 65
~ 0.13			Michaudon 64

TABLE III

$I = \frac{7}{2}-$				
E_0 (eV)	Γ (mV)	$2g\Gamma_n$ (mV)	Γ_γ (mV)	Γ_f (mV)
25.84 ± 0.15	90 ± 33	0.11 ± 0.04	40	50 ± 19
$\lceil\, 26.55 \pm 0.07$	138 ± 34	0.43 ± 0.07	40	98 ± 24
$\lfloor\, 26.48$		0.62 ± 0.03		
$\lceil\, 27.16 \pm 0.07$	74 ± 18	0.18 ± 0.06	40	34 ± 8
$\lfloor\, 27.15$		0.10 ± 0.03		
$\lceil\, 27.86 \pm 0.07$	131 ± 29	0.56 ± 0.06	40	91 ± 20
$\lfloor\, 27.80$	128 ± 15	0.72 ± 0.05	65 ± 17	62 ± 10
$\lceil\, 28.45 \pm 0.09$	97 ± 27	0.17 ± 0.03	40	57 ± 16
$\lfloor\, 28.36$	140 ± 30	0.16 ± 0.02	44 ± 26	96 ± 30
$\lceil\, 28.85 \pm 0.09$	69 ± 25	0.10 ± 0.03	40	29 ± 10
$\lfloor\, 28.69$		0.06 ± 0.03		
$\lceil\, 29.69 \pm 0.09$	62 ± 10	0.19 ± 0.03	40	22 ± 4
$\lfloor\, 29.64$	73 ± 10	0.18 ± 0.01	45 ± 9	28 ± 6
$\lceil\, 30.55 \pm 0.20$	70 ± 21	0.22 ± 0.05	40	30 ± 9
$\lfloor\, 30.59$	150 ± 15	0.21 ± 0.02	73 ± 23	77 ± 18
$\lceil\, 30.86 \pm 0.10$	68 ± 19	0.40 ± 0.09	40	28 ± 8
$\lfloor\, 30.86$	60 ± 6	0.52 ± 0.05	42 ± 8	17 ± 3
$\lceil\, 32.10 \pm 0.09$	96 ± 10	1.97 ± 0.20	50 ± 7	46 ± 6
$\lfloor\, 32.07$	100 ± 10	1.95 ± 0.15	52 ± 11	46 ± 8
$\lceil\, 33.58 \pm 0.09$	65 ± 12	1.7 ± 0.2	40	25 ± 5
$\lfloor\, 33.52$	62 ± 8	1.92 ± 0.12	40 ± 7	20 ± 4
$\lceil\, 34.45 \pm 0.14$	69 ± 14	2.12 ± 0.34	40	29 ± 6
$\lfloor\, 34.34$	85 ± 10	2.20 ± 0.20	47 ± 12	34 ± 7
$\lceil\, 34.90 \pm 0.20$	65 ± 20	0.9 ± 0.3	40	25 ± 8
$\lfloor\, 34.83$	110 ± 30	0.9 ± 0.2	60 ± 30	49 ± 23
$\lceil\, 35.27 \pm 0.10$	114 ± 21	4.5 ± 0.7	40	74 ± 14
$\lfloor\, 35.17$	175 ± 25	4.5 ± 0.3	63 ± 24	107 ± 30
$\lceil\, 38.40 \pm 0.11$	95 ± 27	0.66 ± 0.18	40	55 ± 16
$\lfloor\, 38.31$		0.55 ± 0.06		
$\lceil\, 39.47 \pm 0.11$	73 ± 13	2.6 ± 0.4	40	33 ± 6
$\lfloor\, 39.41$	95 ± 10	2.5 ± 0.2	45 ± 11	47 ± 8

(*Continued*)

$2g\Gamma_n{}^0$ (mV)	θ (Deg)	Spin, etc.	${}^{235}_{92}\text{U}$ $7.13 \times 10^8\ y$ Reference
0.022 ± 0.109			Wang 65
0.08 ± 0.01		$\sigma_0\Gamma_f = 15.3$	Wang 65
0.120 ± 0.006			Michaudon 64
0.035 ± 0.011		$\sigma_0\Gamma_f = 3.9$	Wang 65
0.019 ± 0.006		$\sigma_0\Gamma_f = 4.0 \pm 1.5$	Michaudon 64
0.11 ± 0.01		$\sigma_0\Gamma_f = 17.8$	Wang 65
0.136 ± 0.010		$\sigma_0\Gamma_f = 16.5 \pm 1.0$	Michaudon 64
0.032 ± 0.006		$\sigma_0\Gamma_f = 4.6$	Wang 65
0.030 ± 0.004		$\sigma_0\Gamma_f = 5 \pm 1$	Michaudon 65
0.019 ± 0.006		$\sigma_0\Gamma_f = 2.0$	Wang 65
0.011 ± 0.005			Michaudon 64
0.035 ± 0.006		$\sigma_0\Gamma_f = 2.9$	Wang 65
0.033 ± 0.002		$\sigma_0\Gamma_f = 3.0 \pm 0.6$	Michaudon 64
0.040 ± 0.009		$\sigma_0\Gamma_f = 4.1$	Wang 65
0.038 ± 0.004		$\sigma_0\Gamma_f = 4.6 \pm 0.9$	Michaudon 64
0.072 ± 0.016		$\sigma_0\Gamma_f = 6.8$	Wang 65
0.094 ± 0.010		$\sigma_0\Gamma_f = 6.2 \pm 0.7$	Michaudon 64
0.35 ± 0.04		$\sigma_0\Gamma_f = 37.8$	Wang 65
0.344 ± 0.026		$\sigma_0\Gamma_f = 36 \pm 4$	Michaudon 64
0.29 ± 0.08		$\sigma_0\Gamma_f = 25.7$	Wang 65
0.332 ± 0.020		$\sigma_0\Gamma_f = 24.5 \pm 3.7$	Michaudon 64
0.36 ± 0.06		$\sigma_0\Gamma_f = 31.7$	Wang 65
0.375 ± 0.35		$\sigma_0\Gamma_f = 33 \pm 5$	Michaudon 64
0.15 ± 0.05		$\sigma_0\Gamma_f = 12.8$	Wang 65
0.15 ± 0.03		$\sigma_0\Gamma_f = 15 \pm 5$	Michaudon 64
0.76 ± 0.12		$\sigma_0\Gamma_f = 107.0$	Wang 65
0.758 ± 0.050		$\sigma_0\Gamma_f = 102 \pm 20$	Michaudon 64
0.11 ± 0.03		$\sigma_0\Gamma_f = 13.4$	Wang 65
0.089 ± 0.010		$\sigma_0\Gamma_f = 12 \pm 4$	Michaudon 64
0.42 ± 0.06		$\sigma_0\Gamma_f = 28.7$	Wang 65
0.397 ± 0.032		$\sigma_0\Gamma_f = 41 \pm 4$	Michaudon 64

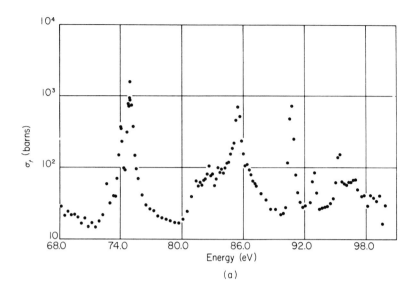

(a)

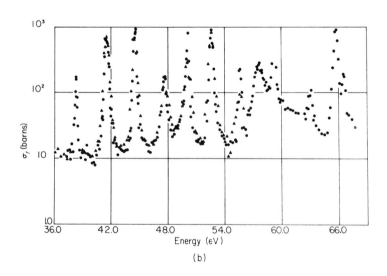

(b)

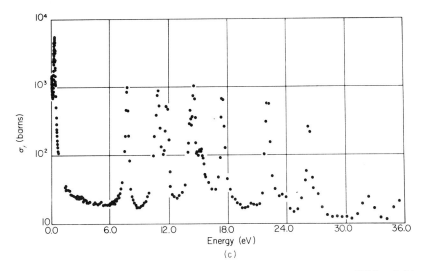

FIG. 20. The total cross section of ²³⁹Pu below 100 eV. Data are those of [(a) and (b), (●)], Uttley (1965) and [(c), (●) and (b), ▲], Bollinger *et al.* (1958). [This figure was supplied by B. A. Magurno of the National Neutron Cross-Section Center at Brookhaven National Laboratory (1967).]

3. ²³⁹PLUTONIUM

The fissile isotope ²³⁹Pu has a low spin, and the resonances are spaced considerably farther apart than in either ²³³U or ²³⁵U. On the average, the resonances also appear to be considerably narrower than in the uranium isotopes. This can be seen in Figs. 20–23 on pp. 410–415, which show the low-energy total and partial cross sections and the energy variation of eta for ²³⁹Pu. Analyses of ²³⁹Pu cross-section data have been reported by Bollinger *et al.* (1958), by Vogt (1960), by Kirpichnikov *et al.* (1964) (see also Ignatiev, 1964), by Derrien *et al.* (1967) and by Farrell (1968). The analysis of Derrien supersedes the preliminary analysis reported by de Saussure *et al.* (1965b). A summary of the resonance parameters is given in Table IV. Spin assignments based on scattering data have been reported by Fraser and Schwartz (1962), by Sauter and Bowman (1965), by Asghar (1967) and by Simpson *et al.* (1968). The results of spin assignments in ²³⁹Pu based on direct measurements are summarized in Table V. Included in parentheses are less definite assignments and assignments based on noninterference of resonance levels in multilevel analysis. It will be noted that in Table V there are very few resonances whose spin has been measured as zero. Indirect spin assignments have been made on the basis of observed fission-fragment mass distribution variations (Cowan *et al.*, 1966) and on fission-fragment kinetic

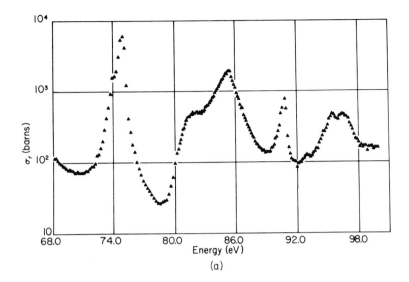

(a)

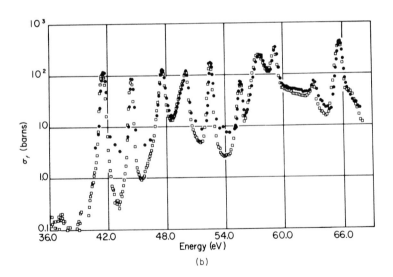

(b)

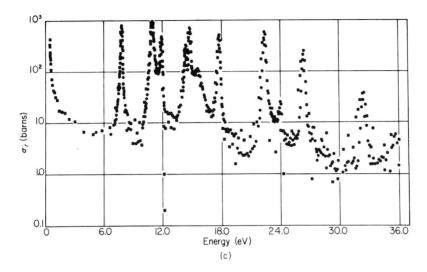

σ_f (barns)

Energy (eV)

(c)

FIG. 21. The fission cross section of ^{239}Pu below 100 eV. Data shown are those of: [(a) ▲, (b) ☐] Shunk *et al.* (1966); [(b) and (c), ●] Derrien *et al.* (1967); and [(c) ■] Bollinger *et al.* (1958). (This figure was supplied by B. A. Magurno of the National Neutron Cross-Section Center at Brookhaven National Laboratory (1967).)

energies (Melkonian and Mehta, 1965). These assignments indicate that there is a definite correlation between spin and the fission width. Resonances having spin of 1$^+$ are generally fairly narrow, while resonances to which 0$^+$ spin is attributed are generally quite low and wide. The resonance neutron scattering from such low, wide resonances is almost unobservable, making direct measurements by this technique very difficult. Michaudon (1967) has reviewed the existing evidence, both direct and indirect, on the spins of resonances in ^{239}Pu, and has concluded that the correlation between the average fission width and resonance spin is well established for this nuclide.

4. ^{241}Plutonium

In ^{241}Pu, there are also observed several low, broad resonances with sharp, narrow resonances superimposed among them. Here, the resonance scattering data have given substantial information on the resonance spins, supporting to a surprising degree the less accurate indirect evidence provided by multilevel analysis.

The low-energy total, fission, and scattering cross sections of ^{241}Pu are shown in Figs. 24–26 on pp. 424–428, for neutron energies below 60 eV. Analyses of the ^{241}Pu data were reported by Simpson and Moore (1961), by Pattenden (1964), by Craig and Westcott (1964), by Moore *et al.* (1964), by

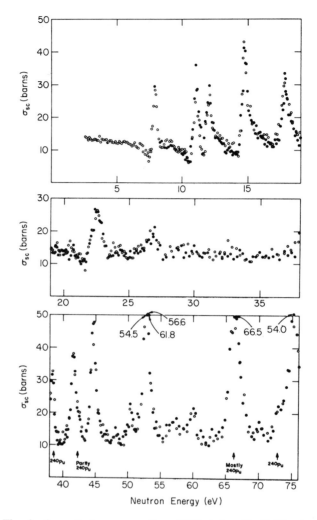

Fig. 22. The elastic scattering cross section of ^{239}Pu below 75 eV. The data are those of Simpson et al. (1968).

James (1965a) and by Simpson et al. (1966). A summary of the parameters obtained is given in Table VI. The results of the analyses indicate that the resonances fall into two groups, with markedly different average widths. Full interference appears to exist among resonances of each group, but not between members of different groups. The two groups were interpreted by Simpson and Moore (1961) and by Moore et al. (1964) as belonging to differ-

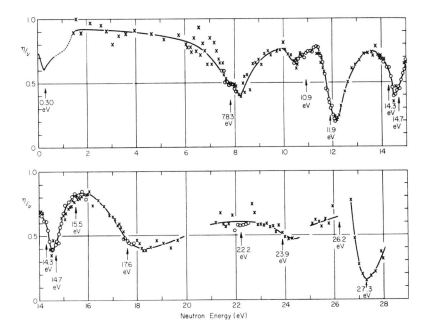

Fig. 23. The number of neutrons emitted per neutron absorbed (eta) for ²³⁹Pu below 30 eV. The data are those of Bollinger *et al.* (1958). (This figure was supplied by Dr. Bollinger.)

ent spin states in the compound nucleus, and the spin assignments by Sauter and Bowman (1968) are consistent with this interpretation.

5. ²²⁹THORIUM AND ²⁴²AMERICIUM

Measurements have also been made on the low-energy total and fission cross sections of ²²⁹Th, the lightest of the odd–even nuclides which is known to be fissile. The total cross section was measured by Coté *et al.* (1961), as shown in Fig. 27 (p. 436), and the fission cross section by Bollinger *et al.* (1963). Since the samples of ²²⁹Th amounted to only a few milligrams, only data over the resonance peaks were obtained. It was concluded that the resonance structure shows the expected large variation in the fission widths, in qualitative agreement with other even–odd fissile targets.

A measurement has also been carried out of the low-energy resonance-fission cross section of the odd–odd isotope ²⁴²Am, which has a 152 year isomeric state. These data, shown in Fig. 28 (p. 437), were obtained by Bowman *et al.* (1968) with a spark chamber detector (Bowman and Hill, 1963).

TABLE IV

$^{239}_{94}$Pu 24360 y			
E_0 (eV)	Γ (mV)	$2g\Gamma_n$ (mV)	Γ_γ (mV)
−1.200			39
−0.26	200 ± 30	0.085 ± 0.030	(40)
⎡0.296			38.6
⎣0.296	101 ± 2	0.121 ± 0.004	39 ± 2
⎡7.82	87 ± 5	1.21 ± .04	39.7 ± 4
7.84		1.33 ± 0.03	(40)
7.90			38
⎣7.83	83 ± 5	1.31 ± 0.03	40.6 ± 3.0
⎡10.93	200 ± 20	2.84 ± .15	55.1 ± 9
10.95		2.15 ± 0.20	(40)
11.0			32
⎣10.93	180 ± 12	2.73 ± 0.10	31.5 ± 9.0
⎡11.89	67 ± 7	1.54 ± 0.10	42.0 ± 4.6
11.95		1.35 ± 0.35	(40)
⎣11.90	64 ± 7	1.60 ± 0.08	40.9 ± 5.0
⎡14.29			40
14.31	102 ± 8	0.91 ± 0.05	
14.25			(40)
⎣14.28	101 ± 11	0.83 ± 0.20	(40)
⎡14.68			40
14.68	70 ± 7	2.88 ± 0.05	37.6 ± 4
14.7			(40)
⎣14.68	79.6 ± 8.0	3.30 ± 0.02	(40)
⎡15.38			40
15.46	700 ± 50	0.98 ± 0.08	
15.5		1.34 ± 0.20	
⎣15.5	800 ± 100	1.20 ± 0.20	(40)
⎡17.65			(40)
17.66	75 ± 7	2.74 ± 0.05	38.7 ± 4.7
17.7		2.2 ± 0.3	(40)
⎣17.6	87 ± 6	2.40 ± 0.10	39.1 ± 5.0
⎡22.26			(40)
22.29	109 ± 9	4.00 ± 0.10	44.3 ± 5.3
22.4		3.3 ± 0.3	(40)
⎣22.2	113 ± 4	3.37 ± 0.15	34.6 ± 4.0
⎡23.91			40
⎣23.94	70 ± 12	0.13 ± 0.01	40

[a] Resonance parameters for (^{239}Pu + n). These parameters are from single-level or multilevel (R-matrix) analysis. Resonance spin determinations are given separately, in Table V. The parameters in this table are given primarily for purposes of comparison. The fission width parameter phases which may be appropriate in a multilevel analysis are shown as signs following the fission widths, to indicate phases of $\pm\pi$. It is expected

Γ_f (mV)	$2g\Gamma_n{}^0$ (mV)	Miscellaneous	$I^\pi = \frac{1}{2}{}^+$ Reference
201+	1.156		Vogt 60
160 $\pm$ 20		$\eta/\nu = 0.80 \pm 0.05$	Bollinger 58
55.4	0.209		Vogt 60
62 $\pm$ 2		$\eta/\nu = 0.61 \pm 0.01$	Bollinger 58
47 $\pm$ 3		$\eta/\nu = 0.54$	Derrien 67
35$-$		$\eta/\nu = 0.46 \pm 0.05$	Ignatiev 64
42*	0.47	*Phase angle = 127°	Vogt 60
41.5 $\pm$ 3.0		$\eta/\nu = 0.49 \pm 0.01$	Bollinger 58
143 $\pm$ 16		$\eta/\nu = 0.72$	Derrien 67
81$-$		$\eta/\nu = 0.67 \pm 0.06$	Ignatiev 64
147$-$	0.83		Vogt 60
146.7 $\pm$ 7.0		$\eta/\nu = 0.73 \pm 0.01$	Bollinger 58
24 $\pm$ 3		$\eta/\nu = 0.37$	Derrien 67
23+		$\eta/\nu = 0.42 \pm 0.05$	Ignatiev 64
22.0 $\pm$ 2.0		$\eta/\nu = 0.34 \pm 0.02$	Bollinger 58
67$-$	0.275		Farrell 68
67 $\pm$ 7		$\eta/\nu = .66$	Derrien 67
34		$\eta/\nu = 0.46 \pm 0.10$	Ignatiev 64
60 $\pm$ 7		$\eta/\nu = 0.60 \pm 0.01$	Bollinger 58
68+	0.644		Farrell 68
30 $\pm$ 3		$\eta/\nu = 0.44$	Derrien 67
23		$\eta/\nu = 0.37 \pm 0.08$	Ignatiev 64
33 $\pm$ 4		$\eta/\nu = 0.45 \pm 0.01$	Bollinger 58
608+	0.268		Farrell 68
650		$\eta/\nu = 0.97$	Derrien 67
1000		$\eta/\nu = 0.9 \pm 0.1$	Ignatiev 64
758 $\pm$ 100		$\eta/\nu = 0.81 \pm 0.01$	Bollinger 58
68.4$-$	0.585		Farrell 68
34 $\pm$ 4		$\eta/\nu = 0.47$	Derrien 67
29		$\eta/\nu = 0.43 \pm 0.05$	Ignatiev 64
46.3 $\pm$ 5.0		$\eta/\nu = 0.46 \pm 0.02$	Bollinger 58
47.2+	0.878		Farrell 68
62 $\pm$ 6		$\eta/\nu = 0.58$	Derrien 67
49		$\eta/\nu = 0.54 \pm 0.02$	Ignatiev 64
75.0 $\pm$ 4.0		$\eta/\nu = 0.58 \pm 0.02$	Bollinger 58
14.8$-$			Farrell 68
55	0.057	$\eta/\nu = 0.79$	Derrien 67

that any single set of parameters will give a more accurate representation of the various cross sections than would an average of the individual resonance parameters. For this reason, no recommended set of parameters has been obtained. A somewhat shortened notation has been used for the referencing in this table; the more complete notation is given in the text. Spin information on resonances in ^{239}Pu is given separately.

TABLE IV

E_0 (eV)	Γ (mV)	$2g\Gamma_n$ (mV)	Γ_γ (mV)
$\lceil$ 23.9	82 ± 10	0.128 ± 0.010	(40)
$\lceil$ 26.25			40
26.24	83 ± 10	2.20 ± 0.15	37 ± 6
26.5		2.2 ± 0.4	(40)
$\lfloor$ 26.2	80 ± 6	2.66 ± 0.20	(40)
$\lceil$ 27.25			40
27.24	42 ± 8	0.12 ± 0.01	34 ± 8
$\lfloor$ 27.3	43 ± 5	0.199 ± 0.040	(40)
$\lceil$ 32.34			40
32.31	153 ± 20	0.42 ± 0.02	41 ± 8
32.5			
$\lfloor$ 32.3	230 ± 50	0.43 ± 0.04	(40)
34.30			40
$\lceil$ 35.47			40
35.50	47 ± 9	0.43 ± 0.02	42.4 ± 8.5
$\lfloor$ 35.3	44 ± 5	0.47 ± 0.04	(40)
$\lceil$ 41.43			40
$\lfloor$ 41.42	52 ± 8	6.20 ± 0.20	43.5 ± 8
$\lceil$ 41.72			40
$\lfloor$ 41.66	105 ± 16	2.02 ± 0.25	49
41.7		4.6 ± 0.5	
41.4	76 ± 10	9.25 ± 0.25	46.8 ± 9.0
$\lceil$ 44.51			40
44.48	58 ± 7	9.97 ± 0.20	46.8 ± 7
44.6		5.5 ± 1.0	
$\lfloor$ 44.5	51 ± 10	9.6 ± 0.2	40.4 ± 10.0
$\lceil$ 47.64			40
47.60	322 ± 25	2.90 ± 0.15	74 ± 16
47.8		3.0 ± 1.0	(40)
$\lfloor$ 47.6	350 ± 50	2.7 ± 0.3	(40)
$\lceil$ 49.60			40
$\lfloor$ 49.71	810 ± 200	2.2 ± 0.2	
$\lceil$ 50.10			40
$\lfloor$ 50.08	57 ± 10	4.55 ± 0.20	42.0 ± 8
49.9		5.1 ± 1.5	(40)
50.0	80 ± 6	6.8 ± 0.4	(40)
$\lceil$ 52.59			40
52.60	68 ± 10	15.70 ± 0.30	48.6 ± 9
52.7		8.7 ± 1.0	(40)
$\lfloor$ 52.6	57 ± 7	22.8 ± 1.5	34.1 ± 5.0
$\lceil$ 55.66			40
$\lfloor$ 55.63	59	2.20 ± 0.12	35

$^{239}_{94}$Pu
24360 y

Γ_f (mV)	$2g\Gamma_n^0$ (mV)	Miscellaneous	$I^\pi = \frac{1}{2}^+$ Reference
42 ± 5		$\eta/\nu = 0.51 \pm 0.03$	Bollinger 58
$44-$	0.540		Farrell 68
44 ± 7		$\eta/\nu = 0.55$	Derrien 67
25		$\eta/\nu = 0.59 \pm 0.12$	Ignatiev 64
37.3 ± 3.0			Bollinger 58
$1.6-$	0.120		Farrell 68
8 ± 4		$\eta/\nu = 0.19$	Derrien 67
3 ± 1		$\eta/\nu = 0.07 \pm 0.02$	Bollinger 58
$99.1+$	0.074		Farrell 68
110 ± 15		$\eta/\nu = 0.73$	Derrien 67
		$\eta/\nu = 0.48 \pm 0.2$	Ignatiev 64
190 ± 50			Bollinger 58
$50-$	0.0006		Farrell 68
$25-$	0.015		Farrell 68
5 ± 2		$\eta/\nu = 0.10$	Derrien 67
4 ± 1		$\eta/\nu = 0.10 \pm 0.02$	Bollinger 58
$39.4+$	0.152		Farrell 68
3		$\eta/\nu = 0.05$	Derrien 67
$79.2-$	0.168		Farrell 68
54		$\eta/\nu = 0.52$	Derrien 67
13		$\eta/\nu = 0.24 \pm 0.05$	Ignatiev 64
10.7 ± 2.0			Bollinger 58
$23.8+$	0.330		Farrell 68
5 ± 1		$\eta/\nu = 0.09$	Derrien 67
5.5		$\eta/\nu = 0.11 \pm 0.03$	Ignatiev 64
4.2 ± 1.0			Bollinger 58
$230+$	0.359		Farrell 68
240 ± 24		$\eta/\nu = 0.76$	Derrien 67
74		$\eta/\nu = 0.59 \pm 0.12$	Ignatiev 64
310 ± 50			Bollinger 58
$900+$	0.268		Farrell 68
690 ± 200		$\eta/\nu = 0.86$	Derrien 67
$23.7-$	0.346		Farrell 68
12 ± 3		$\eta/\nu = 0.22$	Derrien 67
27		$\eta/\nu = 0.65 \pm 0.09$	Ignatiev 64
33.2 ± 4.0			Bollinger 58
$32.4+$	0.608		Farrell 68
9 ± 2		$\eta/\nu = 0.16$	Derrien 67
23		$\eta/\nu = 0.33 \pm 0.04$	Ignatiev 64
7.7 ± 2.0			Bollinger 58
$43.7-$	0.214		Farrell 68
22		$\eta/\nu = 0.40$	Derrien 67

TABLE IV

$^{239}_{94}$Pu 24360 y			
E_0 (eV)	Γ (mV)	$2g\Gamma_n$ (mV)	Γ_γ (mV)
⌈ 57.30			40
| 57.44	500	6.5	
⌊ 57.8			(40)
58.84	1100	5.5	
⌈ 59.22			40
| 59.22	191 ± 16	8.37 ± 0.46	50 ± 10
⌊ 59.8			(40)
⌈ 60.94	6000	15	
⌊ 62.70			40
⌈ 63.16			40
⌊ 63.08	155 ± 17	1.20 ± 0.25	43
⌈ 65.75			40
| 65.71	137 ± 14	18.22 ± 0.50	49.5 ± 8
⌊ 65.9		19.5 ± 4.0	
66.75			40
66.57	180	1.26 ± 0.25	
⌈ 74.19			
⌊ 74.05	71 ± 8	4.75 ± 0.20	37 ± 5
⌈ 74.97			
⌊ 74.95	147 ± 14	33.20 ± 1.50	40.7 ± 6
74.5		29 ± 14	(40)

(Continued)

Γ_f (mV)	$2g\Gamma_n{}^0$ (mV)	Miscellaneous	$I^\pi = \frac{1}{2}^+$
			Reference
1040+	1.304		Farrell 68
			Derrien 67
160		$\eta/\nu = 0.87 \pm 0.09$	Ignatiev 64
			Derrien 67
141−	0.845		Farrell 68
133 ± 15		$\eta/\nu = 0.73$	Derrien 67
160		$\eta/\nu = 0.82 \pm 0.08$	Ignatiev 64
			Derrien 67
4250−	0.844		Farrell 68
54.4−	0.130		Farrell 68
111 ± 37		$\eta/\nu = 0.72$	Derrien 67
127−	1.401		Farrell 68
74 ± 8		$\eta/\nu = 0.60$	Derrien 67
89		$\eta/\nu = 0.61 \pm 0.07$	Ignatiev 64
1355+	0.258		Farrell 68
		$\eta/\nu = 0.57$	Derrien 67
91.9+	0.578		Farrell 68
32 ± 4		$\eta/\nu = 0.47$	Derrien 67
148+	2.840		Farrell 68
84 ± 9		$\eta/\nu = 0.67$	Derrien 67
49		$\eta/\nu = 0.55 \pm 0.10$	Ignatiev 64

TABLE V

SPIN ASSIGNMENTS FOR NEUTRON RESONANCES IN $(^{239}\text{Pu} + \text{n})$

Energy (eV)	Direct determinations spin J				Γ_f (meV)	Indirect determinations	
	Fraser and Schwartz (1962)	Sauter and Bowman (1965)	Asghar (1967)	Simpson et al. (1968)	Derrien et al. (1967)	Cowan et al. (1966)	Melkonian and Mehta (1965)
7.82	1	1		1	47		1
10.93	1	1		1	143		1
11.89	1	1		1	24		1
14.31			(1)	1	67		1
14.69	0	1	1	1	30		1
15.42		(0)			650	(1)	1
17.66	1	1	1	1	34	1	1
22.28	0	1	1	1	62	1	1
26.29			1	1	55	1	(1)
32.38		(0)		(0)	110	(0)	0
35.43		(1)			5		
41.42					3		
41.66	1	1	1	1	54	1	(0)
44.48	0	1	1	1	5	1	(1)
47.60		0	0	(1)	240	1	1
49.71					690	0	(0)
50.08			1	1	12	1	
52.60		1	1	1	9	1	1
57.44				0	$\sim$500	0	1
58.84					$\sim$1100	0	1
59.22			0	1	133	1	
60.94					$\sim$6000	0	
65.71			1	1	74	1	0
66.57					$\sim$140	0	
74.05			1		32		
74.95		1	1	1	84	1	1
81.76					$\sim$2000	0	

[a] Those assignments shown in parentheses are less certain. The values shown on the left represent direct determinations of the spins by resonance neutron scattering. Shown for comparison are values of the fission width as determined by Derrien et al. (1967). At the right are assignments of resonance spins based on the assumption (i) that the mass distribution is spin dependent (Cowan et al., 1966), and (ii) that the fragment kinetic energy distribution is spin dependent (Melkonian and Mehta, 1965). At energies below 50 eV, where the data are best, these two assumptions appear to be valid for ^{239}Pu.

The low-energy resonance cross section shows a very complex structure. The level density appears to be higher than that of ^{233}U, and the average fission width larger. There seem to be pronounced interference effects among the fission resonances, but a conventional resonance analysis appears to be virtually impossible, because of resonance overlap. Nevertheless, it is of interest to note that the data are consistent with what appears to be strong interference, indicating the presence of only a few fission channels.

6. SUMMARY OF RESULTS

The results of analyzing the resonance structure of the common fissile targets may be summarized as follows: Although the analyses are not unique, certain conclusions concerning the fission widths have been drawn which appear to be reasonably independent of the assumptions made. Multilevel analyses have been carried out for four fissile isotopes, ^{233}U, ^{235}U, ^{239}Pu, and ^{241}Pu. In all cases, there is evidence that the resonances show a separation into two groups. This separation is based on the presence or absence of interference effects among various resonances. In all cases, the groups of resonances have different average fission widths. For ^{239}Pu and ^{241}Pu, the average fission widths for the two groups are quite different, For ^{235}U and ^{233}U, the average fission widths may be only slightly different, and the average radiation widths for the two groups also appear to be somewhat different. For ^{239}Pu, direct measurements of the resonance spin and other indirect evidence from fission-fragment mass-yield studies strongly support the idea that the narrow resonances are to be associated with spin 1$^+$, the broad resonances with spin 0$^+$. For ^{241}Pu, the narrow resonances are to be associated with spin 3$^+$, the broad ones with spin 2$^+$.

In the near future, one can expect the analysis of alpha data on ^{233}U, ^{235}U, and ^{239}Pu in the region above that discussed here. These data, obtained by Weston et al. (1967), by de Saussure et al. (1967), and by Gwin et al. (1968), represent simultaneous measurements, at the same high resolution, of both capture and fission cross sections. Their analysis can be expected to give substantially more evidence in the higher energy regions.

Perhaps the most prolific source of new high-resolution data at higher neutron energies is the nuclear explosion. Early measurements, utilizing the nuclear explosion for neutron spectroscopy were pioneered by Cowan et al. (1961), Bame (1963), and Albert (1966). Some of the data obtained from the Petrel experiment were included in Figs. 17, 21, and 25. In this experiment, measurements were made of the fission cross section and capture-to-fission ratio of ^{233}U (Bergen et al., 1966; Bergen and Silbert, 1968) and of the fission cross sections of ^{235}U (Brown et al., 1966), ^{239}Pu

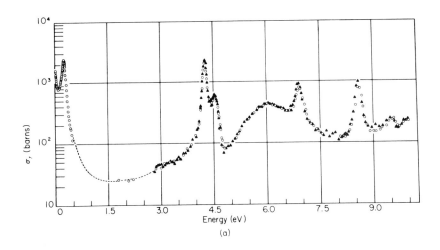

(a)

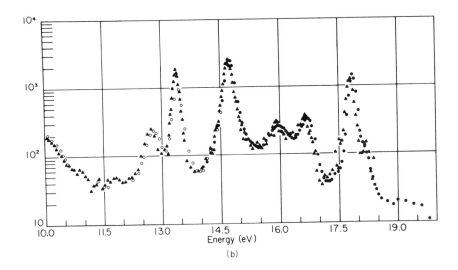

(b)

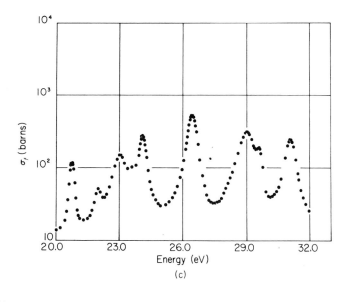

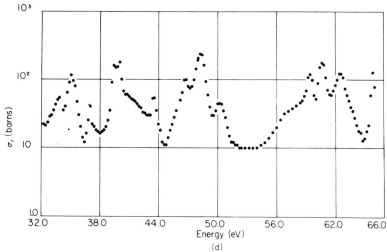

FIG. 24. The total cross section of ^{241}Pu below 60 eV. Data are those of: [(a) and (b) ○] Simpson and Schuman (1961); [(b), (c), (d) ●] Craig and Westcott (1964); and [(a) and (b) ▲] Pattenden and Bardsley (1963). (This figure was supplied by B. A. Magurno of the National Neutron Cross-Section Center at Brookhaven National Laboratory (1967).)

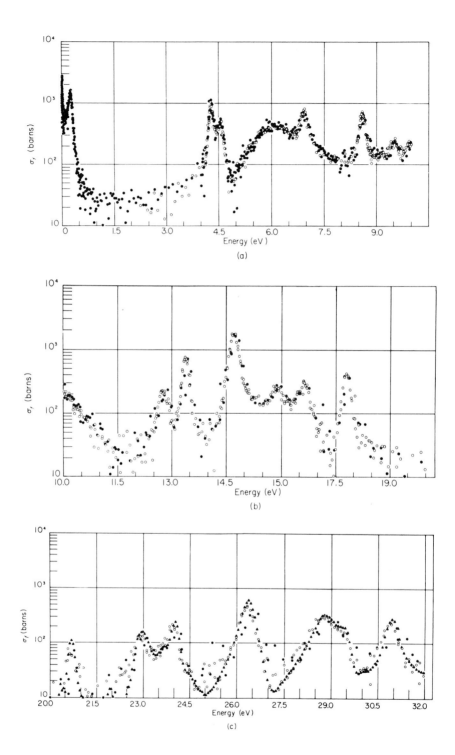

(a)

(b)

(c)

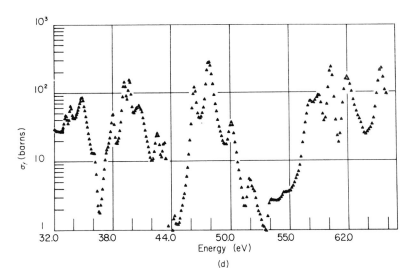

FIG. 25. The fission cross section of ^{241}Pu below 70 eV. Data are those of: [(c) and (d) ▲] Simpson *et al.* (1966); [(a), (b), (c) ●] James (1965a); and [(a), (b), (c) ○] Moore *et al.* (1964). (This figure was supplied by B. A. Magurno of the National Neutron Cross-Section Center at Brookhaven National Laboratory (1967).)

(Shunk *et al.*, 1966; Farrell, 1968), and ^{241}Pu (Simpson *et al.*, 1966). Reported also were measurements of the fission cross sections of ^{241}Am and ^{242}Am, (Seeger *et al.*, 1967), of fission and capture cross sections of ^{240}Pu (Byers *et al.*, 1966), of the capture cross section of ^{238}U (Glass *et al.*, 1966), and of mass-distribution studies of ^{239}Pu resonances (Cowan *et al.*, 1966), all carried out with neutrons from a single nuclear explosion. Since the Petrel experiment, there have been others, notably Persimmon in 1967 and Pommard in 1968. Additional results can be expected in the near future on ^{232}U, ^{233}U, ^{233}Pa, ^{234}U, ^{235}U, ^{236}U, ^{237}U, ^{237}Np, ^{238}Pu, ^{239}Pu, ^{241}Pu, ^{243}Am, and ^{244}Cm from these two events.

IV. FRAGMENT ANGULAR DISTRIBUTIONS AND FISSION NEAR THRESHOLD

A. The Fission Threshold

Slow-neutron spectroscopy permits a separation of the neutron energy into two regions of interest. The region of resolved and partially resolved resonances for heavy nuclei ranges up to perhaps 10^3–10^4 eV. At higher energies, the average properties and systematic effects are of primary interest. For fissionable nuclides, these systematics show important properties of the fission channel spectrum at the top of the fission threshold, or saddle point.

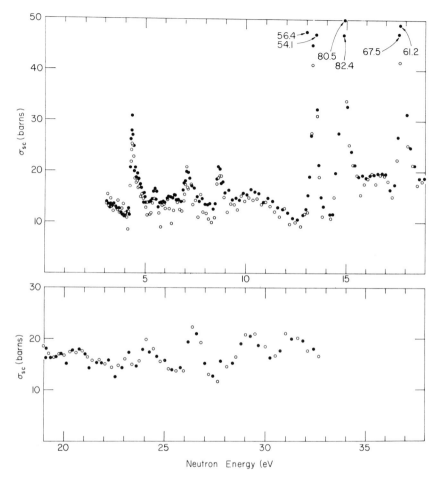

Fig. 26. The elastic scattering cross section of ^{241}Pu below 60 eV. The data are those of Sauter and Bowman (1968). (They were supplied by Dr. Bowman prior to publication.)

If a heavy nucleus absorbs a neutron having kinetic energy in the region below 1 keV, the excitation given to the resulting compound nucleus is almost entirely due to the binding energy of the neutron, and the kinetic energy of the neutron can be neglected. In most cases, such a compound nucleus can lose its excitation energy by the emission of one or many gamma rays (radiative capture), by the reemission of a slow neutron (elastic scattering), or sometimes by fissioning. If fission of the compound nucleus formed by slow-neutron capture is an important mode of de-excitation, then the target nuclide is termed *fissile*. A target nuclide is termed *fissionable** if it can be made to undergo fission by any means, not restricted to the absorption of a slow neutron. In an example of this terminology, one may say that ^{238}U is fissionable by 2 MeV neutrons or by 6 MeV photons and that ^{209}Bi is fissionable by 22 MeV deuterons. One may also say that ^{235}U is fissionable by 0.0253 eV neutrons, or, simply, that ^{235}U is fissile.

These examples are fairly easy to define. However, certain nuclides such as ^{238}Pu and ^{240}Pu show only a small fission component to the resonances. It is generally more convenient to refer to such nuclides as showing *subthreshold fission* under low-energy neutron bombardment.

Since fission is an exoergic reaction, it is not strictly correct to talk about a threshold for fission. The nuclide ^{238}U, for example, shows a detectable spontaneous fission rate, yet it is not usually said to be fissionable by low-energy neutrons, because the rate of fission under slow-neutron bombardment is not detectably higher than the rate of spontaneous fission. The excitation energy contributed by a neutron added to ^{238}U ($\sim$5 MeV) is still low enough that the coulomb barrier for two fission fragments effectively prevents fission of the compound nucleus ^{239}U. It is convenient, however, to define an effective fission threshold which is related to a change in the rate of fission. Figure 29 (p. 438) shows the neutron energy dependence of the fission cross section for several even–even isotopes of uranium (Lamphere, 1956). In ^{238}U, which appears to be the simplest of these, the fission cross-section curve rises smoothly and reaches a more or less constant value at about 2 MeV. This sharp rise is due to the quantum mechanical crossing of the coulomb barrier; an incident neutron with energy above 0.5 MeV contributes enough excitation to the compound nucleus that the probability of crossing the barrier by nascent fission fragments becomes appreciable. One often defines an effective threshold as that energy at which the cross section reaches one-half the value observed at the plateau [cf. Eq. (5), Section I]. If it is assumed that fission well above threshold dominates all other reaction processes, it can be shown that the effective threshold corresponds to that energy at which the average fission width is equal to the sum of the other partial widths. In the low-energy neutron-resonance region, this is the energy at which the average fission width is comparable to the radiative capture width.

The fission threshold is not uniquely defined, then, since it depends on the magnitude of competing processes. Since fission is exoergic, the observation of fission in any given experimental measurement depends to a large

* The distinction between *fissile* and *fissionable* is made in accordance with the definition proposed by the International Standards Organization, as described by Blizard (1961).

TABLE VI

$I = \frac{5}{2}^{+}$			
E_0 (eV)	Γ (mV)	$2g\Gamma_n$ (mV)	Γ_γ (mV)
−0.160			(40)
⌈0.264			(40)
⌊0.25	125		30
⌈4.27			39
4.275			(40)
⌊4.28			(40)
⌈4.57			25
4.580			(40)
⌊4.56			(40)
⌈5.92			40
5.910			(40)
⌊5.91			(40)
⌈6.91			40
6.915			(40)
⌊6.94			(40)
⌈8.57			30
8.585			(40)
⌊8.60			(40)
⌈9.50			25
9.48			(40)
⌊9.56			(40)
⌈10.20			35
10.11			(40)
⌊10.20			(40)
⌈12.74			40
12.75 ± 0.07	235 ± 30		
12.84			(40)
⌊12.77			(40)
⌈13.39			57
13.39 ± 0.08	55 ± 10		
13.45			(40)
⌊13.38			(40)
14.04			(40)
⌈14.70			35
14.75 ± 0.08	150 ± 20		
14.78			(40)
14.73			(40)
⌊14.74 ± 0.05	150 ± 50	6.2 ± 1.0	

[a] Resonance parameters for (^{241}Pu + n). In all cases, these resonance parameters are either single-level or multilevel (R-matrix) parameters. The resonance phases which are appropriate in a multilevel calculation are shown as signs following the fission widths. Recommended values of the parameters are not given; since multilevel analysis is not unique, any single individual set of parameters is expected to give a better representation of the cross section than is an average of several sets which may not have utilized the

Γ_f (mV)	$2g\Gamma_n^0$ (mV)	Spin assignments	$^{241}_{94}$Pu 13 y Reference
60−	0.0725	$J1$	Simpson 61
72+	0.101	$J1$	Simpson 61
95			Schwartz 58
32+	0.327	3	Sauter 68
21+	0.404	$J1$	Moore 64
45+	0.255	$J1$	Simpson 61
142−	0.187	2	Sauter 68
140−	0.204	$J2$	Moore 64
190−	0.194	$J2$	Simpson 61
1330−	0.98	2	Sauter 68
1350−	1.020	$J2$	Moore 64
1350−	1.020	$J2$	Simpson 61
105−	0.258	3	Sauter 68
93−	0.275	$J1$	Moore 64
95−	0.218	$J1$	Simpson 61
85+	0.245	(3)	Sauter 68
70+	0.324	$J1$	Moore 64
70−	0.268	$J1$	Simpson 61
120−	0.063	3	Sauter 68
125−	0.068	$J1$	Moore 64
100+	0.035	$J1$	Simpson 61
990+	0.483	2	Sauter 68
900+	0.47	$J2$	Moore 64
1000−	0.400	$J2$	Simpson 61
250−	0.214	3	Sauter 68
	0.20 ± 0.06		Pattenden 64
266+	0.220	J	James 65
250−	0.22	$J2$	Moore 64
36−	0.552	2	Sauter 68
	0.70 ± 0.04		Pattenden 64
42−	0.596	Not J	James 65
50−	0.5	$J1$	Moore 64
214−	0.048	J	James 65
145+	1.80	3	Sauter 68
	1.67 ± 0.06		Pattenden 64
105+	1.513	J	James 65
135+	1.61	$J2$	Moore 64
	1.6 ± 0.3		Craig 64

same approach in analyzing the data. Unobserved small levels inserted by Simpson *et al.* (1966) to improve the fit have been deleted in this list. Parentheses denote assumed values or less certain values. Spin assignments made on the basis of multilevel interference could have been expected to be highly uncertain. Direct measurements of spin are those of Sauter and Bowman (1968). A shortened notation has been used for the referencing; the more complete notation is used in the text.

TABLE VI

$I = \frac{5}{2}^+$			
E_0 (eV)	Γ (mV)	$2g\Gamma_n$ (mV)	$\Gamma\gamma$ (mV)
15.98			40
15.94 ± 0.07	500 ± 100		
16.06			(40)
16.01			(40)
15.96 ± 0.08	600 ± 150	1.8	
16.68			35
16.6 ± 0.1	210 ± 20		
16.70			(40)
16.65			(40)
16.70 ± 0.08	250 ± 100	1.2	
17.86			80
17.8 ± 0.1	53 ± 20		
17.78			(40)
17.85 ± 0.05	50 ± 90	3.2 ± 0.8	
18.2 ± 0.1	67 ± 10		
20.5 ± 0.1	40 ± 20		
20.63			34
20.7			40
20.7 ± 0.1	90 ± 40		
20.63			(40)
20.75 ± 0.17	90 ± 100	0.32	
21.05			40
21.85			40
21.9 ± 0.1	100 ± 30		
21.99 ± 0.24	200 ± 100	0.13	40
22.88			60
22.99			40
23.0 ± 0.1	310 ± 40		
22.86			(40)
23.04 ± 0.30	600 ± 250	1.5	
23.66			40
23.7 ± 0.1	180 ± 40		
23.97			60
24.04			40
24.0 ± 0.1	190 ± 30		
23.96			(40)
24.12	220 ± 140	1.4 ± 0.9	40
24.57			40
24.7 ± 0.1	60 ± 30		
26.30			30
26.32			40
26.4 ± 0.2	290 ± 30		
26.34			(40)
26.45 ± 0.17	340 ± 100	4.3 ± 1.3	

Γ_f (mV)	$2g\Gamma_n^0$ (mV)	Spin assignments	$^{241}_{94}\text{Pu}$ 13 y Reference
475−	0.386	3	Sauter 68
	0.390 ± 0.015		Pattenden 64
360−	0.344	J	James 65
500−	0.36	$J2$	Moore 64
	0.45		Craig 64
350+	0.47	2	Sauter 68
	0.27 ± 0.02		Pattenden 64
180+	0.323	Not J	James 65
300+	0.36	$J1$	Moore 64
	0.28		Craig 64
37+	0.86	2	Sauter 68
	0.77 ± 0.04		Pattenden 64
80+	0.41	$J1$	Moore 64
	0.76 ± 0.20		Craig 64
	0.037 ± 0.004		Pattenden 64
	0.04 ± 0.02		Pattenden 64
59+		3	Sauter 68
35+	0.17	$J1$	Simpson 66
	0.077 ± 0.009		Pattenden 64
40−	0.08	$J1$	Moore 64
	0.070		Craig 64
300−	0.013	$J1$	Simpson 66
20−	0.010	$J1$	Simpson 66
	0.033 ± 0.007		Pattenden 64
	0.028		Craig 64
335−	0.272	3	Sauter 68
280+	0.245	$J1$	Simpson 66
	0.23 ± 0.02		Pattenden 64
400+	0.24	$J1$	Moore 64
	0.32		Craig 64
450+	0.14	$J2$	Simpson 66
	0.059 ± 0.009		Pattenden 64
185+	0.34	3	Sauter 68
70−	0.26	$J1$	Simpson 66
	0.33 ± 0.03		Pattenden 64
230−	0.31	$J1$	Moore 64
	0.29 ± 0.20		Craig 64
100−	0.013	$J2$	Simpson 66
	0.009 ± 0.007		Pattenden 64
315−	0.817	3	Sauter 68
270+	1.04	$J1$	Simpson 66
	0.88 ± 0.05		Pattenden 64
280+	0.82	$J1$	Moore 64
	0.84 ± 0.24		Craig 64

TABLE VI

E_0 (eV)	Γ (mV)	$2g\Gamma_n$ (mV)	Γ_γ (mV)
$I = \frac{5}{2}^+$			
28.0 ± 0.2	150 ± 50		
$\lceil$ 28.77			60
28.68			40
28.8 ± 0.2	560 ± 60		
28.75			(40)
$\lfloor$ 28.97 ± 0.22	720 ± 100		
$\lceil$ 29.33			35
29.59			40
29.4 ± 0.2	130 ± 30		
29.35			(40)
$\lfloor$ 29.57 ± 0.25	50 ± 50	0.36	
$\lceil$ 30.90			40
30.91			40
30.9 ± 0.2	350 ± 50		
30.88			(40)
$\lfloor$ 31.03 ± 0.21	360 ± 100	2.4 ± 0.6	
33.0 ± 0.2	400 ± 100		
33.27			40
$\lceil$ 33.65			40
$\lfloor$ 33.7 ± 0.2	500 ± 60		
$\lceil$ 34.72			40
35.0 ± 0.1	540 ± 60		
$\lfloor$ 34.90			(40)
$\lceil$ 36.00			40
$\lfloor$ 36.3 ± 0.2	200 ± 50		
37.37			40
$\lceil$ 38.08			40
$\lfloor$ 38.2 ± 0.2	170 ± 30		
$\lceil$ 39.22			40
$\lfloor$ 39.3 ± 0.2	190 ± 30		
$\lceil$ 39.80			40
$\lfloor$ 40.0 ± 0.2	240 ± 30		
$\lceil$ 40.88			40
$\lfloor$ 40.8 ± 0.1	300 ± 70		
41.7 ± 0.2	220 ± 50		
$\lceil$ 42.57			40
$\lfloor$ 42.7 ± 0.1	110 ± 50		
$\lceil$ 43.37			40
$\lfloor$ 43.2 ± 0.2	78 ± 50		
$\lceil$ 46.38			40
$\lfloor$ 46.7 ± 0.2	250 ± 50		
$\lceil$ 47.95			40
$\lfloor$ 48.2 ± 0.2	630 ± 100		
$\lceil$ 50.14			40
$\lfloor$ 50.4 ± 0.2	95 ± 50		

Γ_f (mV)	$2g\Gamma_n{}^0$ (mV)	Spin assignments	$^{241}_{94}$Pu 13 y Reference
	0.22 ± 0.01		Pattenden 64
690+	1.00	2	Sauter 68
750+	1.06	$J2$	Simpson 66
	0.81 ± 0.05		Pattenden 64
750+	1.12	$J2$	Moore 64
	0.93		Craig 64
70+	0.106	(3)	Sauter 68
50−	0.088	$J2$	Simpson 66
	0.15 ± 0.03		Pattenden 64
40−	0.10	$J1$	Moore 64
	0.066		Craig 64
320+	0.489	3	Sauter 68
220+	0.47	$J1$	Simpson 66
	0.55 ± 0.05		Pattenden 64
300+	0.45	$J1$	Moore 64
	0.43 ± 0.10		Craig 64
	0.07 ± 0.02		Pattenden 64
150−	0.044	$J1$	Simpson 66
100−	0.063	$J1$	Simpson 66
	0.20 ± 0.02		Pattenden 64
900−	0.34	$J2$	Simpson 66
	0.34 ± 0.02		Pattenden 64
1200−	0.45	$J2$	Moore 64
500+	0.034	$J2$	Simpson 66
	0.04 ± 0.02		Pattenden 64
500−	0.037	$J1$	Simpson 66
230+	0.09	$J1$	Simpson 66
	0.18 ± 0.04		Pattenden 64
160−	0.25	$J1$	Simpson 66
	0.22 ± 0.02		Pattenden 64
50+	0.26	$J1$	Simpson 66
	0.33 ± 0.04		Pattenden 64
1200−	0.38	$J2$	Simpson 66
	0.13 ± 0.02		Pattenden 64
	0.26 ± 0.04		Pattenden 64
250−	0.053	$J1$	Simpson 66
	0.018 ± 0.007		Pattenden 64
50+	0.034	$J1$	Simpson 66
	0.03 ± 0.01		Pattenden 64
280−	0.31	$J2$	Simpson 66
	0.22 ± 0.04		Pattenden 64
480+	1.05	$J1$	Simpson 66
	0.90 ± 0.70		Pattenden 64
300+	0.09	$J2$	Simpson 66
	0.08 ± 0.02		Pattenden 64

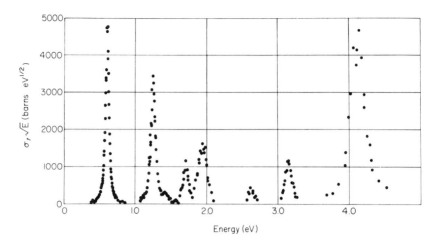

FIG. 27. The total cross section of ^{229}Th as a function of neutron energy below 5 eV, measured by Coté *et al.* (1961) at ANL. The quantity of sample material was limited to a few milligrams, so that only the resonance peaks could be observed. (These data were supplied by Dr. Coté prior to publication.)

extent on the sensitivity of the measuring apparatus. It follows that there is also some arbitrariness about the terms fissionable and fissile since one must decide at what point the fission cross section becomes appreciable, or comparable to that for other processes.

The preceding section has been restricted to a discussion of fission resonances observed in five even–odd heavy targets, ^{229}Th, ^{233}U, ^{235}U, ^{239}Pu, and ^{241}Pu and in one odd–odd target, ^{242}Am. These are among the nuclides which are most likely to be fissile because the addition of a neutron imparts the greatest amounts of binding energy. In addition, certain even-neutron nuclides, those on the neutron-poor side of the beta-stability curve, are expected to show appreciable fission for low-energy neutrons simply because the addition of a neutron to such nuclides contributes a relatively large binding energy. Examples are ^{232}U and ^{238}Pu. However, the systematics of these even–even targets appear to require that they be treated as though they were subthreshold fissionable, and so discussion has been deferred to the present section.

The common fissile nuclides do show the effects of a threshold behavior, as seen in Fig. 30 (Smith, 1962) at neutron energies about 6 MeV. Here the rise to a second plateau is the result of second-chance fission, or the (n,n'f) reaction. Bohr and Wheeler (1939) first discussed this effect: a neutron may be scattered inelastically, leaving the target nucleus with sufficient excitation to undergo fission.

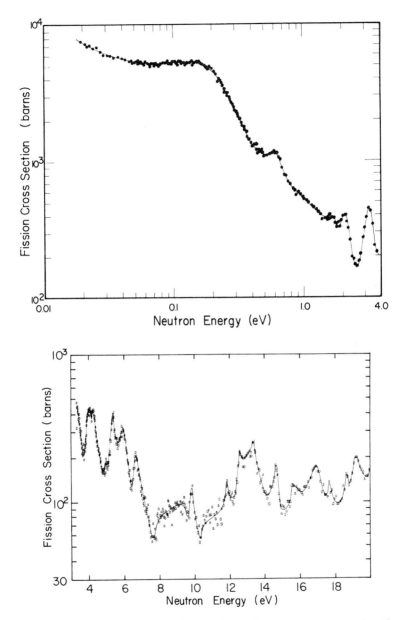

Fig. 28. The fission cross section of the 152 year isomer of ^{242}Am as a function of neutron energy below 20 eV. These data were obtained by Bowman, Hoff, Auchampaugh, and Fultz at LRL, Livermore in early 1964 (Bowman *et al.*, 1968). (This figure was supplied by Dr. Bowman prior to publication.)

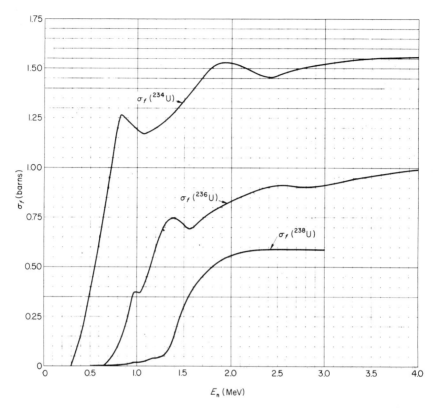

FIG. 29. Fission cross sections of ^{234}U, ^{236}U, and ^{238}U as a function of neutron energy near threshold, as measured by Lamphere (1956). The structure observed in these curves is usually attributed to two effects: the opening of new fission channels, which increases the effective fission cross section, and the opening of neutron inelastic scattering channels (Cranberg and Levin, 1958). These represent channels with which fission is in competition, effectively reducing the fission cross section which would otherwise be observed.

Since the common fissile nuclides show appreciable fission resulting from very low-energy neutron absorption, the first-chance fission thresholds must lie below the excitation afforded by the binding energy of a neutron added to the target nuclides. The fission threshold for fissile nuclides must therefore be studied by some method which gives less energy than the addition of a zero energy neutron, such as a (d,p) reaction with appropriate deuteron and proton energies. This method was first used by Northrop *et al.* (1959) for studying the fission thresholds of the common fissile isotopes. Such studies, when carried out at excitations below the neutron binding energy, are often called studies at negative neutron energies because the

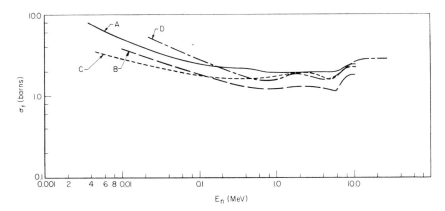

Fig. 30. The fission cross sections of the four common fissile nuclides: (A) $^{233}_{92}$U(——), (B) $^{235}_{92}$U(‐ ‐), (C) $^{239}_{94}$Pu(- -) and (D) $^{241}_{94}$Pu(‐ - ‐), as a function of neutron energy from several kilovolts to 10 MeV, from Smith (1962). Structure near 6 MeV is attributed to "second-chance" fission, the (n,n'f) reaction (see Bohr and Wheeler, 1939). (This figure was supplied by Dr. Smith.)

excitation of the compound nucleus is equal to that which would be attained if neutrons of negative-kinetic energies had been absorbed.

B. Fission-Fragment Angular Distributions in the Resonance Region

Historically, the subject of fission-fragment angular distributions is fundamental to fission theory since it was as an interpretation of the observed fission-fragment anisotropies that Bohr first proposed the channel theory of fission (Bohr, 1956). The idea discussed by Hill and Wheeler (1953) that the saddle point has multiple sheets, each with different quantum numbers and separated in energy, were further developed and made quantitative by Bohr. Bohr first assumed that one of the quantum numbers which could be used to characterize the different saddle-point states or channels is K, the projection of nuclear angular momentum on the nuclear symmetry axis. A second assumption was that the modes of motion which lead to fission are simple ones, analogous to rotations and vibrations which characterize the low-lying states of heavy deformed nuclei near the ground state. The third assumption Bohr made was that fission occurs such that the direction of the fragment emission is along the major axis of the axially symmetric deformed nucleus at the saddle point.

The first assumption is justified on the basis that the deforming of the nucleus to the saddle point is a slow process, compared to the time of rotation. The second and third assumptions, that the fission channels are equivalent to simple vibrations and rotations and that the fragment angular distribution represents the saddle-point distribution, are those which enable one to make quantitative calculations with the Bohr theory. The ground-state rotational and vibrational spectrum of even–even deformed nuclides is an extremely simple one (see, for example, Sheline, 1960; Hyde *et al.*, 1964). There is a ground-state band with $K = 0$ consisting of a series of rotational levels with spins 0^+, 2^+, 4^+, . . . ; with energy spacing proportional to $I (I + 1)$. At least two positive parity excited-state bands are known to exist, built on quadrupole vibrations and rotations. These have been characterized as a gamma-vibrational band with $K = 2$, $I = 2^+$, 3^+, 4^+, . . . , and a beta-vibrational band with $K = 0$, $I = 0^+$, 2^+, 4^+, ... Octupole vibrations are also known. The simplest of these has $K = 0$, $I = 1^-$, 3^-, 5^-, . . . ; and others, with $K = 1$, $I = 1^-$, 2^-, 3^-, . . . , and $K = 2$, $I = 2^-$, 3^-, 4^-, . . . , are expected (Lipas and Davidson, 1961; Huizenga *et al.*, 1965; Griffin, 1965). For odd deformed nuclides, the spectrum is built around single-particle states of the odd nucleon in a deformed potential (Nilsson, 1955; see also Mottelson and Nilsson, 1959). The Nilsson states may in turn be coupled to vibrations of the even–even core. The simplest of these states Bohr (1956) assumed as representing possible fission channels. First, Bohr considered the case of photofission of even–even nuclides. Dipole absorption of a photon by such a target leads to 1^- states with M_z (the projection of angular momentum along the beam direction) equal to ± 1. The lowest lying 1^- states in even–even nuclides have $K = 0$, and the lowest lying 1^- fission channels are also expected to have $K = 0$. Thus the nuclear symmetry axis is perpendicular to the photon beam direction, and the angular distribution of fragments, following the orientation of the nuclear symmetry axis, is of the form $w(\theta) = \sin^2 \theta$, where θ is measured with respect to the photon beam direction. At higher photon energies, one begins to get contributions from other channels, and the angular distribution becomes more nearly isotropic, in agreement with experiment (Winhold and Halpern, 1956). Bohr then went on to develop the form of the angular distribution to be expected for fast neutron fission (see also Griffin, 1959) as well as to comment upon the experiments of interest to slow neutron spectroscopy which were quoted in Section I,A.

One of these experiments is the measurement of fragment anisotropies from slow neutron-induced fission of aligned target nuclei. This experiment was one of the first and most significant tests of the Bohr theory, and shows the importance of the theory to low-energy neutron spectroscopy. The quantitative application of the Bohr theory to this experiment was given by Roberts *et al.* (1957) (see also Roberts *et al.*, 1958; Roberts and Dabbs, 1961). Recent experimental results are those of Dabbs *et al.* (1965). The experiment involved the cooling of a single crystal of $UO_2Rb(NO_3)_3$ with a

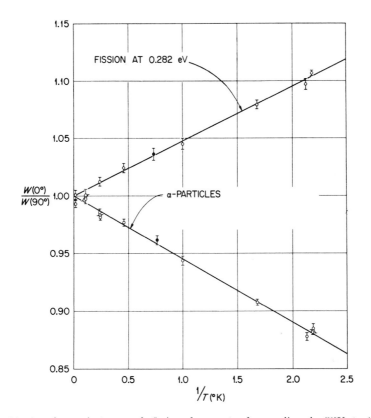

F<small>IG</small>. 31. Angular anisotropy of fission fragments from aligned $(^{235}U + n)$ in $UO_2Rb(NO_3)_3$ at 0.282 eV, as a function of sample temperature using carbon resistance thermometer for $1/T$ values. Fission at 0.282 eV corresponds to $(1 + 0.0307/T)P_2(\cos \theta)$ and particles correspond to $(1 - 0.0575/T)P_2(\cos \theta)$ when corrected for ^{234}U content. [These data were obtained by Dabbs $et\ al.$ (1965).]

layer of the fissile material to be studied on the surface. Cooling of such a crystal leads to orientation of the target nuclei by electric quadrupole coupling. The major axis of a given target nucleus lies preferentially in the plane perpendicular to the c-axis of the crystal. The experiment then consists of the measurement of the anisotropy both of α-particles (from radioactive decay of the oriented nuclei) and of fission fragments (from neutron-induced fission). The anisotropies, some of which are shown in Fig. 31, were found to correspond to an angular distribution given by

$$W(\beta) = 1 + (A/T)P_2(\cos \beta), \tag{21}$$

where $P_2(\cos \beta)$ is the Legendre polynomial of order 2, and T is the temperature of the crystal.

If it is assumed that a separation can be made of the rotational and intrinsic modes of the saddle-point wave function, then the angular distribution of the fragments (assumed to be the same as that of the saddle point) can be related to the initial distribution of aligned target orientations simply by coupling an s-wave neutron to the initial distribution. The largest uncertainty arises in the estimate of the quadrupole coupling constant which gives the degree of target orientation. Estimates of the coefficient A in Eq. (21), based on the measured α-particle anisotropies, are shown in Table VII. Since the nuclear alignment is due to quadrupole coupling, the anisotropy depends on the channel spin J and the absolute value of K, the projection of J on the nuclear symmetry axis, but not on the sign of K. The early experiments were carried out for thermal neutron bombardment of both ^{233}U and ^{235}U targets. The anisotropy for ^{233}U was found to be zero, which suggests that thermal fission of ^{233}U occurs either by way of several channels, or, if only one channel is open, by way of a state having $I = 3$ and $K = 2$. The early ^{235}U data showed that thermal fission contains a substantial contribution of a state having K larger than zero. The measurements on ^{235}U have been extended to various resonance energies as shown in Fig. 32. Here the multilevel curve was calculated according to the analysis of Kirpichnikov et al. (1964) (see also Ignatiev et al., 1964), in which the spins of the 0.28 and 1.1 eV resonances were assumed to be different. One possible interpretation of the data is based on the assumption that the contri-

TABLE VII

ESTIMATES OF THE COEFFICIENT A IN EQ. (21)[a]

K/J	2^+	3^+	3^-	4^-
0	$+0.074$	Parity forbidden	$+0.077$	Parity forbidden
± 1	$+0.037$	$+0.055$	$+0.058$	$+0.065$
± 2	-0.037	0	0	$+0.031$
± 3		-0.092	-0.096	-0.027
± 4				-0.108

[a] From Dabbs et al. (1965). These values are calculated for various possible J and K values appropriate for low energy neutron-induced fission of ^{233}U and ^{235}U. These values depend on the quadrupole coupling constant (determining the degree of nuclear alignment), which was estimated on the basis of alpha particle anisotropies observed in the same experiment. For ^{233}U, the observed anisotropy at thermal neutron energies was very nearly zero. The lowest lying channels for $J = 2$ are expected to have $K = 0$ and 2; those for $J = 3$ should have only $K = 2$. No matter which spin state dominates the ^{233}U cross section at very low energies, the observed low anisotropy indicates that there is a contribution from a channel with $K > 0$. For ^{235}U, only resonances with $J = 3$ should show the effects of a $K = 0$ channel. Thus one could hope to assign $J = 3$ to those resonances which show a larger positive anisotropy. The observed anisotropy as a function of neutron energy is shown in Fig. 32.

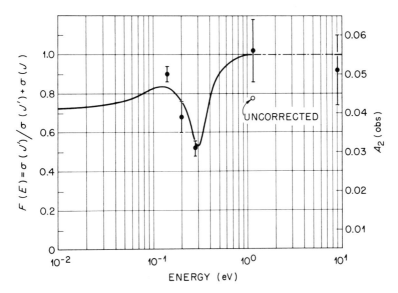

Fig. 32. Variation of fission anisotropy as a function of neutron energy for aligned ^{235}U target nuclei. These data are those of Dabbs *et al.* (1965). The solid curve was derived from the multilevel fit of Kirpichnikov *et al.* (1964), under the assumption that the effective value K which governs the anisotropy depends on the channel spin J.

butions of the various spin states contain a mixture of channels, with the lowest or most probable weighted most heavily. Then the anisotropies shown in Fig. 32 are consistent with spin assignments of $J = 3$ for the 1.1 and 8.8 eV resonances, and of $J = 4$ for the 0.28. These assignments would agree with that of Poortmans *et al.* (1966, 1967) for the 8.8 eV resonance, and with the general conclusions reached by Sailor *et al.* (1966) and Schermer *et al.* (1968) that the 0.28 and 1.1 eV resonances have different spins. However, Sailor and Schermer preferred the opposite spin assignment for these two resonances.

C. Angular Distributions in the Fission of Even–Even Targets Near Threshold

In order for fission fragments to show an angular distribution it is necessary to set up a preferred direction in space. In the resonance neutron experiments of Dabbs *et al.* (1965), this was done by aligning the target nuclei. At higher energies, where the angular momentum of the incoming neutron is different from zero, the neutron beam direction serves as a spatial axis of orientation.

Angular distributions of fast neutrons on ^{232}Th near threshold made by Brolley and Dickenson (1954) were analyzed by Wilets and Chase (1956), according to the ideas suggested by Bohr (1956). This work was extended to other even–even targets by Lamphere (1962, 1965). Lamphere (1962) pointed out that the case of neutrons on even–even (spin zero) targets is an especially simple case because no compound state having a definite spin and parity can be formed by more than one value of the neutron orbital angular momentum. Lamphere noted that the differential cross section, for formation of the compound nucleus with angular-momentum component K along the symmetry axis and with that axis maintaining an angle θ to the neutron beam, can be written as

$$\sigma(\theta) = (\lambda^2/4) \sum_l (2l + 1)^2 (1 - |\eta_l|^2)|D^I_{M,K}|^2, \tag{22}$$

where the $D^I_{M,K}$ are symmetric top functions, given in terms of vector addition coefficients as

$$|D^I_{M,K}|^2 = (-)^{M-K} D^I_{M,K} D^I_{-M,-K}$$
$$= \sum_\nu (-)^{M-K} C(I,I,\nu; M,-M)$$
$$\times C(I,I,\nu; K,-K) P_\nu(\cos\theta), \tag{23}$$

where $P_\nu(\cos\theta)$ is the νth Legendre polynomial. The quantity $1 - |\eta_l|^2$ is defined by

$$1 - |\eta_l|^2 = \sum_{I,j,m} \frac{2I + 1}{(2_j + 1)(2l + 1)} T_{lI} |C(j,l,I; m,0)|^2. \tag{24}$$

Lamphere calculated the neutron transmission coefficients T_{lI} from optical model potentials of Perey and Buck (1962). He found that the resulting angular distributions have the following characteristics: for $K = I = \frac{1}{2}$, the angular distribution is isotropic; for $K = \frac{1}{2}$, $I > \frac{1}{2}$, the angular distribution is peaked at 0 and 180°; and for $K > \frac{1}{2}$, the angular distribution is peaked at 90°.

The most extensive experimental measurements made by Lamphere were on the target nucleus ^{234}U. The cross-section data and measured angular anisotropies are shown in the top half of Fig. 33. The strong forward peaking at 800–1000 keV is indicative of $K = \frac{1}{2}$, $I > \frac{1}{2}$. Since the neutron interactions are expected to be predominantly p-wave in this energy region, Lamphere attributed this strong peaking to contributions from a $K = \frac{1}{2}^-$, $I = \frac{3}{2}^-$ channel. At 500 keV, the strong sideways peaking and the strength of the p-wave interaction suggest $K = \frac{3}{2}^-$. Below 500 keV, the smaller 0–180° peaking was attributed to a lower $K = \frac{1}{2}^+$ band.

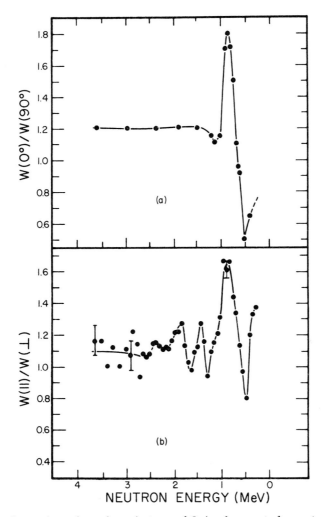

Fɪɢ. 33. Comparison of angular anisotropy of fission fragments for neutron-induced fission of ^{234}U, and for fission of ^{234}U induced by the (d,p) reaction. The neutron studies (a) are those of Lamphere (1962); the (d,pf) work (b) is that of Vandenbosch et al. (1965).

Even though the angular distributions themselves do not give any information on the parity of the saddle-point states, it is difficult to arrive at a conclusion different from the $K = \frac{1}{2}, \frac{3}{2}, \frac{1}{2}$ sequence of Lamphere. The parity is deduced from the observed strengths of the anisotropies and the calculated transmission coefficients for neutrons of definite angular momentum. The analysis of the data is complicated by the presence of neutron-elastic and inelastic scattering, which may compete with fission in certain channels more strongly than in others, but Lamphere (1962) concludes that this effect should be small for channels other than $K = \frac{1}{2}$. Lamphere (1965) reviewed the fast neutron-induced fission anisotropies on other even–even heavy targets, and concluded that the channel sequence $K = \frac{1}{2}$, $\frac{3}{2}, \frac{1}{2}$ is adequate to represent the data for ^{230}Th and ^{232}Th, as well as for ^{234}U, but that ^{238}U and ^{240}Pu targets indicate primarily $K = \frac{1}{2}$ bands, with only slight admixtures of higher K bands.

Fission of the same target nuclides as studied by Lamphere can also be induced by the (d,p) reaction, with appropriate deuteron and proton energies (Vandenbosch et al., 1965). The angular distributions in this case are not expected to be so simple because of the possibility of large angular momentum transfer by the deuteron and proton (see Halpern and Strutinski, 1958; Chaudhry et al., 1962). It is perhaps surprising that the correlation of the data obtained in the two experiments is so good, as shown in Fig. 33 for the anisotropy in ^{234}U. The analysis of these data leads to the same channel sequence, $\frac{1}{2}, \frac{3}{2}, \frac{1}{2}$ for the most prominent structure (Vandenbosch et al., 1965), provided that the barrier curvature is the same for each channel. Vandenbosch (1967) has analyzed the energy variation of the angular distributions, finding that agreement with the data is obtained only if the barrier curvature is different for each channel. This also implies that the sequence of channels may be different from $\frac{1}{2}, \frac{3}{2}, \frac{1}{2}$. Nevertheless, the agreement between the (d,pf) and the (n,f) angular distributions can be taken as evidence that the bands of channels with fairly low K quantum numbers are those with the lowest thresholds for fission.

D. Fragment Angular Distributions for the Fissile Nuclides Near Threshold

The behavior of the neutron-induced fission cross section near threshold has also been measured for the fissile nuclei ^{233}U, ^{235}U, and ^{239}Pu, where the compound nuclei are even–even and the channel spectrum is expected to be much more widely spaced. Here, only charged-particle induced fission can be used; since the fissile nuclides show a large fission cross section at very low neutron energies, their neutron–fission thresholds must lie at somewhat lower excitations than that afforded by the neutron-binding energy, at effectively negative-neutron kinetic energies. By studying deuteron-induced proton-fission coincidences as a function of proton energy, Northrop et al. (1959) determined the fission probability of the compound nucleus near threshold. The results of this experiment are shown in Fig. 34. For ^{233}U and ^{239}Pu targets, multiple thresholds are observed. These have been interpreted as the opening of new fission channels.

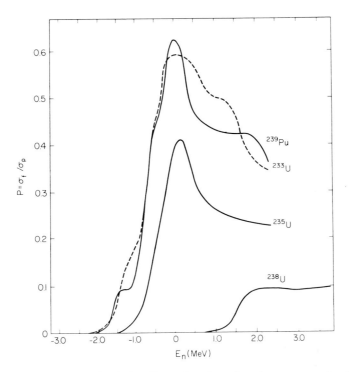

FIG. 34. The probability for fission P as a function of neutron energy, showing the curves of the four nuclides in their proper relative position. This work, by Northrop *et al.* (1959) was the first use of the (d,p) reaction to study fission thresholds at effectively negative neutron energies.

The angular distribution of the fragments is of particular interest. It has been measured by Britt *et al.* (1963b, 1965), and the interpretation of the results follows the ideas discussed by Griffin (1963, 1965). Again it is assumed that the fragment angular distributions are proportional to the square of symmetric top wave functions. For small values of the magnetic quantum number M, the angular distributions are similar to those observed by Lamphere (1962): for small K, the distributions are peaked forward and backward; for large K, they peak sideways. Griffin assumed that well above threshold many channels are open, so that the angular distribution will assume a statistical form

$$W(\theta) \sim \sum_{K=-I}^{+I} |D_{M,K}^{I}|^2 \exp(-K^2/2K_0^2), \qquad (25)$$

where K_0^2 represents the variance of the distribution. The angular distri-

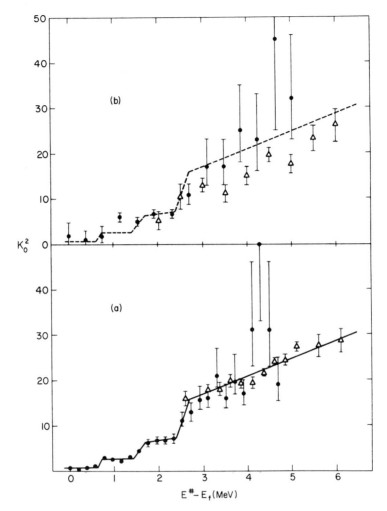

FIG. 35. The (d,pf) experimental results for K_0^2 as a function of energy above thresh-old for (a) a ^{239}Pu target and (to show that the effect is not peculiar to ^{239}Pu) for (b) a ^{233}U target. These data are those of Britt *et al.* (1965). Griffin (1965) points out that near threshold, $K_0^2 \approx 0$, as expected for an even–even fissioning nucleus. At $\sim$2.6 MeV, K_0^2 jumps to a value appropriate for an average over two-quasiparticle states. The smaller jumps at 0.7 and 1.5 MeV may correspond to low-lying collective vibrations in the transition-state spectrum. Triangles are from neutron induced fission, studied by Simmons *et al.* (1965) and Simmons and Henkel (1960).

butions were analyzed in a way which gave the explicit dependence of K_0^2 on the compound nucleus excitation energy, as shown in Fig. 35.

The interpretation of this experiment gives information on the channel structure of even–even fissioning compound nuclei, which is significant to the interpretation of the low-energy neutron-resonance structure of the fissile nuclides. Near threshold, K_0^2 is very near to zero, corresponding to fission through the channel corresponding to the ground-state configuration. At the other extreme, at an excitation of about 2.6 MeV, K_0^2 jumps to a value which is appropriate for an average of two-quasiparticle states, where the excitation is such that it is energetically possible to break apart a pair of neutrons or protons which at lower excitations are coupled together to give a net angular momentum of zero. The mathematical description of this pairing energy gap was reviewed by Gallagher (1964). It was first suggested by Bohr et al. (1958) that the behavior of nuclear matter follows the formal theory of superconductivity proposed by Bardeen et al. (1957). Griffin (1963) (see also Rich and Griffin, 1963; Britt et al., 1963b, 1965), showed that the analysis of fission-fragment angular distribution data by this model implies that the pairing gap at the saddle point is more than twice as large as that appropriate for stable nuclei: the pairing gap for the stable even–even deformed heavy nuclei is only about 1.0 MeV, but for fissioning nuclei deformed to the saddle-point configuration, it appears to be about 2.6 MeV. Griffin suggested that the smaller jumps in K_0^2 at about 0.7 and 1.5 MeV correspond to the opening of channels which correspond to collective vibrations in the transition-state spectrum. By analogy to the spectrum of states observed near the ground state of stable nuclei, one knows how many states or channels to expect, as a function of excitation energy at the saddle point.

E. Subthreshold Resonance Fission

Most even–even and odd–even heavy target nuclides are not considered to be fissile, even though many of them show a small amount of fission in the resonances. Experimentally, these nuclides were first studied by Leonard and co-workers (Leonard and Seppi, 1959; Odegaarden, 1960; Leonard and Odegaarden, 1961). For subthreshold fission one might assume that, even though several fission channels contribute, the cross section is dominated by the lowest. If the Hill–Wheeler (1953) formula is applicable for neutron energies well below the fission barrier, the average fission width may be assumed to vary as

$$\langle \Gamma_f \rangle = (D/2\pi) \exp[-2\pi(E_T - E)/\hbar\omega], \tag{26}$$

where D is the average level spacing, E_T is the energy of the fission "threshold" for the particular channel involved and $\hbar\omega$ is a constant of the barrier. If the energy E_T is known, then a measurement of the fission width should allow one to determine the constant $\hbar\omega$. For subthreshold fission, the ratio Γ_f/D is much less than unity, so the accuracy of the value obtained for $\hbar\omega$ does not depend very strongly on the expected large variation in the fission widths and the resulting inaccuracies in determining $\langle \Gamma_f \rangle$ from one or a few levels.

Some of the results of the measurements of Leonard and Odegaarden (1961) are given in Table VIII. It is interesting to note that $\hbar\omega$ is relatively

TABLE VIII

FISSION PARAMETERS OF RESONANCES OBSERVED IN THE
SUBTHRESHOLD FISSION OF VARIOUS HEAVY NUCLIDES[a]

Nucleus	E_{res} (eV)	σ_{0f} (barns)	Γ_f (eV × 10⁻⁶)	$\hbar\omega$ (MeV)
^{240}Pu	1.06	30	5.2	0.32
	20.4	<535	<200	<0.45
^{242}Pu	2.65	<41	<17	<0.37
^{234}U	5.20	50 ± 9	20	0.33
^{236}U	5.49	<1	<1	<0.45
^{238}U	6.68	<0.15	<0.18	<0.57
	10.2	<0.26	<430	<1.08
^{241}Am	0.307	52	175	0.83
	0.579	14	70	0.73
	1.265	48	170	0.83
^{237}Np	0.49	0.056	0.78	0.31
	1.32	0.097	3.96	0.36
	1.47	0.056	0.59	0.30
^{231}Pa	0.396	0.255	1.5	0.37
	0.496	0.185	8.7	0.43
	0.745	0.075	7.3	0.43
	1.235	<0.12	<6.1	<0.42

[a] From Leonard and Odegaarden (1961).

constant from nuclide to nuclide, with the exception of ^{241}Am. The average value of $\hbar\omega$ obtained from these subthreshold fission measurements is in reasonable agreement with the barrier constant of 0.395 determined from spontaneous fission systematics (Viola and Wilkins, 1966).

If there is only a single fission channel per spin state contributing to subthreshold fission, then interference effects would be expected to be particularly prominent, even though the fission widths are extremely small. However, the existing data in subthreshold fission seem to point to several

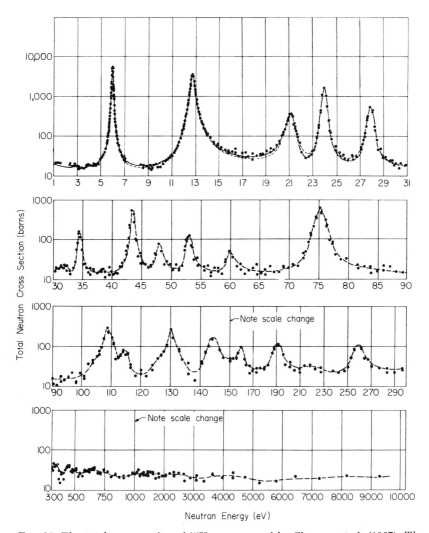

FIG. 36. The total cross section of ^{232}U, as measured by Simpson *et al.* (1967). The fission cross section, measured both by James (1964) and by Auchampaugh *et al.* (1968), confirms the assumption that the observed interference is to be attributed to the fission component of the resonances. MTR fast chopper: —, multilevel 2 channel fit; - - -, multilevel 1 channel fit; – – –, "eyeball" fit above 30 eV.

possible channels instead of just one. This evidence is provided by studies on ^{232}U and ^{238}Pu, which one would expect to be characteristic of even–even fissionable targets. For most even–even nuclides resonance fission is so small that it is considered to be subthreshold fission; the exceptions are

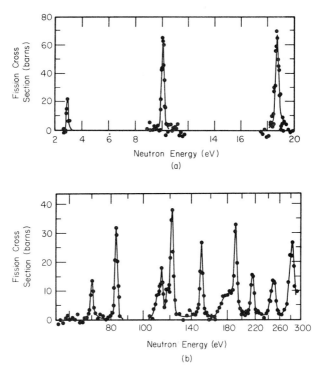

FIG. 37. The fission cross section of ^{238}Pu (a) from 2 to 20 eV, and (b) from 20 to 300 eV, as measured by Stubbins *et al.* (1967). When analyzed and compared with total cross-section data (Young *et al.*, 1962, 1967) these data give evidence that several fission channels are contributing to fission in ^{238}Pu. (See Bowman *et al.*, 1967.)

^{232}U and ^{238}Pu. The low-energy total cross section of ^{232}U is shown in Fig. 36 (Simpson *et al.*, 1967). The fission cross section of ^{232}U has also been measured by James (1964) and by Auchampaugh *et al.* (1968). The fission cross section of ^{238}Pu is shown in Fig. 37 (Stubbins *et al.*, 1967). The total cross section has been measured by Young *et al.* (1967).

Analysis of the ^{232}U data (Simpson *et al.*, 1967; Auchampaugh *et al.*, 1968) showed that the cross section could not be explained by fission in a single channel. Analysis of the ^{238}Pu data (Bowman *et al.*, 1967), which has much less fission in the resonances than does ^{232}U, indicated that perhaps as many as six channels may have appreciable contributions.

The results for both ^{232}U and ^{238}Pu were interpreted (Bowman *et al.*, 1967; Simpson *et al.*, 1967) as experimental evidence for the (n,γf) reaction in even–even heavy nuclides. Lynn (1965b) first suggested that the (n,γf) reaction may be important. He pointed out that all the resonances of fissile nuclei appear to show an appreciable fission component. If, however, the distribution of fission widths follows a chi-squared distribution for one or a few degrees of freedom (Porter and Thomas, 1956), then occasionally one should find a resonance which shows essentially no fission. Such a resonance should be easy to see, in a comparison of total and fission cross sections, but none has ever been observed. There appears to be a minimum value for the fission width of 3 or 4×10^{-3} eV. Lynn suggested that there may be a multichannel process, the (n,γf) reaction, which could account for the paucity of very small fission widths. In the (n,γf) reaction, a gamma-ray transition occurs from a channel through which fission is energetically prohibited to another channel lower in excitation. In this channel, the fission barrier is considerably lower than in the first, and fission is the dominant process. The number of channels for the reaction would be equal to the number of gamma transitions by which the process can occur.

In order for fission to occur by the (n,γf) reaction in ^{232}U or ^{238}Pu, the fission channel sequence must be different from the assumed $K = \frac{1}{2}^+, \frac{3}{2}^-, \frac{1}{2}^-$ sequence postulated by Lamphere (1965). In particular, the $\frac{1}{2}^-$ channel must be enough lower than the $\frac{1}{2}^+$ to permit gamma emission before fission occurs. For ^{238}Pu, the data on fast neutron fission (Vorotnikov et al., 1966) are consistent with such a channel sequence, tending to support the idea that low-energy resonance fission of ^{238}Pu may occur by way of the (n,γf) reaction.

Rae (1965) has suggested that s-wave neutron-induced fission should be generally prohibited by a higher barrier for all heavy even–even targets of uranium and plutonium. Rae pointed to the work of Johansson (1961), who calculated the energies of single particle orbitals appropriate to fissioning nuclei by a model similar to that of Nilsson (1955) which included octupole deformations. Johansson found that the odd parity orbitals are expected to lie considerably lower in energy than those of even parity for deformations corresponding to that of the saddle point. In particular, Rae (1965) pointed out that the lowest $\frac{1}{2}^+$ channel might lie considerably higher ($\sim$500 keV) than the $\frac{1}{2}^-$ and $\frac{3}{2}^-$ channels at the saddle point, thus effectively prohibiting s-wave neutron fission.

There is one more phenomenon in the subthreshold fission resonance region which may have considerable bearing on the detailed shape of the fission barrier. The subthreshold resonance fission of ^{237}Np and ^{240}Pu has been studied up to several hundred electron volts, by Paya et al. (1967) and Migneco and Theobald (1968), respectively. Both groups observed that the fission widths do not seem to be randomly distributed with neutron energy. Instead, there is a pattern of groups of resonances with a large amount of fission, followed by intervals with very little fission. This pattern is very suggestive of intermediate structure, which could arise if the fission barrier were multiply humped, with a rather deep potential well at deformations between the saddle point and the scission point.

Strutinski (1967) postulated a barrier of this nature as the mechanism for the existence of short-lived fissioning isomers. (See Polikanov *et al.*, 1962; Flerov *et al.*, 1965; Lobanov *et al.*, 1965.) Strutinsky calculated fission barrier shapes at large deformations, using Nilsson's approach (Nilsson, 1955) to make shell corrections to the liquid drop energies. He found that secondary maxima and minima are to be expected in the barrier shape. It is interesting to note that there are now good experimental reasons for postulating the existence of fission barriers of this type.

V. THE SCISSION POINT AND NUCLEAR STRUCTURE OF THE FRAGMENTS

A. Mass Distributions and Symmetry Requirements at the Saddle Point

For the common fissile nuclides, the fission-product mass distributions are in general asymmetric, i.e., fission occurs most frequently into two fragments which are unequal in mass. The most probable heavy-fragment mass number is about 139; the most probable light fragment mass is found to vary between 94 and 100, depending on the mass number of the target nuclide. Fission into two fragments of equal mass is much less probable. Under thermal-neutron bombardment, ^{235}U, for example, fissions into two given fragments of roughly equal mass (symmetric fission) only about 0.2% as often as it fissions into the most probable division (asymmetric fission).

The Bohr model (Bohr, 1956), which was so successful in describing quantitatively the behavior of angular distributions of the fragments, does not appear to give a complete description of the mass distributions. Wheeler (1956), on the basis of the Bohr model, suggested that the amount of symmetric fission in ^{235}U should vary from resonance to resonance, depending on the spin state of the resonance. His argument was based on a consideration of the symmetry properties of the wave function of the fissioning nucleus at the saddle point. The ground state spin of ^{235}U is $\frac{7}{2}$, and the parity is negative. The possible spin states in the compound nucleus ^{236}U are 3$^-$ and 4$^-$ since only s-wave neutron absorption is considered. If it is assumed that fission takes place through only a few fission channels, then the lowest-lying channels in each spin state should predominate. The lowest-lying 3$^-$ configuration is based on the rotational band associated with the lowest odd-parity vibrational mode. This has $K = 0$ and is characterized by pear-shaped (octupole) vibration. The antisymmetry of this type of vibration implies zero probability for the symmetric form. If the symmetry properties at the saddle point are the same as at scission, such a vibration implies a low probability for symmetric fission. For 4⁻ resonances, which can be described by octupole vibrations with $K \neq 0$, the lowest lying channel does not restrict symmetric fission in this way. (See Wheeler 1963; Lynn 1965a; Harris, 1966.)

The earliest experimental measurements of the mass distribution of ²³⁵U as a function of energy, by the Los Alamos Radiochemistry Group (1957), showed an effect of the type suggested by Wheeler (1956). The Los Alamos group measured the fission yields of several mass numbers characteristic of both peak (asymmetric fission) and valley (symmetric fission) for thermal and for epicadmium irradiation, and they found a small, but significant, difference for the two types of irradiation. Similar effects were observed for resonances in ²³³U and ²³⁹Pu; for the latter in particular, there is a large change in the symmetric fission yield from thermal-neutron fission to fission at the first resonance (Regier et al., 1959, 1960). For ²⁴¹Pu, no change was found in the mass distribution between thermal and episamarium fission (Regier et al., 1960), implying that the characteristics of the 0.265 eV and negative-energy resonances are the same.

The theoretical arguments for a variation in the mass distributions for ²³³U, ²³⁹Pu, and ²⁴¹Pu were less definite than for ²³⁵U. For ²³⁹Pu, having spin $\frac{1}{2}$ and positive parity, and for ²³³U and ²⁴¹Pu, having spin $\frac{5}{2}$ and positive parity, the symmetry requirement on the vibrational wave function is different since the lowest energy vibrations are expected to be quadrupole. It was argued on this basis that the 0⁺ or 2⁺ possible spin states should perhaps show more symmetric fission than the 1⁺ or 3⁺ spin states, although the spin dependence in this case should have been somewhat weaker (contrary to what was observed in ²³⁹Pu). Nevertheless, for several years the symmetry of the wave function at the saddle point was accepted as a complete explanation for the observed change in the symmetric fission yield from resonance to resonance.

B. Kinetic-Energy Studies and the Variation of Neutron Yield with Mass Number

After the saddle point is crossed, the fissioning nucleus undergoes increasing deformation until at last the two nascent fragments separate, and the nuclear forces between the fragments cease to operate. The point at which this occurs is called the scission point. After the two fragments have separated the forces between them are purely long-range (coulomb) forces

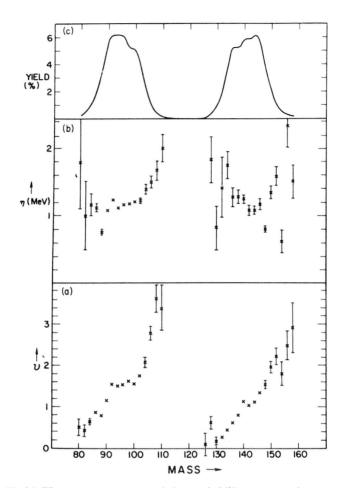

FIG. 38. (a) The average neutron emission probability $\bar{\nu}$, versus fragment mass for $^{233}U + n$. The scale has been arbitrarily normalized to give an average value of 2.50 neutrons per fission. (b) The average neutron kinetic energy η, again arbitrarily normalized to give an average value of 1.2 MeV per neutron. The errors shown are statistical only. (c) The yield of fragments in percent without reference to neutron emission and shown for orientation purposes. The two peaks should be symmetric about mass 117 but have not been folded in the figure so show slight and systematic differences. These data, and similar data for ^{235}U, were obtained by Milton and Fraser (1965). (This figure was supplied by Dr. Milton.)

and the coulomb potential energy eventually is transformed into kinetic energy. Studies of the kinetic energies of the fragments thus give information on the fission process at the scission point.

Complementary to studies of fission-fragment kinetic energies are studies of the prompt neutron yield as it depends on the fission-fragment mass number. The fission energy which does not appear as fragment kinetic energy must appear in prompt neutron and gamma emission, delayed neutron and gamma emission or beta decay (and as binding energy of the fragments). The largest part of this energy appears in prompt neutron emission. Most of the prompt neutrons are emitted after the fragments have been fully accelerated from one another by coulomb forces (see Bowman *et al.*, 1962; Kapoor *et al.*, 1963). By measuring the angular distribution of the prompt neutrons, it is possible for one to tell which fragment emitted these neutrons and, by doing momentum analysis of the fragments, to determine the mass number of the fragment. Measurements of the number of prompt neutrons as a function of fragment mass number have been made by a number of groups (Fraser and Milton, 1954; Stein and Whetstone, 1958; Apalin *et al.*, 1960, 1964, 1965a; Bowman *et al.*, 1963; Milton and Fraser, 1965). Some typical data are shown in Fig. 38 for ^{233}U (Milton and Fraser, 1965). Figure 38 shows the fragment yield, the energy of the emitted neutrons in the center of mass, and the number of neutrons emitted, as a function of fragment mass.

Terrell (1962) showed that the dependence of $\bar{\nu}$ on fragment mass number can be derived from a comparison of mass yield curves before and after neutron emission, i.e., from a comparison of fission fragment time-of-flight data with radiochemical mass yield data. Terrell derived curves of $\bar{\nu}$ versus mass number which agree very well with direct measurements. (Bowman *et al.*, 1963; Milton and Fraser, 1965; see also Schmitt and Konecny, 1966.) The curves derived in this way for various fissioning nuclides are so similar that Terrell proposed a universal curve of $\bar{\nu}$ versus mass number for low-energy fission, as shown in Fig. 39.

The energy distribution of the fission neutrons as a function of mass number is also of interest, as shown in the middle curve in Fig. 38. From studies by Terrell (1959) on the dependence of neutron emission on nuclear temperature, one might expect the neutron energies to follow the neutron emission probabilities (see Terrell, 1965). If the fission-fragment nuclear excitation is high enough to permit large numbers of prompt neutrons to evaporate, one would be led to expect higher neutron energies for such fragments. Instead, as seen in Fig. 38, the neutron-energy distributions as a function of fragment mass are symmetric, with a slight rise in the neutron energy near symmetric fission and near very asymmetric fission. Similar behavior is noted for neutron-energy distributions in ^{252}Cf spontaneous fission (Bowman *et al.*, 1963). One possible explanation for such behavior might be that the neutron emission times are short compared to the time required for reorientation of a highly deformed nascent fragment. Depending on the

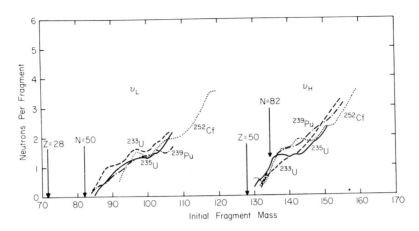

Fɪɢ. 39. Summary of neutron yields derived by Terrell (1962) from cumulative mass yields, as functions of the initial fragment mass. The approximate initial fragment mass corresponding to various magic numbers are also shown, based on constant charge density of the fragments at scission. The similarity of these curves led Terrell to postulate that this is a universal curve; the energy which is expended in neutron emission depends primarily on the structure of the nascent fragments rather than on the fissioning nuclear species.

fragment mass, there is a given amount of energy which is present at scission as potential energy of deformation of one or both fragments. This energy becomes available for excitation as the fragment assumes a less highly deformed shape, and the fission neutrons would be emitted one at a time, as the fragment excitation exceeds the neutron binding energy.

The existence of a "universal curve" for the emission of neutrons by fragments of a given mass implies that $\bar{\nu}$ depends only weakly on such parameters as the excitation energies and the species of the fissioning nuclides, but depends strongly on the nuclear structure of the fission fragments. (Terrell, 1962, 1965.) At the same time, reasoning based on studies of the kinetic energies of the fragments led to the same conclusions. Early measurements of the fragment kinetic energies, made on ^{235}U with solid-state detectors, revealed a striking variation of the symmetric fission yield as a function of total fragment kinetic energy (Safford et al., 1960b). Some of these data are shown in Fig. 40. It is seen that the amount of symmetric fission varies from a reasonably high value at low kinetic energy to very nearly zero at high kinetic energy. Extensions of work of this type led to the discovery of fine structure in the kinetic energy spectra of fission fragments from thermal neutron-induced fission (Gibson et al., 1961; Milton and Fraser, 1961). The energetics of the fission fragment distributions are

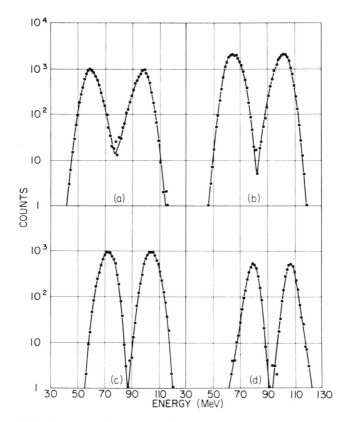

FIG. 40. Pulse-height distributions of kinetic energies of fission fragments from (^{235}U + n) at thermal neutron energies, taken under conditions of constant total kinetic energy, after Safford *et al.* (1960b) (see also Melkonian and Mehta, 1965). The variation of symmetric-fission yield with total kinetic energy can be attributed to a dependence on the amount of energy available to the fragments at the scission point. This energy is in turn related to the number of final states which are energetically permitted to the nascent fragments. (a) E_T = 154 MeV; (b) E_T = 166 MeV; (c) E_T = 178 MeV; (d) E_T = 188 MeV. (These data were supplied by Dr. Safford prior to publication.)

illustrated in Fig. 41, which shows a contour plot of the fission yield as a function of the heavy fragment mass number and of the light fragment kinetic energy (Fraser, 1965). The fine-structure contours are approximately parallel to lines representing constant kinetic energy of the light fragment. (The fine structure can also be noted in pulse-height distributions of correlated fragments, as seen in Fig. 4; Williams *et al.*, 1964.) Vandenbosch and Thomas (1962) (see also Thomas and Vandenbosch, 1964) carried out an analysis of such data, based on a statistical description of the transition

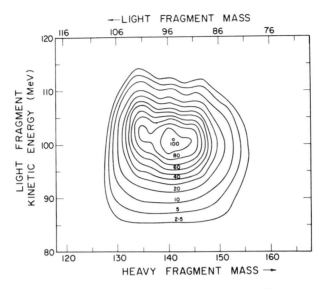

FIG. 41. Probability contours for ^{235}U + n plotted on a grid of light fragment kinetic energy and fragment mass. The numbers adjacent to the contours represent the percentage of the probability at the peak. These data were obtained by Fraser *et al.* (1963). The fine structure contours are apparent at the highest light fragment energies. (This figure was supplied by Drs. J. C. D. Milton and J. S. Fraser prior to publication.)

from the saddle point to scission similar to that suggested by Fong (1956), and showed that the fine structure can be correlated to changes in the nuclear mass surface describing the fragments. Their analysis of the fine structure data on ^{235}U is shown in Fig. 42. The fine structure arises from the larger binding energy of even–even light fragments which correspond to a closed shell of 50 protons and 82 neutrons in the heavy fragment. Such a configuration of fragments involves the lowest excitation energy, and thus appreciable fission can occur at a relatively high coulomb barrier. Thomas and Vandenbosch (1964) carried out a similar analysis of the fine structure observed in the thermal-neutron-induced fission of ^{233}U and ^{239}Pu (Thomas *et al.*, 1965), and obtained reasonable agreement with the data. The results suggest that the fission fragments with high kinetic energy have a minimal excitation energy associated with them. Thus at scission these particular fragments have only small deformations from the shape assumed by stable nuclei in this mass region. The high probability of an asymmetric mass division in the fission of transuranium isotopes is most reasonably explained by shell effects in the fragments. The systematics of the fission mass distribution suggest some of the details of these effects: For fissioning nuclides with masses between about 232 and 252, the position of the cen-

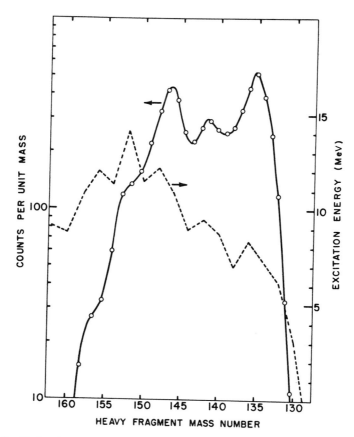

Fig. 42. Fission-fragment mass distribution of the heavy fragment peak from (^{235}U + n) at thermal neutron energies, taken under conditions of constant kinetic energy of the light fragment, after Gibson *et al.* (1961). The dashed line shows the structure due to even–odd effects in the mass surface. The analysis, reported by Thomas and Vandenbosch (1964), showed the importance of the nuclear structure of the nascent fragments to the fission mass distribution ($E_L = 108.7$ MeV).

troid of the heavy peak stays fixed around mass 139, while the position of the light peak depends on the mass of the fissioning heavy nuclide. This suggests the existence of a well-defined core for the heavy peak, presumably consisting of 50 protons and 82 neutrons. The light fragment shows the effects of a closed shell of 50 neutrons, but its structure is less well defined. Whetstone (1959) was one of the first to propose a model to describe the dependence of $\bar{\nu}$ on the nuclear structure of the fragments. According to the Whetstone model, the fissioning nucleus consists of two cores of unequal mass connected by a neck of residual nucleons. The cores, which are due to nuclear shell effects, have already been formed in the time of scission, and the actual mass division is determined by the division of the neck. If the neck divides near the heavier core, the near-symmetric fission is observed, and the energy of deformation of the light core plus the neck becomes available as energy of excitation of the fragment. Similarly, division of the neck near the light core corresponds to a very asymmetric fission, with a high excitation given to and consequently more neutrons emitted from the heavier fragment.

The importance of shell structure and fragment deformation on fission energetics was noted by Brunner and Paul (1960, 1961a,b,c) who showed how the total kinetic energy of the fragments is related to the deformation. The Whetstone model does not describe the details of the variation of fragment kinetic energy with mass number since according to this model the distance between fragment charge centers is very nearly constant. It remained for Vandenbosch (1963) to show that the variation of fragment kinetic energy, as well as the variation in the number of neutrons emitted from the fragment as a function of fragment mass, could be quantitatively related to the deformation of the nascent fragments. Vandenbosch noted that in certain regions of the periodic table, the nascent or primary fission product nuclides are expected to be particularly soft to nuclear deformation. Such nascent fragments at scission will have their effective centers of charge separated far more widely than the more nearly spherical fragments, such that the observed total kinetic energy will be lower than average. In addition, the large deformation will permit large numbers of neutrons to be emitted as the deformation energy is converted to fragment excitation energy.

With the same assumptions about the deformation properties of the nascent fragments at the scission point, Vandenbosch was able to describe the energetics of asymmetric and symmetric fission of elements lighter than thorium, where the behavior is quite different. The fission of bismuth under bombardment of 22-MeV deuterons, for example, shows a single narrow symmetric distribution of fragment mass numbers (Fairhall, 1956). The fission of ^{226}Ra induced by 11-MeV protons (Jensen and Fairhall, 1958) shows a triple-humped curve, presumably due to the presence of both the familiar asymmetric fission shown by (^{235}U + n) and the symmetric fission shown by (^{209}Bi + d). Similar behavior was noted for fission of ^{226}Ra under neutron bombardment (Nobles and Leachman, 1958) and under proton bombardment (Schmitt et al., 1965b); in this case it was found that the symmetric component increases much faster with increasing neutron energy than does the asymmetric component.

Such observations as these had earlier led to a suggestion (Turkevich and Niday, 1951; Fairhall *et al.*, 1958) that two distinct modes of fission may exist, one leading to symmetric division, the other to asymmetric division. The two modes are assumed to have a different dependence on excitation energy (see, for example, Levy *et al.*, 1961). One might expect that there would be two saddle points with different densities of states in the channel spectra. Quadrupole deformations might be expected to lead to a saddle-point characteristic of symmetric fission; while octupole deformations involve a different saddle point which leads to asymmetric fission. This argument was first developed by Businaro and Gallone (1955) and Nossoff (1956), who carried out investigations using the liquid-drop model.

Johansson (1961) included shell effects, following the approach of Nilsson (1955) with octupole deformations. These calculations showed that at large deformations, characteristic of the transition between the saddle point and scission, the energy levels characteristic of octupole deformation can be expected to be considerably lower, leading to an enhancement of asymmetric fission.

The two-mode-of-fission hypothesis has also been invoked to explain the variation in fragment kinetic energy for fission of elements lighter than uranium. Britt *et al.* (1962, 1963c) studied the ^{3}He induced fission of ^{209}Bi and ^{197}Au which shows symmetric fission, and of ^{226}Ra, which shows the characteristic three-humped curve attributed to the presence of both symmetric and asymmetric fission. Britt found that for pure symmetric fission, the variation of the kinetic energy with mass ratio can be described by the functional form

$$E = kZ_1Z_2(A_1^{1/3} + A_2^{1/3}), \qquad (27)$$

where A_1 and A_2 are the masses of the two fragments and Z_1 and Z_2 correspond to the charge division which gives the maximum energy release. For ^{3}He-induced fission of ^{226}Ra, the variation of the kinetic energy with mass ratio for both the asymmetric and symmetric modes was again found to follow the same functional form as Eq. (27), but with different proportionality constants k for the two modes: for the symmetric mode the kinetic energy curve was found to be appreciably lower than for the asymmetric mode. Qualitatively, this is similar to what is observed in ^{235}U + n (see, for example, Milton and Fraser, 1962; Fraser, 1965), in that symmetric fission occurs at lower total kinetic energy. However, for (^{235}U + n), the variation of kinetic energy with mass ratio does not follow Eq. (27), even for the asymmetric fission component.

Schmitt and Konecny (1966), from fragment double energy and velocity measurements, determined the neutron yield as a function of fragment mass from proton-induced fission of radium. They observed a shape qualitatively consistent with the universal curve of Terrell (1962). In analyzing the data on the basis of the two-mode hypothesis, they found that the central symmetric peak could best be described by the liquid-drop model (Nix and Swiatecki, 1965), and the asymmetric peaks by the fragment-structure model (Vandenbosch, 1963; Terrell, 1962, 1965). Very similar conclusions had been reached earlier for high energy (α-particle-induced) fission of ^{230}Th, ^{232}Th, and ^{233}U targets, in separate double-energy and double-energy and double-velocity measurements by Britt and Whetstone (1964) and by Whetstone (1964), respectively, although here the existence of fragment-structure effects in $\bar{\nu}$ was not conclusively shown.

One is led to evaluate the situation as follows: The two-mode-of-fission hypothesis may be the simplest way of describing certain fission phenomena, particularly for elements lighter than thorium and at high excitations. Its use for heavier isotopes at low excitation does not seem to be very fruitful, compared to the fragment-structure model of Vandenbosch and Terrell. In particular, for the description of the low-energy neutron variation of fragment mass yields and kinetic energies, the invoking of completely independent modes for symmetric and for asymmetric division seems to lead to unnecessary complications since it formally requires a different channel or set of channels for asymmetric and for symmetric fission.

C. Ternary Fission

Ternary fission, involving the emission of a long range ($\sim$10–30 MeV) α-particle in addition to two heavy fragments, occurs in about one in five hundred fission events for ^{235}U. Ternary fission of other types is much rarer. Whetstone and Thomas (1965, 1967) have reported on the emission of ^{6}He and other light particles from ^{252}Cf, finding a frequency of 8×10^{-5} per event associated with ^{6}He. Muga (1965) (see also Muga et al., 1967) has found evidence for ternary fission involving three about equally heavy particles. The frequency for such ternary fission is about 1×10^{-6} for ^{252}Cf, or 6×10^{-6} for ^{235}U.

The angular distribution of the fragments from ternary fission in which an α-particle is emitted was studied by Titterton (1951), who observed a strong peaking of α-particles at 82° with respect to the direction of the lighter fragment. The liquid-drop model of fission is useful in explaining ternary fission of this type. According to Hill and Wheeler (1953), the α-particle retains its identity and finds itself in the much lower coulomb potential which exists between the two fragments at the time of scission. The angular distribution is modified by the unequal charge of the two fragments. It may be assumed that ternary fission represents an effective tunnel-

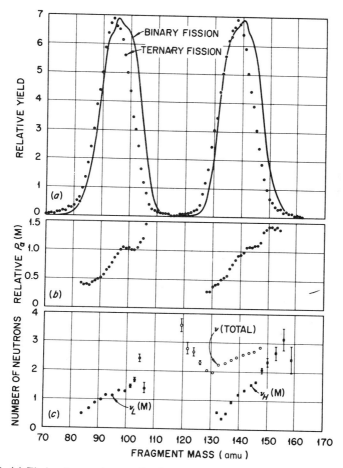

FIG. 43. (a) Fission-fragment mass distributions from energy correlation experiments for binary and ternary fission of ^{235}U, as measured by Schmitt *et al.* (1962). (b) Relative values of the probability of ternary alpha emission as a function of binary fragment mass numbers (Schmitt and Feather, 1964). (c) The number of neutrons emitted by individual fragments as a function of mass number, and the number emitted per fission as a function of heavy fragment mass, as determined by Apalin *et al.* (1960, 1962) (see also Apalin *et al.*, 1965a,b,c).

ing of a charged particle though a potential barrier, such that the relative probability of ternary fission will be larger whenever the barrier is lower. This occurs when the conditions at scission are such that the charge centers of the two nascent heavy fragments are farther apart, i.e., when the fragment deformation is largest. Thus, for a somewhat different reason, the relative probability of ternary fission should show qualitatively the same dependence on fragment mass as does the probability of prompt neutron emission. Detailed measurements by Schmitt *et al.* (1962) (see also Schmitt and Feather, 1964) shown in Fig. 43, demonstrate this effect.

The energetics of ternary fission are consistent with the idea that most of the ternary α-particles are released at the scission point. Schröder *et al.* (1963, 1965) and Schmitt *et al.* (1962) found that the mean total kinetic energy in ternary fission is equal to the mean total kinetic energy in binary fission; i.e., if the mean energy of the α-particle is low, then the mean kinetic energy of the two heavy fission fragments increases by the same amount. Apalin *et al.* (1959) studied the mean number of neutrons emitted when ternary fission occurs. It was found that when an α-particle is emitted, the quantity $\bar{\nu}$ decreases to a value of 1.8 neutrons per fission from the usual value of 2.4 in the case of binary fission of ^{235}U. Furthermore, $\bar{\nu}$ was found to be independent of the energy of the α-particle. These results indicate that the probability of ternary fission depends strongly on the nuclear structure of the fragments at the scission point. The α-particle energy, however, appears to depend primarily on the magnitude of the coulomb potential at the point at which the α-particle is emitted.

The work of Nobles (1962) on ternary-fission systematics shows an interesting effect. Nobles investigated the ternary-fission yield as a function of the fissionability parameter Z^2/A, and as a function of excitation of the fissioning nuclide. He found that the relative ternary-fission yield increases as Z^2/A increases, in a manner similar to the relative symmetric-fission yield. However, he found that as the excitation of the fissioning nucleus increases (up to about 20 MeV), the relative ternary fission yield is found to decrease, unlike the behavior of the symmetric-fission yield. Halpern (1965) has discussed the mechanism of ternary fission, suggesting several possible explanations for this decrease in the ternary-fission probability. Among these is a lowering of the probability of existence of the α-particle cluster inside the nucleus as the nuclear excitation increases.

D. Neutron Spectroscopy of Mass Yields, Kinetic Energies, and Ternary-Fission Yields

Extensive studies of the fission-fragment mass distribution and how it varies with neutron energy have been carried out for both ^{235}U and ^{239}Pu. The earliest results for ^{239}Pu (Regier *et al.*, 1960) indicated that for the lowest observed neutron resonance of ^{239}Pu, at 0.29 eV, the symmetric yield (valley) decreased by a factor of three compared to the valley yield for fission induced by thermal neutrons (near 0.025 eV). Regier assumed that the thermal cross section is composed of a contribution from the 0.29 eV resonance plus a contribution from a negative-energy resonance whose symmetric yield is characteristically different. He concluded that the symmetric yields of the two types of resonances must differ by a factor of five. Cowan et al. (1966) carried out the definitive experiment, with a spinning disc of ^{239}Pu metal and a nuclear explosion (Petrel) as the neutron source. Radiochemical analysis of the fragments as a function of angular position along the disc gave the results which have been shown in Table V. Two groups of resonances were observed, with characteristically different relative symmetric yields. Cowan assumed that the observed effect depends only on the resonance spin, and inferred the spin from the relative number of levels observed. Direct measurements of the spins generally confirmed Cowan's assignments (see Table V).

Melkonian and Mehta (1965) reported on studies of the variation of kinetic energy and of ternary-alpha yield for ^{239}Pu. In these experiments the existence of a variation in kinetic energy was determined from a measurement of the relative yield of fission fragments from a thick sample to that from a thin sample.* An increase in the relative thick-sample yield was interpreted as an increase in the average fragment kinetic energy. Melkonian and Mehta assumed that the kinetic energy and the yield of ternary-alpha particles depended on the resonance spins, and inferred the spins from their data. The results of the measurements have also been shown in Table V. In general, the agreement of the resonance spin assignments made by these indirect methods (mass distributions and kinetic energy variations) is good enough to permit one to conclude that for ^{239}Pu, the observed effects depend primarily on the resonance spin.

For ^{235}U, the experimental data do not permit this simple interpretation. The most extensive mass distribution studies are again those of Cowan et al. (1961, 1963). The fission-fragment kinetic energy variations in the same energy region were again determined by Melkonian and Mehta (1963, 1965) (see also Mehta, 1963). A comparison of the results is shown in Table IX. It is seen that correlations do exist, but the separation into two groups of resonances with two distinctly different types of behavior is not as conclusive for ^{235}U as for ^{239}Pu.

The yield of ternary fission, involving the emission of a high-energy α-particle, has also been shown to show changes from resonance to resonance for ^{235}U. (Mehta, 1963; Mostavoya et al., 1964; Michaudon, 1964; Michaudon et al., 1965b; Melkonian and Mehta; 1965, Kvitek et al., 1965.) The validity of earlier data on the variation of the yield of ternary α-particles

---

* Here the terms *thick* and *thin* refer to a comparison with the range of a typical fission fragment in plutonium, which amounts to 5 to 10 mg/cm². All such samples would be considered thin to neutrons.

TABLE IX

COMPARISON OF RESULTS ON $^{235}U^a$

Resonance no.	Resonance energy (eV)	Ratio of thick U to thin U^b	Relative yield of High energy fragments^b	Relative symmetric fission yield^c
1	8.78	0.730 ± 0.012	Low	(Low)
2	15.40	0.912 ± 0.090	High	Low
3	16.09	0.801 ± 0.083	High	Low
4	16.67	0.768 ± 0.063	Uncertain	Low
5	18.05	0.769 ± 0.079	Uncertain	(Low)
6	19.3	0.810 ± 0.022	High	Low
7	21.1	0.823 ± 0.039	High	Low
8	22.98	0.850 ± 0.084	Uncertain	Low
9	23.5	0.894 ± 0.081	High	Low
10	25.6	0.838 ± 0.081	Uncertain	High
11	27.87	0.881 ± 0.077	High	(Low)
12	32.13	0.886 ± 0.043	High	(Low)
13	33.6	0.687 ± 0.040	Low	(Low)
14	34.45	0.749 ± 0.034	Low	(High)
15	35.25	0.758 ± 0.021	Low	High
16	39.5	0.740 ± 0.040	Low	High

^{a} Comparison of results on ^{235}U of the relative yield of high kinetic energy fragments (Melkonian and Mehta, 1965), and of the relative yield of fragments of symmetric fission (Cowan et al., 1961, 1963). Parentheses denote assignments which are to some degree uncertain. For resonances where definite results were obtained in both measurements (resonances 2, 3, 6, 7, 9, 15, and 16), the correlation is readily apparent.

^{b} Melkonian and Mehta (1965).

^{c} Cowan et al. (1961, 1963).

has been questioned because of the possibility of confusing the ternary-fission reaction with α-particles from the reaction ^{235}U (n,α) ^{232}Th (Sowinski et al., 1963). In more recent studies, the possibility of observing the (n,α) reaction has been eliminated by coincidence methods or high biasing, and these data show that a variation in the ternary-fission yield exists. The data also show some correlation with the high symmetric fission yield as given in Table IX although there is not a 1:1 correspondence of high ternary fission with high symmetric fission. The ternary-fission yield appears to be related to the sizes of the fission resonances. According to Mehta (1963), those resonances with a high value of the product $\Gamma_{\lambda n}^0 \Gamma_{\lambda f}$ almost invariably show a lower ternary fission yield. (Here $\Gamma_{\lambda n}^0$ is the reduced neutron width, $\Gamma_{\lambda n}/E^{\frac{1}{2}}$, and $\Gamma_{\lambda f}$ is the fission width for level θ.) This effect cannot be attributed to variations in $\Gamma_{\lambda f}$ alone since the values of $\Gamma_{\lambda n}^0$ for the resonances showing low ternary fission are much larger than $\langle \Gamma_{\lambda n}^0 \rangle$, the average

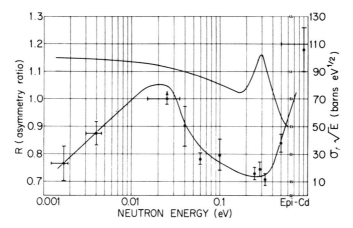

Fig. 44. Variation of the asymmetry ratio R, the ratio of asymmetric to symmetric fission mass yields normalized to the ratio at 0.025 eV, as a function of neutron energy below 0.5 eV for ^{235}U, as measured by Faler and Tromp (1963). The mass distribution variations do not follow the structure in the fission cross section, which is shown in the figure. They do appear to be correlated with changes in the ternary fission yield (Deruytter and Neve de Mevergnies, 1965), and with changes in the binary-fission kinetic-energy distribution (Miller and Moore, 1967). It is possible that changes in the angular distribution of fragments emitted in the neutron-induced fission of oriented ^{235}U targets (Dabbs et al., 1965) also show this kind of energy dependence, rather than that given by the solid line in Fig. 32.

neutron width. The relative ternary-fission yield also appears to vary with the energy of the long range α-particle (Michaudon, 1964; Michaudon et al., 1965b).

For ^{235}U it also seems that the variation of the mass and kinetic-energy distributions with neutron energy does not necessarily follow the observed resonance structure. This has been most extensively studied at very low energies, in the region below 1 eV, as shown in Fig. 44. The variation in the mass distribution (Faler and Tromp, 1963) does not follow the resonance structure. Deruytter and Neve de Mevergnies (1965) found evidence that the ternary-alpha yield shows the same anomalous behavior, following the symmetric-fission yield in this region.

The relative changes in both the mass distribution and in the kinetic-energy distribution are more complex than has been so far described. Croall and Willis (1965) studied the complete mass distribution of fragments from fission of ^{239}Pu irradiated by thermal and epicadmium neutrons. They showed that the dramatic decrease in the symmetric fission yield (Regier et al., 1960) is accompanied by an increase in the relative yield of very asymmetric fragments. The data obtained by Miller and Moore (1967) suggest that a similar effect exists for changes in the kinetic-energy distribution in binary fission. An increase in the yield of the highest energy heavier fragment is accompanied by a decrease in the yield of the highest energy lighter fragment. These results may imply the same sort of skewing of the mass and kinetic-energy distributions as that observed by Senchenko *et al.* (1967) for fast-neutron irradiation of ^{233}U.

Of considerable interest are studies of correlations of the prompt neutron yield from resonance to resonance, carried out by Weinstein (1968) (see also Weinstein and Block, 1968). For ^{239}Pu, Weinstein found a small, but significant, difference in $\bar{\nu}$ for the two spin states. These changes in $\bar{\nu}$ appear to be in qualitative agreement with the changes in fragment kinetic energy.

E. Channel Effects in Fission Induced by *p*-Wave Neutrons

In the region of 0.1 to 1 MeV neutron energy the individual resonance widths become larger than the average level spacings, and only the averages of the fission width, the neutron width, or the capture width are significant. The fission cross sections in this energy region might be expected to show structure which could be related to the opening of additional fission channels. However, the dependence of the fission cross section is not easy to interpret, for several reasons. Low excited states of the fissile target nuclei lie a few, or tens of, kilovolts above the ground state. Neutron inelastic scattering feeding these levels constitutes a channel which does not lead to fission, causing a decrease in the fission cross section as the neutron energy is raised. Absorption of neutrons of higher angular momenta can introduce additional fission channels, as well as additional channels for processes which do not lead to fission. Fission induced by *p*-wave neutrons may be important as low as 10 keV neutron energy and is expected to be comparable to the *s*-wave contribution at 100 keV. Fission by *d*-wave neutrons may be important at 500 keV, and higher partial waves enter in the million electron volt region (see Rae *et al.*, 1958; Auerbach and Perey, 1962). Uttley *et al.* (1967) have determined strength functions from total cross-section measurements, and find good agreement with calculations based on the optical potentials of Perey and Buck (1962).

Where the neutron kinetic energy becomes significant compared to the binding energy, the quantity $\bar{\nu}$ can no longer be treated even as approximately constant. A monotonic increase in $\bar{\nu}$ with increasing neutron energy would be consistent with the view that the primary dependence of $\bar{\nu}$ is on the excitation energy of the fragments (Terrell 1959, 1962). Meadows and Whalen (1962, 1967), Mather *et al.* (1964, 1965), and Hopkins and Diven (1963) have reported measurements of the dependence of $\bar{\nu}$ with incident

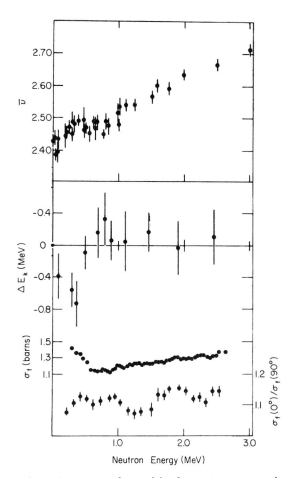

Fig. 45. Comparison of structure observed in the neutron energy dependence of $\bar{\nu}$, the fragment kinetic energy, the fragment angular anisotropy, and the fission cross section for ($^{235}U + n$), as presented by Blyumkina et al. (1964).

neutron energy for various fissionable materials. Blyumkina et al. (1964) have noted various changes in the slope of $\bar{\nu}$ at energies below 1 MeV, which they have interpreted as structure due to the opening of p-wave neutron channels. Blyumkina also pointed out a correlation of structure in the energy dependence of the fission cross section, the fission-fragment anisotropy, the fragment kinetic energy, and the number of neutrons emitted for ^{233}U and ^{235}U, as shown in Figs. 45 and 46. Blyumkina cited this correlation as evidence that both variations in the fragment kinetic energy and the

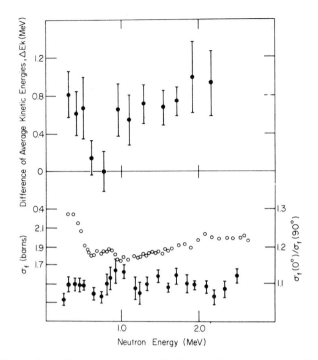

Fig. 46. Comparison of structure observed in the neutron energy dependence of the fragment kinetic energy, the fission-fragment angular anisotropy, and the fission cross section of ^{233}U after Blyumkina *et al.* (1964).

variation in $\bar{\nu}$ are due to varying contributions from saddle-point transition states or fission channels as defined by Bohr (1956).

Blyumkina interpreted the experimental results according to channel theory, requiring but one additional assumption: that the rotational and vibrational energy associated with the various channels at the saddle point eventually reappears as fragment kinetic energy. It is thus not available to the fragments as excitation or deformation energy at scission, and so $\bar{\nu}$ for fission through such channels is relatively lower. Blyumkina assumed that the effective positive parity saddle-point states are lower in energy than the negative parity states by about 0.8 MeV, which is about the size of the observed change in kinetic energy. The target nuclide ^{235}U has negative parity, and so only fission through the negative parity saddle-point states is permitted with s-wave neutrons. As soon as p-wave neutrons give an appreciable contribution, the kinetic energy drops because of fission through the now opened positive parity channels, and $\bar{\nu}$ increases to a larger value. When the p-wave contribution ceases to dominate because of d-wave and higher-angular-momentum neutron absorption and because of appreciable fission through higher saddle-point states of both parities, the kinetic energy release again increases.

The same assumption appears adequate to explain the ^{233}U data as well. Here the target nucleus has positive parity, and p-wave neutron absorption requires that the higher energy negative-parity channel dominate at a few hundred kilovolts. When d-wave neutrons are permitted, near 1 MeV, positive parity channels again lower the kinetic energy to the thermal or s-wave value. Higher channels and higher angular momenta again cause the increase above 1 MeV.

Mass distribution studies have also been made for p-wave neutron induced fission of ^{235}U and ^{239}Pu targets (Cuninghame et al., 1961, 1966). The first of these experiments, on ^{235}U, was done in order to test the suggestion of Wheeler (1956) that symmetric fission is prohibited for the 3^{-} state in the configuration (^{235}U + n) because of the antisymmetric character of the octupole vibration. If fission can be induced by p-wave neutrons incident on ^{235}U, the possible spin states are 2, 3, 4, and 5; all with positive parity. In the 2^{+} and 4^{+} channels, the wave function for vibrational motion is not required to be antisymmetric, so a net enhancement of symmetric fission might be expected for p-wave neutron-induced fission. Based on calculations of Rae et al. (1958), it was estimated that fission by p-wave neutrons is expected to be important compared to s-wave fission ($\sim$30% of total fission) at $\sim$50 keV. The peak-to-valley ratios determined by Cuninghame et al. (1961) are shown in Fig. 47. These data show an effect which can be attributed to p-wave fission; however, contrary to expectation, there is considerably less symmetric fission in this region than at thermal neutron energies. Cuninghame et al. (1966), in an attempt to resolve the questions which had been raised, carried out a similar study of ^{239}Pu. This nuclide has spin $\frac{1}{2}$ and positive parity, so that absorption of an s-wave neutron will lead to possible spins of 0^{+} and 1^{+}. The wave function for the 1^{+} state is permitted to be antisymmetric (Griffin, 1965) although that for the lowest 0^{+} state is not. Absorption of a p-wave neutron permits states of negative parity, and certain of these could also have an antisymmetric wave function describing the saddle-point configuration. Cuninghame et al. (1966) chose to interpret their mass distribution data in this way, according to the expected symmetry properties of the energetically permitted saddle-point wave function. They were able to show that the ^{239}Pu data are in qualitative agreement with such a scheme although the ^{235}U data shown in Fig. 47 still show a discrepancy.

As an alternative approach, one can assume that the variations in the mass distributions of Cuninghame et al. (1961) should be simply related to

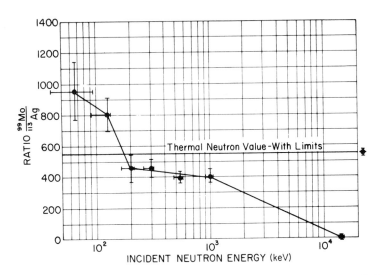

FIG. 47. Ratio of asymmetric to symmetric fission yields for ^{235}U + n (Harwell data), as determined by Cuninghame et al. (1961). The shape of the curve is consistent with what would be expected from a comparison with Fig. 46. (The symmetric yield should follow the yield of prompt neutrons.) However, the curve appears to be inconsistent because of its position with respect to the thermal neutron value.

the variations in fragment kinetic energy and $\bar{\nu}$ as observed by Blyumkina et al. (1964). One expects qualitatively that if the scission configuration is such that less kinetic energy is released, the extra energy can go into fragment excitation and the symmetric-fission yield will be enhanced. The data show just the opposite behavior, however. If one is led to conclude that a mechanism other than symmetry requirements on the wave function at the saddle point is responsible for the symmetry of the mass distribution, then it must also be granted that this mechanism is somewhat more complex than a simple argument based on the energy available to the nascent fragments at scission.

F. Fission Gamma and Electron Emission

Prompt and delayed fission-gamma radiation has been reviewed by Johansson and Kleinheinz (1965) and by Maier–Leibnitz et al. (1965a) Although the gamma-radiation studies have revealed a considerable body of information about the fission process at the scission point, there has been virtually no effort made to study fission-gamma systematics as a function of neutron energy. Perhaps this is because the gamma-ray spectra are so complex that even the highest resolution lithium-drifted germanium detectors probably do not resolve more than the strongest of the individual gamma-ray peaks (see Bowman *et al.*, 1964, 1965).

The energy spectrum of the gamma rays emitted from ^{235}U was first measured by Maienschein *et al.* (1958) (see also Peelle *et al.*, 1962). The gamma spectrum consists of fairly soft radiation: the average energy of a prompt fission-gamma ray is about 1 MeV (Hopkins and Diven, 1962). Since there is an average of about 7–9 MeV released in fission in the form of prompt gamma radiation, the multiplicity must be high. The large multiplicity and the short lifetime ($\lesssim 10^{-11}$ sec) of the gamma-emitting states (Johansson, 1964) suggests that the radiation consists of successive fast E2 transitions. The angular correlation of the gamma rays with respect to the direction of motion of the fission fragments is consistent with the interpretation that these are E2 transitions, as well as that the initial spin of the nascent fragments is high, oriented perpendicular to the direction of flight. (See Hoffman, 1964; Kapoor and Ramanna, 1964, Graff *et al.*, 1965.) The comparatively successful competition of gamma radiation with neutron emission also implies that the initial spins of the nascent fragments must be large. All the data seem to be consistent with a spin of about $I = 7$ to 10 for each fragment. The high spin of the primary fragments also implies that there is a large orbital angular momentum between the fragments (up to 10 to 20 units, depending on the mass split). Such a large orbital angular momentum leads one to expect that the fragments may possess as much as 1–2 MeV or more of kinetic energy at scission (Maier–Leibnitz *et al.*, 1965a). The dependence of the number of fast prompt gamma rays on the fragment mass shows the same saw-tooth structure as the prompt neutron emission probability (Johansson, 1964; Maier–Leibnitz *et al.*, 1965b). The number of slower prompt gammas (emission times of 10^{-8} to 10^{-7} sec) shows peaking in the region of closed shell nuclei, which Johansson (1964) has interpreted as vibrational transitions in these nuclei. Conversion electron and prompt x-ray spectra have been studied by Bowman *et al.* (1965). The yields are also consistent with that which would be expected from internal conversion of prompt E2 transitions.

Almost nothing has been done on the neutron spectroscopy of prompt fission-gamma radiation. Moore and Spencer (1964) attempted to find evidence for the (n,γf) reaction in ^{233}U by comparing the fission-gamma spectra at 0.06 and at 1.8 eV. This nuclide is not the best one to choose for such a study since the effect should not be very pronounced for ^{233}U (see Lynn, 1965b), and the experimental results were inconclusive. The experiment should perhaps be repeated, with higher resolution detectors, for ^{239}Pu or ^{235}U, or perhaps, as an ideal nuclide, ^{238}Pu.

VI. THE QUESTION OF SPIN DEPENDENCE OF THE MASS DISTRIBUTION AND THE FISSION-WIDTH DISTRIBUTION

A. Fission Widths and the Fission-Channel Spectrum

In order to discuss the possibility of a spin dependence of the fission-fragment mass distribution and of the fission widths, it is necessary to summarize the details of the stages in fission, the pertinent experimental data, and the various models of fission which apply. We have already introduced the idea that fission induced by the resonance absorption of low-energy neutrons is a multistep process. The formation of a compound nucleus is followed by the crossing of the saddle point. Then the nucleus undergoes still further deformation, perhaps slowly at first, and more quickly as it approaches the scission point. It is at this point that the nuclear forces cease to operate, and so the fission-product mass distribution must have been determined at this or some earlier stage. Finally, the nascent fission fragments separate, gaining kinetic energy as a result of coulomb repulsion, and, in reorienting themselves from the deformed configuration at the scission point, emit most or all of the neutrons and prompt fission gamma radiation.

The fission-width distribution is related to the conditions at the saddle point, while the mass and kinetic-energy distributions are related to the conditions at scission. The crossing of the saddle point is slow, and it is generally accepted that the fission process at this stage has a collective nature (Bohr and Mottleson, 1953; Hill and Wheeler, 1953). The nature of the transition from the saddle point to scission is open to question. The large number of fission-product masses indicates that the transition takes place slowly while preserving statistical equilibrium (Fong, 1956). The angular distribution of the fragments, and the apparent correlation of symmetric-fission-product mass yields with the properties of the saddle-point states would seem to require that the transition take place rather quickly, without any appreciable changes occurring in nuclear orientation or wave function symmetry. (Wilets, 1959, 1964; Cuninghame et al., 1966.) However, as Wilets (1959) has pointed out, if one only assumes conservation of the quantum number K throughout the process, the angular-distribution data are consistent with a slow transition in which statistical equilibrium is maintained.

The properties of the fission-width distribution can be discussed in detail by considering only the spectrum of transition states at the saddle point. The type of interference observed in fission and the manner in which the fission widths are distributed are characteristic of a few-channel process.

According to the Bohr model (Bohr, 1956), such behavior is expected if the transition-state spectrum at the saddle point is relatively widely spaced, allowing only a few fission channels to be energetically permitted. It seems reasonable to assume that the fission channels revealed by the fission-width distribution and by the interference among the fission resonances are in fact the same saddle-point channels revealed by fragment angular distribution studies. One is then able to make some statements about what kind of behavior is to be expected in the low-energy resonances. Bohr suggests that the potential surface corresponding to low-lying states in the compound nucleus will be preserved as the deformation increases, and at the saddle point these potential surfaces constitute the channel spectrum. Thus some information on the possible modes of oscillation of the highly deformed compound nucleus at the saddle point can be obtained from the energy levels observed near the ground state of this compound nucleus.

The compound nucleus formed by the addition of a neutron to any of the common fissile isotopes is even–even. In particular, ^{229}Th, ^{233}U, and ^{241}Pu have spin $\frac{5}{2}$ and even parity. Spin states of 2^+ and 3^+ are formed in the compound nucleus by adding an s-wave neutron to any of these isotopes. For ^{235}U, having a spin and parity of $\frac{7}{2}^-$, the spin states of interest are 3^- and 4^-. For ^{239}Pu ($\frac{1}{2}^+$), spin states of 0^+ and 1^+ are formed.

Hyde et al. (1964) (see also Sheline, 1960) have reviewed the level systematics of deformed even–even nuclei, on the basis of the unified model (Bohr and Mottelson, 1953). The description of rotational states given by this model provides a convenient reference frame for discussion.

The low-lying level systematics of even–even heavy nuclides are indicated in Fig. 48. Most of the data are those reviewed by Hyde et al. (1964). According to the collective model, in even–even deformed nuclides, low-lying positive parity levels are attributed to rotational bands built upon the 0^+ ground-state band head and upon quadrupole vibrations whose lowest members are 0^+ (sometimes called β-vibrations) and 2^+ (γ-vibrations). If one neglects band mixing, one can think of the ground state and β-vibrational bands, consisting of the even spin sequences 0^+, 2^+, 4^+, . . . , as a "breathing mode" vibration along the nuclear symmetry axis. At the saddle point such a vibration is unbound since there are no restoring forces, and the β-vibration might be considered to be degenerate with the motion corresponding to the ground-state deformation (Huizenga et al., 1965). The γ-vibrational band consists of the levels in the spin sequence 2^+, 3^+, 4^+, . . . , and if pure, can be thought of as an axially asymmetric vibration perpendicular to the symmetry axis. Low-lying negative parity levels are attributed to rotational motions built upon octupole vibrations with spin sequences 1^-, 3^-, 5^-, . . . , and 2^-, 3^-, 4, . . . , sometimes called b- and

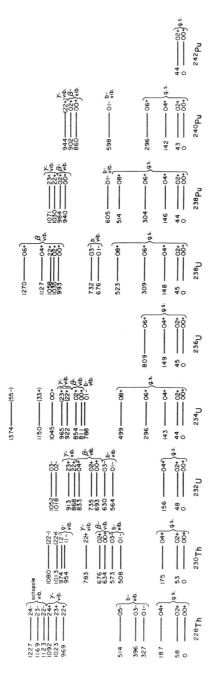

Fig. 48. Level systematics of even–even heavy nuclei. The level schemes shown were constructed from those reviewed by Hyde *et al.* (1964). The nuclides of primary interest are ²³⁰Th, ²³⁴U, ²³⁶U, ²⁴⁰Pu, and ²⁴²Pu, which correspond to compound nuclei formed by adding a neutron to the most common fissile nuclides. The low-lying levels of ²³⁶U and ²¹²Pu have not been extensively studied, but they are not expected to be very different from those of the other even–even nuclides in the region. In this figure the notation "g.s." refers to the rotational structure associated with the ground state; the *b*-vibrational band is an octupole band with $K = 0$; the β-vibrational band is a quadrupole band with $K = 0$; the γ-vibrational band is a quadrupole band with $K = 2$; and the *g*-vibrational band is an octupole band with $K = 1$. Of particular interest is the highest structure shown for ²²⁸Th. This may represent an octupole band with $K = 2$, or, as Arbman *et al.* (1960) suggest, a gamma vibration superimposed on the $K = 0$ octupole band, which is very low in ²²⁸Th. Also of considerable interest is the 0^+ state at 1045 keV in ²³⁴U, which perhaps represents the magnitude of the pairing energy gap for stable even–even nuclides in this region. (See Bjørnholm and Nilsson, 1963.)

g-vibrations, respectively, in analogy to the β- and γ-vibrations (Lipas and Davidson, 1961). The spectra of levels of all the heavy, deformed, even–even nuclides are very similar, and show structure characteristic of that described above. There is additional structure in ^{234}U consisting of an 0^+, 2^+, . . . , sequence near 1 MeV. This may be the lowest structure resulting from the breaking of a nucleon pair, which would represent the onset of an extremely complex set of intrinsic states based on quasiparticle coupling (Gallagher et al., 1961; Gallagher, 1962; Bjørnholm and Nielsen, 1963). According to the superfluid model of nuclear structure at nuclear excitations less than this energy gap of 1 MeV, the nucleus can be treated as if it were a superconductor, where the only excited states possible will consist of collective excitations and rotations (see Bohr et al., 1958; Griffin, 1963; Gallagher, 1964).

For purposes of this discussion, let us consider the spectrum of low-lying levels observed in these even–even nuclides. On the basis of the Bohr model, it can readily be seen how the fission-width distribution could be spin dependent. The nuclides ^{233}U and ^{241}Pu, for example, allow the formation of resonances of spin 2^+ and 3^+. If band mixing is neglected, the simple motions described above (ground-state and γ-vibrations) would allow only one low-lying channel characteristic of 3^+ spin, but at least two such channels for 2^+ spin.

Following Bohr and Wheeler (1939) (see also Wheeler, 1956), one expects that the average fission width should be proportional to the number of open fission channels. It can in this way be argued that the average fission width should be different for 2^+ and 3^+ resonances. This is consistent with the results of multilevel analysis of ^{241}Pu fission resonances and possible also for ^{233}U, where one finds that there are two groups of interfering resonances with different average widths, and that these resonances are characteristic of different spins. (See Moore et al., 1964; Sauter and Bowman, 1968.)

It is evident that the modes of motion which lead to fission are not restricted to these simple ones described since on the basis of this model, no mechanism exists for the observed fission of ^{239}Pu in the 1^+ spin state. One possible answer to this puzzle was suggested by Griffin (1965). From studies of the fragment angular distributions from fission of even–even compound nuclei (Britt et al., 1963b, 1965), the spectrum of saddle-point states was inferred as shown in Fig. 49. At the saddle point, the onset of two-quasiparticle channels does not appear until 2.6 MeV above the lowest configuration. This permits the possible modes of motion to consist not only of the simple quadrupole and octupole vibrations but also of double vibrations, states whose wave function consists of products of the simple vibrations. Griffin (1965) suggested that the 1^+ resonances in ^{239}Pu do, in fact,

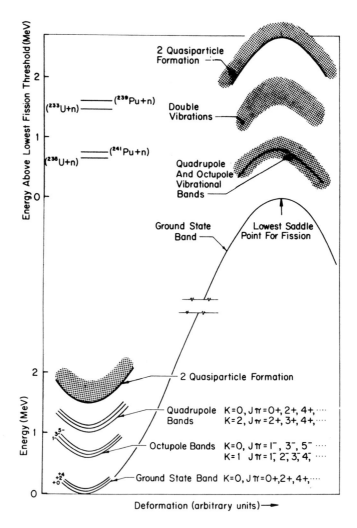

FIG. 49. Relation between level structure of even–even stable nuclei and fission-channel structure at the saddle point of fissioning even–even compound nuclei. The addition of a slow neutron to any of the common fissile isotopes leads to an excitation which lies well above the lowest fission threshold. This energy excess is indicated in the upper left-hand corner of the figure. It will be noted that some fission of (^{235}U + n) and (^{241}Pu + n) is expected by way of channels corresponding to single vibrations, while some fission of (^{233}U + n) and (^{239}Pu + n) is expected through channels corresponding to double vibrations as well.

TABLE X

CLASSIFICATION OF THE SEQUENCE OF MODES OF MOTION OR
FISSION CHANNELS EXCEPTED AT THE SADDLE POINT[a]

Energy of transition state above fission barrier (MeV)	K	I_π	Description	Remarks
0	0	0^+	Ground state	
0.011		2^+	+ Rotation	
0.035		4^+	+ Rotation	
~0.5	0	1^-	1 quantum mass asymmetry vib^n	May lie even lower in energy.
		3^-	+ rotation	
		5^-	+ rotation	
~0.7	2	2^+	1 quantum gamma vib^n	These are assumed to be the levels observed by Britt *et al.* (1963b, 1965).
		3^+	+ rotation	
		4^+	+ rotation	
		5^+	+ rotation	
~1.0	1	1^-	1 quantum bending vib^n	May be considerably lower in energy.
		2^-	+ rotation	
		3^-	+ rotation	
		4^-	+ rotation	
		5^-	+ rotation	
~1.2	2	2^-	1 qu. mass asymmetry + 1 qu. gamma	
		3^-	+ rotation	
		4^-	+ rotation	
		5^-	+ rotation	
~1.5	1	1^+	1 qu. mass asymmetry + 1 qu. bend	Suggested by Griffin (1965).
		2^+	+ rotation	
		3^+	+ rotation	

[a] These states are those suggested by Wheeler (1963), and extended by Lynn (1965a). Here K is the projection of angular momentum I on the nuclear symmetry axis. In each band, one expects the individual rotational states shown to be separated by roughly the same energy differences indicated for the ground state band.

fission through one or more of the channels whose wave functions represent double vibrations.

The allowable modes of motion have been summarized by Wheeler (1963; see also Lynn, 1965a; Harris, 1966; Cuninghame *et al.*, 1966). These modes are shown in Table X (Lynn, 1965a). For the low-energy resonances in fissile nuclei, the available channels (those which will be open) depend on the excitation energy. The excitation of the compound nuclei formed by adding a slow neutron to the four fissile target nuclides is also indicated in Fig.

49 (Northrop *et al.*, 1959; Britt *et al.*, 1966). Even though there are fairly large uncertainties in the fission-threshold measurements as well as in the saddle-point channel-spectrum data, one can conclude that the excitation of ^{235}U and ^{241}Pu is such that some contributions from single vibrations may be expected. For ^{233}U and ^{239}Pu, the channels corresponding to single vibrations are undoubtedly fully open, and some contributions from double vibrations are expected. Table XI shows the expected sizes of the average fission width for the four common fissile nuclides. These values are based on estimates of the number of channels as given in Table X and Wheeler's estimate of the fission width as given in Eq. (3) (Bohr and Wheeler, 1939; Wheeler, 1956). Table XI also shows the average fission widths determined

TABLE XI

COMPARISON OF ESTIMATED AVERAGE NEUTRON RESONANCES OF FISSILE NUCLIDES[a]

Target	Channel spin	No. of channels	D (eV)	$\langle \Gamma_f \rangle_{\text{est}}$ (eV)	$\langle \Gamma_f \rangle_{\text{obs}}$ (eV)
^{233}U	2^+	2.5	1.4	0.57 ± 0.11 ⎱	
^{233}U	3^+	1.5	1.0	0.24 ± 0.08 ⎰	0.403
^{235}U	3^-	1.5	1.4	0.33 ± 0.11 ⎱	
^{235}U	4^-	0.5	1.1	0.08 ± 0.08 ⎰	0.052
^{239}Pu	0^+	1.5	10.0	2.4 ± 0.8	0.2
^{239}Pu	1^+	0.5	3.3	0.27 ± 0.27	0.05
^{241}Pu	2^+	1.5	2.5	0.60 ± 0.20	0.51
^{241}Pu	3^+	0.5	1.8	0.14 ± 0.14	0.16

[a] In estimating the fission widths the level spacing D was assumed to vary inversely as $2J + 1$ per spin state. The average spacing of the 0^+ levels of ^{239}Pu were calculated under the assumption that all the narrow observed levels have spins of 1^+. The quantity $\langle \Gamma_f \rangle / D$ was assumed to have a maximum value of $n/2\pi$ where n is the number of channels energetically permitted. As is noted in Fig. 49, there is in all cases a set of channels which may be partially open. In these cases, it was arbitrarily assumed that just one of these channels is half open, with an uncertainty in the partial width of 100%.

from neutron spectroscopy in the resonance region. One concludes that the sizes of fission widths are consistent with the values expected from the saddle-point spectrum of transition states.

B. The Transition from Saddle Point to Scission

In order to be able to say what effect the saddle-point configuration has on the fission-product mass distribution, it is necessary to know something about the transition from saddle point to scission. In effect, this question hinges on whether or not the time from saddle point to scission is long enough for statistical equilibrium to be attained. A study on this subject was made by Wilets (1959), who described a surface-coupling mechanism based on the collective model for evaluating the time required for statistical equilibrium. Both the liquid-drop model (Bohr and Wheeler, 1939) and the statistical model proposed by Fong (1956) assume statistical equilibrium at scission. The channel theory of Bohr (1956) requires that certain constraints, K (the angular-momentum projection on the nuclear symmetry axis) and π (the parity), be conserved in the transition from saddle point to scission. However, as Wilets (1959) points out, this model is otherwise consistent with almost any descent from the saddle point; i.e., statistical equilibrium at the scission point is not a requirement.

It is of interest to examine the statistical model of fission in somewhat more detail. As originally proposed by Fong (1956), the relative probability of a given mode in the mass distribution is determined by the number of final quantum states which can lead to that mode. Thus the mass distribution is related to the total energy of excitation of the two nascent fragments at the point of scission as well as to the level density. Fong examined the nuclear level density empirically to take account of shell effects, and was able in this way to calculate a mass distribution curve for thermal neutron-induced fission of ^{235}U which agree well with the observed data. Perring and Story (1955) applied Fong's method to the case of ^{239}Pu but were unable to achieve a fit to the observed mass-distribution curve. Newton (1956) and Cameron (1958) refined Fong's theory on the basis of more realistic expressions for the density of final states, and concluded that a purely statistical approach is inadequate to describe the observed mass-distribution curve. Cameron was able to reproduce the curve for (^{235}U + n) by allowing the radius parameter at scission to vary as a function of both the energy of excitation and the mass ratio (perhaps foreshadowing present ideas of the importance of fragment deformation). Newson (1961) extended the statistical model to include a description of the symmetric fission of (^{209}Bi + d) and the three-humped mass distribution curve resulting from charged particle irradiation of ^{226}Ra (Jensen and Fairhall, 1958, 1960). Newson's calculations required the use of two additional parameters, one to correct for the excitation required for the break-up of the doubly-magic, but possibly distorted, core ^{132}Sn and the other a free "smearing" parameter to account for unknown nonstatistical effects at the instant of scission. These models of fission have been summarized by Wilets (1964).

Brunner and Paul (1960, 1961a,b,c) emphasized the formation of magic number clusters in the fissioning nucleus. They used the surface tension concept of the liquid-drop model, and showed that the increased surface tension of magic-number nuclei implies a much lower deformation for magic-number fragments. By assuming a mass-ratio dependent two-body nuclear potential between the two clusters before fission, they were able to fit the observed mass distribution for low-energy neutron fission of ^{235}U. They found a dependence on the interaction radius very similar to that required by Cameron (1958). In addition, the correct mass-ratio dependence of the mean kinetic energy of the fragments was given; the symmetric deuteron-induced fission of ^{209}Bi was correctly predicted; and the qualitative dependence of the symmetric fission yield on neutron energy in the million electron volt region was described.

The work of Thomas and Vandenbosch (1964), in correlation the characteristics of the mass distribution to the amount of excitation energy available, is perhaps one of the most significant contributions to neutron spectroscopy (see Fig. 42). Their work was based on a statistical approach, but also demonstrated the importance of the magic number cores to the available excitation energy and thus to the mass distribution. Vandenbosch (1963) developed the idea that a large deformation of the nascent fragments at the scission point can describe the variation of $\bar{\nu}$ and the ternary-fission yield with mass number. Fong (1961, 1964), following these ideas, has modified his original statistical model (Fong, 1956) to include the effects of deformation of the nascent fragments. With the modified theory, he has been able to describe the kinetic-energy distribution of fission fragments and the dependence of $\bar{\nu}$ on mass number.

All of the models described above have in common an emphasis on the nuclear structure of the primary fragments at the scission point as the determining factor for such properties of fission as the mass and kinetic-energy distributions. They have as an underlying basis the assumption that statistical equilibrium is maintained over most of the way in the transition from the saddle point to scission. Many of the successful theoretical approaches used in recent years have treated the transition as statistical over the entire distance, such that minimizing the potential energy at scission is valid (Terrell, 1965; Vandenbosch, 1963; Ramanna et al., 1965). Ramanna estimated that the transition time is long enough, compared to the time required for a nucleon to traverse the nucleus, to permit about 500 nucleon transfers between the nascent fragments to take place during the transition from saddle point to scission.

One must also require that certain constraints, including the projection of angular momentum on the nuclear symmetry axis and the parity, are conserved in the transition from saddle point. Wilets (1964) has pointed out that many of the dynamical variables which one usually regards as constants of the motion are generally not so under deformation. Included among these, for certain types of deformation, are the intrinsic angular momentum of the system and the parity. On this basis, it would appear that the maintaining of such constraints is only possible if one assumes that the transition from the saddle point to scission is sudden rather than statistical. Yet most of the evidence concerning the fission-fragment mass and kinetic-energy distributions clearly supports the idea that in the transition statistical equilibrium is maintained.

C. The Fission-Channel Spectrum and the Fission-Product Mass Distribution

The fission-width distribution has been assumed above to be determined by the conditions at the saddle point. In particular, the spin dependence of the fission-width distribution seems to be consistent with the idea that only a few fission channels or modes of motion are energetically allowed as the fissioning nucleus crosses the saddle point. The fission-product mass distribution does not appear to be determined primarily by the conditions at the saddle point but rather by the conditions at the scission point: in particular, the deformation and excitation of the primary or nascent fragments. However, it is important to bear in mind that the mass distribution has been observed to vary in a systematic way from resonance to resonance. Since the resonances are properties of the compound nucleus, some property of the compound nucleus must still be in effect at the point at which the mass distribution is determined. Somehow, the deformation and excitation of the fragments at scission must depend on such a property of the resonances of the fissioning nucleus.

It has been commonly accepted that the observed variation in the mass distribution from resonance to resonance is characteristic of the resonance spin state in the compound nucleus. If the characteristics of the mass distribution are determined by the amount of energy available to the fragments for deformation and excitation at scission, then in order for a spin dependence to exist, the height of the coulomb barrier at the scission point (and thus the energy available to the fragments) must be different for the different modes of motion which occur in the transition from saddle point to scission. The simplest and most obvious mechanism is that proposed by Blyumkina et al. (1964): that the potential energy of collective motion at the saddle point is maintained in the transition, and eventually appears as kinetic energy. It is not available as deformation energy of the fragments at scission, and as a consequence the mass distribution changes in a systematic way. Another mechanism might be that the relative separation of the nascent fragments at scission could be different for different types of quadrupole or octupole vibrations. Such a difference should be observed if the shape or curvature of the barrier were different for the different channels (Vandenbosch, 1967). There is evidence that neither of these views is completely satisfactory (Croall and Willis, 1965; Cuninghame *et al.*, 1966; Miller and Moore, 1967), but the exact mechanism which affects the mass distribution is not important to the argument. It is sufficient to postulate that some constraint is maintained in the transition from saddle point to scission, such that various modes of motion lead to configurations where different amounts of excitation and deformation energy are available to different fragments.

It may be noted that a constraint of this nature implies not only a spin dependence of the fission-fragment mass distribution but also a fission-channel dependence as well. As an example, let us consider s-wave neutrons on the target nucleus ^{235}U, where the channel spins of interest are 3^- and 4^-. The channel spectrum will be assumed to be restricted to the octupole vibrations. The lowest of these with $K = 0$ and a spin sequence 1^-, 3^-, 5^-, . . . , is a channel through which resonance fission of 4^- resonances cannot occur. Other octupole channels with higher K values contain both 3^- and 4^- in the spin sequences. If the mass distribution depends on spin, it must also depend on the fission channel, i.e., for 3^- resonances, fission occurring through a channel having $K = 1$ or 2 might be expected to show a mass distribution resembling that of 4^- resonances, but fission occurring through channels having $K = 0$ could show a different mass distribution. This type of channel dependence of the mass distribution could readily be demonstrated. The fission vectors (partial-fission-width parameters as defined by Vogt, 1958, 1960) in different channels are not expected to be correlated with respect to phase. Thus if the mass distribution is channel-dependent, a plot of the ratio of relative symmetric fission to total fission across a resonance could show striking interference effects. Small effects of this kind appear to have been experimentally observed in the kinetic-energy variation of certain fragments (Moore and Miller, 1965). These effects are found to be correlated qualitatively to changes in the symmetric-fission yield (Faler and Tromp, 1963), and perhaps to the variation of the fragment angular distribution from fission of aligned nuclei as well (Dabbs *et al.*, 1965).

At higher energies, the existence of such a correlation has not yet been established. Cuninghame *et al.* (1961) and Blyumkina *et al.* (1964) found evidence for changes in the mass distribution and in the fragment angular distribution, respectively, for p-wave fission of ^{235}U, but the observed changes do not seem to show the expected correlation. Vandenbosch *et al.* (1965) sought, but failed, to observe a dependence of the mass asymmetry on angular anisotropy for $^{234}U(d,pf)$ in the energy region where marked structure is seen in the angular anisotropy. On the other hand, Kapoor *et al.*

(1965) found evidence that a correlation between angular anisotropy and mass asymmetry does exist for 4 MeV neutrons on ^{235}U. The direct evidence for a correlation between the saddle-point states and the mass asymmetry is so far inconclusive.

When Bohr first proposed the channel theory of fission (Bohr, 1956), he suggested several experiments which would be appropriate to the field of low-energy neutron spectroscopy (see Section I). We have seen that most of these experiments have been done. In general, they show that channel theory can be applied to the description of low-energy neutron induced fission. We conclude that the average fission width does depend on the spin of the compound nucleus and appears to be proportional to the number of available or open fission channels. The fragment mass distribution also depends on the compound nucleus spin. It appears to vary in a way which suggests that it too is characteristic of the fission channel, rather than depending on spin alone. The fission-fragment kinetic energies and the relative ternary-fission yield also appear to be related to the resonance spin by way of the appropriate saddle-point state. The variation of the mass distribution, kinetic-energy distribution, and ternary-fission yield is probably not a consequence of the symmetry properties of the wave function at the saddle point, however, but instead seems to be related to one of the constraints which may exist in the relatively slow transition from the saddle point to scission. The exact nature of such a constraint is not understood, and until it can be theoretically justified, its existence will be open to question.

A variation of $\bar{\nu}$, the number of neutrons emitted per fission, is also expected in the low-energy region. Since $\bar{\nu}$ depends on fragment mass and kinetic energy, and the fragment mass yield varies with incident neutron energy, so must $\bar{\nu}$. Measurements by Weinstein (1968) have confirmed that this variation exists. One expects also to find variations in the yield and energy of prompt fission gammas and conversion electrons with neutron energy since they too depend on the fragment mass.

The above are only a few of the many experiments which suggest themselves in the neutron spectroscopy of fissile nuclides. Although there has been a reawakening of interest in fission with the development of high-resolution fission detectors, the understanding of the fission process is far from complete. The dependence of the average fission width and of the fission-product mass distribution on spin has been established for a few levels in $(^{239}\mathrm{Pu} + \mathrm{n})$ and $(^{241}\mathrm{Pu} + \mathrm{n})$. For other cases, the fission-width dependence rests upon indirect evidence—that of noninterference between groups of wide and narrow levels. As soon as the spin dependence of the fission characteristics of resonance levels has been determined, one can anticipate experiments which will connect these fission characteristics with the properties of the fission channels in a given spin state.

APPENDIX

In order to describe the resonance structure of fissile nuclides, it is necessary to use a multilevel formula to take into account the interference in fission among the resonances. There are in general use three different formulas which were developed explicitly for the fission problem. (Reich and Moore, 1958; Vogt, 1958; Adler and Adler, 1962.) All of these can be derived from the general dispersion formalism of Wigner and Eisenbud (1947). A review of the general theory of nuclear reactions has been made by Lane and Thomas (1958). A review of the specific approaches to the fission problem was made by Adler and Adler (1966b), to which the interested reader is referred for greater detail. The two earlier formulas differ only in the computational procedures and in the convenience of using them. The formula developed by Reich involves a summation over levels and a matrix inversion with respect to channels. The formula developed by Vogt (1958) involves a summation over channels and a matrix inversion with respect to levels. The cross sections calculated with either formula from a given set of parameters are the same. One very slight difference is that the Reich–Moore formula requires only that the off-diagonal matrix elements describing interference in the radiative-capture channels must vanish. The Vogt formula requires in addition that the level–level interference in the neutron channel be equal to zero; whereas the Reich–Moore formula treats the latter effect explicitly. For the fissile nuclei, this difference in treatment is almost always of negligible importance.

The Adler formula (Adler and Adler, 1962), although derivable from the R-matrix theory of Wigner and Eisenbud (1947), corresponds to the general formulation of Kapur and Peierls (1938) (see also Lane and Thomas, 1958). The derivation requires a redefinition of the level parameters. Formally the redefined level parameters are related to the conventional R-matrix parameters, where the relation involves a complex matrix transformation. In practice, the redefined parameters are usually taken as primary parameters, and their relation to the nuclear parameters is ignored.

The relationships among the various approaches can be seen by considering the general theory and how it reduces to the special case of neutron-induced fission. The low-energy neutron reaction cross sections σ_{nc} and the total cross section σ_{nT}, which are due to resonance levels of the same spin and parity in the compound nucleus are related to the elements U_{nc} of the collision matrix by expressions of the form

$$\sigma_{nc} = \pi \lambda_n{}^2 g |\delta_{nc} - U_{nc}|^2 \qquad \text{(A-1a)}$$

and

$$\sigma_{nT} = 2\pi \lambda_n{}^2 g \, \text{Re}(1 - U_{nn}). \qquad \text{(A-1b)}$$

Here the subscript n refers to the neutron channel, the channel c to any other single channel, λ_n is the neutron wavelength, g is a statistical weighting factor for the spin state involved, and δ_{nc} is the Kronecker delta. The specifically nuclear properties of the interaction are described by the derivative matrix $\mathbf{R}$, in the following form:

$$\mathbf{R} = \sum_\lambda \frac{\gamma_\lambda \times \gamma_\lambda}{E_\lambda - E}, \qquad \text{(A-2)}$$

where the notation $\gamma_\lambda \times \gamma_\lambda$ denotes the direct product of the vectors γ_λ, and the subscript λ refers to levels. The elements $\gamma_{\lambda c}$ of these vectors are the reduced-width parameters for the level λ in channel c. In notation following Lane and Thomas (1958), the collision matrix $\mathbf{U}$ can be related to the derivative matrix $\mathbf{R}$ by an expression of the form

$$\mathbf{U} = \mathbf{0}^{-1}(\mathbf{1} - \mathbf{RL})^{-1}(\mathbf{1} - \mathbf{RL}^*)\mathbf{0}^*, \qquad \text{(A-3)}$$

where the diagonal matrices $\mathbf{O}$ and $\mathbf{L}$ describe the properties of the external region, and are thus assumed to be known quantities. In particular, the matrix $\mathbf{L}$ has real and imaginary parts, $\mathbf{L} = \mathbf{S} + i\mathbf{P}$, where the elements S_c are level shift parameters for channel c, and the P_c are penetrability parameters. In the special case of low-energy neutron-induced fission, it is customary to choose the boundary conditions such that the real components of $\mathbf{L}$, the shift parameters S_c, are formally allowed to vanish. Otherwise, the characteristic energies E_λ do not correspond to the observed resonance energies, in the absence of interference among levels.

The difficulty in using this formulation lies in performing the matrix inversion indicated in Eq. (A-3). Very early attempts to analyze fission resonances were based on an approximation valid where the levels are well separated. In this case, following Feshbach *et al.* (1954), one can rewrite Eq. (A-2), in the region of the ith resonance, as

$$\mathbf{R} = \frac{\gamma_i \times \gamma_i}{E_i - E} + \mathbf{g}, \qquad \text{(A-4)}$$

where

$$\mathbf{g} = \sum_{\lambda \neq i} \frac{\gamma_\lambda \times \gamma_\lambda}{E_\lambda - E}.$$

The second term $\mathbf{g}$ in Eq. (A-4) has no singularities near $E = E_i$ and is of the order of $\langle \Gamma \rangle / D$, the ratio of the average level width to the average spacing. When the quantity **g** is small, the matrix inversion can be performed by means of a power series expansion in **g**. The result of this approach is an approximate multilevel formula, valid for cases in which the level spacing is much greater than the widths. Such an approximation, however, is not found to be justified by most of the fissile nuclides.

The approach used by Vogt (1958) was to convert the problem of the matrix inversion with respect to channels into one of matrix inversion with respect to levels. The technique, described in detail by Wigner (1946) and by Lane and Thomas (1958), consists of assuming an expansion of the form

$$(1 - \mathbf{RL})^{-1} = \left[1 - \sum_{\lambda} \frac{\boldsymbol{\gamma}_\lambda \times \boldsymbol{\beta}_\lambda}{E_\lambda - E} \right]^{-1} = 1 + \sum_{\mu\nu} A_{\mu\nu}(\boldsymbol{\gamma}_\mu \times \boldsymbol{\beta}_\nu), \quad \text{(A-5)}$$

where the vector $\boldsymbol{\beta}_\lambda = \mathbf{L}\boldsymbol{\gamma}_\lambda$, and the subscripts λ, μ, and ν all refer to levels. In order that the inverse have the form indicated, the $A_{\mu\nu}$ must satisfy the following equation (Wigner, 1946):

$$\delta_{\mu\nu} = A_{\mu\lambda}(E_\lambda - E) - i \sum_{\nu} A_{\mu\nu}(\boldsymbol{\gamma}_\nu \cdot \boldsymbol{\beta}_\lambda), \quad \text{(A-6)}$$

where

$$(\boldsymbol{\gamma}_\nu \cdot \boldsymbol{\beta}_\lambda) \equiv \sum_c \gamma_{\nu c}\beta_{\lambda c}$$

and the summation c is now over all channels. The inverse of the Wigner level matrix A is then given by

$$(A^{-1})_{\mu\lambda} = (E_\lambda - E)\,\delta_{\mu\lambda} - \sum_c \gamma_{\mu c}\beta_{\lambda c} \quad \text{(A-7)}$$

and the components of A are obtained by inverting the matrix $(\mathbf{A}^{-1})$ in Eq. (A-7), under the assumption that the matrix $(\mathbf{A}^{-1})$ is finite (i.e., that all but a few levels can be neglected).

Reich preferred to eliminate the radiative-capture channels, by partitioning the matrix $(1 - \mathbf{RL})$ into a 2×2 matrix, each element of which is itself a matrix, in a manner similar to the channel elimination method of Teichmann and Wigner (1952) see also Lane and Thomas (1958):

$$(1 - \mathbf{RL}) \equiv \mathbf{a} \equiv \begin{pmatrix} \mathbf{a}_{k \times k} & \mathbf{a}_{k \times n} \\ \mathbf{a}_{n \times k} & \mathbf{a}_{n \times n} \end{pmatrix}, \quad \text{(A-8)}$$

where the matrix $\mathbf{a}_{k \times k}$ describes the neutron channel and the $(k - 1)$ fission channels, and the matrix $\mathbf{a}_{n \times n}$ describes the n radiative-capture channels. The matrix inversion of $\mathbf{a}$ is then straightforward for any finite number of fission channels, provided that the inversion of the matrix $\mathbf{a}_{n \times n} = [\mathbf{1} - \mathbf{RL}]_{n \times n}$ can be done (for the radiative capture channels only). This is again accomplished by the technique described in Eq. (A-4). In the present case, however, an important simplification results: in the summation over radiative-capture channels in Eq. (A-7), the off-diagonal elements ($\mu \neq \lambda$) vanish, and the sum becomes merely half the radiation width $\Gamma_{\lambda\gamma}$ for the level λ. Thus the inversion of ($\mathbf{A}^{-1}$) has a particularly simple form if only the radiative-capture elements are involved:

$$A_{\mu\lambda} = \delta_{\mu\lambda}/(E_\lambda - E - \tfrac{1}{2}i\Gamma_{\lambda\gamma}), \qquad (A\text{-}9)$$

and the inversion of $[\mathbf{1} - \mathbf{RL}]^{-1}$ reduces to an inversion of a matrix of rank k, where $k - 1$ is the number of fission channels. The explicit summation over levels is retained, so that the interfering contribution from distant levels or from direct interaction can be handled very simply as an additive constant.

The multilevel approach of Reich and Moore (1958) is thus convenient to use in the case of a few interfering channels and many levels; the approach of Vogt (1958) is more convenient in the case of many interfering channels and few levels. In Vogt's formulation, the number of interfering channels is not explicitly stated. An estimate of this number can be obtained from the average angle between the fission vectors γ_λ in a multidimensional channel space (see Vogt, 1958, 1960). Up to the present time, the fitting of fission data has required no more than two fission channels per spin state. However, as Vogt points out, this does not necessarily mean that only two channels are open, but merely that the data which presently exist are not sensitive to the effects of more than two channels.

Both formulas discussed above (Reich and Moore, 1958; Vogt, 1958) are cumbersome to use. For each neutron energy, some kind of matrix inversion must be performed, with respect to channels or with respect to levels, in order to evaluate the interference effects. The Adler formula (Adler and Adler, 1962) introduces parameters which are to a first approximation energy independent, and in which the matrix inversion is handled implicitly. The collision matrix is written in terms of the Wigner level matrix A, whose inverse is given in Eq. (A-7), as

$$U_{cc}{}' = \exp[i(\phi_c + \phi_c{}')]\left[\delta_{cc}{}' + i\sum_{\lambda\mu} \gamma_{\lambda c}\beta_{\mu c}{}'A_{\lambda\mu}\right], \qquad (A\text{-}10)$$

where the subscripts c and c' refer to channels and λ and μ to levels. (The phase factors $\exp(i(\phi_c))$ and $\exp(i\phi_{c'})$ are related to properties of the external region.) The elements of the matrix A can also be written as

$$A_{\lambda\mu} = \sum_c \frac{S_{\lambda c} S_{\mu c}}{a_c - E}, \tag{A-11}$$

where the elements $S_{\lambda c}$ and $S_{\mu c}$ are the complex and energy dependent elements of the transformation matrix which diagonalizes $\mathbf{A}^{-1}$, i.e.,

$$\mathbf{D}^{-1} = \mathbf{S}^{-1}\mathbf{A}^{-1}\mathbf{S}, \tag{A-12}$$

where $\mathbf{D}^{-1}$ is diagonal. The term a_c in Eq. (A-11) is also complex and, in general, energy dependent.

However, Adler and Adler have shown that for the particular case of low-energy neutron-induced fission, the elements $S_{\lambda c}$ and the terms a_c, while still complex, are not dependent on the neutron energy, at least to a very good approximation. In Eq. (A-7), the summation $\Sigma_c \gamma_{\lambda c} \beta_{\mu c}$ does contain terms involving the energy since the neutron reduced widths $\beta_{\mu n}$ are energy dependent. However, for low-energy neutron-induced fission, the neutron widths are invariably so much smaller than the fission and radiation widths that their energy dependence is negligible in the sum over channels.

Adler and Adler showed that to this approximation the elements of $\mathbf{A}$ depend on neutron energy only as explicitly shown in Eq. (A-11). The resulting form of the collision matrix consists of Breit–Wigner type terms, of the form

$$U_{cc'} = e^{i(\phi_c + \phi_{c'})} \left[\delta_{cc'} + i \sum_k (a_k - E)^{-1} \sum_{\lambda\mu} \gamma_{\lambda c} \beta_{\mu c'} S_{\lambda k} S_{\mu k} \right]. \tag{A-13}$$

One can then treat the complex parameters defined in Eq. (1-13) as fictitious level widths and energies, parameters to be extracted by fitting the data. The necessity of performing a matrix inversion at each value of the neutron energy is obviated.

Perhaps the only disadvantage of this approach is that the parameters obtained, while adequate to describe the data, are not simply related to the nuclear level parameters, which are the quantities of physical interest. Nevertheless, the inherent simplicity of the formulation is very attractive. It lends itself readily to techniques of Doppler broadening and to least-squares extraction of the level parameters (Adler and Adler, 1962, 1963, 1966a,b; see also Adler et al., 1964, 1965; Fluharty et al., 1966).

ACKNOWLEDGMENTS

The author would like to express his appreciation to R. G. Fluharty for his continuing interest and for many helpful discussions on low-energy neutron-induced fission, as well as to all those who have supplied drawings and figures and to those who have consented to the presentation of their data, in many cases prior to publication. In this latter connection, particular thanks are due to C. D. Bowman, G. F. Auchampaugh, and G. D. Sauter of the Lawrence Radiation Laboratory at Livermore; to L. M. Bollinger and A. B. Smith of the Argonne National Laboratory; to J. W. T. Dabbs, G. deSaussure, L. W. Weston, and H. W. Schmitt of the Oak Ridge National Laboratory; to A. J. Deruytter of the Belgian Center for Nuclear Energy Studies at Mol; to J. S. Fraser and J. C. D. Milton of the Atomic Energy of Canada Ltd., Chalk River; to B. R. Leonard of the Pacific Northwest Laboratory at Hanford; to G. J. Safford of the Union Carbide Nuclear Corporation at Tuxedo Park, to V. L. Sailor of the Brookhaven Nationla Laboratory, to T. D. Thomas of Princeton University and R. Vandenbosch of the University of Washington, and to S. Weinstein of the Rensselaer Polytechnic Institute.

Many of the figures showing the most recent cross-section data in the resonance region were supplied by B. A. Magurno and the staff of the National Neutron Cross-Section Center at Brookhaven National Laboratory, from data which were obtained prior to 1967. It is impossible to give adequate recognition and thanks to all those with whom helpful discussions and communications were held. One of the most valuable of these was with J. E. Lynn of the Atomic Energy Research Establishment at Harwell, who also provided a prepublication manuscript of his review (Lynn, 1968). Mrs. J. M. Moore was invaluable in proofreading the manuscript, checking references, and preparing tables. The contributions of the staff of the Nuclear Physics Branch of the Idaho Nuclear Corporation, particularly, K. T. Faler, L. G. Miller, F. B. Simpson, and O. D. Simpson, and the hospitality of the Nuclear Engineering and Science Department of the Rensselaer Polytechnic Institute while this manuscript was in preparation, are gratefully acknowledged.

REFERENCES

ADLER, D. B., and ADLER, F. T. (1962). *Trans. Am. Nucl. Soc.* **5,** 53.

ADLER, D. B., and ADLER, F. T. (1963). *Proc. Conf. Breeding, Economics Safety Large Fast Power Reactors* p. 695, USAEC Rept. ANL-6792.

ADLER, D. B., and ADLER, F. T. (1966a). USAEC Rept. TID-23396; USAEC Rept. BNL-50045.

ADLER, F. T., and ADLER, D. B. (1966b). *Proc. Conf. Neutron Cross Section Technol., Washington, D.C.* p. 873. USAEC Rept. CONF-660303.

ADLER, F. T., ADLER, D. B., and GOODWIN, W. A. (1964). *Trans. Am. Nucl. Soc.* **7,** 86.

ADLER, D. B., ADLER, F. T., and LEWIS, E. E. (1965). *Proc. Conf. Appl. Computing Methods Reactor Problems.* USAEC Rept. ANL-7050, p. 457.

ALBERT, R. D. (1966). *Phys. Rev.* **142,** 778.

ANDRITSOPOULOS, G. (1967). *Nucl. Phys.* **A94,** 537.

APALIN, V. F., DOBRYNIN, YU. P., ZAKHAROVA, V. P., KUTIKOV, I. E., and MIKAELYAN, L. A. (1959). *At. Energ.* **7,** 375; *Soviet J. At. Energy* **7,** (*English Transl.*) 855.

APALIN, V. F., DOBRYNIN, YU, P., ZAKHAROVA, V. P., KUTIKOV, I. E., and MIKAELYAN, L. A. (1960). *At. Energ.* **8,** 15; *Soviet J. At. Energy* (*English Transl.*) **8,** 10 (1961).

APALIN, V. F., GRITSYUK, YU. N., KUTIKOV, I. E., LEBEDEV, V. I., and MIKAELYAN, L. A. (1962). *Nucl. Phys.* **38,** 193.

APALIN, V. F., GRITSYUK, YU, N., KUTIKOV, I. E., LEBEDEV, V. I., and MIKAELYAN, L. A. (1964). *Nucl. Phys.* **55,** 249.

APALIN, V. F., GRITSYUK, YU. N., KUTIKOV, I. E., LEBEDEV, V. I., and MIKAELYAN, L. A. (1965a). *Nucl. Phys.* **71,** 553.

APALIN, V. F., GRITSYUK, YU. N., KUTIKOV, I. E., LEBEDEV, V. I., and MIKAELYAN, L. A. (1965b). *Nucl. Phys.* **71,** 546.

APALIN, V. F., GRITSYUK, YU. N., KUTIKOV, I. E., LEBEDEV, V. I., and MIKAELYAN, L. A. (1965c). *In* "Physics and Chemistry of Fission," Vol. I, p. 587. IAEA, Vienna.

ARAS, N. K., MENON, M. P., and GORDON, G. E. (1965). *Nucl. Phys.* **69,** 337.

ARBMAN, E., BJØRNHOLM, S., and NIELSEN, O. B. (1960). *Nucl. Phys.* **21,** 406.

ASGHAR, M. (1967). *Nucl. Phys.* **A98,** 33; see also "Nuclear Data for Reactors," Vol. II, p. 185. IAEA, Vienna.

AUCHAMPAUGH, G. F., BOWMAN, C. D., and EVANS, J. E. (1968). *Nucl. Phys.* **A112,** 329.

AUCLAIR, J. M., LANDON, H. H., and JACOB, M. (1955). *Compt. Rend. Acad. Sci. Paris* **241,** 1935.

AUCLAIR, J. M., LANDON, H. H., and JACOB, M. (1956). *Physica* **22,** 1187.

AUERBACH, E. H., and PEREY, F. G. J. (1962). USAEC Rept. BNL-765.

AXTMANN, R. C., and KEDEM, D. (1965). *Nucl. Inst. Methods* **32,** 70.

BABCOCK, R. V. (1961). *IRE Trans. Nucl. Sci.* **NS-8,** No. 1, 98.

BAME, S. J. (1963). USAEC Rept. CONF-481-71. Text of invited paper, *Bull. Am. Phys. Soc. Ser. II* **9,** 76.

BARDEEN, J., COOPER, L. N., and SCHRIEFFER, J. R. (1957). *Phys. Rev.* **108,** 1175.

BELL, G. I. (1953). *Phys. Rev.* **90,** 548.

BELL, G. I. (1966). *Proc. Conf. Neutron Cross Section Technol., Washington, D.C.* p. 454. USAEC Rept. CONF-660303.

BENNETT, W. R., JR. (1962). *Ann. Phys. (N.Y.)* **18,** 376.

BERGEN, D. W., and SILBERT, M. G. (1968). *Phys. Rev.* **166,** 1178.

BERGEN, D. W., SILBERT, M. G., and PERISHO, R. C. (1966). *Proc. Conf. Neutron Cross Section Technol. Washington, D.C.* p. 895. USAEC Rept. CONF-660303.

BERGSTRÖM, I., and DOMEIJ, B. (1966). *Nucl. Inst. Methods* **43,** 146.

BERRETH, J. R. (1965). *Nucl. Appl.* **1,** 230.

BERTOLINI, G., DEPASQUALI, G., and FANTECHI, R. (1965). *Nucl. Instr. Methods* **32,** 355.

BETHE, H. A. (1950). *Rev. Mod. Phys.* **22,** 213.

BETHE, H. A., and ASHKIN, J. (1953). *In* "Experimental Nuclear Physics" (E. Segre, ed.), Vol. I, p. 166. Wiley, New York.

BIRKS, J. B. (1964). "Theory and Practice of Scintillation Counting." MacMillan, New York. See especially p. 570.

BJØRNHOLM, S., and NIELSEN, O. B. (1963). *Nucl. Phys.* **42,** 642.

BLACKBURN, D. H., and HALLER, W. (1965). *Rev. Sci. Inst.* **36,** 901.

BLANKENSHIP, J. L., and BORKOWSKI, C. J. (1960). *IRE Trans. Nucl. Sci.* **NS-7,** No. 2–3, 190.

BLIZARD, E. P. (1961). *Nucl. Sci. Eng.* **9,** No. 3, i.

BLOCK, R. C., SLAUGHTER, G. G., and HARVEY, J. A. (1959). *Bull. Am. Phys. Soc. Ser. II* **4,** 34.

BLOCK, R. C., SLAUGHTER, G. G., and HARVEY, J. A. (1960). *Nucl. Sci. Eng.* **8,** 112.

BLOCK, R. C., SLAUGHTER, G. G., WESTON, L. W., and VONDERLAGE, F. C. (1961). *In* "Neutron Time-of-Flight Methods" (J. Spaepen, ed.), p. 203. Euratom, Brussels.

BLYUMKINA, YU. A., BONDARENKO, I. I., KUZNETSOV, V. F., NESTEROV, V. G., OKOLOVITCH, V. N., SMIRENKIN, G. N., and USACHEV, L. N. (1964). *Nucl. Phys.* **52,** 648.

BOHR, A. (1952). *Kgl. Danske Videnskab. Selskab Mat. Fys. Medd.* **26,** No. 14.

BOHR, A. (1956). *Proc. Intern. Conf. Peaceful Uses At. Energy* **2,** 151.

BOHR, A., and MOTTELSON, B. R. (1953). *Kgl. Danske Videnskab. Selskab Mat. Fys. Medd.* **27,** No. 16.

BOHR, A., MOTTELSON, B. R., and PINES, D. (1958). *Phys. Rev.* **110**, 936.

BOHR, N. (1948). *Kgl. Danske Videnskab. Selskab Mat. Fys. Medd.* **18**, No. 8.

BOHR, N., and WHEELER, J. A. (1939). *Phys. Rev.* **56**, 426.

BOHR, N., and LINDHARD, J. (1954) *Kgl. Danske Videnskab. Selskab Mat. Fys. Medd.* **28**, No. 7.

BOICOURT, G. P., and BROLLEY, J. E., JR. (1954). *Rev. Sci. Instr.* **25**, 1218.

BOLLINGER, L. M. (1957). *Proc. Tripartite Conf. Cross Sect. Fissile Nuclei* (N. J. Pattenden, ed.), p. 21. United Kingdom Atomic Energy Authority Rept. AERE NP/R 2076 (Rev.).

BOLLINGER, L. M., COTÉ, R. E., HUBERT, P., LEBLANC, J. M., and THOMAS, G. E. (1956). *Bull. Am. Phys. Soc. Ser. II* **1**, 165.

BOLLINGER, L. M., COTÉ, R. E., and THOMAS, G. E. (1958). *Proc. Intern. Conf. Peaceful Uses At. Energy 2nd* **15**, 127.

BOLLINGER, L. M., DIAMOND, H., and GINDLER, J. E. (1963). *Bull. Am. Phys. Soc. Ser. II* **8**, 370.

BOWMAN, C. D., and HILL, R. W. (1963). *Nucl. Instr. Methods* **24**, 213.

BOWMAN, C. D., AUCHAMPAUGH, G. F., FULTZ, S. C., MOORE, M. S., and SIMPSON, F. B. (1966). *Proc. Conf. Neutron Cross Sect. Technol. Washington, D.C.* p. 1004. USAEC Rept. CONF-660303.

BOWMAN, C. D., AUCHAMPAUGH, G. F., STUBBINS, W. F., YOUNG, T. E., SIMPSON, F. B., and MOORE, M. S. (1967). *Phys. Rev. Letters* **18**, 15.

BOWMAN, C. D., AUCHAMPAUGH, G. F., FULTZ, S. C., and HOFF, R. W. (1968). *Phys. Rev.* **166**, 1219.

BOWMAN, H. R., THOMPSON, S. G., MILTON, J. C. D., and SWIATECKI, W. J. (1962). *Phys. Rev.* **126**, 2120.

BOWMAN, H. R., MILTON, J. C. D., THOMPSON, S. G., and SWIATECKI, W. J. (1963). *Phys. Rev.* **129**, 2133.

BOWMAN, H. R., THOMPSON, S. G., and RASMUSSEN, J. O. (1964). *Phys. Rev. Letters* **12**, 195.

BOWMAN, H. R., THOMPSON, S. G., WATSON, R. L., KAPOOR, S. S., and RASMUSSEN, J. O. (1965). *In* "Physics and Chemistry of Fission," Vol. II, p. 125, IAEA Vienna.

BREIT, G., and WIGNER, E. P. (1936). *Phys. Rev.* **49**, 519.

BRITT, H. C., and WEGNER, H. E. (1963a). *Rev. Sci. Instr.* **34**, 274.

BRITT, H. C., and WEGNER, H. E. (1963b). *Rev. Sci. Instr.* **34**, 627.

BRITT, H. C., and BENSON, G. C. (1964). *Rev. Sci. Instr.* **35**, 842.

BRITT, H. C., and WHETSTONE, S. L., JR. (1964). *Phys. Rev.* **133**, B603.

BRITT, H. C., WEGNER, H. E., and GURSKY, J. (1962). *Phys. Rev. Letters* **8**, 98.

BRITT, H. C., WEGNER, H. E., and WHETSTONE, S. L., JR. (1963a). *Nucl. Instr. Methods* **24**, 13.

BRITT, H. C., STOKES, R. H., GIBBS, W. R., and GRIFFIN, J. J. (1963b). *Phys. Rev. Letters* **11**, 343.

BRITT, H. C., WEGNER, H. E., and GURSKY, J. C. (1963c). *Phys. Rev.* **129**, 2239.

BRITT, H. C., GIBBS, W. R., GRIFFIN, J. J., and STOKES, R. H. (1965). *Phys. Rev.* **139**, B354.

BRITT, H. C., NEWSOME, R. W., JR., and STOKES, R. H. (1966). *Bull. Am. Phys. Soc. Ser. II* **11**, 30.

BROLLEY, J. E., JR., and DICKINSON, W. C. (1954). *Phys. Rev.* **94**, 640.

BROOKS, F. D. (1959). *Nucl. Instr. Methods* **4**, 151.

BROOKS, F. D. (1961). *In* "Neutron Time-of-Flight Methods" (J. Spaepen, ed.) p. 131. Euratom, Brussels.

BROOKS, F. D. (1966). *In* "Reactor Physics in the Resonance and Thermal Regions" (A. J. Goodjohn and G. C. Pomraning, eds.), Vol. II, p. 193. MIT Press, Cambridge, Massachusetts.

BROOKS, F. D., PRINGLE, R. W., and FUNTS, B. L. (1960). *IRE Trans. Nucl. Sci.* **NS-7,** No. 2–3, 35.

BROWN, W. K., BERGEN, D. W., and CRAMER, J. D. (1966). *Proc. Conf. Neutron Cross Sect. Technol., Washington, D.C.* (P. B. Hemmig, ed.) p. 971. USAEC Report CONF-660303.

BRUNINX, E. and RUDSTAM, G. (1961). *Nucl. Instr. Methods* **13,** 131.

BRUNNER, W., and PAUL, H. (1960). *Ann. Physik* (7) **6,** 267.

BRUNNER, W., and PAUL, H. (1961a). *Ann. Physik* (7) **7,** 326.

BRUNNER, W., and PAUL, H. (1961b). *Ann. Physik* (7) **7,** 333.

BRUNNER, W., and PAUL, H. (1961c). *Ann. Physik* (7) **8,** 146.

BUNEMANN, O., CRANSHAW, T. E., and HARVEY, J. A. (1949). *Can. J. Res.* **27A,** 191.

BUSINARO, U. L., and GALLONE, S. (1955). *Nuovo Cimento* **1,** 629.

BYERS, D. H., DIVEN, B. C., and SILBERT, M. G. (1966). *Proc. Conf. Neutron Cross Sect. Technol., Washington, D.C.* (P. B. Hemmig, ed.) p. 903. USAEC Report CONF-660303.

CAMERON, A. G. W. (1958). *Rev. Mod. Phys.* **30,** 553; and in *Proc. United Nations Conf. Peaceful Uses At. Energy 2nd* **15,** 425.

CARSWELL, D. J., and MILSTED, J. (1957). *J. Nucl. Energy* **4,** 51.

CHANG, W. Y., and ROSENBLUM, S. (1945). *Phys. Rev.* **67,** 222.

CHAUDHRY, R., VANDENBOSCH, R., and HUIZENGA, J. R. (1962). *Phys. Rev.* **126,** 220.

COHEN, S., and SWIATECKI, W. J. (1962). *Ann. Phys. (N.Y.)* **19,** 67.

COTÉ, R. E., and BOLLINGER, L. M. (1961). *Phys. Rev. Letters* **6,** 695.

COTÉ, R. E., DIAMOND, H., and GINDLER, J. E. (1961). *Bull. Am. Phys. Soc.* **6,** 417.

COWAN, G. A., TURKEVICH, A., BROWNE, C. I., and Los Alamos Radiochemistry Group (1961). *Phys. Rev.* **122,** 1286.

COWAN, G. A., BAYHURST, B. P., and PRESTWOOD, R. J. (1963). *Phys. Rev.* **130,** 2380.

COWAN, G. A., BAYHURST, B. P., PRESTWOOD, R. J., GILMORE, J. S., and KNOBELOCH G. W. (1966). *Phys. Rev.* **144,** 979.

CRAIG, D. S., and WESTCOTT, C. H. (1964). *Can. J. Phys.* **42,** 2384.

CRAMER, J. D. (1967). *Bull. Am. Phys. Soc. Ser. II* **12,** 521.

CRANBERG, L., and LEVIN, J. S. (1958). *Phys. Rev.* **109,** 2063.

CRANSHAW, T. E., and HARVEY, J. A. (1948). *Can. J. Res.* **26A,** 243.

CROALL, I. F., and WILLIS, H. H. (1965). *In* "Physics and Chemistry of Fission." Vol. I, p. 355. IAEA, Vienna.

CUNINGHAME, J. G., KITT, G. P., and RAE, E. R. (1961). *Nucl. Phys.* **27,** 154.

CUNINGHAME, J. G., FRITZE, K., LYNN, J. E., and WEBSTER, C. B. (1966). *Nucl. Phys.* **84,** 49.

CURRAN, S. C., and WILSON, H. W. (1965). *In* "Alpha, Beta and Gamma Ray Spectroscopy" (K. Siegbahn, ed.), Vol. I, p. 303. North Holland Publ., Amsterdam.

CURRAN, S. C., COCKROFT, A. L., and ANGUS, J. (1949). *Phil. Mag.* **40,** 929.

DABBS, J. W. T., WALTER, F. J., and PARKER, G. W. (1965). *In* "Physics and Chemistry of Fission," Vol. I. p. 39. IAEA, Vienna.

DAR, Y., LOEBENSTEIN, H. M., and DEBOER, J. (1964). *Nucl. Instr. Methods* **27,** 327.

DEARNALEY, G. (1961). United Kingdom Atomic Energy Authority Rept. AERE-R3874.

DEARNALEY, G., and WHITEHEAD, A. B. (1961). *Nucl. Instr. Methods* **12,** 205.

DEARNALEY, G., and NORTHROP, D. C. (1963). "Semiconductor Counters for Nuclear Radiations." Wiley, New York.

DECARVALHO, H. G., CORTINI, G., MUCHNIK, M., RINZIVILLO, R., and SASSI, E. (1964). *Nucl. Phys.* **53,** 345.

DECARVALHO, H. G., DASILVA, A. G., ALVES, R. N., BÖSCH, R., and WÖLFLI, W. (1965). *In* "Physics and Chemistry of Fission." Vol. II, p. 343. IAEA, Vienna.

DERRIEN, H., BLONS, J., EGGERMANN, C., MICHAUDON, A., PAYA, D., and RIBON, P. (1967). *In* "Nuclear Data for Reactors." Vol. II, p. 195. IAEA, Vienna.

DERUYTTER, A. J. (1961). *J. Nucl. Energy, Pt. A B, (Reactor Sci. Technol.)* **15,** 165.

DERUYTTER, A. J., and NEVE DE MEVERGNIES, M. (1965). *In* "Physics and Chemistry of Fission." Vol. II, p. 429, IAEA, Vienna.

DERUYTTER, A. J., SCHRÖDER, I. G., and MOORE, J. A. (1962). Unpublished Data.

DE SAUSSURE, G., WESTON, L. W., GWIN, R., RUSSELL, J. W., and HOCKENBURY, R. W. (1965a). *Nucl. Sci. Eng.* **23,** 45.

DE SAUSSURE, G., BLONS, J., JOUSSEAUME, C., MICHAUDON, A., and PRANAL, Y. (1965b). *In* "Physics and Chemistry of Fission." Vol. I, p. 205. IAEA, Vienna.

DE SAUSSURE, G., WESTON, L. W., GWIN, R., INGLE, R. W., TODD, J. H., HOCKENBURY, R. W., FULLWOOD, R. R., and LOTTIN, A. (1967). *In* "Nuclear Data for Reactors." Vol. II, p. 233. IAEA, Vienna.

DESHPANDE, R. Y. (1967). *Nucl. Instr. Methods* **46,** 255.

DIVEN, B. C. (1958). *Proc. United Nations Conf. Peaceful Uses At. Energy 2nd* **15,** 60.

DIVEN, B. C. (1965). *In* "Nuclear Structure Study with Neutrons (M. Neve de Mevergnies, P. Van Assche, and J. Vervier, eds.), p. 441. North Holland Publ., Amsterdam.

DIVEN, B. C. (1966). *Proc. Conf. Neutron Cross Sect. Technol. Washington, D.C.* p. 1051. USAEC Rep. CONF-660303.

DIVEN, B. C., and HOPKINS, J. C. (1961). *In* "Neutron Time-of-Flight Methods" (J. Spaepen, ed.), p. 407. Euratom, Brussels.

DIVEN, B. C., MARTIN, H. C., TASCHEK, R. F., and TERRELL, J. (1956). *Phys. Rev.* **101,** 1012.

DIVEN, B. C., TERRELL, J., and HEMMENDINGER, A. (1958). *Phys. Rev.* **109,** 144.

DIVEN, B. C., TERRELL, J., and HEMMENDINGER, A. (1960). *Phys. Rev.* **120,** 556.

DRAWBAUGH, D. W., and GIBSON, G. (1966). *Proc. Conf. Neutron Cross Sect. Technol. Washington, D.C.* p. 939. USAEC Rept. CONF-660303.

DRAWBAUGH, D. W., and GIBSON, G. (1967). *In* "Nuclear Data for Reactors," Vol. II, p. 251. IAEA, Vienna.

EGGLER, C., and HUDDLESTON, C. M. (1956). *Nucleonics* **14,** No. 4, 34.

EVANS, J. E., and FLUHARTY, R. G. (1960). *Nucl. Sci. Eng.* **8,** 66.

FAIRHALL, A. W. (1956). *Phys. Rev.* **102,** 1335.

FAIRHALL, A. W., JENSEN, R. C., and NEUZIL, E. F. (1958). *Proc. Intern. Conf. Peaceful Uses At. Energy 2nd* **15,** 452.

FALER, K. T., and TROMP, R. L. (1963). *Phys. Rev.* **131,** 1746.

FANO, U. (1946). *Phys. Rev.* **70,** 44.

FANO, U. (1947). *Phys. Rev.* **72,** 26.

FANO, U. (1963). *Ann. Rev. Nucl. Sci.* **13,** 1.

FAROUK, M. A., NASSEF, M. H., EL-BEHAY, A. Z., and ZALOUBOVSKY, I. I. (1965). *Nucl. Instr. Methods* **35,** 210.

FARRELL, J. A. (1968). *Phys. Rev.* **175,** 1371.

FESHBACH, H., PORTER, C. E., and WEISSKOPF, V. F. (1954). *Phys. Rev.* **96,** 448.

FLEISCHER, R. L. (1966). *Rev. Sci. Instr.* **37,** 1738.

FLEISCHER, R. L., PRICE, P. B., and WALKER, R. M. (1965). *Ann. Rev. Nucl. Sci.* **15,** 1.

FLEROV, G. N., POLIKANOV, S. M., MIKHEEV, V. L., PERELYGIN, V. P., and PLEVE, A. A. (1963). *In* "Proceedings of the Third Conference on Reactions between Complex Nuclei" (A. Ghiorso, R. M. Diamond, and H. E. Conzett, eds.), p. 219. Univ. of California Press, Berkeley, California.

FLEROV, G. N., OGANESYAN, YU. TS., LOBANOV, YU. V., KUZNETSOV, V. I., DRUIN, V. A., PERELYGIN, V. P., GAVRILOV, K. A., TRETIAKOVA, S. P., and PLOTKO, V. M. (1964). *Phys. Letters* **13,** 73.

FLEROV, G. N., PLEVE, A. A., POLIKANOV, S. M., IVANOV, E., MARTALOGU, N., POENARU, D., and VILCOV, N. (1965). *In* "Physics and Chemistry of Fission," Vol. I, p. 307. IAEA, Vienna.

FLUHARTY, R. G., MARSHALL, N. H., and SIMPSON, O. D. (1966). *Proc. Conf. Neutron Cross Sect. Technol. Washington, D.C.* (P. B. Hemmig, ed.) p. 985. USAEC Rept. CONF-660303.

FONG, P. (1956). *Phys. Rev.* **102,** 434.

FONG, P. (1961). *Phys. Rev.* **122,** 1543.

FONG, P. (1964). *Phys. Rev.* **135,** B1338.

FOOTE, H. L. (1958). *Phys. Rev.* **109,** 1641.

FRANKEL, S., and METROPOLIS, N. (1947). *Phys. Rev.* **72,** 914.

FRASER, J. S. (1965). *In* "Physics and Chemistry of Fission," Vol. I, p. 451. IAEA, Vienna.

FRASER, J. S., and MILTON, J. C. D. (1954). *Phys. Rev.* **93,** 818.

FRASER, J. S., and SCHWARTZ, R. B. (1962). *Nucl. Phys.* **30,** 269.

FRASER, J. S., and MILTON, J. C. D. (1966). *Ann. Rev. Nucl. Sci.* **16,** 379.

FRASER, J. S., MILTON, J. C. D., BOWMAN, H. R., and THOMPSON, S. G. (1963). *Can. J. Phys.* **41,** 2080.

FRISCH, O. R. (1994). British Atomic Energy Rept. BR-49.

FRÖHNER, F. H., HADDAD, E., LOPEZ, W. M., and FRIESENHAHN, S. J. (1966). *Proc. Conf. Neutron Cross Sect. Technol.* (P. B. Hemmig, ed.) p. 55. USAEC Rept. CONF-660303.

FULBRIGHT, H. W. (1958). *In* "Encyclopaedia of Physics" (S. Flügge and E. Creutz, eds.), Vol. 45, p. 1. Springer, Berlin.

GALLAGHER, C. J., JR. (1962). *Phys. Rev.* **126,** 1525.

GALLAGHER, C. J., JR. (1964). *In* "Selected Topics of Nuclear Spectroscopy" (B. J. Verhaar, ed.), p. 133. North Holland Publ., Amsterdam.

GALLAGHER, C. J., JR., NIELSEN, H. L., and NIELSEN, O. B. (1961). *Phys. Rev.* **122,** 1590.

GARRISON, J. D. (1966). *In* "Reactor Physics in the Resonance and Thermal Regions" (A. J. Goodjohn and G. C. Pomraning, eds.), Vol. II, p. 3. MIT Press, Cambridge, Massachusetts.

GETOFF, N., and BILDSTEIN, H. (1965). *Nucl. Instr. Methods* **36,** 173.

GETOFF, N., BILDSTEIN, H., and PROKSCH, E. (1967). *Nucl. Instr. Methods* **46,** 305.

GIBBONS, J. H., MACKLIN, R. L., MILLER, P. D., and NEILER, J. H. (1961). *Phys. Rev.* **122,** 182.

GIBSON, W. M., THOMAS, T. D., and MILLER, G. L. (1961). *Phys. Rev. Letters* **7,** 65.

GIBSON, W. M., MILLER, G. L., and DONOVAN, P. F. (1965). *In* "Alpha, Beta, and Gamma Ray Spectroscopy" (K. Siegbahn, ed.), Vol. I. p. 345. North Holland Publ., Amsterdam.

GINDLER, J. E., and HUIZENGA, J. R. (1968). *In* "Nuclear Chemistry" (L. Yaffe, ed.). Academic Press, New York.

GLASS, N. W., THEOBALD, J. K., SCHELBERG, A. D., WARREN, J. H., and TATRO, L. D. (1966). *Proc. Conf. Neutron Cross Sect. Technol. Washington, D.C.* (P. B. Hemmig, ed.) p. 766. USAEC Report CONF-660303.

GOULDING, F. S. (1966). *Nucl. Instr. Methods* **43**, 1.

GRAFF, G., LA TAI, A., and NAGY, L. (1965). *In* "Physics and Chemistry of Fission," Vol. II, p. 163, IAEA, Vienna.

GRIFFIN, J. J. (1959). *Phys. Rev.* **116**, 107.

GRIFFIN, J. J. (1963). *Phys. Rev.* **132**, 2204; see also (1964). *Phys. Rev.* **135**, No. 7AB, 2.

GRIFFIN, J. J. (1965). *In* "Physics and Chemistry of Fission," Vol. I, p. 23. IAEA, Vienna.

GWIN, R., WESTON, L. W., DESAUSSURE, G., and HOCKENBURY, R. W. (1968). Preliminary Results on Measurements of Alpha for ^{239}Pu (To be published).

HADDAD, E., FRÖHNER, F. H., LOPEZ, W. M., and FRIESENHAHN, S. J. (1966). *In* "Reactor Physics in the Resonance and Thermal Regions" (A. J. Goodjohn and G. C. Pomraning, eds.), Vol. II, p. 125. MIT Press, Cambridge, Massachusetts.

HAHN, O., and STRASSMANN, F. (1939). *Naturwiss.* **27**, 11.

HAINES, E. L., and WHITEHEAD, A. E. (1965). *Rev. Sci. Instr.* **36**, 1385.

HAINES, E. L., and WHITEHEAD, A. E. (1966). *Rev. Sci. Instr.* **37**, 190.

HALPERN, I. (1959). *Ann. Rev. Nucl. Sci.* **9**, 245.

HALPERN, I. (1965). *In* "Physics and Chemistry of Fission," Vol. II, p. 369. IAEA, Vienna.

HALPERN, I., and STRUTINSKI, V. M. (1958). *Proc. United Nations Conf. Peaceful Uses At. Energy. 2nd* **15**, 408.

HARRIS, D. R. (1966). *Proc. Conf. Neutron Cross Sect. Technol. Washington, D.C.* (P. B. Hemmig, ed.) p. 823. USAEC Rept. CONF-660303.

HARVEY, J. A. (1966a). *In* "Reactor Physics in the Resonance and Thermal Regions" (A. J. Goodjohn and G. C. Pomraning, eds.), Vol. II, p. 103. MIT Press, Cambridge, Massachusetts.

HARVEY, J. A. (1966b). *Proc. Conf. Neutron Cross Sect. Technol. Washington, D.C.* (P. B. Hemmig, ed.) p. 31. USAEC Report CONF-660303.

HAVENS, W. W., JR., MELKONIAN, E., RAINWATER, L. J., and ROSEN, J. L. (1959). *Phys. Rev.* **116**, 1538.

HEMMENDINGER, A. (1965). *Phys. Today* **18**, No. 8, p. 17.

HEMMENDINGER, A. (1967). *In* "Nuclear Data for Reactors," Vol. II, p. 219. IAEA, Vienna.

HEMMENDINGER, A., SILBERT, M. G., and MOAT, A. (1965). *IEEE Trans. Nucl. Sci.* **NS-12**, 304.

HEMMENDINGER, A., DIVEN, B. C., BROWN, W. K., ELLIS, A., FURNISH, A., and SHUNK, E. R. (1967). USAEC Rept. LA-3478, Part I.

HILL, D. L., and WHEELER, J. A. (1953). *Phys. Rev.* **89**, 1102.

HOFFMAN, M. M. (1964). *Phys. Rev.* **133**, B714.

HOLLOWAY, M. G., and LIVINGSTON, M. S. (1938). *Phys. Rev.* **54**, 18.

HOPKINS, J. C., and DIVEN, B. C. (1962). *Nucl. Sci. Eng.* **12**, 169.

HOPKINS, J. C., and DIVEN, B. C. (1963). *Nucl. Phys.* **48**, 433.

HORROCKS, D. L. (1963). *Rev. Sci. Instr.* **34**, 1035.

HUDDLESTON, C. M. (1958). *Proc. Intern. Conf. Peaceful Uses At. Energy 2nd* **14**, 298.

HUDDLESTON, C. M. (1960). *In* "Fast Neutron Physics" (J. B. Marion and J. L. Fowler, eds.), p. 441. Wiley (Interscience), New York.

HUIZENGA, J. R., and VANDENBOSCH, R. (1962). *In* "Nuclear Reactions" (P. M. Endt and P. B. Smith, eds.), Chapter II, p. 42. North Holland Publ., Amsterdam.

HUIZENGA, J. R., UNIK, J. P., and WILKINS, B. D. (1965). *In* "Physics and Chemistry of Fission," Vol. I, p. 11. IAEA, Vienna.

HUMBLET, J., and ROSENFELD, L. (1961). *Nucl. Phys.* **26**, 529.

HYDE, E. K. (1960a). USAEC Rept. UCRL-9036.

HYDE, E. K. (1960b). USAEC Rept. UCRL-9065.

HYDE, E. K. (1964). "The Nuclear Properties of the Heavy Elements, Fission Phenomena," Vol. III. Prentice-Hall, Englewood Cliffs, New Jersey.

HYDE, E. K., PERLMAN, I., and SEABORG, G. T. (1964). "The Nuclear Properties of the Heavy Elements," Vols. I, II. Prentice-Hall, Englewood Cliffs, New Jersey.

IGNATIEV, K. G., KIRPICHNIKOV, I. V., and SUKHORUCHKIN, S. I. (1964). *At. Energ.* **16**, 110; *Soviet J. At. Energy (English Transl.)* **16**, 121.

JAMES, G. D. (1964). *Nucl. Phys.* **55**, 517.

JAMES, G. D. (1965a). *Nucl. Phys.* **65**, 353.

JAMES, G. D. (1965b). *In* "Physics and Chemistry of Fission," Vol. I, p. 235. IAEA, Vienna.

JENSEN, R. C., and FAIRHALL, A. W. (1958). *Phys. Rev.* **109**, 942.

JENSEN, R. C., and FAIRHALL, A. W. (1960). *Phys. Rev.* **118**, 771.

JESSE, W. P., and Sadauskis, J. (1955). *Phys. Rev.* **97**, 1668.

JOHANSSON, S. A. E. (1961). *Nucl. Phys.* **22**, 529.

JOHANSSON, S. A. E. (1964). *Nucl. Phys.* **60**, 378.

JOHANSSON, S. A. E., and KLEINHEINZ, P. (1965). *In* "Alpha, Beta, and Gamma Ray Spectroscopy" (K. Siegbahn, ed.), Vol. I, p. 805. North Holland Publ., Amsterdam.

KAPOOR, S. S., and RAMANNA, R. (1964). *Phys. Rev.* **133**, B598.

KAPOOR, S. S., RAMANNA, R., and RAMA RAO, P. N. (1963). *Phys. Rev.* **131**, 283.

KAPOOR, S. S., NADKARNI, D. M., RAMANNA, R., and RAMA RAO, P. N. (1965). *Phys. Rev.* **137**, B511.

KAPUR, P. L., and PEIERLS, R. (1938). *Proc. Roy. Soc. (London)* **A166**, 277.

KIRPICHNIKOV, I. V., IGNATIEV, K. G., and SUKHORUCHKIN, S. I. (1964). *At. Energ.* **16**, 211; *Soviet J. At. Energy* **16**, 251.

KLINGENSMITH, R. W. (1961). *IRE Trans. Nucl. Sci.* **NS-8**, No. 1, 112.

KOBISK, E. H. (1965). USAEC Rept. ORNL-3829.

KOCH, L. (1960). Doctoral dissertation, Univ. of Paris, Centre d'Etudes Nucleaires de Saclay Rept. CEA-1532.

KONECNY, E., and HETWER, K. (1965). *Nucl. Instr. Methods* **36**, 61.

KONSHIN, V. A., MATUSEVICH, E. S., and REGUSHEVESKII, V. I. (1965). *In* "Physics and Chemistry of Fission," Vol. II, p. 349. IAEA, Vienna.

KVITEK, I., POPOV, YU. P., and RYABOV, YU. V. (1965). *In* "Physics and Chemistry of Fission," Vol. II, p. 439. IAEA, Vienna.

LAMB, W. E., (1939). *Phys. Rev.* **55**, 190.

LAMPHERE, R. W. (1956). *Phys. Rev.* **104**, 1654.

LAMPHERE, R. W. (1960). *In* "Fast Neutron Physics" (J. B. Marion and J. L. Fowler, eds.), p. 449. Wiley (Interscience), New York.

LAMPHERE, R. W. (1962). *Nucl. Phys.* **38**, 561.

LAMPHERE, R. W. (1965). *In* "Physics and Chemistry of Fission," Vol. I, p. 63. IAEA, Vienna.

LANE, A. M., and THOMAS, R. G. (1958). *Rev. Mod. Phys.* **30**, 257.

LASSEN, N. O. (1949). *Kgl. Danske Videnskab. Selskab Mat. Fys. Medd.* **25**, No. 11.

LASSEN, N. O. (1951). *Kgl. Danske Videnskab. Selskab Mat. Fys. Medd.* **26**, No. 5.

LAUER, K. F., and VERDINGH, V. (1963). *Nucl. Instr. Methods* **21**, 161.

LEONARD, B. R., JR. (1955). Unpublished data, reported by V. L. Sailor in *Proc. Intern. Conf. Peaceful Uses At. Energy* **4**, 199.

LEONARD, B. R., JR. (1956). *Proc. Intern. Conf. Peaceful Uses At. Energy* **4**, 193.

LEONARD, B. R., JR. (1962). *In* "Neutron Physics" (M. L. Yeater, ed.), p. 3. Academic Press, New York.

LEONARD, B. R., JR. (1967). *In* "Plutonium Handbook, A Guide to the Technology" (O. J. Wick, ed.), Vol. I., Gordon and Breach, New York.

LEONARD, B. R., JR., and FRIESENHAHN, S. J. (1959). USAEC Rept. HW-62727, p. 19.

LEONARD, B. R., JR., and SEPPI, E. J. (1959). *Bull. Am. Phys. Soc. Ser. II* **4**, 31; see also USAEC Rept. HW-56919, p. 62.

LEONARD, B. R., JR., and ODEGAARDEN, R. H. (1961). USAEC Rept. HW-67219, p. 4.

LEONARD, B. R., JR., SEPPI, E. J., and FRIESEN, W. J. (1956). USAEC Rept. HW-44525, p. 47.

LEVY, H. B., HICKS, H. G., NERVIK, W. E., STEVENSON, P. C., NIDAY, J. B., and ARMSTRONG, J. C., Jr. (1961). *Phys. Rev.* **124**, 544.

LINDHARD, J. (1965). *Kgl. Danske Videnskab. Selskab Mat. Fys. Medd.* **34**, No. 14.

LINDHARD, J., and THOMSEN, P. V. (1962). *Proc. Symp. Radiation Damage Solids Reactor Materials, Venice* IAEA, Vienna.

LINDHARD, J., and WINTHER, A. (1964). *Kgl. Danske Videnskab. Selskab Mat. Fys. Medd.* **34**, No. 4.

LINDHARD, J., NIELSEN, V., SCHARFF, M., and THOMSEN, P. V. (1963a). *Kgl. Danske Videnskab. Selskab Mat. fys. Medd.* **33**, No. 10.

LINDHARD, J., SCHARFF, M., and SCHIÖTT, H. E. (1963b). *Kgl. Danske Videnskab. Selskab Mat. fys. Medd.* **33**, No. 14.

LIPAS, P. O., and DAVIDSON, J. P. (1961). *Nucl. Phys.* **26**, 80.

LIVINGSTON, M. S., and BETHE, H. A. (1937). *Rev. Mod. Phys.* **9**, 245.

LOBANOV, YU. V., KUZNETSOV, V. I., PERELYGIN, V. P., POLIKANOV, S. M., OGANESYAN, YU. TS., and FLEROV, G. N. (1965). *Yadern. Fiz.* **1**, 67; *Soviet J. Nucl. Phys. (English Transl.)* **1**, 45.

Los Alamos Radiochemistry Group (1957). *Phys. Rev.* **107**, 325.

LOVE, L. O., PRATER, W. K., SCHEITLIN, F. M., WHITEHEAD, T. W., JR., DAGENHART, W. K., TRACY, J. G., JOHNSON, R. L., and BANIC, G. M. (1965). *Nucl. Instr. Methods* **38**, 148.

LUBITZ, C. R. (1967). Private communication; see also Francis, N. C., Lubitz, C. R., Reynolds, J. T., and Slaggie, E. L. (1967). *In* "Nuclear Data for Reactors," Vol. II, p. 267. IAEA, Vienna.

LYNN, J. E. (1964). *Phys. Rev. Letters* **13**, 412.

LYNN, J. E. (1965a). *In* "Nuclear Structure Study with Neutrons" (M. Neve de Mevergnies, P. Van Assche, and J. Vervier, eds.), p. 125. North Holland Publ., Amsterdam.

LYNN, J. E. (1965b). *Phys. Letters* **18**, 31.

LYNN, J. E. (1967). *In* "Nuclear Data for Reactors," Vol. II, p. 89. IAEA, Vienna.

LYNN, J. E. (1968). "The Theoretical Interpretation of Neutron Spectroscopic Data." Oxford University Press, London.

MAGLEBY, E. H., SMITH, J. R., EVANS, J. E., and MOORE, M. S. (1956). USAEC Rept. IDO-16366.

MAIENSCHEIN, F. C., PEELLE, R. W., ZOBEL, W., and LOVE, T. A. (1958). *In Proc. Intern. Conf. Peaceful Uses At. Energy, 2nd* **15,** 366.

MAIER-LEIBNITZ, H., ARMBRUSTER, P., and SPECHT, H. J. (1965a). *In* "Physics and Chemistry of Fission," Vol. II, p. 113. IAEA, Vienna.

MAIER-LEIBNITZ, H., SCHMITT, H. W., and ARMBRUSTER, P. (1965b). *In* "Physics and Chemistry of Fission," Vol. II, p. 143. IAEA, Vienna.

MANLEY, J. H. (1962). *Nucl. Phys.* **33,** 70.

MATHER, D. S., FIELDHOUSE, P., and MOAT, A. (1964). *Phys. Rev.* **133,** B1403.

MATHER, D. S., FIELDHOUSE, P., and MOAT, A. (1965). *Nucl. Phys.* **66,** 149.

MAURETTE, M., and STEPHAN, C. (1965). *In* "Physics and Chemistry of Fission," Vol. II, p. 307. IAEA, Vienna.

MEADOWS, J. W., and WHALEN, J. F. (1962). *Phys. Rev.* **126,** 197.

MEADOWS, J. W., and WHALEN, J. F. (1967). *J. Nucl. Energy* **21,** 157.

MEHTA, G. K. (1963). Doctoral dissertation, Columbia Univ.

MEITNER, L., and FRISCH, O. R. (1939). *Nature* **143,** 239.

MELKONIAN, E. (1961). *Nucl. Instr. Methods* **11,** 307.

MELKONIAN, E., and MEHTA, G. K. (1963). *Trans. Am. Nucl. Soc.* **6,** 262.

MELKONIAN, E., and MEHTA, G. K. (1965). *In* "Physics and Chemistry of Fission," Vol. II, p. 355. IAEA, Vienna.

MENTI, W., MARTIN, M., BAS, E. B., and VOGT, O. (1964). *Nucl. Instr. Methods* **31,** 25.

MEYER, M. A. (1963). *Nucl. Instr. Methods* **23,** 277

MICHAUDON, A. (1963). *Phys. Letters* **7,** 211.

MICHAUDON, A. (1964). Doctoral dissertation, Univ. of Paris, Centre d'Etudes Nucleaires de Saclay Rept. CEA-R2552.

MICHAUDON, A. (1965). Unpublished lecture at an informal symposium on fission at ORNL, August 1965.

MICHAUDON, A. (1967). *In* "Nuclear Data for Reactors," Vol. II, p. 161. IAEA, Vienna.

MICHAUDON, A., DERRIEN, H., RIBON, P., and SANCHE, M. (1965a). *Nucl. Phys.* **69,** 545.

MICHAUDON, A., LOTTIN, A., PAYA, D., and TROCHON, J. (1965b). *Nucl. Phys.* **69,** 573.

MIGNECO, E., and THEOBALD, J. R. (1968). *Proc. Conf. Neutron Cross-Sect. Technol., 2nd, Washington, D.C.* NBS Spec. Pub. 299 Vol. 1, p. 527.

MILLER, G. L., GIBSON, W. M., and DONOVAN, P. F. (1962). *Ann. Rev. Nucl. Sci.* **12,** 189.

MILLER, L. G. (1963). Unpublished studies on radiation damage by fission fragments to surface barrier detectors.

MILLER, L. G. (1965). Unpublished studies on radiation damage by fission fragments to diffused junction detectors.

MILLER, L. G., and MOORE, M. S. (1957). Unpublished data on ^{233}U fission cross section.

MILLER, L. G., and MOORE, M. S. (1967). *Phys. Rev.* **157,** 1055.

MILTON, J. C. D., and Fraser, J. S. (1958). *Phys. Rev.* **111,** 877.

MILTON, J. C. D., and FRASER, J. S. (1961). *Phys. Rev. Letters* **7,** 67.

MILTON, J. C. D., and FRASER, J. S. (1962). *Can. J. Phys.* **40,** 1626.

MILTON, J. C. D., and FRASER, J. S. (1965). *In* "Physics and Chemistry of Fission," Vol. II p. 39. IAEA, Vienna.

MOAK, C. D., DABBS, J. W. T., and WALKER, W. W. (1966). *Rev. Sci. Instr.* **37,** 1131.

MOAK, C. D., LUTZ, H. O., BRIDWELL, L. B., NORTHCLIFFE, L. C., and DATZ, S. (1967). *Phys. Rev. Letters* **18,** 41.

MOORE, M. S., and REICH, C. W. (1960). *Phys. Rev.* **118,** 718.

MOORE, M. S., and SIMPSON, F. B. (1962). *Nucl. Sci. Eng.* **13,** 18.

MOORE, M. S., and SPENCER, R. R. (1964). USAEC Rept. IDO-17025.

MOORE, M. S., and MILLER, L. G. (1965). *In* "Physics and Chemistry of Fission," Vol. I, p. 87. IAEA, Vienna.

MOORE, M. S., and SIMPSON, O. D. (1966). *Proc. Conf. Neutron Cross Sect. Technol. Washington, D.C.* (P. B. Hemmig, ed.) p. 840 USAEC Rept. CONF-660303; and in "Reactor Physics in the Resonance and Thermal Regions" (A. J. Goodjohn and G. C. Pomraning, eds.), MIT Press, Cambridge, Massachusetts.

MOORE, M. S., and MILLER, L. G. (1967). *Phys. Rev.* **157**, 1049.

MOORE, M. S., MILLER, L. G., and SIMPSON, O. D. (1960). *Phys. Rev.* **118**, 714.

MOORE, M. S., SIMPSON, O. D., WATANABE, T., RUSSELL, J. E., and HOCKENBURY, R. W. (1964). *Phys. Rev.* **135**, B945.

MOSTOVAYA, T. A., MOSTOVOI, V. I., and YAKOVLEV, G. V. (1964). *At. Energ.* **16**, 3; *Soviet J. At. Energy (English Transl.)* **16**, 1.

MOTTELSON, B. R., and NILSSON, S. G. (1959). *Kgl. Danske Videnskab. Selskab Mat. fys. Skrifter.* **1**, No. 8.

MUGA, M. L. (1965). *In* "Physics and Chemistry of Fission," Vol. II, p. 409. IAEA, Vienna.

MUGA, M. L., RICE, C. R., and SEDLACEK, W. A. (1967). *Phys. Rev. Letters* **18**, 404.

MULÁS, P. M., and AXTMANN, R. C. (1966). *Phys. Rev.* **146**, 296.

NEWSON, H. W. (1961). *Phys. Rev.* **122**, 1224.

NEWTON, T. D. (1956). *Proc. Symp. Phys. Fission* (G. C. Hanna, J. C. D. Milton, W. T. Sharp, N. M. Stevens, and E. A. Taylor, eds.) p. 307. Atomic Energy of Canada Ltd. Rept. CRP-642-A.

NIFENECKER, H. (1964). *J. Phys.* **25**, 877.

NIFENECKER, H., and PERRIN, G. (1965). *In* "Physics and Chemistry of Fission," Vol I. p. 245. IAEA, Vienna.

NIFENECKER, H., MICHAUDON, A., and FAGOT, J. (1961). *In* "Neutron Time-of-Flight Methods" (J. Spaepen, ed.), p. 413. Euratom, Brussels.

NIFENECKER, H., PAYA, D., and FAGOT, J. (1963). *J. Phys. Radium* **24**, 254.

NILSSON, S. G. (1955). *Kgl. Danske Videnskab. Selskab Mat. Fys. Medd.* **29**, No. 16.

NIX, J. R. (1966). *Nucl. Phys.* **81**, 61.

NIX, J. R., and SWIATECKI, W. J. (1965). *Nucl. Phys.* **71**, 1.

NOBLES, R. A. (1956). *Rev. Sci. Instr.* **27**, 280.

NOBLES, R. A. (1962). *Phys. Rev.* **126**, 1508.

NOBLES, R. A., and LEACHMAN, R. B. (1958). *Nucl. Phys.* **5**, 211.

NORTHCLIFFE, L. C. (1963). *Ann. Rev. Nucl. Sci.* **13**, 67.

NORTHROP, J. A., and NOBLES, R. A. (1956). *Nucleonics* **14**, No. 4, 36.

NORTHROP, J. A., GURSKY, J. M., and JOHNSRUD, A. E. (1958). *IRE Trans. on Nucl. Sci.* **NS-5**, No. 3, 81.

NORTHROP, J. A., STOKES, R. H., and BOYER, K. (1959). *Phys. Rev.* **115**, 1277.

NOSHKIN, V. E., JR. (1963). Thesis, Clark Univ., Worcester, Massachusetts.

NOSSOFF, V. G. (1956). *Proc. Intern. Conf. Peaceful Uses At. Energy* **2**, 205.

ODEGAARDEN, R. H. (1960). USAEC Rept. HW-64866, p. 4.

OLEKSA, S. (1958). *Phys. Rev.* **109**, 1645.

OWEN, R. B. (1958). *IRE Trans. on Nucl. Sci.* **NS-5**, No. 3, 198.

PALEVSKY, H., HUGHES, D. J., ZIMMERMAN, R. L., and EISBERG, R. M. (1956). *J. Nucl. Energy* **3**, 177.

PARKER, W., and FALK, R. (1962). *Nucl. Instr. Methods* **16**, 355.

PARKER, W. C., and SLÄTIS, H. (1965). *In* "Alpha, Beta, and Gamma Ray Spectroscopy" (Kai Siegbahn, ed.), Vol. I, p. 379. North Holland Publ., Amsterdam.

PARKER, W., BILDSTEIN, H., and GETOFF, N. (1964). *Nucl. Instr. Methods* **26**, 55.

PATRICK, B. H., SCHOMBERG, M. G., SOWERBY, M. G., and JOLLY, J. E. (1967). *In* "Nuclear Data for Reactors," Vol. II, p. 117. IAEA, Vienna.

PATTENDEN, N. J., and HARVEY, J. A. (1960). *Proc. Intern. Conf. Nucl. Structure, Kingston, Ontario* (D. A. Bromley and E. W. Vogt, eds.) p. 882. Univ. of Toronto, Toronto, Canada.

PATTENDEN, N. J., and BARDSLEY, S. (1963). *Proc. Intern. Conf. Nucl. Phys. Reactor Neutrons* p. 369. USAEC Rept. ANL-6797.

PATTENDEN, N. J., and HARVEY, J. A. (1963). *Nucl. Sci. Eng.* **17**, 404.

PATTENDEN, N. J., and BARDSLEY, S. (1964). Unpublished preliminary analysis reported in USAEC Rept. BNL-325, Second ed., Supplement 2, Vol. III (1965); see also United Kingdom Atomic Energy Authority Rept. AERE-PR/NP 6, p. 10.

PAYA, D., DERRIEN, H., FUBINI, A., MICHAUDON, A., and RIBON, P. (1967). *In* "Nuclear Data for Reactors." Vol. II, p. 128. IAEA, Vienna.

PEELLE, R. W., MAIENSCHEIN, F. C., ZOBEL, W., and LOVE, T. A. (1962). *In* "Pile Neutron Research in Physics." p. 273. IAEA, Vienna.

PEREY, F., and BUCK, B. (1962). *Nucl. Phys.* **32**, 353.

PERRING, J. K., and STORY, J. S. (1955). *Phys. Rev.* **98**, 1525.

POLIKANOV, S. M., DRUIN, V. A., KARNAUKHOV, V. A., MIKHEEV, V. L., PLEVE, A. A., SKOBELEV, N. K., SUBBOTIN, V. G., TER-AKOPYAN, G. M., and FOMISHEV, V. A. (1962). *JETF* **42**, 1464; *Soviet Phys. JETP (English Transl.)* **15**, 1016.

POORTMANS, F., CEULEMANS, H., and NEVE DE MEVERGNIES, M. (1966). *Proc. Conf. Neutron Cross Sect. Technol., Washington, D.C.* (P. B. Hemmig, ed.), Vol. II, p. 755. USAEC Rept. CONF-660303.

POORTMANS, F., CEULEMANS, H., and NEVE DE MEVERGNIES, M. (1967). *In* "Nuclear Data for Reactors," Vol. II, p. 211. IAEA, Vienna.

PORTER, C. E., and THOMAS, R. G. (1956). *Phys. Rev.* **104**, 483.

PRICE, P. B., and WALKER, R. M. (1962a). *Phys. Rev. Letters* **8**, 217.

PRICE, P. B., and WALKER, R. M. (1962b). *J. Appl. Phys.* **33**, 3400.

PRICE, P. B., and WALKER, R. M. (1962c). *J. Appl. Phys.* **33**, 3407.

PRICE, P. B., and WALKER, R. M. (1962d). *Phys. Letters* **3**, 113.

PRICE, P. B., FLEISCHER, R. L., WALKER, R. M., and HUBBARD, E. L. (1963). *Proc. Conf. Reactions Complex Nuclei, 3rd* (A. Ghiorso, R. M. Diamond, and H. E. Conzett, eds.), p. 332. Univ. of California Press, Berkeley, California.

RAE, E. R. (1965). *In* "Physics and Chemistry of Fission," Vol. I, p. 187. IAEA, Vienna.

RAE, E. R., MARGOLIS, B., and TROUBETSKOY, E. S. (1958). *Phys. Rev.* **112**, 492.

RAMANNA, R., SUBRAMANIAN, R., and AIYER, RAJU, N. (1965). *Nucl. Phys.* **67**, 529.

REGIER, R. B., BURGUS, W. H., and TROMP, R. L. (1959). *Phys. Rev.* **113**, 1589.

REGIER, R. B., BURGUS, W. H., TROMP, R. L., and SORENSEN, B. H. (1960). *Phys. Rev.* **119**, 2017.

REICH, C. W., and MOORE, M. S. (1958). *Phys. Rev.* **111**, 929.

REINES, F., COWAN, C. L., JR., HARRISON, F. B., and CARTER, D. S. (1954). *Rev. Sci. Instr.* **25**, 1061.

RETHMEYER, J., BOERSMA, H. J., and JONKER, C. C. (1961). *Nucl. Instr. Methods* **10**, 240.

RICH, M., and GRIFFIN, J. J. (1963). *Phys. Rev. Letters* **11**, 19.

RICHMOND, R., and PRICE, B. T. (1956). *J. Nucl. Energy* **2**, 177.

ROBERTS, L. D., and DABBS, J. W. T. (1961). *Ann. Rev. Nucl. Sci.* **11**, 175.

ROBERTS, L. D., DABBS, J. W. T., and PARKER, G. W. (1957). USAEC Rept. ORNL-2430, p. 51.

ROBERTS, L. D., DABBS, J. W. T., and PARKER, G. W. (1958). *Proc. Intern. Conf. Peaceful Uses At. Energy, 2nd* **15**, 322.

ROELAND, L. W. (1958). Thesis, Univ. of Amsterdam. USAEC Div. of Techn. Information Rept. NP-10428.

ROELAND, L. W., BOLLINGER, L. M., and THOMAS, G. E. (1958). *Proc. Intern. Conf. Peaceful Uses At. Energy* **15**, 440.

ROSENZWEIG, N. (1958). *Phys. Rev. Letters* **1**, 24.

ROSENZWEIG, N. and PORTER, C. E. (1960). *Phys. Rev.* **120**, 1698.

ROSSI, B., and STAUB, H. (1949). "Ionization Chambers and Counters." McGraw-Hill, New York.

ROUSH, M. L., WILSON, M. A., and HORNYAK, W. F. (1964). *Nucl. Instr. Methods* **31**, 112.

RYCE, S. A. (1966). *Nature* **209**, 1343.

RYCE, S. A., and WYMAN, R. R. (1964). *Can. J. Phys.* **42**, 2185.

SAFFORD, G. J., and HAVENS, W. W., JR. (1961). *Nucl. Sci. Eng.* **11**, 65.

SAFFORD, G. J., HAVENS, W. W., JR., and RUSTAD, B. M. (1959). *Nucl. Sci. Eng.* **6**, 433.

SAFFORD, G. J., HAVENS, W. W., JR., and RUSTAD, B. M. (1960a). *Phys. Rev.* **118**, 799.

SAFFORD, G. J., SCHRÖDER, I. G., and MOORE, J. A. (1960b). Unpublished data; see also Melkonian and Mehta, *In* "Physics and Chemistry of Fission," Vol. II, p. 358. IAEA, Vienna.

SAILOR, V. L., BRUNHART, G., PASSELL, L., REYNOLDS, C. A., SCHERMER, R. I., and SHORE, F. J. (1966). *Bull. Am. Phys. Soc. Ser. II*, **11**, 29.

SANDERS, J. E., SKARSGARD, H. M., and KENWARD, C. J. (1957). *J. Nucl. Energy* **5**, 186.

SAPLAKOGLU, A. (1958). *Proc. United Nations Conf. Peaceful Uses At. Energy, 2nd* **16**, 103.

SAPLAKOGLU, A. (1961). *Nucl. Sci. Eng.* **11**, 312

SAUTER, G. D., and BOWMAN, C. D. (1965). *Phys. Rev. Letters* **15**, 761.

SAUTER, G. D., and BOWMAN, C. D. (1968). *Phys. Rev.* **174**, 1413.

SAYRES, A., and WU, C. S. (1957). *Rev. Sci. Instr.* **28**, 758.

SCHERMER, R. I., PASSELL, L., BRUNHART, G., REYNOLDS, C. A., SAILOR, V. L., and SHORE, F. J. (1968). *Phys. Rev.* **167**, 1121.

SCHMITT, H. W., and LEACHMAN, R. B. (1956). *Phys. Rev.* **102**, 183.

SCHMITT, H. W., and FEATHER, N. (1964). *Phys. Rev.* **134**, B565.

SCHMITT, H. W., and KONECNY, E. (1966). *Phys. Rev. Letters* **16**, 1008.

SCHMITT, H. W., and PLEASONTON, F. (1966). *Nucl. Instr. Methods* **40**, 204.

SCHMITT, H. W., NEILER, J. H., WALTER, F. J., and SILVA, R. J. (1961). *Bull. Am. Phys. Soc. Ser. II* **6**, 240.

SCHMITT, H. W., NEILER, J. H., WALTER, F. J., and CHETHAM-STRODE, A. (1962). *Phys. Rev. Letters* **9**, 427.

SCHMITT, H. W., GIBSON, W. M., NEILER, J. H., WALTER, F. J., and THOMAS, T. D. (1965a). *In* "Physics and Chemistry of Fission," Vol. I, p. 531. IAEA, Vienna.

SCHMITT, H. W., DABBS, J. W. T., and MILLER, P. D. (1965b). *In* "Physics and Chemistry of Fission," Vol. I. p. 517. IAEA, Vienna.

SCHMITT, H. W., KIKER, W. E., and WILLIAMS, C. W. (1965c). *Phys. Rev.* **137**, B837.

SCHRÖDER, I. G., MOORE, J. A., and DERUYTTER, A. J. (1963). *J. Phys.* **24**, 900.

SCHRÖDER, I. G., DERUYTTER, A. J., and MOORE, J. A. (1965). *Phys. Rev.* **137**, B519.

SCHWARTZ, R. B. (1958). *Bull. Am. Phys. Soc. Ser. II* **3**, 176. See also: Stehn *et al.* (1965).

SEEGER, P. A., and BERGEN, D. W. (1967). USAEC Rept. LA-3478, Part 2.

SEEGER, P. A., HEMMENDINGER, A., and DIVEN, B. C. (1967). *Nucl. Phys.* **A96**, 605.

Sechenko, V. I., Tarasko, M. Z., Mikhailov, V. B., and Kuzminov, B. D. (1967), *Yadern. Fiz.* **5**, 514; *Soviet J. Nucl. Phys. (English Transl.)* **5**, 362.

Sheline, R. K. (1960). *Rev. Mod. Phys.* **32**, 1.

Sher, R., and Felberbaum, J. (1962). USAEC Rept. BNL-722.

Sher, R., and Felderbaum, J. (1965). USAEC Rept. BNL-918.

Shore, F. J., and Sailor, V. L. (1958). *Phys. Rev.* **112**, 191.

Shunk, E. R., Brown, W. K., and La Bauve, R. (1966). *Proc. Conf. Neutron Cross Sect. Technol., Washington, D.C.* (P. B. Hemmig, ed.) p. 979. USAEC Rept. CONF-660303.

Silbert, M. G. (1968). Unpublished studies on fission fragment damage.

Silbert, M. G. and Moat, A. (1964). Unpublished studies on response of solid state detectors to high intensity ion beams.

Silk, E. C. H., and Barnes, R. S. (1959). *Phil. Mag.* (8) **4**, 970.

Simmons, J. E., and Henkel, R. L. (1960). *Phys. Rev.* **120**, 1985.

Simmons, J. E., Perkins, R. B., and Henkel, R. L. (1965). *Phys. Rev.* **137**, B809.

Simpson, F. B., Miller, L. G., Moore, M. S., Hockenbury, R. W., and King, T. J. (1968) (to be published).

Simpson, O. D., and Moore, M. S. (1961). *Phys. Rev.* **123**, 559.

Simpson, O. D., and Schuman, R. P. (1961). *Nucl. Sci. Eng.* **11**, 111.

Simpson, O. D., Moore, M. S., and Simpson, F. B. (1960). *Nucl. Sci. Eng.* **7**, 187.

Simpson, O. D., Marshall, N. H., and Young, R. C. (1962). *Nucl. Instr. Methods* **16**, 97.

Simpson, O. D., Fluharty, R. G., Moore, M. S., Marshall, N. H., Diven, B. C., and Hemmendinger, A. (1966). *Proc. Conf. Neutron Cross Sect. Technol., Washington, D.C.* (P. B. Hemmig, ed.), **II**, 910. USAEC Rept. CONF-660303.

Simpson, O. D., Moore, M. S., and Berreth, J. R. (1967). *Nucl. Sci. Eng.* **29**, 415.

Skarsgard, H. M. and Kenward, C. J. (1958). *J. Nucl. Energy* **6**, 212.

Smith, A. B. (1962). *In* "Physics of Fast and Intermediate Reactors," Vol. I, p. 29. IAEA, Vienna.

Smith, J. R., and Magleby, E. H. (1957). USAEC Rept. IDO-16373. p. 39.

Smith, J. R., and Magleby, E. H. (1958). USAEC Rept. IDO-16505. p. 54.

Smith, J. R., and Fast, E. (1966). *Proc. Conf. Neutron Cross Sect. Technol., Washington, D.C.* (P. B. Hemmig, ed.) **II**, 919. USAEC Rept. CONF-660303.

Solbrig, A. W., Jr. (1961a). *Am. J. Phys.* **29**, 257.

Solbrig, A. W., Jr. (1961b). *Nucl. Sci. Eng.* **10**, 167.

Sowinski, M., Dakowski, M., and Piekarz, H. (1963). *Phys. Letters* **6**, 321.

Staub, H. H. (1953). *In* "Experimental Nuclear Physics" (E. Segre, ed.), Vol. I, p. 1. Wiley, New York.

Stehn, J. R., Goldberg, M. D., Wiener-Chasman, R., Mughabghab, S. F., Magurno, B. A., and May, V. M. (1965). USAEC Rept. BNL-325, Second ed., Supp. 2, Vol. III.

Stein, W. E. (1957). *Phys. Rev.* **108**, 94.

Stein, W. E., and Whetstone, S. L., Jr. (1958). *Phys. Rev.* **110**, 476.

Strutinski, V. M. (1967). *Nucl. Phys.* **A95**, 420.

Stubbins, W. F., Bowman, C. D., Auchampaugh, G. F., and Coops, M. S. (1967). *Phys. Rev.* **154**, 1111.

Swiatecki, W. J. (1956a). *Phys. Rev.* **101**, 97.

Swiatecki, W. J. (1956b). *Phys. Rev.* **104**, 993.

Swiatecki, W. J. (1958). *Proc. United Nations Intern. Conf. Peaceful Uses At. Energy, 2nd* **15**, 248.

SWIATECKI, W. J. (1965). *In* "Physics and Chemistry of Fission," Vol. I, p. 3. IAEA, Vienna.

TEICHMANN, T., and WIGNER, E. P. (1952). *Phys. Rev.* **87**, 123.

TERRELL, J. (1959). *Phys. Rev.* **113**, 527.

TERRELL, J. (1962). *Phys. Rev.* **127**, 880. See also *Phys. Rev.* **128**, 2925.

TERRELL, J. (1965). *In* "Physics and Chemistry of Fission," Vol. II, p. 3. IAEA, Vienna.

THOMAS, T. D., and VANDENBOSCH, R. (1964). *Phys. Rev.* **133**, B976.

THOMAS, T. D., GIBSON, W. M., and SAFFORD, G. J. (1965). *In* "Physics and Chemistry of Fission." Vol. I, p. 467. IAEA, Vienna.

TITTERTON, E. W. (1951). *Nature* **168**, 590.

TURKEVICH, A., and NIDAY, J. B. (1951). *Phys. Rev.* **84**, 52.

UTTLEY, C. A. (1965). Unpublished data reported in USAEC Rept. BNL-325, Supplement No. 2, Vol. III (1965). See also: "Nuclear Structure Study with Neutrons" (M. Neve de Mevergnies, P. Van Assche, and J. Vervier, eds.), p. 535. North Holland Publ., Amsterdam.

UTTLEY, C. A., NEWSTEAD, C. M., and DIMENT, K. M. (1967). "Nuclear Data for Reactors," Vol. I, p. 165. IAEA, Vienna.

VAN AUDENHOVE, J. (1965). *Rev. Sci. Instr.* **36**, 383.

VAN AUDENHOVE, J. ESCHBACH, H. L., and MORET, H. (1963). *Nucl. Instr. Methods* **24**, 465.

VANDENBOSCH, R. (1963). *Nucl. Phys.* **46**, 129.

VANDENBOSCH, R. (1967). *Nucl. Phys.* **A101**, 460.

VANDENBOSCH, R., and THOMAS, T. D. (1962). *Bull. Am. Phys. Soc. Ser. II* **7**, 37; also USAEC Rept. BNL-5876.

VANDENBOSCH, R., UNIK, J. P., and HUIZENGA, J. R. (1965). *In* "Physics and Chemistry of Fission," Vol. I, p. 547. IAEA, Vienna.

VERDINGH, V., and LAUER, K. F. (1964). *Nucl. Instr. Methods* **31**, 355.

VIOLA, V. E., JR., and O'CONNELL, D. J. (1965). *Nucl. Instr. Methods* **32**, 125.

VIOLA, V. E., JR., and WILKINS, B. D. (1966). *Nucl. Phys.* **82**, 65.

VOGT, E. (1958). *Phys. Rev.* **112**, 203.

VOGT, E. (1960). *Phys. Rev.* **118**, 724.

VOROTNIKOV, P. E., DUBROVINA, S. M., OTROSHCHENKO, G. A., and SHIGIN, V. A. (1966). *Yadern. Fiz.* **3**, 479; *Soviet J. Nucl. Phys. (English Transl.)* **3**, 348.

WALTER, F. J. (1964). *IEEE Trans. Nucl. Sci.* **NS-11**, 232.

WALTER, F. J., DABBS, J. W. T., and ROBERTS, L. D. (1960). *Rev. Sci. Instr.* **31**, 756.

WANG SHIH-TI, WANG YUNG-CH'ANG, DERMENDZHIEV, E., and RYABOV, YU. V. (1965). *In* "Physics and Chemistry of Fission," Vol. I, p. 287. IAEA, Vienna.

WATANABE, T., and SIMPSON, O. D. (1964). *Phys. Rev.* **133**, B390.

WEINSTEIN, S. (1968). Doctoral dissertation, Rensselaer Polytechnic Institute.

WEINSTEIN, S. and BLOCK, R. C. (1968). *Proc. Neutron Cross Sect. Technol. Conf., 2nd Washington, D.C.* NBS Spec. Pub. 299, 635.

WESTCOTT, C. H., EKBERG, K., HANNA, G. C., PATTENDEN, N. J., SANATANI, S., and ATTREE, P. M. (1965), *At. Energy Rev.* **3**, No. 2, p. 3.

WESTON, L. W., GWIN, R., DE SAUSSURE, G., INGLE, R. W., FULLWOOD, R. R., and HOCKENBURY, R. W. (1967) USAEC Rept. ORNL-TM-1751.

WHALING, W. (1958). *In* "Encyclopaedia of Physics" (S. Flugge and E. Creutz, eds.) Vol. 34, p. 193. Springer, Berlin.

WHEELER, J. A. (1956). *Physica* **22**, 1103.

WHEELER, J. A. (1963). In "Fast Neutron Physics" (J. B. Marion and J. L. Fowler, eds.), Part 2, p. 2051. Wiley (Interscience), New York.

WHETSTONE, S. L., JR. (1959). *Phys. Rev.* **114**, 581.

WHETSTONE, S. L., JR. (1963). *Phys. Rev.* **131**, 1232.

WHETSTONE, S. L., JR. (1964). *Phys. Rev.* **133** B613.

WHETSTONE, S. L., JR., and THOMAS, T. D. (1965). *Phys. Rev. Letters* **15**, 298.

WHETSTONE, S. L., JR., and THOMAS, T. D. (1967). *Phys. Rev.* **154**, 1174.

WIGNER, E. P. (1946). *Phys. Rev.* **70**, 606.

WIGNER, E. P., and EISENBUD, L. (1947). *Phys. Rev.* **72**, 29.

WILETS, L. (1959). *Phys. Rev.* **116**, 372.

WILETS, L. (1962). *Phys. Rev. Letters* **9**, 430.

WILETS, L. (1964). "Theories of Nuclear Fission." Oxford Univ. Press (Clarendon) London and New York.

WILETS, L., and CHASE, D. M. (1956). *Phys. Rev.* **103**, 1296.

WILKINSON, D. H. (1950). "Ionization Chambers and Counters." Cambridge Univ. Press, London and New York.

WILLIAMS, C. W., SCHMITT, H. W., WALTER, F. J., and NEILER, J. H. (1964). *Nucl. Instr. Methods* **29**, 205.

WINHOLD, E. J., and HALPERN, I. (1956). *Phys. Rev.* **103**, 990.

YAFFE, L. (1962). *Ann. Rev. Nucl. Sci.* **12**, 153.

YEATER, M. L., HOCKENBURY, R. W., and FULLWOOD, R. R. (1961). *Nucl. Sci. Eng.* **9**, 105.

YOUNG, T. E., SIMPSON, F. B., and COOPS, M. S. (1962). *Bull. Am. Phys. Soc. Ser. II* **7**, 305.

YOUNG, T. E., SIMPSON, F. B., BERRETH, J. R., and COOPS, M. S. (1967). *Nucl. Sci. Eng.* **30**, 355.

Author Index

The numbers in italics refer to pages on which the complete references are listed.

Z

Subject Index

A

Activation techniques, 221

²⁴¹Americium
  fission cross section of, 51–52

²⁴²Americium
  fission cross section of, 415, 423, 437

Analysis of cross sections of fissile nuclides, *see also* individual nuclides
  area, 162
  individual nuclides, 387–423
  multilevel nuclides, 383–386
  summary of results, 423, 427

Analysis of gamma-ray spectra, 253–258
  Ge-diode data, of, 256
  NaI data, of, 253–257
  response function, 253–256

Analysis of transmission data, *see also* Area analysis of transmission data
  area method, by, 124–135
  average cross sections, 135–139
  Doppler effect, 118–120, 126
  shape method, by, 116–124
  thick–thin area method, 128–129, 132–133

Angular distributions
  capture gamma-rays, of, 331
  fission fragments, of, 439–449
  scattered neutrons, of, 47–48, 224

Angular distributions of fission fragments
  deuteron induced fission, from, 445–449
  fissile nuclides, from, 51–52, 439–443, 446
  fissionable nuclides, from, 443–446
  resonance region, in, 439–443
  theory, 439–442, 444–446, 447–449
  threshold, near, 443–446

Area analysis, *see also* Area analysis of transmission data
  area function in, 125–131, 160–162, 174
  capture data, of, 161–165, 196–197

combination of partial areas, of, 163–165
  partial data, of, 159, 161–165
  scattering data, of, 161–165, 172–175
  self-indication and transmission data, of, 22–25, 141–143
  transmission data, of, 124–135, 158–159

Area analysis of transmission data, 124–135
  area function, 125–131
  Doppler broadening, with, 127–129, 131–135
  interference, with, 129–135
  self-indication detector, using, 139–143
  thick and thin sample limits, 126, 158–159
  thick–thin sample method, 128–129, 132–133

Average cross sections, *see also* Analysis of transmission data
  capture, 225–226
  formula for total, 135–136, 225
  measurements of partial, 166–167
  strength functions from, 136–138

B

Backgrounds, *see also* Detectors
  black sample technique for, 111, 167
  capture measurements, in, 196–197
  scattering measurements, in, 176, 180–181, 183–187
  total cross section measurements, in, 110–111

Bismuth
  transmission of 790-eV resonance, 129–130

¹⁰Boron
  (n,α) cross section, 47–48

Breit–Wigner single-level formulas
  capture cross section, 157
  limitations of, 156, 379–380, 383–386

T

U